Migration

Migration

The Biology of Life on the Move

Hugh Dingle

New York   Oxford

OXFORD UNIVERSITY PRESS

1996

Oxford University Press

Oxford   New York
Athens   Auckland   Bangkok   Bombay
Calcutta   Cape Town   Dar es Salaam   Delhi
Florence   Hong Kong   Istanbul   Karachi
Kuala Lumpur   Madras   Madrid   Melbourne
Mexico City   Nairobi   Paris   Singapore
Taipei   Tokyo   Toronto

and associated companies in
Berlin   Ibadan

Copyright © 1996 by Oxford University Press, Inc.

Published by Oxford University Press, Inc.,
198 Madison Avenue, New York, New York 10016

Oxford is a registered trademark of Oxford University Press

Library of Congress Cataloging-in-Publication Data
Dingle, Richard Douglas Hugh, 1936–
Migration: the biology of life
on the move / Hugh Dingle.
p.   cm. Includes bibliographical references.
ISBN 0-19-508962-6 (cl.)
ISBN 0-19-509723-8 (pbk.)
1. Animal migration.
I. Title.
QL754.D515   1995
591.52'5—dc20   94-39538

9 8 7 6 5 4 3 2 1

Printed in the United States of America
on acid-free paper

Preface

Migration is a wonderfully diverse biological phenomenon that has fascinated scientist and layman alike for centuries. Most attention has focused on the "charismatic megafauna" of the larger and more spectacular animals that undertake it, but it is not limited to these. Enormous numbers of tiny, fragile, and inconspicuous animals migrate as well, and—as I shall argue in this book—so do many plants and fungi when, as propagules, they move to new germination sites. In spite of the diversity of migratory phenomena, however, most books on the subject are confined to a particular taxon. Thus we have a plethora of excellent single-authored books and symposium volumes on bird migration, several on the movements of fish, some on mammals, some on butterflies, a few on insects in general, and a smattering of others, including books that cover a range of taxa. Although there have recently been some first-rate, very attractive popular books on migration that are not taxon limited, the past two decades have seen only two major scientific publications with broad taxonomic perspective. R. R. Baker's huge 1978 work is not only dated now, but its highly eclectic approach has had limited acceptance among biologists. The outstanding volume edited by M. A. Rankin (1985), that derived from a 1983 symposium at Port Aransas, Texas, revealed clearly that the important "how" and "why" questions concerning migration transcended taxonomic boundaries, but its papers are based on information that is now more than 10 years old.

For these reasons it seems a good time for a book on migration that is not taxon limited and that focuses on the similarities among migrants as well as on the differences. I have long felt that those of us working on the many varieties of migrating organisms had much to teach each other. This book is my attempt to find a common ground and to focus on migration as a biological phenomenon, rather than on the migrating portion of a group that I study. This has often led me to explore the biology of taxa that I do not know very well, and I hope I have not committed too many errors in the process. Pursuit of my aim will have been well worth that risk, however, if the result is a greater understanding of an important event in the lives of many organisms. I hope readers will inform me of errors, misinterpretations, or faulty logic they may find.

I have written this book both for students of migration and for those biologists who are generally interested in the functioning of whole organisms. I have tried to keep the discussion at a level that can be understood by anyone with a good background in biology, and especially by beginning graduate students or even upper division undergraduates, because future studies of migration will be in their hands.

In the genesis and writing of this book, I am indebted to many people. First and foremost is the late Prof. John S. Kennedy F.R.S., whose influence on my thinking about migration will be evident throughout. I spent a postdoctoral year with John in Cambridge in 1962–63, and his stimulating ideas, advice, and encouragement launched me on my own studies of not only migration but also other aspects of behavior. John's clear thinking and uncanny ability to focus on the heart of a problem have been inspiration to me ever since. This book is dedicated to his memory.

Much of the writing was done while I was on sabbatical leave from the University of California at Davis, and I am extremely grateful for the enlightened policy that made the time for learning, thinking, and writing available. Three institutions provided much-appreciated logistical and facilities support during my sabbatical. The Department of Entomology, CSIRO, in Canberra under the directorship of Max Whitten sponsored my visit to Australia, and Roger Farrow kindly provided office space. In conjunction with my CSIRO appointment, the Australian National University appointed me a Visiting Fellow sponsored by Dave Shaw of the Research School of Biological Sciences. In Germany space, facilities, and support were provided by the Vogelwarte Radolfzell under the directorship of Peter Berthold, and I was financially supported by a Senior Scientist Award from the Alexander von Humboldt Foundation. My own research on migration has been supported for some 30 years by the U.S. National Science Foundation.

Portions of the manuscript have been read by Peter Berthold, Alistair Drake, Roger Farrow, Gavin Gatehouse, Sid Gauthreaux, and Wayne Rochester. All were also most generous in sharing their expertise about various aspects of migration and in providing me with manuscripts or illustrations. I acknowledge Alistair Drake, in particular, for his help in understanding meteorological events. Others who were most helpful in various ways were Elizabeth Arias-Tobar, Joe Cech, Peter Fullagar, Fran James, Arja and Veijo Kaitala, Bernd Leisler, Peter Moyle, Marc Mangel, Randy Snyder, Eugene Spaziani, Rob Toonen, and John and Jill Thompson. The students in a Population Biology Center "Monte Carlo" seminar, especially Greg Lowenberg and Eric Nagy, were helpful in introducing me to the plant literature. Thanks also to all those who permitted use or modification of illustrations.

In preparing the manuscript I was aided immensely by Mitch Baker and Chuck Fox, who prepared most of the illustrations. Mitch's artwork in particular has increased their attractiveness and clarity. The entire final draft of the manuscript, and several preceding drafts, were cheerfully and expertly produced by Nancy Dullum. Kirk Jensen guided the book through the publication process. Finally, special thanks goes to Jeri Dingle who got the laptop working, typed several early chapter drafts, read the entire manuscript, tried to ensure that words and meanings stayed together, and made often heroic efforts to prod a recalcitrant and procrastinating author to write.

Davis, Calif.   H. D.
November 1994

Contents

Introduction and Plan of the Book

Migration is one of the most fascinating of all behaviors. For sheer drama and beauty it is hard to match the movements of millions of individuals over what are often sizable portions of the surface of our planet. For sheer intrigue it is hard to match the question: Why do they do it? It is little wonder that migration in all its spectacular variety is a perpetual subject for coffee-table books, television programs, and even the daily press. Its "mystery" and its "marvels" are responsible for attracting much of the attention the public devotes to matters of natural history.

The conspicuousness of migration in the lives of organisms also signals the importance of its role in life histories, and it is this role that is the focus of this book. In view of the many books written about migration, a word is due here regarding why I have written another one. Broadly, there are two reasons. First, with but only one or two exceptions, previous books on migration have focused on a single taxon. Biologists have as a result tended to view migration from the perspective of a single group and indeed have often defined migration in terms of its properties within that group. Thus ornithologists and other vertebrate biologists have stressed migration as a round-trip movement, even though round-trips are characteristic only of long-lived organisms like vertebrates; the vast majority of migrants, with insects being a prime example, migrate only one way. One of the major points I make is that if one compares across a spectrum of organisms, it is apparent that physiological and behavioral characteristics define migration, not the particular route followed. A reason for writing this book then, is to consider migration across a range of taxa to determine its common properties and the way natural selection has molded those properties to fit the life histories of the organisms concerned.

A further major departure from previous books on migration is the inclusion of plants and, to a minor degree, protists and fungi among the organisms considered. As I read through the literature, it seemed apparent that the properties evolved in these organisms to ensure their movement from one place, host, or habitat to another shared much in common with the more traditionally considered migratory movements of animals. I have tried, therefore, to include and discuss cases from plants, especially, where appropriate. My own knowledge of plants is

3

limited, so I may well have made errors in the interpretation of individual cases. I would maintain, however, that the general point that plants display migratory behavior is valid, and I hope my argument in this book is a convincing one.

The second reason for writing this book follows from the first. Because students of migration generally tend to be taxon-oriented, there is much confusion not only over terminology but also over the conceptual basis for the classification of animal movements. I have met more than one biologist who has thrown up his or her hands in the face of attempts to define migration because the arrays of behaviors to which the term has been attached have so little in common. Frequent among the sources of confusion has been the confounding of population processes with individual behaviors, as I discuss in detail in Chapter 3. I therefore felt there was a need to sort out migration from other movements so that biologists of whatever taxonomic focus could discuss it from a base on common ground. Once that need is met, there is a trove of physiological and ecological adaptations and evolutionary pathways that can be more profoundly understood by comparing the tactics and strategies of diverse organisms living in different habitats and displaying a variety of life styles. Thus one of the aims of the book is to show how comparisons across taxa can illuminate migratory life cycles and the relation of migration to other movements, once migration is firmly placed on common conceptual ground. At the same time, I hope the book leads also to asking new and important questions about movement behavior.

The book is divided into four parts, each with a brief introductory section. In Part I, I define migration, give examples, and place it in the spectrum of movement behaviors, concluding with a chapter on methods for its study. Part II focuses on proximate mechanisms, including physiology, morphology, and the constraints associated with them, the interactions between migration and wind and current patterns, and finally, the various orientation and navigation mechanisms by which migrants find their way about. The longest section of the book is Part III, on the evolution of migratory life histories. This subject has received less attention in most other books on migration (another reason for writing this one), but I devote a major portion to it here; the evolutionary and ecological basis for migration—although much more difficult in study than proximate factors—is of the greatest importance to understanding the roles of migration not only in the lives of individual organisms, but also in the ecological communities in which they live. Those roles are important also for their impact on the interactions between migrants and humans, so a final section of the book, Part IV, is devoted to a brief consideration of migration and its relation to conservation and pest management.

I have not intended the book to be a comprehensive review; this, with the explosion of the literature on migration, would be far too daunting a task. Also, there are several excellent books and symposia reviewing particular aspects of migration or describing it in various groups. Rather, I have tried to stress major conceptual points and to illustrate those points with specific examples of migratory behavior or ecology. Where data from comparisons were available, I have also made liberal use of these to add conceptual generality to particular issues. Each chapter closes with a section I have called a "Summing Up," where the intent is not only to summarize and draw conclusions, but also to speculate and reflect briefly on the major issues raised. Finally, at the end of the book, there is a brief chapter in which I try to draw together the major threads and to offer some ideas on future directions research on the biology of migration might take.

PART ONE

Migration and Methods for Its Study

Even to the most casual observer of nature, it is obvious that organisms make many different kinds of movements. Mostly these are local in scale and are linked to the daily and other short-term activities contributing to growth, survival, and reproduction. On occasion, however, movements occur on a larger scale and with an often distinct physiology; these movements take the organism to a different habitat, serving for both escape and colonization. The terminologies applied to different forms of movement vary widely, reflecting origins from and applications to the groups of organisms with which biologists using the terminologies are most familiar. Thus, early on the term "migration" came to be associated with the to-and-fro movements of birds and other vertebrates. Unfortunately this ornithocentric view often led to the confounding of issues and to the failure of biologists working on different groups to communicate about common features of movement behavior, including migratory movements that were not to-and-fro but that did include functions and physiologies similar to those of round-trip journeys.

The opening two chapters of this first section of the book deal with the problems of defining different types of movements. In Chapter 1, I outline a taxonomy that places migration in the context of movement behaviors in general. Starting with stasis and the simplest forms and rates of turning, I describe briefly and define a set of movements that are directly related to the resources an organism encounters in the course of its local and short-term activities. Inputs arising from those resources directly influence the onset and cessation of these movements, which include kineses, foraging, and other activities that I designate broadly as station-keeping. Similar, but also involving a search for a new home range, are movements that I call "ranging" because they usually include a specifically exploratory component; or at least we presume they do, because it is also apparent that we know relatively little of the details of any of the described movement behaviors. Finally I introduce migration as a type of movement that differs both qualitatively and quantitatively from the others. Migration is not a proximate response to resources and serves not to keep an organism in its habitat, but to remove it to another.

Having dealt with movement in general in Chapter 1, I describe and define migration in detail in Chapter 2. I first review the history of how biologists have viewed it and of how this view has changed over the past few decades. This discussion sets the stage for a definition based on the physiological characteristics of migration and a description based on its ecological outcome. The definition is based on the experimental work of J. S. Kennedy on the black bean aphid, *Aphis fabae*; this work is outlined in some detail in order to make clear the basis for the definition. The chapter includes some examples of migration from different taxa to illustrate the main points of the definition and to point out its general applicability, even though it is based in research on aphids. I distinguish between migration as an individual behavior and dispersal and aggregation as population processes. I also make the point that migrants occur in many taxa.

The rich diversity of migratory patterns is illustrated more fully in Chapter 3. I begin with the familiar form of round-trip journeys, but show that even these can take a multiplicity of forms—from the globe-straddling excursions of arctic terns to insect movements to and from diapause sites, which may require travel of only a few tens or hundreds of meters. Other patterns are far less regular, as illustrated by the fact that denizens of habitats dependent on erratic rainfall display routes that are different each time migration is undertaken. At that, such diversity of pathways is also a characteristic of many organisms that migrate only once in a lifetime. Other migrants may make round-trips, but each leg may occur at a different stage of the life cycle or even in a different generation. The point is made that it is the characteristics of the behavior and not of the pathways followed that determine whether a particular organism is a migrant. Examples throughout the chapter are drawn from a variety of taxa as well as a variety of patterns and pathways.

Chapter 4 concludes Part I. Here I address the problem of how one studies migration, whether in the field or in the laboratory. In the field the first problem is to identify migrants. Sometimes this can be accomplished with just a knowledge of natural history, but a number of ways have also been devised to mark individuals and, in some cases, groups. The chapter includes a brief survey of how migrating individuals may be counted, and also of capture and marking methods so that routes can be estimated from recapture data. A second major problem is how to follow migrants on their journeys. Advancing technologies have produced some ingenious and sophisticated ways of accomplishing this, including radar and satellite telemetry. The limitations of these methods are also assessed. Some aspects of migratory behavior can best be studied in the laboratory. This can be a real challenge to investigators, and some of the ways those challenges have been met are also surveyed in this chapter. Overall the summary of methods is intended to provide enough understanding to allow appreciation of the problems faced in gathering the data that will provide insights into the behavior and ecology of migration.

1

A Taxonomy of Movement

This is a book about behavior and its ecological consequences, and specifically it is a book about a particular kind of behavior known as *migration*. Migration is one of several types of movement, all of which sum over the course of an organism's life cycle to produce a *lifetime track* (Baker 1978). Before proceeding to a full discussion of migratory behavior and its evolution and ecology, I shall briefly define and describe movements and related behaviors that organisms display, including some examples and an adumbration of their roles in life cycles (Table 1-1). This taxonomy should serve to place migration in the context of the lifetime track and to indicate its significance to life histories. I then consider in greater detail a definition and description of migration in Chapter 2.

Most movements occur within a relatively circumscribed area or *home range* over which an organism travels to acquire the resources it needs for survival and reproduction. The sizes of home ranges depend on the habitat and the size and powers of movement of their residents. Most cover only a small area relative to the range of the organism, but some, like those of pelagic seabirds, may extend over considerable portions of the earth's surface. The kind of area within which the home range occurs and which produces the required resources for each discrete phase of the life cycle is the *habitat* (Southwood 1981). Usually in anticipation of declining resources and in contrast to the other behaviors outlined in Table 1-1, migration takes an individual out of its current habitat to a new one often at some distance elsewhere. The focus of this book is the behavior that leads to that ecological consequence.

The form of the lifetime track and its partitioning into constituent movement (and stationary) elements are determined by natural selection. They are a function of the organism's size, its life history traits, its powers of movement, its geographic range, and its habitat. All of these factors will be discussed with respect to migration in this book. In turn the dynamics of populations and communities are greatly influenced in both time and space by the lifetime tracks of the organisms of which they are composed. Below and in Table 1-1 I outline the movements that in their various ways contribute to lifetime tracks.

Table 1–1. A Taxonomy of Movement Behaviors Contributing to the *Lifetime Tracks* of Organisms

Movement	Characteristics	Examples
Movements home range or resource directed		
Stasis	Organism stationary	Corals, trees, hibernating mammals
Station keeping	Movements keeping organisms in home range	
Kineses	Changes in rate of movement or turning	Planarian in shadow; moth in a pheromone "plume"
Foraging	Movement in search of resources; movement stops when resource encountered	Movement in search of food or oviposition site (animals); modular growth (plants, corals)
Commuting	Movement in search of resources on a regular short-term basis, usually daily; ceases when resource encountered	Albatross foraging; diel "migration"; vertical "migration" in plankton
Territorial behavior	Movement and agonistic behavior directed toward neighbors and/or intruders; stops when intruder leaves	Many examples across taxa
Ranging	Movement over an area so as to explore it; ceases when suitable habitat/ territory located	"Dispersal" of some mammals; "natal dispersal" of birds; fungal spore ejection; parasite host seeking
Movement not directly responsive to resources or home range		
Migration	Undistracted movement; cessation primed by movement itself. Responses to resources/ home range suspended or suppressed.	Annual journeys of birds to and from breeding grounds; flight of aphids to new hosts; transport of some seeds to germination sites
Movement not under control of organism		
Accidental displacement	Organism does not initiate movement. Movement stops when organism leaves transporting vehicle.	Storm vagrancy

Stasis

Before proceeding to the various forms of movements, it is worth noting that many organisms spend most of their lifetimes without moving at all. The most obvious of these occur in plants, fungi, or coral colonies with extreme cases being species of tree like the bristlecone pines (*Pinus aristata*) in the southwestern United States.

These conifers may live fixed in one place for thousands of years with but a brief period of movement in the life cycle when, as seeds, they leave their parents for new germination sites. Among corals, fixed colonies of the same genotype (genets) may date from the Pleistocene (Potts 1984), although individual polyps may have a life span of only days or weeks. Highly mobile terrestrial animals such as insects may spend a period in suspended development known as *diapause*, which can last for several years (e.g., Danks 1987), and some small mammals spend months in hibernation (escape in time rather than in space). Most animals, however, come to more or less complete rest only when basking or sleeping and then only for a portion of the daily cycle. These latter two behaviors combine with a set of localized movements to keep an animal in place (home range) within a habitat or even more locally within a particular habitat patch (Hassel and Southwood 1978).

Station Keeping

Activities and movements that keep an animal in a home range have been called *station keeping* (Hassel and Southwood 1978; Kennedy 1985), and this seems a useful term to describe a number of behaviors. These include an array of interactions with both the abiotic and biotic environment, all of which can be characterized as "vegetative functions" or activities that exploit available resources to promote growth and reproduction (Kennedy 1985). These resources include not only food but also others required for the maintenance of individuals and species including shelter, mates, nest sites, landmarks, microclimate, and so forth, all requirements for existence usually incorporated in the organism's home range. A salient characteristic of station-keeping movement behaviors is that they cease when a resource is located. A predator stops hunting when it kills its prey and begins to feed, and a male moth stops orienting to female sex pheromones when it locates a mate. As we shall see, this cessation of activity upon contact with a vegetative stimulus is not a characteristic of migration.

Among the simplest of station-keeping movements are *kineses* or variations in the rate of movement. These can be either a change in the speed of movement (orthokinesis) or a change in the rate of turning (klinokinesis), with slower movement and more frequent (random) turns tending to keep the animal in place. The classic biology laboratory example of the planarian that slows down and increases its turning per unit of distance to remain in shadow is a case in point. A more complex example of kineses is the counterturning, or execution of alternating right and left turns, exhibited by many insects when responding to windborne odor (Bell and Tobin 1982; Kennedy 1983; Bell 1991). On contact with odor the insect turns upwind and zigzags progressively toward the source in alternating turns of less than 180°. If it loses contact with the odor, it begins casting with turns approximating 180° that do not result in progression upwind but do increase the probability of relocating the odor. When the odor is relocated, upwind zigzagging is resumed. These behaviors have been most extensively studied in male moths in the presence of a pheromone "plume," but they occur as foraging responses in most if not all insects, and in many other organisms as well, while locating mates and food sources.

Kineses grade into a set of behaviors, which may be termed *foraging*, by which organisms seek resources by moving within a habitat. Foraging includes related behaviors such as searching (Bell 1991) and scanning and can be defined as "reiterative . . . activity that is readily interrupted by an encounter with a resource item of a particular kind" (Kennedy 1985). Foraging usually takes place within habitats or habitat patches and encompasses the search not only for food but also for other resources. A good description of foraging movements within habitat patches and transitions between patches is given by Zalucki and Kitching (1982) for the monarch butterfly, *Danaus plexippus*. Both sexes were more directional and flew faster when flying in the open than within patches of food plants. Females foraging for oviposition sites, however, exhibited more localized turning (kineses) than males, presumably because males were seeking females. The type and distribution of a resource can thus influence the partitioning among behaviors. The "tactics" of partitioning have been analyzed extensively under the rubric of "optimal foraging" theory (Stephens and Krebs 1986), and the ecological outcome of these station-keeping responses has been studied within the framework of the "resource concentration hypothesis" (Kareiva 1983; Root and Kareiva 1984). The "arms race" between foragers and their biotic resources forms the basis for some of the more intensively studied coevolutionary relationships as occur, for example, between butterflies and the plants on which they oviposit (Thompson and Pellmyr 1991).

One of the more important developments in plant biology in recent years has been the understanding of plant growth as a modular process (Lopez et al. 1994 and references therein). Individual plants are seen as populations of iterative units or modules, each with its own growth and development coupled to the demography of the whole plant. This approach allows growth analysis in relation to resource acquisition as an expression of the foraging behavior of plants. Clonal plants (and also clonal corals, tunicates, and bryozoans) have provided useful insights into the relations among morphological structure, resource abundance and distribution, and consequent foraging strategies. There are many interesting parallels between the foraging behaviors of plants and animals that illustrate how natural selection has acted to produce equivalent solutions to similar problems, suggesting common biological processes (Lopez et al. 1994). Analyses of these commonalities free of particular organism bias should be an exciting area of future research.

Some foraging movements in animals can be extensive, covering tens, hundreds, and in a few instances even thousands of kilometers of home range. Two examples in insects are the desert locust, *Schistocerca gregaria* (Rainey 1989; Farrow 1990) and the cabbage butterfly, *Pieris rapae* (Baker 1968a,b). Locusts aggregate into large swarms, which may move up to thousands of kilometers across the countryside, stopping to feed and lay eggs in suitable areas (Chapter 11). The devastation they can cause to crops has been noted since biblical times (e.g., in the

Book of Exodus). In the British Isles a *P. rapae* female may oviposit over distances of 100 or more kilometers and apparently uses a sun compass (Chapter 8) to orient in a straight line while doing so. On long days the females move north and so lay eggs as the season advances; they reverse direction with the shortening days of late August. Baker calls these long oviposition pathways migration, but because the females are responding to suitable host plants, extended foraging seems more appropriate, although there may be bouts of migration when they do not respond to host plants (see below). When host plants occur in limited patches, the oviposition foraging of *P. rapae* incorporates much turning (kineses) so that the butterflies remain in a limited area (Jones 1977). In birds there are several examples of long-distance foraging. Many oceanic species including storm petrels, shearwaters, and albatrosses fly over wide areas of ocean during the nonbreeding season (Alerstam 1990) and North American wood warblers (Parulinae) may move about extensively when overwintering in Central and South America (Keast and Morton 1980). Likewise many Palearctic migrants move over extensive areas of Africa (Moreau 1972). In spite of the distances traversed, however, these movements can be classified as foraging (or ranging, see below) because they are apparently readily terminated, even if only temporarily, by suitable resources.

In much of the insect literature, the term "trivial" has been used to identify within-habitat movements in search of resources, with the movements assumed to be relatively short distance and therefore trivial relative to long-distance migratory movements (Southwood 1962; Johnson 1969). There is an element of circularity in the use of the term because it was defined in terms of the habitat, which in turn was the area within which trivial movements took place. An additional problem is that there are no good behavioral criteria to distinguish trivial movements from whatever "nontrivial" movements might be. Foraging and other station-keeping behaviors seem to be what was originally intended, provide more information about what actually occurs, and avoid the connotation of insignificance—which of course these behaviors are not. For these reasons, I have not included "trivial movements" in my behavioral taxonomy outlined in Table 1–1; they are included within the set of station-keeping responses.

Some foraging pathways take the form of more or less round-trip excursions, which seem to be most logically designated *commuting* (Kennedy 1985) by analogy with the well-known daily round-trips of humans in urban societies. The most studied of these are the daily vertical "migrations" of many marine and freshwater organisms. These include the microanimals and plants of the plankton and also fish and larger invertebrates (summarized in Dingle 1980; chapters in Rankin 1985; and in many texts on marine and freshwater biology). The general pattern in zooplankton involves a rise to the surface at night and a descent, often to quite considerable depths, during the day, with the behavior proximately controlled by light levels. The best evolutionary explanations seem to be the avoidance of predation by rising to the surface at night when largely invisible to predators and the exploitation of resources in different water layers or both. Phytoplankton can move vertically by regulating buoyancy or, if possessing flagella, actively swimming; their vertical movements appear to be responsive to light and nutrient gradients, enabling cells to move into regions conducive to growth or to avoid deleterious conditions. Among terrestrial animals, the caterpillars of many lepidopterans make vertical excursions to the tops of plants at night to feed but shelter at ground level during the day, and soil organisms undertake vertical transits on both a daily and seasonal basis.

There are also several animals of both marine and terrestrial communities that make more or less short-term horizontal commutes. A good example here is the occurrence of diel foraging movements in both tropical reef and temperate nearshore fishes so that diurnal and nocturnal fish communities can be quite different (Ogden and Quinn 1984; Quinn and Brodeur 1991). Some marine birds make long-distance daily or even longer duration commutes from nesting colonies to foraging areas. The most extreme example of such long-distance commuting is the wandering albatross (*Diomedea exulans*). Satellite tracking has revealed that these birds may make foraging trips of up to 33 days for distances of 3600 to 15,000 kilometers when their mates are incubating and for 3 days and over 300 km after the chicks have hatched (Jouventin and Weimerskirch 1990). In the Serengeti Plains of East Africa, spotted hyenas (*Crocuta crocuta*) establish clan territories, but when the migratory herds of ungulates on which the hyenas feed move away, clan members commute an average of 40 km to feed on the nearest migratory grazers. In this way clan sizes are maintained above the apparent carrying capacities of the territories (Hofer and East 1993). At the interface between localized foraging and daily commuting to foraging areas are the composite foraging pathways of butterfly fish (*Chaetodon* spp.) on reefs in the tropical Pacific. Their excursions may involve repeated daily circuits with stops to feed on specific coral heads (Reese 1989).

All the commuting movements I have described are basically extended foraging excursions, readily terminated by immediate responses to inputs signaling resource availability. Because they include round-trips, a criterion often used to define migration (Chapter 2), these commutes have been referred to as migratory, especially in the case of the so-called "vertical migrations" of plankton. Migratory movements, however, can be characterized by an accompanying physiology that includes undistractability by those resources that readily elicit responses from commuters and other nonmigrants, as will be discussed below and in detail in Chapter 2. The term "commuting" thus seems best to describe to-and-fro movements primarily for the purpose of securing resources.

In the case of many animals the resources for maintenance, growth, and reproduction are acquired on a defended territory. From our perspective, those activities required for establishing and defending territories also constitute station-keeping responses. These might include patrolling territorial boundaries, fending off intruders, and where necessary, attracting a mate to the territory. As with other station-keeping activities these are primarily resource directed. They differ somewhat, however, in that they are often directed agonistically against other individuals, namely those that intrude on the territory. This agonistic component means the difference between territorial behaviors and the other station-keeping movement responses is qualitative and based in physiology, as is also the difference between migration and station keeping. Because of its obvious role with respect to resource acquisition and home range, it nonetheless seems reasonable to discuss territoriality here in the context of station keeping and in contrast to the movements generated by migrations. Territorial behaviors are characteristically station-keeping responses in their constant monitoring of environmental inputs by the organism and in their rapid reaction to changes by adjusting to maintain the integrity of the resource space.

Ranging

At certain times in its life cycle an organism may leave its home range and seek another one. There are two movement behaviors that accomplish this change in home range, which I shall call *ranging* and migration. Migration will be discussed in the next section and, especially, in the next chapter. Ranging, like migration, occurs in a great variety of organisms and is characterized by a departure from the current habitat patch (emigration), the seeking of a new patch, and the occupation of the first available and suitable habitat patch discovered (immigration). Following Jander (1975) and Bell (1991) I refer to this travel between resources or patches as ranging because it involves "movement over an area so as to explore it" (*Webster's Collegiate Dictionary*, Tenth Edition). Ranging differs from migration in that responses to sensory inputs from resources and orientation to those resources are not inhibited or suppressed. Rather the organism responds to the resources or other inputs signaling a suitable habitat patch as soon as they are encountered, and in this way ranging is similar to station-keeping behaviors.

I use ranging in preference to what in much of the literature on movement is called "dispersal." Dispersal is commonly used to mean movement from a natal site to a site where reproduction takes place (Howard 1960; Holekamp and Sherman 1989; Stenseth and Lidicker 1992). I shall discuss extensively difficulties with the term in Chapter 2 where I analyze it in the context of migration. Suffice it to say here, there are two major problems with it. First, in many contexts and in accordance with its usual dictionary definition, dispersal indicates the breaking up of a group of objects or individuals and the process of increasing distances among them. The use of dispersal to mean what I here call ranging confounds a process that occurs in populations with the behavior of the individuals in those populations. In fact a population of individuals ranging or migrating (again individual behavior) can either disperse or aggregate (Taylor 1986a,b) and many descriptions of so-called "dispersal" in fact describe aggregation. Second, none of the definitions of dispersal indicate what the organism is doing behaviorally. Ranging includes within it the behavioral notion of exploration. In addition I add to it a physiological nexus of response to inputs indicating resources, to contrast with the physiology of migration, in which those responses are inhibited or suppressed (Kennedy 1985).

A few examples indicate the enormous variety ranging behavior can take. It is common among mammals and birds where it can occur either as natal ranging ("natal dispersal") or as reproductive (breeding) ranging ("breeding dispersal"). In the former it is young individuals that seek new home ranges and in the latter it is the movement of adults to new breeding locations. Ranging tends to be sex-biased toward males in mammals and toward females in birds, apparently because the largely polygynous breeding systems in mammals favor philopatric (site faithful) females, while the greater degree of monogamy in birds favors male philopatry (Cockburn 1992). Other examples are at least some of the ballistic devices of fungi and plants used to propel spores and seeds, respectively, out of the immediate area (Simons 1992). Particularly interesting cases occur among parasites. The various life cycle and behavioral devices by which parasitic organisms from several phyla move from host to host or locate new hosts are probably best considered ranging movements, because they usually involve responses to the new host (resource) with the first contact, while also including a change in home range (the host's body).

It should be noted, however, that in many cases distinctions between ranging and foraging on the one hand or ranging and migration on the other will be difficult to make. In part the reason is that we lack sufficient information because specific distinctions of this sort have not been looked for. I suspect, for example, that many mammal "dispersers" are migrating rather than ranging, but because these behaviors have not been distinguished in studies of mammals, except in some obvious round-trip migrations of large grazers, we have little information on which to make a determination of which behavior is actually occurring. It is also the case, however, that there are gradations across the spectrum of foraging through ranging to migration, so that in some cases absolute distinctions will simply not be possible. Defining categories in a behavioral taxonomy of movement is nevertheless still useful because the taxonomy can indicate the direction natural selection has taken the organization of that behavior and so point out what critical elements of an organism's biology should be a focus for study (see below).

Migration

Movement behavior that constitutes migration differs in important ways from station keeping and ranging. These differences will be discussed in depth in Chapter 2, and I shall outline them only briefly here so that contrasts with other movements are evident. The most distinctive characteristics of migratory behavior are the undistractability of the individual displaying it (if an animal) and the special physiological mechanisms in both animals and plants assuring that it takes place. Animal migrants do not respond to sensory inputs from resources that would readily elicit responses in other circumstances, the presence of food or a shelter, for instance. I well remember the behavior of migrant seabirds on a pelagic birdwatching trip in Monterey Bay, California, which illustrates this point quite well. Bits of fish thrown from the stern of the boat attracted a milling throng of gulls, terns, some shearwaters, and even a South Polar skua. This food was completely ignored, however, by migrating arctic terns and Sabine's gulls, which flew by without hesitating, much to the disappointment of the assembled birders. Many plant seeds similarly "ignore" resources during migratory transit because hard seed coats (which may require partial digestion in a gut) or other constraints prevent germination until completion of a journey. In both animals and plants the stimuli that eventually terminate migration do so only after a considerable period spent migrating and may be primed by the migration itself. This contrasts with station-keeping and foraging movements, which terminate as soon as stimuli from suitable resources are encountered. Other characteristics of migration include responses to environmental signals, photoperiod for example, that initiate departure before environmental deterioration and specialized departing and arriving behaviors. Departing migrant birds illustrate this specialization in their often rapid climb to heights well above those of their usual daily journeys (Alerstam 1985; Richardson 1990). Distance traveled does not per se determine whether a journey is migratory, although for a given organism migration is apt to occur over a greater distance and for a longer time than single station-keeping or ranging movements. Note also that migration does not necessarily include a to-and-fro pathway.

Thus, on a proximate scale migration is not driven by resources, although ultimately it certainly is. Functionally, and over the long term, migration reduces the environmental heterogeneity experienced by the organism (Leggett 1984) and serves to place it under optimal conditions throughout its life. Because migration so often is a move beyond current habitat as well as beyond current home range, natural selection has molded it in such a way that it does not depend on response to resource deterioration. If departure were delayed until resources were depleted, it would often be difficult if not impossible for the organism to acquire the energy necessary for the journey; once on the journey, immediate response to resources could result in premature termination in a place where in turn resources would decline. There are thus ultimately logical reasons why migrants would time departure by inputs forecasting resource deterioration and suppress responses to resources while in transit. Within that framework natural selection has produced an often astonishing variety of migratory syndromes.

Accidental Displacement

One additional type of movement needs mention before moving on to a more complete characterization of migration in Chapter 2. This is accidental movement, in the sense that no overt behavior on the part of the organism leads to the movement path in question. Sometimes migrants may be displaced in unpredictable fashion. The ornithological literature is replete with examples of birds being carried beyond or outside their usual ranges by storms or periods of prevailing winds in the "wrong" direction (e.g., Alerstam 1990). Many terrestrial insects may also be blown off course; one of the most frequently cited examples is the large fallout of aphids on the snows of Spitzbergen reported by Elton (1927) as a consequence of strong winds blowing from the Kola Peninsula (Johnson 1969). Spores, seeds, small insects and other organisms are also no doubt on occasion simply plucked from their habitats by storms or high winds. By far the majority of accidental travelers undoubtedly perish, as with Elton's aphids or the Australian locusts studied by Farrow (1975a) that were carried out to sea on the winds that were transporting them; and for migrants this mortality must be considered a cost of migration, balanced by the success of those travelers that succeed in colonizing new habitats. In a few cases, accidentally displaced individuals may colonize a completely new region, as has evidently happened with the Australian grey teal (*Anas gibberifrons*) which is now established in New Zealand (Frith 1982). Accidental movements can thus be of considerable ecological importance under some circumstances (Taylor 1986a,b). They differ from migration in two ways: first, they are involuntary, and in fact there may be behaviors or devices such as adhesive appendages or holdfast organs specifically evolved to counter them; second, they cease once the organism is deposited by the transporting vehicle rather than as a result of specific behavioral and physiological processes.

Summing Up

Broadly speaking, there are two types of movement behavior (Table 1–1). The first type concerns a range of movements associated with resources and home ranges. Physiologically, these movements can be characterized by a tendency to be proximately triggered by inputs coming from resources necessary for growth and maintenance. A shortage stimulates movement until the resource in question is encountered, whereupon inputs from the resource cause the movement to cease. Thus an animal forages when it is hungry but stops foraging and begins to feed when it encounters a food item of an appropriate kind. The modular growth form of many plants and corals can be considered a foraging "movement" that accomplishes the same end. A young mammal driven from its home range by social interactions ranges over the habitat until it encounters cues informing it of a new suitable home range that it can occupy; it then stops ranging. The ecological outcome of these station-keeping and ranging movements is to keep the organism in the habitat where it acquires the resources necessary for given stages of the life cycle.

Migration, the second type of movement behavior, is different. It is a behavior that takes an organism out of its current home range and habitat. Physiologically, migration is characterized by a suppression or inhibition of responses to proximate stimuli emanating from resources required for vegetative functions. Movement is usually triggered by inputs that forecast a shortage of resources and/or by the organism's own endogenous rhythms. Migration can be distinguished from other forms of movement because inputs that would cause an organism to cease ranging or station keeping do not cause it to stop migrating. Rather migration ceases as a result of physiological changes brought about by the movement itself. The ecological consequence of migration is movement to habitats different from the one the organism leaves and usually, although not always, at some distance from it. All these factors will be considered in detail in Chapter 2.

The various kinds of movement do not partition into watertight categories. This is especially true of station-keeping and ranging responses where intergradations between categories may be extensive. Defining the categories nevertheless should provide a useful guide as to how to go about studying them. What is obvious in trying to define the categories is that all too often we really know very little about the behaviors in question. This is basically for two reasons. The first is that they are often difficult to study, and the second is that we have largely lacked a framework that classified movements without regard to which particular group of organisms was performing them. Too often this has meant that important similarities and differences in physiology and ecology have been ignored and that potentially interesting and important questions have therefore simply not been asked. Asking whether a parasite, say, is ranging or migrating when moving between hosts can allow the formulation of hypotheses about important basic and applied aspects of its biology.

In spite of the sometimes vague boundaries between movement categories, distinctions are important because natural selection will act differently on the different behaviors. Our ability, or lack of it, to distinguish among the categories will in part reflect our ignorance, but it will also reflect the strength of natural selection in promoting the evolution of distinct behaviors. It is important to understand why selection might sometimes act to make movement behaviors of different kinds very distinct and why under other circumstances it may act to render them extensively overlapping. I hope the taxonomy outlined will provide some guidelines not only for distinguishing among behaviors contributing to lifetime tracks, but also for determining how natural selection has produced the many patterns observed.

2

Migration: A Definition

Historical Background

Perhaps no other behavior has suffered so much from confusion of definition as migration. What types of movement should be considered migration and how these types relate to one another are complex questions that have resulted in a definition barrier impeding synthesis and generalization across systems (Rankin 1985b). To a great extent this is because students of the phenomenon have tended to focus on a single taxon and its peculiarities, with the result that definitions depended very much on the particular group under study (Baker 1978; Taylor 1986a,b; Stenseth and Lidicker 1992a,b). Ornithologists paid little attention to the work of entomologists and vice versa and neither paid much attention to the movements of aquatic organisms. There has also been a strong tendency to concentrate effort on the readily observed and often spectacular movements of long distance to-and-fro migrants among birds, butterflies, large mammals, or fish. Return movements were established early on by vertebrate biologists as the criterion for "true migration" (Thomson 1926; Heape 1931). Thomson described such "true" migrations as "changes of habitat, periodically recurring and alternating in direction, which tend to secure optimal environmental conditions at all times" (1926, p. 3). Because return movements were unknown, with but few exceptions, among insects and most mammals, most of these animals were specifically excluded. Later authors, summarizing the movements of various vertebrates, adopted similar definitions and often emphasized a distinction between migration and other movements that, although sometimes extensive, did not involve to-and-fro passage (Heape 1931; Dorst 1962; Harden Jones, 1968; Orr 1970; Stenseth and Lidicker 1992a,b; for an exception, see Gaines and McClenaghan 1980). These latter were referred to as "emigration," "nomadism," or in the case of the many small organisms such as parachuting seeds, aphids, spiders, and marine plankton that were transported by wind or currents, "passive dispersal" (Heape 1931; Hardy 1958; Williams 1958).

The difficulty with limiting definitions of migration to movements with a to-and-fro pattern is that many organisms display patterns that, although they do not involve round-trips, have the same function as the so-called true migrations. In other words they clearly work "to secure optimal environmental conditions at all times" by allowing exploitation of different habitats by different ontogenetic stages or exploitation of habitats that change seasonally, successively, or as a result of resource depletion. Many of these movements would be those described as "on the whole a rather quiet, humdrum process . . . taking place all the time as a result of the normal life of animals" by the pioneering ecologist Charles Elton (1927). Some, however, are quite dramatic, such as the large-scale, long-distance movements of armyworm moths (*Spodoptera* spp.) (Rose et al. 1987) and many butterflies. Conversely, many round-trip movements can clearly be seen as foraging or commuting rather than migration (Chapter 1).

Furthermore, there are strong behavioral similarities between many long-distance one-way and round-trip movements. Both involve emphasis on locomotion not seen at other times of the life cycle. In long-lived organisms, round-trips can take place within a generation; in short-lived forms, one-way movements may eventually become round-trips, but only after several generations have elapsed. Even the purportedly passive movements on winds or currents involve active embarkation by the partaking organisms. The takeoff behavior of spiderlings and the photopositive upward swimming of marine larvae entering the plankton are cases in point. These "passive" journeys can occur over considerable distances in vehicles of long duration and consistent direction (Drake and Farrow 1988 and Chapter 5). These (and other) movements are also called "migration," and the question thus arises as to whether it is possible to arrive at a conceptual definition of the term based on common features irrespective of variation in distance, direction, or probability of return.

Developments over the past few decades have further focused the issue. An important step occurred around 1960 when there was a major paradigmatic shift among entomologists in thinking about insect migration. Until that time attitudes about insect movements basically reflected those concerning birds, and only large powerful flyers such as butterflies, dragonflies, or locusts were considered migrants, and—at least for butterflies and dragonflies—it was usually assumed that the journeys involved a round-trip. This outlook was typified by the great summary works of C. B. Williams (1930, 1958). But with the publication of the important papers of Johnson (1960, 1963), Kennedy (1961), and Southwood (1962), the perspective changed quite dramatically from concentration on the geographical description of movements to their behavioral and physiological bases (Johnson and Kennedy) and to the selective factors acting upon them (Southwood).

Johnson took as his primary point of departure the observation that it was mostly young adult insects that migrated, usually just after the teneral period (from *tener*, Latin for soft) of cuticle hardening. As he noted, this was before reproduction, with the consequence that migration was linked to reproductive physiology; Johnson called this tendency to depart before the onset of reproduction the "oögenesis-flight syndrome." The ecological consequence was that insects would produce their young in the habitat they reached at the end of migration, thus placing migration firmly in the context of life histories. These themes were developed further in Johnson's comprehensive book on insect migration and dispersal (Johnson 1969; see also Rankin et al. 1986 for a discussion in relation to recent developments in insect physiology).

Kennedy, working with aphids, developed the theme that migratory behavior was integrated with the settling responses that ended movement. The aphid experiments will be discussed more fully in the next section, but Kennedy's basic concept was that there was a reciprocal interaction between migratory behavior and some portion of the settling responses (what he called the "vegetative" responses promoting feeding, growth, and reproduction). Each inhibited the other in Kennedy's scheme, with the consequences of the inhibition being that each could later "rebound" once the inhibition was removed. Migration, by initially suppressing settling, thus promoted its stronger manifestation later and vice versa. Both Johnson and Kennedy stressed that migration represented specialized behavior.

That behavior was further placed into an ecological and evolutionary context by Southwood, who summarized evidence that migration was characteristic of insects living in temporary habitats, such as vernal ponds or early successional fields, whereas insects living in more permanent situations like lakes or forests were nonmigratory, and many were even wingless. This pattern of migration in response to shifting environments was later formalized with the notion of a habitat "templet" expressed in the ratio H/τ, with H the duration of the habitat and τ the generation time of the insect. More migration occurs as this ratio shrinks toward unity (Southwood 1977).

Parallel developments in conceptualizing the evolutionary and ecological milieu of migration were also occurring among students of other taxa. In the 1950s ornithologists began developing more sophisticated notions of migration as an evolved behavior integral to life histories (Kalela 1954; Lack 1954; Salomonsen 1955). The stage had been set by Elton (1927, 1930), who clearly included animal movements in his considerations of population properties. Kalela, Lack, and Salomonsen put them in the context of what we would now call behavioral ecology, explicitly outlining the evolutionary significance of migration as a behavior molded by natural selection to ensure breeding at the right place and time. More recently ichthyologists have formulated a model of fish migration as a life history syndrome involving energetic trade-offs between movement patterns and reproductive output (Schaffer and Elson 1975; Leggett 1977). There has thus been a steady shift toward attempts to understand the behavioral characteristics and the ecological and evolutionary context of migration in addition to describing its course and timing. These changes are evident in recent symposia on migratory movements (McCleave et al. 1984; Rankin 1985a; Gwinner 1990; Rainey et al. 1990; Dingle and Gauthreaux 1991).

As a result of these developments, it has become apparent to many students of migration that any useful definition can no longer be restricted to round-trip movements. At the same time it is necessary to avoid a conceptualization that is so broad as to be devoid of biological meaning or heuristic content. We thus return to the question raised earlier: Can we arrive at a definition that encompasses all (or most) of the phenomena called migration and that focuses on the characteristics of the migrant?

Defining Migration

Migration as Specialized Behavior

The work of Johnson, Kennedy, and Southwood clearly focused attention on the behavioral characteristics of migratory individuals. This concentration on what migrants actually do has revealed the ways they differ from organisms that may be moving, but are not engaged specifically in migratory behavior (Chapter 1). The characteristics are summarized by Kennedy (1985) and Southwood (1981), and they fall into five major categories that seem to apply in varying degrees across all taxa of migrants, regardless of the particular size of the organism, the medium in which it travels, or its mode of locomotion. Migrants may fly, swim, walk, float, or balloon on parachutes or silk threads, but all show at least some of the five characteristics of migration even though distance, duration, timing, frequency, and final destination may all vary.

The first characteristic of migrants is persistent movement. This activity carries the migrant beyond its original habitat where it obtained resources to a new one in which it also gathers resources; there may, in fact, be new and different resources gleaned at the destination, as well as ones similar to those at the departure point, although this isn't necessarily so. An insect or bird, for example, may both feed and reproduce at the termination of a migratory flight (the site of egg laying or nesting being a new resource) whereas it only fed in the site of origin. A habitat can thus be considered "the area that provides the resource requirements for a discrete phase of an [organism's] life" to follow Southwood's (1981) convenient and ecologically meaningful definition.

These migratory movements *between* habitats are quite different from movements *within* a single habitat. As was noted in Chapter 1, the within-habitat movements of station keeping and ranging are focused on the available resources and cease when a resource of a particular kind is encountered. Migratory movements are characterized by the temporary suppression of responses to resources. In the course of movement to a new habitat, an organism usually covers much greater distances than it does while performing station keeping or ranging activities. Many of these can be impressive indeed. The arctic tern may travel nearly 20,000 kilometers between Arctic breeding grounds and Antarctic feeding areas, and even tiny aphids may traverse 1000 km or more when migrating to a new host plant (Alerstam 1990; Johnson 1969; and Chapters 3 and 5).

The second characteristic of migratory behavior is that it is straightened out, in contrast to station keeping in particular in which there may be much turning or backtracking. In self-powered animals such as birds, fish, or whales, which make one or more round-trip journeys within a lifetime, this straightening may take a specified direction whose maintenance requires sophisticated orientation and navigation mechanisms (Chapter 8). In insects, plant seeds, and planktonic marine larvae, direction may or may not be specific and is determined by the wind or currents in which migration take place. Often there seem to be mechanisms available, however, to select transporting agents most likely to assure arrival in a suitable habitat (Drake and Farrow 1988 and Chapter 5). Furthermore, even the most feeble flyer, swimmer, or balloonist usually exerts some control over its direction and its entry to and exit from the transporting vehicle. Entry into the appropriate wind or current stream can assure that migrants maintain a consistent direction for hundreds or even thousands of kilometers.

Third, migrant organisms are undistracted by those stimuli that would arrest their movements were they station keeping or ranging (Chapter 1). Responses to inputs arising from resources promoting growth and maintenance are evidently inhibited or suppressed during migration. Except when they have depleted fat reserves, migrant birds will not stop and feed even when they could easily do so, and migrant black bean aphids (*Aphis fabae*) will not respond to the young leaves of desirable host plants which would ordinarily induce probing with the mouthparts, feeding, and larviposition(Kennedy and Booth 1963b). More interestingly, whereas flight initially inhibits settling in these aphids, the settling responses later become stronger than they would have been without flight, so that flight actually "primes" settling (Kennedy and Booth 1964) as will be discussed in more detail below. Such priming also occurs in birds; the *Zugenruhe* or migratory restlessness of caged migrants declines more rapidly in birds that have flown than in those kept continuously caged in the laboratory (Gwinner and Czeschlik 1978, Berthold 1993, and Chapter 6). In some migrants there may be alternation between response to and suppression of inputs from resources. Both birds and butterflies will feed en route if it becomes necessary to increase energy stores for the next part of the journey (Brower 1985; Bairlein 1990), and some birds may even hold feeding territories temporarily (Rappole and Warner 1976).

Fourth, distinct behaviors of leaving and arriving are characteristic of migrants. Most migrant birds, for example, become hyperphagic before departure and may increase food intake by as much as 40% above normal, with the excess stored as lipid fuel. This condition was recognized as long ago as 1928 by Groebbels and termed zugdisposition or "travel disposition" (Groebbels 1928; see also Berthold 1975; Rappole and Warner 1976). Phoretic mites actively seek out and attach themselves to an insect for transport and then release and resume foraging behaviors (inhibited during transport) after a period of travel (Binns 1982; Houck and O'Connor 1991). First instars of the salt marsh aphid, *Pemphigus treherni*, which lives on the roots of the sea aster, become photopositive when ready to migrate and move to the soil surface where they are picked up by the tide and distributed over the marsh; after 30 minutes of floating, they become photonegative and move down to the roots when they encounter a new host plant (Foster 1978; Foster and Treherne 1978). Similarly, *Aphis fabae* is sensitive to blue light from the sky on takeoff, but becomes sensitive to yellow light reflected from the immature leaves of host plants at the termination of a migratory flight (Kennedy et al. 1961).

Fifth, migrants reallocate energy specifically to support movement. Thus birds may double their body weight in subcutaneous fat, insects vastly increase the size of the fat body, and plants allocate fat to the embryo in a departing seed (Chapter

6). There are similarities here to the energy storage that accompanies the many forms of dormancy and hibernation, and in fact, in organisms such as insects and plant seeds, migration and some form of dormancy (reproductive diapause in insects) may occur simultaneously. Because of trade-offs between migration and reproduction in the allocation of resources, migration is an important life history trait.

In summary there are five attributes that stand out as characteristics identifying migration as a distinct and specialized behavior. These are (1) persistent movement of greater duration than occurs during the station keeping or ranging movements of the same organism; (2) straightened-out movement that differs from the relatively frequent turning that occurs during ranging and station keeping, especially the latter; (3) initial suppression or inhibition of responses to stimuli that arrest other movements but with their subsequent enhancement; (4) activity patterns particular to departure and arrival; and (5) specific patterns of energy allocation to support movement. Not all migrants should be expected to display all five attributes all the time—natural selection will see that they are apportioned appropriately in each case—but migrants can be expected to show some if not most of the traits most of the time. Any definition of migration must encompass that fact.

Most definitions of migration focus on its outcome rather than on its behavioral characteristics. As a result, they are liable to ambiguities, and they do not provide a ready means to tell whether a particular individual is a migrant, which the above characteristics do allow. The most complete definition of migration incorporating the special traits is the one proposed by J. S. Kennedy first in 1961 (Kennedy 1961) and then later (Kennedy 1985) in a somewhat modified form as follows:

> Migratory behavior is persistent and straightened-out movement effected by the animal's own locomotory exertions or by its active embarkation on a vehicle. It depends on some temporary inhibition of station-keeping responses, but promotes their eventual disinhibition and recurrence.

With some modifications, which I shall discuss below, this is the definition of migration I use in this book.

Some points about this definition are worth mentioning. First, it does not depend on the length of the path or the repetition of the route but focuses on the characteristics of the behavior itself. Secondly, a key word stressed by Kennedy is the "some" modifying "temporary inhibition." Not all station-keeping (or ranging) responses are inhibited or suppressed all the time or 24 hours a day. As examples, many birds migrate at night and may feed during the day, and the large milkweed bug, *Oncopeltus fasciatus*, segregates migratory and vegetative activity into separate parts of the diel cycle by means of a circadian "clock" (Caldwell and Rankin 1974). Thirdly, although Kennedy restricted his definition to animals, I extend it to include at least plants and fungi as well. These organisms have evolved anatomical and physiological mechanisms that produce, in effect, specific departure and transport behaviors such as pappi that act as parachutes, arals that guarantee carriage in avian guts (Howe and Smallwood 1982) or devices for expelling spores (van der Pijl 1969; Pedgley 1982; Simons 1992) all of which "inhibit," in the sense that they prevent, station keeping. If one modifies Kennedy's definition slightly by substituting "organism" for "animal" and interprets inhibition to include those aspects of plant or fungal physiology that prevent premature germination of migrants, many of the movement behaviors of fungi and plants can be readily included. This should become more apparent as I discuss various examples and aspects of migration in the remainder of the book.

With this modification to expand the life forms covered, the Kennedy definition has major advantages over all other definitions appearing in the migration literature. By focusing on the behavior itself, on what the organism is actually doing to move and not simply on the geographic coordinates of the movement, it is predictive, rather than merely descriptive. Implicit in the definition are tests for determining whether a particular individual is a migrant. If it is, it should manifest reciprocal interactions between ranging and station keeping responses on the one hand and migratory responses on the other, and these should, at least in principle, be detectable with properly designed behavioral or physiological experiments. Regrettably these sorts of behavioral tests have rarely been applied or even considered, so that reciprocal interactions have not been looked for. I hope this book prompts much greater investigation of these largely unexplored interacting behaviors; it should be an exciting area for future research. Further, by explicitly focusing on the interaction between growth and reproductive behaviors and movement behavior, the Kennedy definition puts migration firmly in the context of life histories. Finally, it defines a set of variables and parameters subject to the action of natural selection in the molding of migration–life history syndromes (cf. Dingle 1985, 1989). No other definition is as complete; and it is my argument, therefore, that Kennedy's definition is the best characterization of migration across taxa unless or until its specific tenets are invalidated by suitable test.

The Migratory Behavior of *Aphis fabae*

Kennedy formulated his definition of migration in large part based on his detailed analysis of the migration of the black bean aphid, *Aphis fabae*, perhaps the most thorough study of the migratory behavior of any organism (Kennedy 1958 et seq; Kennedy and Booth 1963a,b, 1964; Kennedy and Ludlow 1974). I shall first describe Kennedy's system in some detail, and then in the light of his results, discuss some examples of migration in other organisms in the next section. *Aphis fabae*, like most aphids of temperate climates, produces both sexual and partheno-genetic forms (holocycly); it also alternates between a primary woody host, in this case the burning bush or spindle (*Euonymus europaea*), and summer herbaceous hosts of the legume family, including the fava bean, *Vicia faba*. In the spring aphids hatch from zygotic eggs and spend two generations on *Euonymus*. They then produce migrants that fly to bean fields and produce apterous parthenogenetic young by vivipary, and these offspring and their descendants continue the parthenogenetic part of the cycle throughout the summer. When apterous females become crowded, they produce winged alates via a maternal effect (Mousseau and Dingle 1991), and these fly to new hosts. In the autumn sexuals are produced, which generate the overwintering zygotic eggs, and the cycle repeats. Kennedy studied the migratory behavior of the summer viviparous parthenogenetic alates that fly to a new host plant before larvipositing their young.

Migration was studied in a laboratory flight chamber that allowed the aphids to fly freely (see Chapter 4 for details of the design). Aphids were induced to fly by placing them under a bright light. As they flew upward their rate of rise was controlled by a downward windstream, regulated by a butterfly valve attached to a recording device. In this way the wind speed and the rate of climb could be matched, and hence the latter continuously recorded. Settling was induced by presenting a leaf below the flying aphid; this leaf was attached to a lever arm that could also be twisted to shake off a settled individual in order to induce further flight. Aphids would fly continuously in this chamber for periods of up to 2 to 3 hours.

A typical record of an aphid flight is illustrated in Figure 2–1A. The aphid starts out with a high rate of climb of over 40 cm·sec^{-1}, but as flight proceeds, it enters a "cruising" phase, and the rate of climb declines. In the example in the figure, the aphid was presented with a host leaf after approximately 130 minutes of flight, on which it landed and undertook settling responses (see Table 2–1). The aphid was then induced to takeoff by twisting the lever arm (natural takeoffs also occur), and it climbed at a much higher rate than recorded just prior to landing. Not only is this climbing rate greater than pre-landing, however; on many occasions it was also greater even than the rate on initial takeoff. This increased strength of flight after landing strongly resembles the enhanced response of a reflex (or even of a single nerve cell) after nervous inhibition, prompting Kennedy to postulate that the process inducing strengthened flight was similar to the alternating "rebound" of responses occurring with antagonistic reflexes, discovered by Sherrington (1906). Kennedy termed the process "antagonistic induction" to indicate its similarity to the "successive induction" that Sherrington described for spinal reflexes. The strength of flight rebound increased with longer stays on the leaf and with greater suitability of the leaf as a host, as determined by the probability of larviposition. The strengthened flight did not simply reflect recovery from fatigue because the same "rest" periods induced different rates of climb in subsequent flight, depending on leaf quality.

Additional evidence that flight rebound did not simply represent recovery from fatigue was provided by an experiment comparing continuous uninterrupted flight with flight periodically interrupted by the presentation of a leaf for settling (Fig. 2–1B). The experiment was analogous to the "touch-and-go" landings practiced at airports by pilots in training. Aphids were divided into two groups. In one, individuals were allowed to fly continuously for 20 minutes. In the second, aphids also flew for 20 minutes, but once every minute they were presented with a leaf and allowed to "touch down," but were then shaken off to resume flying before

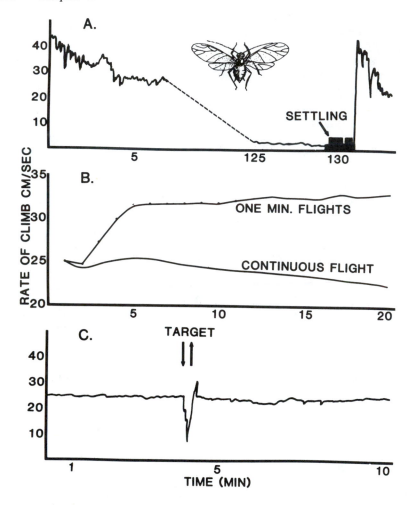

Figure 2–1. Responses of free-flying aphids in the Kennedy aphid flight chamber. *A.* Continuous flight interrupted by settling (upper level of settling consisted of aphid probing): note rebound of flight after settling. *B.* Boosting of flight following a series of touchdowns at one-minute intervals as compared to continuous flight. *C.* Response of aphid to target presented for 10 seconds at lateral edge of visual field. (From Dingle 1985, after Kennedy 1966 and Kennedy and Ludlow 1974.)

ensuing settling responses could take place. The repeated touch-and-go landings did not allow rest and recovery (if anything they should have induced greater fatigue because of the repeated climbs following takeoff), but they nevertheless still "boosted" flight to even higher levels of activity, often to the point where the aphids would totally ignore leaves that were otherwise powerful landing stimuli.

Reciprocal effects of flight on settling were also analyzed and are summarized

Table 2–1. Influence of Flight Duration in a Laboratory Flight Chamber on Subsequent Settling Behavior of *Aphis fabae*

Flight Duration	Settling Behavior	Percentage of Individuals	
15–30 sec	Settling without probes*	50	(17)[†]
	One probe only	50	(50)
	>1 probe	—	(33)
	Going below[‡]	—	(—)
	Larviposition	—	(—)
60–75 sec	Settling without probes	47	(27)
	One probe only	12	(67)
	>1 probe	29	(6)
	Going below	12	(—)
	Larviposition	—	(—)
6 min	Settling without probes	5	(11)
	One probe only	21	(89)
	> 1 probe	16	(—)
	Going below	37	(—)
	Larviposition	21	(—)
40–60 min	Settling without probes	8	(5)
	One probe only	8	(80)
	> 1 probe	8	(10)
	Going below	22	(5)
	Larviposition	54	(—)

*Probe = apparent attempt to insert stylets into leaf
[†]First number = percentage when settling on mature bean leaf; second number = percentage when settling on fuchsia
[‡]Movement of aphid to underside of leaf (where larviposition takes place)
SOURCE: From Kennedy and Booth 1963b.

in Table 2–1. Longer flights promoted enhanced levels of settling behavior, culminating in larviposition on the underside of a leaf if flights were long enough. The influence of host plant suitability is also apparent. Settling responses after long flights were stronger on a mature bean leaf (a moderately suitable host) than on a fuchsia leaf, a totally inappropriate host on which these aphids do not survive or reproduce. On fuchsia there is, in fact, little effect of flight length on settling responses, demonstrating an interaction between the expression of priming by flight and host plant suitability.

Reciprocal coordination occurred not only between flight and settling responses, but between different kinds of flight behavior; an example is illustrated in Figure 2–1C. During cruising migratory flight, in which the aphid oriented toward the overhead light source, a leaf-shaped yellow card was presented to one side of the flying aphid. This card induced a "targeted" flight, differing from cruising in that it involved a horizontally directed approach to the presented object that included a depressed rate of climb. Longer migratory flights produced more persistent horizontal approach flights directed at the target and a greater depression of climbing rate. If the target was presented repeatedly, climbing rates increased after the target was withdrawn in a manner similar to the increased rates induced by the touch-and-go experiments involving landing responses to a leaf. There were thus reciprocal effects between the two types of flight.

Finally, to complicate the situation still further, there was antagonistic depression between flight and settling in addition to antagonistic induction. If flight was weak, or if settling was strong at a previous landing, settling could have a depressing rather than a boosting effect on flight. Once again host leaf suitability was involved because stronger settling responses were more likely when there was a preferred host. When flight was particularly strong, as after much boosting by intermittent landings, flight could depress settling.

The whole migratory system of the aphid thus is a finely organized combination of reciprocally exciting and inhibiting effects between flight and settling and between migratory cruising and horizontally directed target flight. Migrants can be identified by the fact that they will ignore sensory inputs that will later cause them to settle, feed, and deposit young, *but* eventually these responses will be disinhibited and will return stronger than before. These aphid results are the data base from which Kennedy derived his definition of migratory behavior, and show clearly that migration in these insects consists of a complex behavioral and physiological syndrome incorporating interactions between movement behavior and the "vegetative" behaviors that eventually arrest that movement. These interactions include initial inhibition by flight of responses that later terminate flight when suitable conditions are encountered, but eventual lowering of the response thresholds to those conditions. Migratory flight thus differs physiologically from other types of behavior, including other types of flight. We must now ask whether the aphid case is unique or whether the sorts of reciprocal interactions observed might also characterize other migrants.

Migratory Behavior

One of the difficulties inherent in attempting to compare migration behaviors of other organisms with those of aphids is that the sorts of patterns that Kennedy observed have essentially never been specifically looked for elsewhere. Nevertheless, in recorded observations of migrants the major characteristics do appear, and they do so in examples from across the spectrum of migrating taxa. Comparisons reveal the similarity in performance characteristics in the face of wide variation in modes of locomotion, life history stage, or the coordinates of the route traveled. We saw some examples of characteristics of migratory behavior paralleling those of aphids in the prelude leading up to Kennedy's definition of migration above. Here I summarize a few more examples to make the parallels more explicit.

An example of a parallel occurring in birds is the behavior of transoceanic migrants flying from eastern North America over the Atlantic to South America described by Williams and Williams (1990). The flight occurs in three phases, each distinguished by different means of orientation. At takeoff, these migrants are sensitive to local wind and weather allowing departure under suitable conditions.

Rates of climb and often the ascent to high altitudes (Alerstam 1990) also distinguish these takeoffs from those occurring at other times. Once on the transoceanic journey, the migrants appear to assume a cruising or programmed phase during which they are insensitive to local weather or landmarks. In the final phase of the oceanic crossing, migrants are once again responsive to local weather and landmarks as they descend for landfall. Note the parallels to the phases of flight observed by Kennedy in the black bean aphid (Fig. 2–1). Leggett (1985) describes similar changes in responsiveness in American shad (*Alosa sapidissima*) entering rivers on their spawning migrations. These fish are highly selective of water temperature when they begin their runs, entering the rivers only when outflows reach 15°C, a temperature that predicts favorable spawning conditions upstream. Once in the river, however, they are insensitive to temperature fluctuations. Interactions between the upstream journey and responsiveness to cues promoting spawning at the breeding sites are also likely but remain to be investigated.

Marine larvae respond to specific stimuli to ensure entry into the plankton to migrate to new settling sites, as illustrated by the larvae of the stomatopod crustacean *Gonodactylus bredini* (Dingle 1969). These animals live in rock cavities in the tropical marine littoral. Females lay their eggs in the cavities and show considerable maternal behavior, including holding, cleaning, and defending the eggs (Dingle and Caldwell 1972). Newly hatched larvae remain in the cavity with the female through the first three stadia, surviving on egg yolk. During this time they are negatively phototactic and highly thigmotactic, clinging to the walls of the chamber or any loose debris. Precisely at the completion of molt to the fourth stadium, they cease thigmotaxis and become positively phototactic, leaving the cavity and entering the plankton. They then spend several stadia in the plankton, but at the time of molt to the settling postlarva, they again become thigmotactic and descend to the substrate seeking shelter in cavities or interstices. Stomatopods are not unique in these behaviors. A survey of 141 species of marine invertebrates with planktonic larvae found that the juveniles of 82% of the species displayed photopositive behavior when entering their pelagic phases and later photonegativity when settling to the substrate (Thorsen 1964). Settling behavior itself can be complex as shown by the larvae of the red abalone, *Haliotis rufescens*. These exhibit a characteristic behavior called "bottom hopping" when they complete the pelagic migratory period and become ready to take up life on the bottom (Morse 1991). They alternately sink to the substrate and then rise again until they eventually contact red coralline algae on which they settle in response to the presence of a small peptide molecule, related to the neurotransmitter gamma amino-butyric acid (GABA), on the algal surface.

Equally characteristic of migratory behaviors are those of ballooning spiders and phoretic mites. Young wolf spiders (*Pardosa* spp.) climb first to the top of plants or other objects, a behavior also seen in the black bean aphid. They then attach their walking threads or "drag lines" to the substrate and free themselves from attachment by breaking them with a jerk. Following this they stand on "tiptoe" and extrude six to eight silk threads from the spinnerets and are lifted off the substrate by the wind when the threads reach a length of about 70 cm (Richter 1970). McCook (1890) gives a wonderful description of specialized landing responses as the spiderlings gather in their silk lines to facilitate descent and touch down. Phoretic mites in migration behave in a manner paralleling spiders and aphids as summarized by Binns (1982) and Houck and O'Connor (1991). While waiting to attach to a suitable transporting insect, the mites may develop into "waiting stages" that exhibit "questing" behavior with the first two pairs of legs raised vertically while the body is anchored to the substrate by caudoventral suckers. This behavior is often undertaken after ascending an object, as was the case with spiders, and in some cases a mite may actually jump 2 to 5 cm onto a passing host. Similar behavior of scale insects is described in Chapter 5. Once attached, the mites ride without feeding. Growth-promoting or "vegetative" responses are thus suppressed both during the lead in to attachment and once attachment occurs, consistent with our definition of migration. The detachment behavior of the mites has proved much more difficult to study and virtually nothing is known about its characteristics.

Analogues to spider ballooning and mite phoresy occur among plants and fungi (see reviews by van der Pijl 1969, Howe and Smallwood 1982, Fenner 1985, and Simons 1992). Ballistic systems may assist embarkation of seeds (these also occur in fungi, mosses, and liverworts for the release of spores), or fleshy sweet-tasting pulp or arals may attract vertebrates to eat the fruits and hence transport the seeds which are regurgitated or passed through the gut. *Virola sebifera*, a large fruiting tree of New World tropical rainforests whose fruits possess fleshy arals, is a good example of a plant well adapted to vertebrate transport of the seeds (Howe 1981). Wings, tufts, and pappi have evolved in different taxa of plants to facilitate wind-aided transport. There may even be special devices such as the hygroscopic bristles of the "cranesbill" of *Geranium* to assist in movement to suitable germination sites once landing occurs. Plants do not have nervous systems, so the interaction between sets of behaviors, in the sense that they occur in animals, are not possible, but in other respects the "behavioral" adaptations of plants for migration are remarkably parallel to those in animals. The physiological mechanisms accompanying seed transport, such as the requirement for partial digestion of a tough seed coat before germination is possible, can be considered analogues of mechanisms priming settling in aphids. The search for more such analogues, from the perspective that seeds in some cases are the migratory phase of the plant, would seem to pose interesting questions for future research.

This brief survey of examples indicates the presence of interacting sequences of behaviors during migration and thus parallels with the migratory behavior of Kennedy's aphids. There is persistent movement and active embarkation, and the movements tend to be straightened out. During migration, activities related to growth and development are suppressed. Finally, there are distinct behaviors associated with the beginning and ending of migration, although our general lack of knowledge about the latter is one of the major lacunae in our understanding of migratory syndromes. Although this and other gaps exist as targets for future research the commonalities observed among the very different sorts of organisms suggests that the definition of migration derived from aphids does indeed cover migration in its many forms.

The Problem of "Dispersal"

If migration suffers from a confusion of definition, no word has contributed more to that confusion than the term "dispersal." The term is most often used for presumptively one-way movements and usually applied to individuals moving to breeding locations away from the place they were born. Thus in Howard's classic use for small mammals it is "the permanent movement an individual makes from its birthsite to the place where it reproduces or would have reproduced had it survived and found a mate" (Howard 1960). More recent definitions such as "a complete and permanent emigration from an individual's home range" (Holekamp and Sherman 1989) or the movement from one home site to another (Stenseth and Lidicker 1992b) are briefer but basically similar to Howard's. The term is also used more loosely as a synonym for one-way movement, as a synonym for movement in general, and interchangeably with emigration (e.g., Berger 1987). Watson (1992) argues that dispersal is a vague term meaning different things to different workers and even different things to the same workers and prefers instead to use emigration and immigration. Migrants, however, also immigrate and emigrate, terms which simply indicate arrival or departure.

The real difficulty with the term "dispersal" is that it confuses individual behavior with population processes. I have looked up "dispersal" in several dictionaries, including dictionaries of biological terminology, and all definitions include a component of scattering or spreading. In biological terms, that translates to an increase in the mean distance between individuals and so is a characteristic of populations (Baker 1978; Southwood 1981; Kennedy 1985). It is individuals that move, and their movements may or may not result in dispersal of the population. To the contrary, movement, whether it be station keeping, ranging, or migrating, can lead to convergence and often does (Baker 1978). Insects, birds, or parachuting seeds migrating on winds, for example, may be concentrated by converging winds in storms (Drake and Farrow 1988 and Chapter 5). Convergence and dispersal may also occur at different parts of the life cycle; in the ladybird beetle, *Hippodamia convergens*, concentrations occur following migration to montane diapause sites, but dispersal is more prevalent when migrating to fields of new crops in the spring (Hagen 1962). Even in the case of phoretic mites, where movement is by attachment to an insect host, the behaviors of both host and mite can lead to benefits that would be reduced or even eliminated by dispersal (Houck and O'Connor 1991). Many classical to-and-fro migrations act specifically to prevent dispersal because they facilitate return to specific sites for overwintering, diapausing, and/or breeding. The ornithological and entomological literatures abound with examples.

A particularly interesting case of alternative aggregating and dispersing occurs in the marine polychaete tubeworm *Hydroides dianthus* (Toonen and Pawlik 1995). These worms produce two distinct types of larvae. The first type, or founding settlers, settle indiscriminately on uninhabited substrata usually within the first 12 days of larval life. These larvae disperse the population by occupying scattered substrates and so increasing mean distance among individuals. The second type, or gregarious settlers, will settle only in response to cues associated with conspecific adults and so cause aggregation of the population. The individual behaviors of these two types of migrating larvae thus produce two quite different population consequences, dispersal and convergence.

Using foraging and migration as examples, I have summarized the relationships between movement and spatial change in Figure 2–2, which is modeled after diagrams in Solbreck (1978) and Southwood (1981). On the vertical axis is the change in mean distance among individuals in the population; that is, the degree of scattering or dispersal versus the degree of convergence. Convergence can result in aggregation if the organisms are actually attracted to each other and over-dispersal can result if there is mutual repulsion. Foraging movements keep individuals "here" in the present habitat, and migratory behavior takes them "elsewhere" to a new habitat as indicated on the horizontal axis. If the local environment contains concentrated resources, covergence can occur through for-

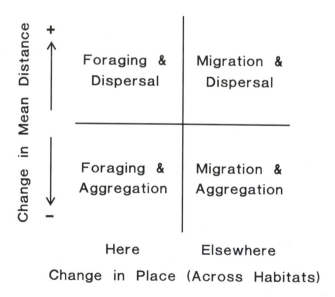

Figure 2–2. Relation between two individual movement behaviors, foraging and migration, and the population processes of dispersal and aggregation, which are functions of the increase or decrease in the mean distance among individuals. Both foraging, which takes place within the home range or "here," and migration, which takes the individual elsewhere, can result in either dispersal or aggregation.

aging; similarly, if suitable habitat is not widespread or site philopatry is strongly developed, convergence can result from migration. Widespread distribution of resources and habitats produces dispersal. Finally, no net change may result as might be the case with similar sized territories from year to year in a migratory bird or the simple filling of existing habitat lacunae in a migrating small mammal. The point to be made is that both migration and foraging, which can be distinguished behaviorally at the individual level, can result in dispersal, covergence, or neither at the population level. The distribution of the population over space, its dispersion, is the *outcome* of migratory, foraging, and other movement behaviors.

Confusion thus arises from the use of the term "dispersal" because it is often, if not usually, used to describe movements that do not increase the mean distance among individuals and may even reduce it. The reason the term comes into the literature as an alternative to the term "migration" seems to be historical. So long as biologists were constrained to see migration as a movement that was defined as self-propelled, long distance, and to-and-fro, any movement not satisfying those criteria required separate terminology (Kennedy 1985). This was particularly true for insects and other terrestrial arthropods, seeds and spores, planktonic organisms, and small mammals. In the latter, almost any movement away from a natal area is termed dispersal; the actual behavior involved during movement or the responses terminating moving are almost never treated; perhaps, as Taylor (1986b) indicates, because of a preoccupation with the consequences of movement for population cycling in time rather than for the distribution in space. Although the mapping of territories and home ranges of small mammals is commonplace, mammalogists have given little attention to the effects of different scales of movement on the patterns of distribution. Nor have they given much effort to learning about the actual behavior of the "dispersers." In part this is because of the difficulty of observing small mammals in the field, but it is also in part a conceptual problem. One doesn't find what one doesn't look for. What behavior and physiology have been described suggests that some small mammals may deposit fat prior to departure, ignore inputs from resources, and otherwise show signs of being migrants (Gaines and McClenaghan 1980). In the case of birds, "dispersal" is widely applied to movements between breeding sites that do not involve to-and-fro migration (ranging in my taxonomy). The most frequent of these movements occur in young birds ("natal dispersal"). Even though they warn of the consequences of density and dispersion in their major review of "dispersal" in birds, however, Greenwood and Harvey (1982) still do not address the consequences of increasing or decreasing the distances among individuals, yet cite examples of both.

The essence of dispersal in the view expressed here is not that an individual leaves its living place, as Lidicker claims in the Introduction to Lidicker and Caldwell (1982), but concerns the degree of scattering that occurs in the *population* as the individuals move about. The problem with using the term migration as a subset of dispersal as Lidicker and many entomologists and marine biologists do, even if for historical reasons, is that it confounds several issues of importance to the biology of the organism. With that perspective one does not attempt to distinguish behaviorally between migration and other movements, and one does not seek to determine if a movement increases or decreases the space among individuals. There are good reasons to think that natural selection will act to discriminate between the behaviors and to differentiate among their outcomes. In other words, different biological functions are involved. Calling migration or any other movement by an individual "dispersal" simply confounds all these issues. With the broadening of the definition of migration and focus on its behavioral rather than topographical characteristics, a term such as dispersal to cover non–round-trip movements becomes unnecessary and focus can now center on whether, indeed, dispersal is an ecologically important process occurring during migratory, station keeping, or ranging behavior. I therefore, advocate that "dispersal" be dropped as a synonym for one-way movement and restored to its use as a description of the *process* leading to a distributional outcome of movement. Population dispersion is that outcome.

Ecological Definitions of Migration

The use of "dispersal" to indicate one-way movements is one manifestation of attempts to define migration ecologically. The leading and most eloquent advocate of this latter approach is L. R. Taylor (1986a,b). The crux of Taylor's argument is that "migration is a life function of whole individuals . . . and can only be defined ecologically" (Taylor 1986b). The reasons for this conclusion reside in the observations, derived from long-term studies of insect movements in the British Isles (Taylor 1986a), that insect populations are constantly shifting in space and time and that these shifts must be incorporated into models of population dynamics. A population was envisaged as an anastomosing fern stele with the vertical axis time and the horizontal axis space (Taylor and Taylor 1977). From the perspective of the population, it is immaterial how individuals move; the "accidental" transport by storm or surge is equivalent to self-controlled behavior. However, the movements involved in population fluctuations were then seen as density-dependent, changing, mutually attracting and repelling interactions between conspecifics. Taylor's model has been criticized on both mathematical and logical grounds (Dye 1983; Thorarinsson 1986; Routledge and Swartz 1991) and, most importantly for our discussion here, because it confounds behavioral and ecological levels by attributing population changes to particular locomotory responses and interactions among individuals in the face of the claim that no special behaviors are necessarily involved. It thus suffers from an internal contradiction (Kennedy 1985).

Migration under Taylor's scheme (Taylor 1986a,b) becomes a matter of movement between the geographic coordinates of the birthplace and those of the next generation, by what means being ecologically immaterial. Taylor then reinforces this conceptual view by noting that the Oxford English Dictionary (and other dictionaries) defines four types of migration: the action of moving from place to place, the transfer of residence, periodical movements, and alteration of position within the body. Combining these with population concepts, Taylor categorizes four types of migration among organisms: passive migration, dynamic (one-way) migration, homeostatic (two-way) migration, and finally social migration within a population by analogy with movement within a body (Table 2 in Taylor 1986b). One again, however, in order to define these types, he must resort to behavior (Gatehouse 1987b) as, for example, in his "Migration of the second kind" (i.e., one-way) which includes "*actively initiated* emigration; *volition* continued through transmigration. . ." (italics mine) so that again ecology and behavior are conflated. Indeed, in defining migration as an action, dictionaries are explicit in labeling it a behavior. It thus makes more sense to define migration as a *behavioral process with ecological consequences* as Gatehouse (1987b) does following Rogers (1983) and Kennedy (1985). Migration is then seen as one of the movements (Table 1—1) by which organisms get about (Gatehouse 1987b, 1989). Accepting migration as a behavioral process also brings it into focus as the result of evolution by natural selection; selection does not and cannot by definition influence accidental transport (Dingle 1980). Further, the only way that an accidental vagrant can have ecological import is to arrive, equally by chance, in an environment favorable for its reproduction and growth (Gatehouse 1987b).

Other attempts to define migration ecologically also suffer because they focus on outcomes rather than the behavior that produces those outcomes. This is the flaw in defining migration as a to-and-fro pathway as ornithologists (Alerstam 1990) and ichthyologists (Harden Jones 1984) are still wont to do. Defining migration in this way ignores the behavior that leads to these circuit movements and can inhibit looking at migratory behavior in new and interesting ways as Leggett (1984), especially, notes. C. G. Johnson's (1969) definition of migration as a change in breeding habitat suffers the same problems, even though Johnson, like Taylor, could not avoid including a behavioral component, the oögenesis-flight syndrome (Chapter 6). Baker's (1978) definition of migration as "the act of moving from one spatial unit to another" includes essentially all movement, so obscuring important differences. In including accidental or unintentional movement, it also includes components that cannot be responsive to natural selection (Dingle 1980). I attempted to include behavior and natural selection in my definition of migration as "specialized behavior especially evolved for the displacement of the individual in space" (Dingle 1980), a definition also adopted by Kerlinger (1989) among ornithologists, but as Kennedy (1985) has indicated, this definition does not differentiate migratory from nonmigratory behavior that also causes displacement, the problem again of incorporating outcome.

In spite of their inadequacies, however, ecological definitions, especially Taylor's comprehensive version, do illustrate two important points. The first is that it is necessary to distinguish hierarchical levels when discussing migration. Ecology and physiology must be kept distinct, but both are important to a full understanding of migratory phenomena. Gatehouse (1987b) characterizes the situation well when he states that migration can only be *defined* in behavioral terms, although it can be *described* in terms of its ecology, which determines its evolutionary course. Thus the evolution, genetics, ecology, and physiology of migration are critical elements of this book. The second point is that the issues are not merely semantic. What migrants actually do, the ways in which they do it, and the outcomes of their actions are biologically fundamental. It matters to be able to distinguish a migrant from a forager because the outcomes of these two behavior patterns can be quite different. Similarly, it might be desirable to know if an organism far out of its normal range (a "vagrant") is an off-course migrant or simply the victim of accidental transport. Without a behavioral definition of migration this isn't possible, even in principle, because there is no specific hypothesis to test for making the distinction. Important, too, is the distribution of organisms in space, whether they are dispersed or aggregated, for that continuum of states has profound implications for the ecology of populations and the fitness of individuals. As I hope I have made clear, the reason for adopting Kennedy's definition for this book is that it sets out in bold relief the issues we must consider to fully understand migration as a biological phenomenon.

Summing Up

Historically studies of migration have focused on a particular taxon with definitions circumscribed by the characteristics of the taxon in question. This has led to the obscuring of commonalities of function and behavior that are characteristic of migration in a great variety of organisms. In the new view arising in studies of the biology of migrants, characteristics once thought to define migration, such as to-and-fro pathways, are now seen as subsets of a much broader spectrum of behavior. This poses the question of whether we can find a definition of migration that encompasses that spectrum and provides a guidepost for future studies. It is my contention that such a definition should focus on behavior and not its ecological outcome. Outcomes are important for determining how natural selection influences the behavior, but not in defining it.

A survey of a broad array of organisms suggests there are five characteristics of migration that distinguish it from other forms of movement. These are (1) it is persistent, (2) it is straightened out, (3) it is undistracted by resources that would ordinarily halt it, (4) there are distinct departing and arriving behaviors, and (5) energy is reallocated to sustain it. Not all migrants will display all of these characteristics all of the time, but most will display most of them at least part of the time during which they are migrating. A definition that incorporates these characteristics is that of J. S. Kennedy (1985), which stresses that migrants inhibit or suppress inputs from resources that terminate other types of movement and that migratory behavior itself primes later responses to those same inputs. An advantage of the definition is that it provides hypotheses to test to determine if a given individual is a migrant. Although derived from observations on the behavior of aphids, the definition seems to apply also to other migratory organisms, at least insofar as there is evidence to make a judgment. There is clearly much research that needs doing on the behavioral characteristics of migrants within the guidelines this definition provides.

There have been various attempts to define migration ecologically, but the difficulty with such efforts is that they confuse behavior with its outcome. Such is the case with the term "dispersal," which is commonly used to describe what in most cases are probably one-way migrations. Dispersal means a scattering or spreading of the constituent elements of a group or population. In biological terms this means increasing the mean distance among individuals; it is, in other words, a population phenomenon, not a characteristic of an individual. Dispersal and its opposites, convergence and aggregation, are all possible population processes that can occur during movement, including but not limited to migratory movement. Dispersion is then the distributional outcome of the population over the landscape.

The definition of migration is not simply a semantic problem. How we define a behavior determines how we view its biology and therefore how we study it. Defining migration in behavioral terms and describing its ecology allows us to formulate hypotheses about both the process of natural selection acting on it and, in this case, the result of the process in terms of behavior. It is not always possible to make hard and fast distinctions between migration and other movement behaviors. Indeed this is to be expected. Given the complexities of the balance between chance and certainty in the lives of most organisms, there will be times when behavioral alternatives are not obvious. Those situations are themselves interesting and should be incorporated in the analysis of migratory syndromes. The framework provided by the definition I am advocating should provide the necessary conceptual perspective.

3

Patterns in Migratory Journeys

The definition elaborated in Chapter 2 explicitly excludes the pathway followed as a criterion for migration. Thus freed from the restrictions imposed by cartography, we can examine the diverse ways a similar behavioral organization can be modified by natural selection to serve a multiformity of ecological requirements. In the following sections I outline a number of different migratory pathways and their spatial and temporal elements. These should serve to illustrate the richness of pattern generated in the life cycles of organisms by the synergism between migratory behavior and the exigencies of the environment and set the stage for a more detailed examination of both proximate and ultimate factors in migration in the remainder of this book.

To-and-Fro Migrations

Many to-and-fro migrations are ones with which we are all likely to be familiar. They include the great seasonal movements that bring songbirds back to Temperate Zone forests and gardens, the ducks and geese to northern lakes and marshes, the great whales swimming past our coasts, and ladybird beetles from mountain refuges to lowland fields and flower beds. Many of these movements occur on a large scale over major portions of the planet; others are on a much smaller scale—sometimes, as in the case of many insects, over distances of only a few hundreds or thousands of meters. All involve breeding at one end of the migratory pathway and most include refuging at the other to escape seasonal harshness in the breeding area. As with all migratory journeys great variety is evident, as the following few examples will illustrate.

I shall begin with the arctic tern (*Sterna paradisaea*), for perhaps no other organism makes such a spectacular journey or so captivates the imagination of ornithologists and lay persons alike. Arctic terns have a circumpolar breeding distribution in the northern hemisphere, nesting along the northern rims of both the Old and New World continents and on Arctic and subarctic islands. In the autumn they depart on the longest migratory journey known for any bird (Alerstam 1985, 1990) to feeding areas at the edge of the pack ice off the coast of Antarctica (Fig. 3–1). They spend their time in the Antarctic between approximately 30° west and

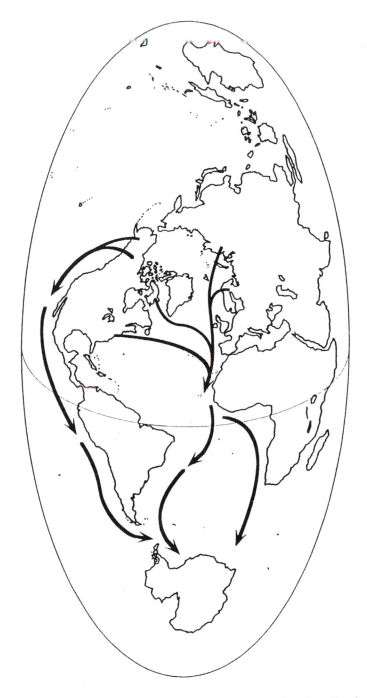

Figure 3–1. Southward migration of the Arctic tern. Populations breeding in both the eastern Nearctic and Palearctic converge in the eastern Atlantic. Off Africa the population divides to migrate down both sides of the Atlantic to Antarctic regions. Birds from the western Nearctic migrate south along the Pacific coast of the Americas.

90° east longitude, or roughly the region due south of Africa and most of Asia (Salomonsen 1967; Alerstam 1990). The terns breeding in northern Europe and along the north coast of Siberia and its outlying islands migrate along the western borders of Europe and then Africa to the Cape of Good Hope; here they enter the prevailing westerly winds of the southern oceans (discussed fully in Chapter 5) for a southeastward flight to the margins of the pack ice. This means a considerable *westward* journey before turning south for birds breeding in the eastern parts of Siberia.

Terns breeding in the New World divide between the Atlantic and the Pacific for their southward journeys. Those from eastern North America, Greenland and Iceland fly first eastward across the North Atlantic to the coast of Europe. This transoceanic journey covers a distance of over 3000 kilometers, during which "they stop to feed only on the rarest occasions (once seen) and never by any chance settle on the water" (Wynne-Edwards 1935; see also pp. 16–17). On reaching the waters off Britain, they join the streams of their Old World counterparts in the movement south. A part of this joint stream branches off at the great bend of West Africa and crosses the Atlantic to Brazil, moving on down the coast of South America before entering the westerly winds for an eastward course to the pack ice. The third group of terns from populations west of Hudson's Bay migrates down the Pacific coasts of North and South America and around Cape Horn in the prevailing westerlies for the final eastward flight to their destination.

The absence of migrating terns on the eastern sides of the continents presumably reflects the availability of food in the surface waters of the oceans in which the terns must feed en route. Cold currents rich in nutrients flow equatorward on the western rims of continents along the eastern edges of ocean basins (see Chapter 5). Conversely, warm currents with less nutrients flow along the eastern continental margins. The terns thus follow routes that take them over the richest waters. Their return northward begins in March and by and large follows the same nutrient-rich routes north, although there is some indication that a portion of the eastern North American birds cross the South Atlantic to reach the East Coast of North America.

Juvenile birds may spend more than one year in the Southern Hemisphere, making the migratory cycle of the arctic tern more complex than just a simple annual to-and-fro journey. Some first-year birds stop first for up to a year in the cold Humboldt Current off western South America or in the vicinity of the southern tip of Africa. The former may contribute the bulk of the birds arriving via the Pacific each year when in their second year they rejoin the southward stream. Other juveniles may circumnavigate the South Pole over the course of a year before again returning north, and sometimes turn up on the southern coast of Australia. The Southern Hemisphere movements of the arctic tern are still known rather poorly and much remains to be learned about the migration patterns of this champion of long-distance flyers.

Although many birds undertake migratory journeys almost as spectacular as that of the arctic tern, many do not but rather make only short trips between breeding and nonbreeding areas. Nevertheless these journeys still function to place birds in the most favorable possible environments throughout the year. The swift parrot (*Lathamus discolor*) is a case in point. This small (24 cm) green parrot breeds on the eastern two thirds of the island of Tasmania, approximately 300 km across the Bass Strait from the southeast coast of mainland Australia, and migrates in the southern autumn (February to April) to southeastern Australia (Blakers et al. 1984 and Fig. 3–2). It forages on the blossoms and nectar of *Eucalyptus* spp. and on the sugary exudates of lerp (scale) insects, all of which are in short supply during the winter in Tasmania. On the Australian mainland the parrots wander widely in search of flowers and lerp south and east of a slightly concave line that extends from Adelaide in the west to Brisbane in the north and east. In this they are similar to arctic terns, which also wander widely in their nonbreeding range. The overall pattern of to-and-fro seasonal movement as seen in the swift parrot and arctic tern is in fact quite typical of most avian migration patterns, albeit varying in scale.

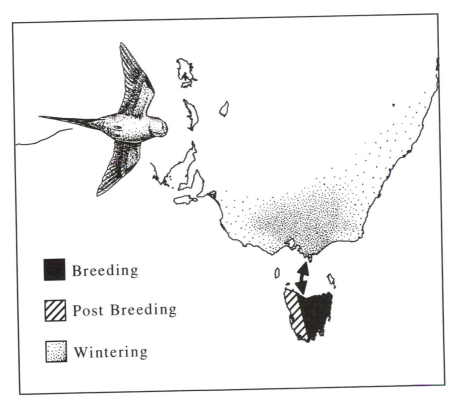

Figure 3–2. Migration of the swift parrot in southeastern Australia. The birds breed in Tasmania and winter on the mainland roughly from Adelaide in the west to Brisbane in the northeast, with greatest concentrations toward the center of the wintering range.

On a smaller scale yet are the seasonal to-and-fro movements of the Eurasian milkweed bug, *Lygaeus equestris*, on the island of Gotland in the Baltic Sea between Sweden and Finland (Solbreck 1971, 1972, 1976; Solbreck and Kugelberg 1972; Solbreck and Sillén-Tullberg 1981 and summarized in Figure 3–3). Their migration is an example of movements to special habitats, discussed more fully in Chapter 10. These insects feed on the seeds of the milkweed *Vincetoxicum hirundinaria*, and a generation of bugs from eggs laid in the spring matures during the summer. In the autumn the bugs fly to sheltered overwintering sites in stone outcrops or in various interstices in and around stone buildings (old Viking lighthouses are a particularly favored spot; see Fig. 3–4). At this time the ovaries of females are undeveloped, the fat body is large and prominent, and the bugs enter a diapause that carries them through the winter. This diapause can be readily demonstrated by bringing the bugs into the laboratory in the early winter, giving them a supply of the seeds on which to feed, and warming them to summer temperatures. Under these conditions it is still some 60 days before females begin to lay eggs. In contrast, if the experiment is repeated in February or March, oviposition begins after only 15 days, demonstrating that diapause has ended in anticipation of spring warming. The fat body supplies energy for overwintering and declines throughout the winter. In May the bugs leave the diapause sites, fly out to the surrounding fields to feed on the developing milkweeds, and produce the next generation. Similar patterns are seen in the related *L. kalmii* (Caldwell 1974) and the coccinellid ladybird beetle *Coleomegilla maculata* (Solbreck 1974) both in North America; all these round-trip flights take place over only a few kilometers at most.

Two points emerge from consideration of these bird and insect movements. The first is a reiteration of the fact that it is not distance traveled that determines whether a specific movement pattern is migratory. Arctic terns fly quite literally from one end of the earth to the other, the swift parrot only a few hundred kilometers, and the European milkweed bug perhaps only a few hundred meters, but all share common features of migratory behavior and common ecological function. During migration and overwintering other functions such as reproduction are discontinued, and the journeys all serve to place the organisms in the most suitable habitats throughout the year. The second point, deriving from the first, is that the behavior is closely linked to physiology. Not only do the animals move in a specific way, but physiological functions like reproduction are suppressed or inactivated during migration activity and the non-breeding portion of the annual cycle. An active process of suppression is indicated by the fact that, for example, milkweed bugs do not reproduce in the autumn months even when placed in suitable conditions in the laboratory. I shall have more to say about physiological interactions in Chapter 6.

The three instances of migration described so far all take place by flight, but migratory journeys also take place in the oceans, freshwaters, and over land. Perhaps the best known of aquatic migrations are those of salmon, which take place between fresh and salt water (diadromy), in this case with breeding in freshwater

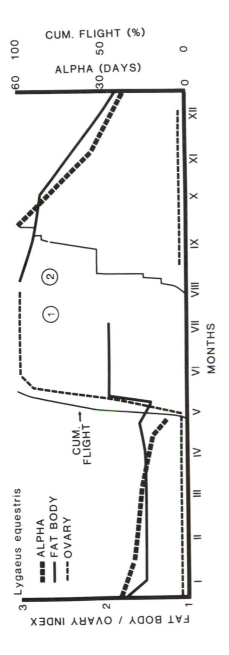

Figure 3–3. Annual life cycle of the milkweed bug, *Lygaeus equestris* in Sweden. Migration to diapause sites occurs in August and September as indicated by the line for cumulative flight beginning in month VIII. At this time, age at first reproduction (ALPHA) occurs about 60 days after bugs are warmed up in the laboratory. The fat body is large in the autumn but shrinks during the winter as reserves are used up. By May age at first reproduction is only 6 days when warmed in the lab; this is about the time bugs are leaving diapause sites and seeking host plants. Circled numbers indicate generations. (From Dingle 1985, based on data from Solbreck 1972, 1976 and Solbreck and Sillén-Tullberg 1981.)

Figure 3–4. Photograph of a cluster of *Lygaeus kalmii* basking in the early spring at a diapause site (the doorway of a Viking lighthouse). (Photo by Christer Solbreck.)

and one or more years maturing in the ocean. (This is anadromy; the opposite cycle, as seen in eels, which breed in the ocean but mature in freshwater, is termed catadromy.) The migration of the sockeye salmon (*Oncorhynchus nerka*) illustrates both the general pattern of salmonid migration and the considerable variation present (Quinn 1985; McDowall 1988). The sockeye is one of six species of Pacific salmon (*Oncorhynchus*), all of which spawn only once (semelparity); two other genera of anadromous salmonids, the Atlantic salmon and several species of trout (*Salmo*) and the lake trout and char (*Salvelinus*), are repeat spawners (iteroparity). Sockeye populations occur on both sides of the Pacific.

Like other salmonids, the eggs of sockeyes hatch in freshwater, in this case in the upper reaches of streams. Unlike their counterparts the pink (*O. gorbuscha*) and chum (*O. keta*) salmon whose fry migrate immediately to the sea, the fry of the sockeye migrate first to nearby lakes where they feed and grow for 2 to 4 years. Chinook (*O. tshawytscha*) and coho (*O. kisutch*) are intermediate to this pattern, but do not use lakes; the former spends up to a year in streams before entering the sea and the latter up to 3 years. Following their stay in a lake, sockeye fry undergo a transformation into the *smolt*, and in this stage undertake a second migration to the sea. In the ocean they wander widely, although there is some evidence from tagging to suggest that the populations from the Eastern and Western Pacific have different oceanic distibutions (but with some overlap). After 2 to 3 years at sea

the now mature sockeye return to their home streams (usually) and migrate upstream to the spawning areas, with gonads maturing as they do so. After spawning,
they die.

This several-year round-trip from stream to lake to ocean and back to stream
is the typical anadromous pattern in the sockeye, but there are two major deviations
from this theme. First, some individuals, called "residuals," remain in the
freshwater lakes, mature there, and make their upstream spawning migrations from
there; they can and do breed with fish returning from the ocean. Secondly, there
are whole populations or "stocks" that are landlocked and apparently have been for
a considerable period. These are different enough in appearance to be given a
separate name, the kokanee. The ability to migrate to and survive in the sea is not
lost, however, as offspring of kokanee parents can still go to sea and return as
typical anadromous fish (Foerster 1947). There thus seems to be much flexibility
built into the sockeye phenotype. Landlocked populations of the Atlantic salmon are
also known, and the overall behavioral flexibility of salmonids has allowed them
to colonize most of the accessible streams and coldwater lakes of the Northern
Hemisphere.

The final example of classic to-and-fro migrations is that of the mule deer, the
Rocky Mountain race of the North American white-tailed deer, *Odocoileus
virginianus* (McCullogh 1985). The general pattern of migration is an annual
movement from high-altitude summer ranges to lower winter ranges of milder
climate and, probably more importantly, less accumulated snow. Deer occupy
individual home ranges, and the annual migrations are often from quite specific
summer ranges to equally specific ranges in the winter, as confirmed by marking
studies of individual deer. Migration is not simply the result of moving the shortest
distance along an elevational gradient between summer and winter areas. Rather,
the routes taken are often quite lengthy and complex, involving considerable
distances, with the maximum recorded route being 155 km and involving the
crossing of six mountain ranges in Idaho. There is thus a major lateral component
to migration as well as an altitudinal one. Routes are individual specific and often
cross the routes of other deer; thus, if mapped, there is an elaborate criss-crossing
network. A similar pattern is seen in races from eastern North America where
there is little or no altitudinal component. Further, deer at a particular site in one
season may scatter (disperse) to a number of different sites for the other, before
reassembling again the following year, a consequence of the fact that deer travel
individually or in small groups and remain faithful to learned home ranges.
Individual migration is the accumulation of generations of learned routes and home
ranges, modified by the fact that young are driven out of the home ranges of the
parents and can scatter widely. It is thus no wonder that there is an enormous
amount of individual variation. As new areas are colonized, following clear-cutting
of forests for example, new migration routes are developed, but deer often return
to the range of origin in the winter. These deer thus have a flexible and effective
migration system that allows them to exploit habitats only seasonally available,
because of heavy winter snows, and to colonize habitats that become newly

available. The large populations of *O. virginianus* across North America are testament to the success of this migration system.

Modifications of the To-and-Fro Pattern

Perhaps the most vivid and well known of insect migrants is the New World monarch butterfly, *Danaus plexippus*, whose movements and ecology have been studied for several years by Lincoln Brower and his students and colleagues (summarized in Brower and Malcolm 1991 and Malcolm et al. 1993) and by F. A. Urquhart (1987). This butterfly is a member of the tropical and subtropical subfamily Danainae; it is unique among the 157 known species in undertaking an extraordinary migratory journey of some 3000 km. This journey occurs in the autumn in the population in Eastern North America and is necessitated because the butterfly cannot overwinter in the Temperate Zone, even in diapause. The species is, however, very variable in its migratory behavior, with populations in tropical Central and South America, the West Indies, and Hawaii (where it is introduced) being sedentary, and populations in western North America and Australia (where it is also introduced and called the wanderer) moving for considerably shorter distances (Malcolm and Zalucki 1993).

Monarchs specialize on milkweeds (Asclepiadaceae), and their migratory patterns have evolved to exploit the often abundant but seasonal milkweed food supply. The eastern North American population was long thought to have a single round-trip migration with a flight south in the fall and a return flight to northern breeding areas in the spring (the single sweep hypothesis). The situation has now been shown to be more complicated. In the fall these monarchs do indeed make a single journey south; their reproduction is suppressed by short photoperiods and cool temperatures (Barker and Herman 1976). The bulk of the population overwinters in spectacular aggregations of millions of butterflies at specific sites in the fir (*Abies religiosa*) forests in mountains of the Transvolcanic Range in central Mexico, with a few butterflies clustering at sites in peninsular Florida. These overwintering sites are evidently just cool enough to permit stored fat to carry the butterflies through the winter and just warm and moist enough to prevent death from freezing or dehydrating (Chapter 6). Beginning in late February and early March, the butterflies in the aggregations become more active, mating occurs, and the monarchs begin moving northward.

Contrary to the original descriptions of monarch migration, however, the individuals leaving the clusters do not re-invade the original habitats in the northeastern United States and southern Canada. Rather they stop along the rim of the Gulf of Mexico to breed, feeding on the early spring milkweeds emerging there. It is their offspring of subsequent generations that invade the later emerging milkweed patches farther north, thus confirming the successive brood hypothesis of monarch migration (Fig. 3–5). This was demonstrated by chemically identifying the cardenolides in the milkweeds that the butterflies ate as caterpillars. Milkweeds

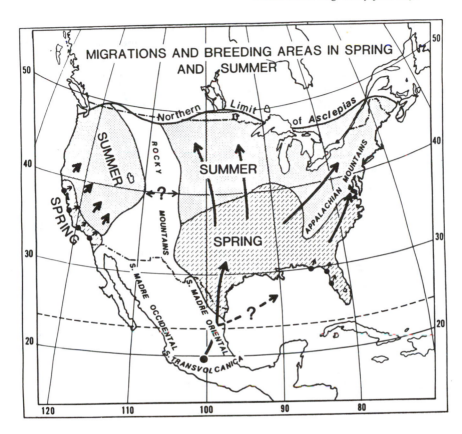

Figure 3-5. Spring migration of the monarch butterfly in North America. Eastern populations overwinter in Mexico and at scattered sites in Florida (*filled circles*). The northward movement takes place over at least two generations. In western North America overwintering takes place on the coast of California with both movement inland and some breeding near overwinter sites. (From Brower and Malcolm 1991.)

contain species specific arrays of cardenolides, which are nauseating heart poisons to vertebrates, and these can be separated chromatographically, allowing each butterfly to be "fingerprinted" according to the milkweed species on which it fed as a larva. Butterflies overwintering in Mexico all have the cardenolides of northern milkweeds, especially the most common species, *Asclepias syriaca*. In contrast, butterflies reentering northern areas in the spring have fingerprints characteristic of the Gulf Coast milkweeds, *A. viridis* and *A. humistrata*, thus conclusively proving that the return northward movement is at least a two-generation process. Recent evidence suggests even more generations may be involved. Eastern monarchs thus resemble many agriculturally important species, such as the noctuid moths *Mythimna separata* in China and *Heliothis punctigera* in Australia that have a poleward migration in the spring lasting from two to several generations with a single generation return movement more or less equatorward in the autumn (Fitt 1989; Han and Gatehouse 1991).

The population movements of the monarch in western North America cover much shorter distances but are more complex (Nagano et al. 1993; Wenner and Harris 1993). Overwintering aggregations of up to several thousand (rather than millions) occur at several places along the coast of California from just north of San Francisco to the vicinity of San Diego (Fig. 3–5). Milkweeds can start sprouting in the foothills of the Coast Ranges as early as late January (personal observations; Wenner and Harris 1993), the common species being *A. californica* in the south and *A. cordifolia* in the north. The first generation of butterflies breeds on these. In April and May, *A. fascicularis* emerges along the coast and in coastal valleys and *A. speciosa* emerges inland so the butterflies actually move in both directions and one or more summer generations are reared on both host plants. Autumn brings a return movement to the coast of California. The pattern of the monarch (wanderer) migrations in Australia is similar, with overwintering sites near Sydney and Adelaide and summer movement inland, but some breeding continuing in the overwintering areas. The subtropical climate near Brisbane to the north allows continuous breeding throughout the year (Smithers 1977; James 1984, 1988, 1993). The overall migration pattern of the monarch is thus not quite the straightforward to-and-fro north to south single-generation movement it was originally thought to be; rather, several generations are involved (eastern North America), or there may be mixtures of quasi-residents and migrants over the range (western North America and Australia).

Another modification of the round-trip scheme of migration occurs in the red-billed quelea (*Quelea quelea*) of Africa. These birds are major pests because they eat great quantities of grain; because of their propensity to appear suddenly wherever grain crops were ripening and establish large breeding colonies, they were originally thought to be nomadic with little pattern to their breeding. Analysis of the seasonality of grass growth and seed production, however, combined with banding data and other observations revealed that queleas undertake regular seasonal migrations that are variable and complex (Ward 1971; Jones 1989). These migrations are tied to the passage of the Inter-Tropical Convergence Zone (ITCZ) where northern and southern air masses meet and bring rainfall (see Chapter 5). The ITCZ moves back and forth over the equator with the north and south passages of the sun and is preceded by a band of rainfall. The pattern of quelea movements and its relation to rainfall are illustrated in Figure 3–6.

During the dry season, queleas feed on the fallen seeds of several species of annual grass that have ripened during the previous rainy period. At this time they often form huge roosts which ebb and flow in size and location as a function of seed abundance. With the coming of the rains, the remaining uneaten seeds germinate, eliminating the food supply. For a short time the birds can move ahead of the rains, but at best this is a delaying tactic because the seed supply will inevitably disappear. A second tactic is to feed on the swarms of migrating termites

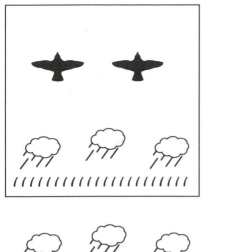

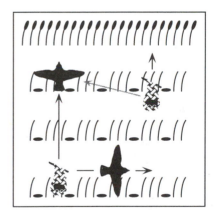

Figure 3–6. Migrations of the red-billed quelea in East Africa. The birds first move ahead of the advancing rain front (upper left) and then fly back over the front to breeding where grasses are seeding (*upper right*). Birds not breeding at this time follow behind the front to breed near later seeding grasses (*lower left*). Birds that have bred once may also move to new areas and breed again (*lower right*). (Redrawn from Jones 1989.)

that are triggered by the first light rainfalls, but again this swarming occurs for only a brief period. Faced with an acute food shortage, the queleas depart in an "early-rains migration," laying down fat before they do so. The amount of fat deposited is proportional to the distance traveled in each geographic race, but how this is triggered or accomplished is unknown. Also unknown is the source of energy intake because fattening can occur with or without feeding on termites, an obvious energy-rich source (Ward and Jones 1977; see also Chapter 6). The course of this early-rains migration is through (or over) the advancing rain front, followed by a flight of at least 300 km over the area behind the front, where grass is sprouting but has not yet set seed, to a final destination in an area of ripe grass heads and abundant insect larvae. These latter will be the result of sprouting and hatching with the passage of the rain front up to several weeks previously. Because the rain front is apt to be variable in time of passage and amount of moisture produced, the quelea movements can also be highly variable from year to year. Indeed this is what made the migrations initially difficult to sort out.

Birds that have completed the prenuptial molt and have mature gonads establish nesting colonies on arrival in areas with ripe seeds, but many birds are apparently still not ready for breeding. The earliest breeding episodes are therefore likely to be on a small to medium scale. The initial nonbreeders move along the zone of ripening grass behind the rain front; when their gonads are mature and they have completed the molt to breeding plumage, they stop moving and form breeding colonies. In the meantime, the birds that bred early are likely to have completed the breeding cycle, which is very rapid in queleas, lasting only 5 to 6 weeks. These birds now depart the original breeding areas, leaving their fledglings behind to finish the maturation process on their own, and migrate to catch up with the rain front. When they reach a suitable area of seeds and available insect larvae, they stop and breed again. This pattern of itinerant breeding may involve two and perhaps in some cases even more nesting episodes during the same rainy season in places as much as 500–700 km apart and simultaneously distributed over more than 300 km (Jaeger et al. 1986). The differences in timing and spacing are determined by the pattern of the seeding of annual grasses.

The juveniles left behind when their parents depart the breeding area remain in the vicinity of the colony for several weeks feeding on fallen grass seeds. As these become harder to obtain, the juveniles may move into cereal crops where seeds are readily available. Seed size dominates food selection, with the birds preferentially taking small seeds in the approximate order grass > millet > sorghum. Eventually the young birds move away and probably assume a wandering life during the dry season. Similarly, the adult breeders, after the final breeding episode, initially concentrate at the start of the dry season at the place where they last bred; later they also wander widely. There seems to be little mixing of adults and juveniles during the dry season, but the reasons why are unclear. The arrival of the next rainy season starts the migratory cycle again. There are thus three major phases of quelea movment: dry season wandering, a more or less straight early-rains migration over or through the rain front to some distance beyond it, and sporadic movements and itinerant breeding behind the rains.

A somewhat different migratory pattern keyed by rainfall occurs in another African animal, the wildebeest (*Connochaetes taurinus*) in the Serengeti-Mara region of Kenya and Tanzania (Talbot and Talbot 1963; Pennycuick 1975; Inglis 1976). The basic pattern is illustrated in Figure 3–7. The Serengeti wildebeest spend the wet months of December through April in the grassy plains of the

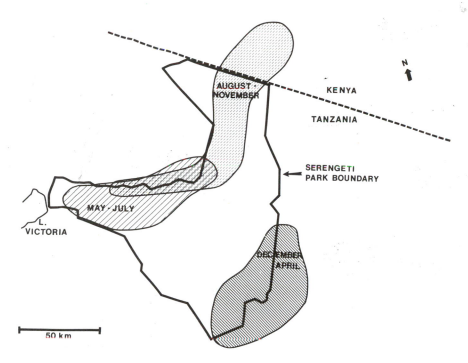

Figure 3–7. Wildebeest migration in the Serengeti Plains of East Africa. Movement is, first, from rainy season breeding areas (December–April) in the southeast to wet areas near Lake Victoria. This is followed by movement to the northeast to meet the earliest rains. The final movement is back to the southeast. (From Dingle 1980.)

southeastern part of the region where their preferred food of short grass is green and high in nutrient content. Most calving takes place at this time. As the grass dries out in May and June, the behavior of the animals changes, and they form into large herds that migrate northwestward to the wetter scrubland near Lake Victoria. This movement may occur in spectacular fashion with single files of wildebeest extending for miles. Toward the end of July or beginning of August as the dry season approaches its end, the herds move northeastward to spend the next 2 or 3 months in the northern part of the region where rains arrive with the ITCZ before they do further south. In November and December there is a return movement to the southeast and the breeding area more or less behind the rain front. There is year to year variation in the time of occupancy of a given area, strongly correlated with the annual variation in onset and amount of rainfall.

There is also variation among wildebeest individuals and populations. Some old bulls, for example, may not join the migratory herds, but rather hold territories throughout the year in the southeastern breeding area, dependent upon the presence of some green vegetation in wetter sites and on access to a permanent water source.

These sites apparently confer some reproductive advantage, as the males holding them are ready to mate immediately with females returning from migration. In the nearby Ngorongoro Crater, the wet and dry seasons are less extreme than in the Serengeti, and the wildebeest population is largely sedentary, although there may be some facultative movement out of the Crater in the more extreme dry years. As in the maintenance of year-round territories in the Serengeti, the establishment of sedentary populations requires a water source and the availability of at least some food throughout the year. Migrations, and the variability present in these wildebeest populations, are thus a response to the resource pattern.

The same is true for the migratory circuits of such oceangoing fishes as the herring (*Clupea harengus*), the cod (*Gadus morhua*), and the plaice (*Pleuronectes platessa*) as described by Harden Jones (1968) based on extensive fishing and other records (see also Dingle 1980). Here again the different populations of each species display considerable variation, even to breeding in different seasons. In the herring, for example, some populations breed off Norway in the spring and others on the Dogger Bank in the North Sea in the fall. The basic pattern for all these fishes, however, is a movement between spawning sites, feeding areas, and wintering areas. Juveniles may spend some time in a nursery area, to which the eggs and newly hatched larvae have drifted on currents, before entering the primary migration circuit upon reaching maturity. The movement among different areas allows these fishes to exploit an extensive resource base, and Nikolsky (1963) has suggested that this may in part account for the abundance of these species, allowing their exploitation by commercial fisheries. Be that as it may, the major point to be noted is that these fishes, like the monarch butterfly, the sockeye salmon, the red-billed quelea, and the wildebeest all migrate on more or less regular circuits, but with considerable modification of the "classic" to-and-fro pattern. Nevertheless, the behavior in each case meets the criteria for migration outlined in Chapter 2.

Nomadic and Opportunistic Migrations

A number of birds become irruptive or nomadic under the right conditions, but it is not clear for many of these whether the movements represent migration or extended foraging in search of scattered food. Perhaps the best known are northern birds such as crossbills (*Loxia* spp.) and waxwings (*Bombycilla garrulus*) that wander widely, especially under conditions of food shortages during the winter (Chapter 1). It is the case, however, that crossbills at least display many of the physiological properties of true migrants such as pre-movement fat deposition (Berthold 1993). But the bird that perhaps most epitomizes the nomadic habit and that does appear to display truly migratory journeys to breeding sites is the banded stilt (*Cladorhynchus leucocephalus*) of Australia. The species was first described in 1816 from a skin collected between 1800 and 1804, but it was not until 1930 that the first observation of breeding was made. This was in spite of the fact that the bird is quite common throughout southern Australia roughly from Melbourne westward to Perth (Blakers et al. 1984; Robinson and Minton 1989). In 1930 some 40,000 nests were found at Lake Grace in the state of Western Australia and about 27,000 nests at Lake Callabonna in South Australia. These are somewhat remote desert salt pans that are usually dry, but fill to form large shallow lakes after heavy rains. These lakes and the brine shrimp they contain are the key to the life cycle of the banded stilt. When breeding does take place, it can involve enormous numbers of birds. The Lake Barlee breeding event in 1980, for example, involved 179,000 nests and probably about 500,000 eggs. Lake drying killed about half the eggs and chicks, but still left some 250,000 chicks to set out from the colonies (Pringle 1987).

During the nonbreeding times, which can extend for several years, the stilts congregate at salt lakes and salt marshes in large numbers. I saw several thousand at the salt works north of Adelaide in November 1991. Stilts may also range widely over at least several hundred kilometers in search of suitable habitat. Breeding is triggered by heavy rainfall that fills the large shallow salt lakes and causes the hatching of dense concentrations of brine shrimp upon which the adults and newly hatched young feed. Lakes with fishes or other breeding birds do not attract the stilts, perhaps because of competition for the brine shrimp food supply (Robinson and Minton 1989).

The major breeding event on islands in Lake Torrens, South Australia in 1989 has provided considerable data on breeding and its relation to the nomadic migratory cycle in this species (Robinson and Minton 1989). In March 1989, heavy rains fell in the vicinity of the Flinders Ranges in South Australia, including 676 millimeters at one station, the highest monthly rainfall ever recorded there. The runoff from this torrential rainfall filled Lake Torrens to the west of the Flinders Ranges to form a salt lake some 200 km long by 30 km wide but only about half a meter deep over most of its area. By the middle of April, 4 weeks after the heaviest of these rains, the first nesting colony of several thousand pairs had been established on an island in the lake, to be followed quickly by more colonies on nearby islands. Overall up to 100,000 birds nested. Stilts do not normally occur in this area because there is no permanent standing water, and a major question is thus just how they found the breeding site. The successive start-ups of nesting colonies implies that it was not found by all birds at once. A logical hypothesis is that major frontal systems bringing rain, likely to occur over a wide area, trigger migration by the stilts, leading to the location of breeding sites, if present. Pairing and mating seem to occur en route so this is a likely scenario. It remains to be tested, however, and this will not be easy to do, given the infrequency of breeding and the difficulty of reaching the breeding sites, especially following rains when roads are usually impassable.

Once breeding begins, it is rapid, much in the manner of the red-billed quelea. The arriving mated pairs scrape out a nesting hollow to which they may add a few sticks and grass stems. The three or four eggs are large and produce large, very precocious young in 19 to 21 days (Pringle 1987). As soon as the chicks dry following hatching, they leave the nest, and within 2 days they proceed to the lake either singly or in small groups accompanied by up to several adults. Once in the water they immediately begin feeding. At this stage they often form into large "creches," accompanied by some adults, and swim in a moving wall feeding as they go. The track of a feeding group is readily visible from the air, indicating the rate at which the chicks must be consuming the "soup" of brine shrimp. They can move up to 100 km from the hatching site.

Following the post-hatching departure of the chicks from the nest site, adults mate and immediately begin nesting again. At Lake Torrens the first eggs were laid on the 15th of April and hatched over the 2 weeks from the 4th to the 19th of May. The second brood was begun on the 18th of May and all had hatched by mid-June. Two complete nesting cycles were thus compressed into 12 weeks, an astonishing performance for a bird of this size (circa 800 grams). Apparently even more broods may sometimes be produced, but heavy predation (about 95% by gulls at the end of the Lake Torrens second brood) and lake drying bring the breeding cycle to a close. The rapid production of huge numbers of offspring is clearly an adaptation to the nomadic lifestyle with breeding often years apart.

A number of Australian ducks and other waders are also nomadic and opportunistic in their breeding. Examples include the freckled duck (*Stictonetta naevosa*) and the grey teal (*Anas gibberifrons*). The former requires lignum, a dense matted vine, for nesting, and in particular lignum overhanging water. This situation occurs only after heavy rains and flooding and then only infrequently, as lignum can be some distance from the water source. Its life cycle is thus not unlike that of the banded stilt. The teal is less stringent in its requirements and so breeds more frequently. Its movement may be more akin to extended foraging or ranging. In years of very severe drought, however, major irruptions can occur, such as the one that resulted in the colonization of far northern Australia and New Zealand in 1957–58. These latter events seem more likely to be migratory. In northern Australia the teal colonies did not persist, but there has been breeding in New Zealand ever since (Blakers et al. 1984). Because of the frequent widespread dry periods, most Australian waterbirds are more or less nomadic and opportunistic in their migration and breeding patterns; the relation between the timing and the life histories of the birds is virtually aseasonal and quite different from that observed in the Northern Hemisphere (P.J. Fullagar, personal communication).

An insect with an opportunistic migratory and breeding pattern is the African armyworm moth, *Spodoptera exempta* (Gatehouse 1986, 1987, 1989; Rose et al. 1987; Pedgley et al. 1989). In many ways the armyworm migratory pattern is similar to that of the red-billed quelea; both are dependent on sprouting grasses following rainfall, for example. But they also differ in that the armyworm breeds throughout the year, and its migratory tendency is influenced by density. The similarities indicate the absence of hard and fast differences in the types of migratory patterns observed in organisms. The year-round breeding of the armyworm is possible because even in the dry season sporadic rains occur, especially along the coast of East Africa, and provide the necessary moisture to maintain some suitable habitat for the moths in the form of green grass. The tendency to migrate is increased by high density (part of the *phase polyphenism*, or change in morphology and behavior, caused by density effects in these moths that will be discussed in detail in Chapter 13), but even in low densities some genetically programmed migration occurs. In the dry season this migration allows moths to find suitable although patchy habitats. These migrations tend to occur downwind with a net inland movement. In the dry season many of these migrating moths undoubtedly perish, but with the coming of the rains produced by the southward moving Inter-Tropical Convergence Zone (ITCZ), the situation changes. Now the downwind migrating moths are likely to encounter rainstorms with convergent winds, which concentrate the moths, in contrast to the dispersive effects of winds during the dry season. Furthermore, the rain tends to wash the moths out of the air. The net result is an often very high concentration of moths laying eggs in an area that is favorable because rains have just occurred there. These concentrations often result in primary outbreaks of the "armyworm" larvae feeding voraciously on local crops.

At high densities these larvae undergo a "phase transformation" to a dark caterpillar that is highly visible because, in contrast to the green low-density larva, it feeds in the open at the tops of grass stems. As an adult the migratory genotype is much more readily expressed, even to possessing more fat to use as fuel, and each night large numbers of moths take off on downwind migrations that can cover considerable distances because the moths will fly for long periods. The greater migratory tendency allows moths to escape overcrowded conditions where they are susceptible to disease and predation. The general result is that the moths, as a consequence of their migration, will disperse to any habitats where recent rains have made grasses favorable for reproduction and growth of the larvae. At the conclusion of migration females also need water to begin egg laying. With large numbers of moths settling in a favorable area, a *secondary outbreak* can occur. Eventually, however, the dry season returns, and the moth population shrinks back to a few isolated favorable patches. The whole migratory system is genetically programmed and phenotypically modified to allow the moths to find suitable habitats, where crowding can occur, but to escape crowding with an enhanced migratory exodus. The frequencies of the different levels of migratory capability are set by varying intensities of natural selection. As with the red-billed quelea, the life cycle of the armyworm moth is a consequence of adaptation to the changing wet and dry seasons of the eastern half of sub-Saharan Africa.

One-Way Migrations

One-way migrations are those in which there is no apparent circuitry in the pattern of movements, either by individuals or by populations. This does not necessarily mean that a given area is abandoned completely by migrants because there may be immigration by individuals originating elsewhere even as those reared in an area are emigrating. In late spring monarch butterflies, for example, individuals make only a single one-way migration, but the overall pattern for the population is a return, at least more or less, to the situation prevailing at some arbitrary point at the "beginning" of the cycle such as the site of overwintering aggregations. In this section I shall consider some cases where such population return movement and cyclicity is not evident.

A classic example of a one-way migrant, if for no other reason than its fame in ballads and legends of the American West, is the tumbleweed or Russian thistle (*Salsola kali*) as described by Young (1991). Tumbleweeds are common weeds of roadsides, waste areas, and degraded open habitats generally and were introduced from the Old World to the New presumably with European settlement. Their migratory capabilities are a major factor contributing to their now widespread distribution over the drier areas of North America. In late summer and autumn, tumbleweeds dry out and break away from their roots (Fig. 3–8). This process, as well as seed maturation, is accelerated by short days; there are specialized cells in the stem that self-destruct under the influence of short days, allowing the upper portion of the plant to separate from the roots. This portion is globular, and flowers and seeds, rather than being on the outer surface, are contained within it and protected from abrasion. This feature is an important part of the migratory mechanism because it prevents the seeds from being quickly lost as soon as the plant separates and strikes the ground surface. The rounded canopy also facilitates

A

B

C

Figure 3–8. Migration of the tumbleweed. *A.* Windrow of tumbleweeds at the edge of a field in Davis, California. *B.* Individual plant showing its globular structure. *C.* Close-up showing specialized suture at the point where the upper portion of the plant separates from the roots. (Photos by H. Dingle.)

the rolling and tumbling that later shakes the seeds loose as the plant rolls along, effectively distributing them over a wide area. The timed-release mechanism, the globular form, and the protection of the seeds that delays their release constitute an effective adaptive combination for migrating and then scattering seeds in windy areas, like the American West and the steppes of Central Asia, where the plant is native. Both places have broad flat expanses of land with little to obstruct the prevailing winds. It is little wonder that windrows of tumbleweeds along fence rows are a common sight in western North America in late summer and autumn.

One-way migrations are also common in insects in which, unlike the monarch butterfly, one-way paths do not sum to any particular pattern when viewed topographically. Usually the one-way migrations are colonization events, with individuals departing a habitat that is deteriorating or responding to some signal such as photoperiod or crowding that indicates that habitat deterioration is imminent. Such are the migrations of most aphids, an example of which was discussed in detail in Chapter 2. Other particularly striking examples of migration to colonize a new site are the swarms of social insects. In ants and termites these are the winged sexuals produced by the colony that usually emerge to migrate at certain seasons. In much of the tropics the swarming of termites at the beginning of the rainy season is an annual event. A somewhat special case, and one of the best studied because of the economic significance of the insects involved, is the swarming behavior of the honeybee, *Apis mellifera*.

The migratory portion of the colony cycle of the European race of the honeybee, the race kept by Temperate Zone beekeepers, can be summarized as follows (Seeley 1985). Usually in the spring the workers in an established colony begin the construction of queen cells, which are larger than the cells constructed for the rearing of workers. The resident queen then lays eggs in 20 or more of these "queen cups," and the workers feed the hatched larvae royal jelly, which ensures that they will mature into new queens. During this time when the new queens are developing, the colony queen is fed less by the workers, her abdomen shrinks, and her egg production declines precipitously. The workers also periodically shake the queen, a behavior that has been interpreted as forcing the queen to keep walking about the hive, presumably further contributing to her slimming and weight loss. Shortly after the first cell of the newly developing queens is capped as the larva enters pupation, the old queen departs the colony on a migratory flight, accompanied by some 10 to 15 thousand workers. After flying a short distance from the hive, the swarm pauses, usually as a large cluster on the limb of a tree. Scout bees leave the swarm and explore cavities in the surrounding area for potential nest sites (ranging). Eventually one is selected, the swarm is signaled that a colony site is available, and the cluster breaks up to fly to the new home, ending the migratory period.

European bees are quite predictable in their swarming and migration, but the same is not true for the African race. This well-publicized bee has invaded the New World via Brazil, hybridized with local varieties, and is seemingly on the verge of invading the United States. One of the potential problems the Africanized bees present for beekeepers is that an entire colony will abscond and migrate elsewhere when disturbed, as when a hive is moved from one field or orchard to another during the flowering season to pollinate crops. This absconding behavior is apparently a reflection of adaptation to a more uncertain environment, one likely to be influenced by unpredictably severe dry periods as occur over much of Africa. The absconding is interesting because it can take two forms (Winston et al. 1979): short distances of around 10 km or less and long distances which can be up to 100 km. The former seem to be changes in nest site due to local factors such as predation, too small a nest cavity, or excessive exposure to the elements. This local change in nest site is probably a form of ranging as I have defined it in Chapter 1. The long-distance movements seem to involve migration from a poor or deteriorating environment to one richer in resources; there are several examples of these movements, especially into areas where flowers rich in nectar and pollen have come into bloom (Fletcher 1978). These long-distance swarm migrations are the most dramatic examples of the migratory behavior of honeybee colonies and among the most dramatic of one-way movements.

Migration with the Aid of Devices

Many migrants move not under their own power, but with a device that allows them to embark on a transporting vehicle. In the case of terrestrial organisms, the focus of my discussion, that vehicle is either the wind or a mobile animal on which or in which the migrant takes passage. The journeys undertaken fall roughly into four categories. First, there are those made by spinning a silken thread that is then used for transport by the wind, a method usually referred to as "ballooning." This method is used by some lepidopteran larvae and by arachnids, both spiders and mites. Second, there are seeds that are attached to fluffy "parachutes" or pappi (singular pappus), which also function for wind transport; the milkweeds (Asclepiadaceae) and many composites (Asteraceae) such as the common dandelion (*Taraxicum officinale*) are well-known examples. Third, there are seeds with fleshy arals, embedded in a fleshy fruit, or bearing a fleshy attachment such as an eliasome, all to facilitate animal transport. Arals and fruits ensure that the seeds are swallowed by vertebrates, which then regurgitate them or pass them through the gut, in most cases depositing them at some distance from the parent plant. Eliasomes are fleshy seed attachments attractive to ants, which then transport the seeds to suitable germination sites in and around the ant colonies. As might be expected, eliasomes occur chiefly in small plants of the forest floor like the violets and their relatives. Examples of seed transport by animals are extensively discussed by Howe and Wesley (1988). Finally, there is transport by phoresy, or the superficial attachment to a mobile organism so that the migrant is itself transported. This mode of migration is common in pseudoscorpions (Weygoldt 1969; Zeh and Zeh 1992) and mites (Binns 1982; Houck and O'Conner 1991), and these tiny animals are frequently found attached to flies and beetles. The esoteric and charming nature of many phoretic associations continually delight students of co-evolution. Many plants of open areas also have "hitchhiking" seeds that bear spines or similar structures allowing them to adhere to the fur of mammals for transport to new sites.

The behavior of ballooning spiders is an excellent illustration of the point that migration is an active behavioral process even when the agent of transport is external. (As we shall see in Chapter 5, propulsion and wind can also act synergistically during migration.) A good example involves young wolf spiders, *Pardosa* spp. (Richter 1970). To initiate migration by ballooning, the spiderlings first climb to the tops of plants to gain maximum exposure to the wind. Next they attach their walking threads or "draglines" to the substrate and break them with a jerk so that they are free to become airborne. They then stand on "tiptoe" and extrude six to eight silk threads from the spinnerets, allowing these to waft in the wind. When these threads reach about 70 mm in length, more or less depending on wind velocity, the spiderlings are lifted off the substrate and carried downwind. After remaining aloft for some time, the spiderlings may facilitate their own landing by gathering in the silk lines to reduce lift (McCook 1890). If winds are too strong, the spiders delay takeoff. During certain periods of the summer in the Temperate Zones of both hemispheres, the air may be full of ballooning spiders. Because of the constraints of this method of transport (see Chapter 7), most ballooning spiders are small; in samples from Australia and the United States the majority were below 1.0 mg, although the largest weighed 25.5 mg (Greenstone et al. 1987). Not surprisingly, therefore, the best represented family was the Linyphiidae, a group of very small species, and many other families were represented primarily by juveniles.

Size is also limited by the transport mechanisms of seeds that parachute or attach to mammalian fur. Seeds that become airborne are, like spiderlings, prisoners of the drag forces required for takeoff and those that attach are restricted by adhesive strength. Phoretic pseudoscorpions and mites are also small, not only for ease of transport by the insect carrying them, but also, presumably, so that they won't be detected and removed. Seeds transported in the guts of vertebrates are less constrained by size and some can be quite large. All that is required is that they can be handled by the trophic apparatus and the digestive tract of the transporting animal. Many have evolved thick or hard seed walls to prevent damage by chewing or tearing and some have become specialized to the extent that passage through a vertebrate gut is necesary before germination can occur (Howe and Smallwood 1982; Howe and Wesley 1988). This is one more example of how in plants, as well as animals, specialized physiological adaptations have evolved to facilitate the migration process.

Summing Up

Migration is a behavior of great variety, and it frequently reflects awesome performance. It is not, however, defined either by its magnitude or the peculiarities of its routes. Rather its defining characteristics of active embarkation and locomotory behavior and the temporary suppression of maintenance, trophic, and reproductive activities occur in movements as disparate as the short round-trips of Swedish milkweed bugs to and from diapause shelters and the transcontinental wanderings of Australian banded stilts in search of their remote and ephemeral breeding sites. Nor can migration be distinguished on the basis of active versus passive movements, because even when migrants are borne ballooning on the wind or in the gut of an unwitting host, they got there through precise behavioral or physiological mechanisms that assured their transport. In spite of the detail in the variety, the common thread of specialized behavior to leave an old home well behind and seek a new is woven throughout the migration phenomenon. In the tiny and short-lived, the journey occurs only once; in the large and long-lived it may occur tens of times throughout the life span.

Equally as important as the behavior itself, in all the examples I have outlined, is its ecological outcome. In the classic to-and-fro migrants like the arctic tern or the swift parrot, the journey means a predictable safe haven at each end of the route. One of these is for breeding in a region where food is abundant, but only for a relatively brief period; the other is a place where foraging can be sustained throughout the nonbreeding season. In many Temperature Zone bird migrants this latter may in fact be a tropical ancestral home (Chapter 9). Other versions of round-trip migration may include stops at three or more sites to exploit resources that also may be available only for a limited time. Such circular or quasi-circular patterns occur in monarch butterflies, Serengeti wildebeest, and North Sea herring. The seemingly erratic migrations of the banded stilt also have their ecological logic, allowing exuberant reproduction in those years and in those places where the itinerant rains briefly soak the landscape. The ecological requirements of each species select for the timing of migration and the particular pattern of each migratory cycle. Underlying each, however, is a physiological syndrome producing a similar well-defined behavior. The ultimate outcome of that behavior is survival and reproduction in an environment that varies greatly over space and time.

Thus the examples in this chapter reflect the fact that migration is a characteristic behavior with specific physiological properties molded by natural selection to yield specific ecological outcomes. It is the properties that must be considered when defining migration, and the outcomes that must be considered when determining its ecological and evolutionary significance.

4

Methods for Studying Migration

The methods used for studying migrating organisms range from simple counts to the use of complex electronic devices to track migrants across space and time. The recent rapid advances in technology from miniaturization to the development of sophisticated computers has permitted the counting or tracking of migrants in ways that often provide startling insights into the details of migratory pathways. At a different level, new technologies are also allowing extensive measurements on migrants under controlled laboratory conditions, yielding new understanding of physiological mechanisms underlying migratory behavior. Sophisticated technology, however, is not always necessary, for there is still much to be learned from relatively simple methods of marking and counting.

In the following sections I briefly describe some of the means used to study migration. These are divided into field and laboratory methods, with each used to address particular issues. Field techniques are used to determine what migrants are actually doing under natural conditions; laboratory methods may be necessary to answer questions about physiology or life history. All have major advantages and disadvantages, and so with all there will be trade-offs involved. Whatever the method it will be apparent that because migration often takes place over large scales of space and time relative to the size and life span of the migrating organism, it is one of the most difficult of behaviors to study. The design and use of adequate methods thus presents a formidable challenge to the ingenuity of investigators.

Field Methods

Direct Visual Observations and Counts

By far the most common method for studying migration in the field involves counting migrants as they move past a lookout. The use of this technique is especially frequent among ornithologists, both professional and amateur (Kerlinger 1989; Alerstam 1990), but it has also been used for insects (Johnson 1969), for whales passing a headland or promontory, and for assorted other organisms in various circumstances. Most often observers take advantage of the fact that topography such as ridges, coastlines, peninsulas, and isthmuses will funnel or collect large numbers of migrants. Well-known features where bird and insect counts have taken place are the Falsterbo Peninsula in southern Sweden; various Alpine passes; Gibraltar and the Bosporus at the two ends of the Mediterranean; Point Pelee, Ontario, on the northern shore of Lake Erie; Cape May Point in Southern New Jersey; Hawk Mountain on the Kittatinny Ridge in eastern Pennsylvania; and the Isthmus of Panama. The passage of aerial migrants at such lookouts can be quite spectacular as anyone who has seen the clouds of broad-winged and Swainson's hawks (*Buteo platypterus* and *B. swainsonii*) and turkey vultures (*Cathartes aura*) over Panama City, Panama, or the streams of monarch butterflies flying south at Point Pelee, Ontario, can testify. Over 500,000 raptors, for example, pass over Balboa on the south side of the Isthmus of Panama in a season (Smith 1980), and some 1.8 million birds of the 25 most numerous species pass by the observatory at Falsterbo, Sweden (Alerstam 1990). Many locations consistently record over 10,000 raptors in passage. Counts of migrants at topographic features have the very real advantage of being easy methods to use because they require no elaborate or expensive equipment. If conducted over long periods as they have been at Falsterbo and Hawk Mountain, for example, they can provide a great deal of information about long-term trends in populations and in the seasonal timing of migration at the particular location.

There are also, however, rather severe limitations in what can be accomplished by simple visual observations at a topographic feature. Many flying migrants, including both birds and insects, migrate at considerable altitude where they are invisible to observers even with the aid of binoculars. Migrant birds, for example, generally fly at 300 meters or above where they are most likely to take advantage of geostrophic winds, and many insects do likewise (Drake and Farrow 1988; Gauthreaux 1991 and Chapter 5). Even very large animals like whales will be quite invisible if they are migrating far offshore. Night migrants, which predominate in some groups of passerine birds and in moths, will also be invisible to ground observers. Further, because observations are made only at a single point, it is not possible without additional information to determine sources of variation. Do two hawk species that differ in the time of peak passage do so because they start at different times or because they originate at different distances from the observation point? Does annual variation in numbers arise because of differences in population sizes as a consequence of high mortality or breeding failures or because migrants vary in their routes from year to year in response to weather conditions? Kerlinger (1989) gives a good discussion of the kinds of problems encountered as they relate to hawk migration, and much of his critique is generally applicable.

Various methods have been used, some quite ingenious, to overcome the limitations of simple observations and counts. One of the earliest was moon-watching, which came into regular use around 1950. The idea here is to observe with binoculars or telescope the passage of migrants across the face of the moon. The grandest use of moonwatching took place in North America on four consecutive nights at the beginning of October 1952 (Lowery and Newman 1966). More than 1000 birders and astronomers at 265 observation posts watched the passage of migrants across the face of the moon. Although in some places the moon was obscured by cloud, the results were very interesting. They showed, first, that the most intensive migration occurred in areas where north or northwesterly winds were blowing, providing a tailwind for these southward moving migrants (see Chapter 5). Secondly, they showed what enormous numbers of birds were moving. On peak nights of migration, extrapolations of the moonwatch counts yielded estimates of from 10,000 to 50,000 birds moving south per kilometer of front per hour. Over a 200-kilometer front this would be equal to roughly 50 million migrants on a night's passage, assuming an average of 25,000 birds·km·hr^{-1}, an impressive number to say the least (Alerstam 1990).

A further modification of simple sight observations has been used by Walker (1985a) to estimate numbers of three species of butterfly migrating through peninsular Florida and other locations in the southeastern United States. The three species—the cloudless sulphur (*Phoebis sennae*), the gulf fritillary (*Agraulis vanillae*), and the long-tailed skipper (*Urbanus proteus*)—migrate in large numbers at ground level and are relatively large and conspicuous. They can, therefore, be visually tracked fairly easily. The butterflies were counted as they flew between two poles 3 m in height and spaced 45 m apart in the midst of an 8-hectare grassy field, giving observers an unobstructed view of the butterflies as they flew past. The poles were placed so that the line between them ran ENE-WSW or perpendicular to the axis of the Florida peninsula, as it had been determined that the butterflies flew roughly parallel to this axis (see Chapter 8). Butterflies were counted for successive 15-minute intervals three times a week from the end of August to the middle of November, the period encompassing fall migration. The counts could be used to determine the distribution of migrants over the course of the day, and from these data total migrants per day over the period, and over a 100-km front for the whole season, could be estimated. The latter estimates were 11.3 million *A. vanillae*, 29.2 million *P. sennae*, and 234.5 million *U. proteus*, which appear to be at least order of magnitude estimates of the numbers of these butterflies entering Florida. Walker's results with traps will be discussed below.

Observations at times or places where migrants could not have been produced locally can also provide useful information on timing and extent of movements (Drake 1991). Observations from offshore islands or islands of habitat surrounded by inhospitable terrain can indicate how far a species can move (e.g., Farrow 1984). Immigration can often be detected in areas where a species does not overwinter, as is the case for many birds and butterflies, or early in the season where a locally breeding population has not yet emerged, as in the case of the *Heliothis* moths studied by Fitt and Daly (1990). These methods can be extended by using "synoptic surveys" or monitoring over wide areas to identify breeding and emergence sites as well as sites of immigration. These surveys are likely to be particularly important for forecasting the migrations of major pest species such as *Heliothis* (Joyce 1981 and Chapter 15). They can be further enhanced by using development models (Oku 1983a; Farrow and McDonald 1987) or regional population models (Stinner et al. 1986) to interpret and integrate data gathered from a wide area. Needless to say, for these survey methods to be effective, they require the cooperation and coordination of several observers (see also Chapter 15).

The Use of Natural Markers

Natural markers of a variety of sorts can be used to identify migrants, to distinguish migrants from local resident populations, to determine the source or route of migrants, or any combination of these. The best markers allow identification of individuals, and markings on flukes, heads, or bodies have been used very successfully to identify individual whales and map their routes over long distances, often with the help of extensive photographic libraries (Lockyer and Brown 1981; Bigg et al. 1986; Ford 1991). More usually, however, it is only populations that can be separated. Among birds, migrants are often subspecifically distinct in plumage patterns and sometimes sufficiently so that they can be distinguished in the field. The Tasmanian race *lateralis* of the silvereye *Zosterops lateralis* with its rufous flanks and white throat can be separated from mainland races with either pale buff or gray flanks and yellow throats. Tasmanian birds are absent in the summer from mainland Australia, but in the winter populations can be recognized as far north as southern Queensland; thus they are clearly migratory (Blakers et al. 1984). Different populations of the white-crowned sparrow (*Zonotrichia leucophrys*) of North America sing locality-specific songs and can be identified by these dialects (Baptista and Petrinovich 1984). By this means birds wintering in the Sacramento Valley of California can be identified by their songs (which they sing on winter territories) as belonging to a population that breeds in Alaska (*Z. l. gambeli*). Populations of migratory moths can be identified by specific genetic markers such as allozymes identified by electrophoresis (Pashley and Bush 1979). Use of the latter method does, however, up the cost considerably and requires capture of individuals to obtain blood or tissue samples for analysis. At a higher cost yet, but still more precise, are specific sequences identified in DNA. Chemicals in the food can also be "fingerprinted" as we have seen in the case of the monarch butterfly in Chapter 3, in which identification of species-specific milkweed cardenolides conclusively demonstrated that northward migration in the spring is at least a two-generation process.

One of the more interesting uses of natural markers is the generation of evidence from parasites to deduce the patterns of movement of the albacore tuna (*Thunnus alalunga*) in the South Pacific (Jones 1991). Parasites were collected from over 400 fish, and of these, 12 species could be accurately recognized and so used in the analysis. Fish caught near New Zealand showed a decrease in the occurrence of digenic trematodes with increasing fish length up to 70–79 cm. Subsequently these parasites increased in larger fish. These changes in parasite load were consistent with the notion that the juvenile fish migrate south and lose parasites in cooler waters. As the fish mature they first migrate back to tropical areas to spawn, are re-infected, and then return south again to feed and grow further. In southern areas they appear to move along a subtropical convergence zone as revealed by the decline in parasite load eastward from New Zealand along the convergence. Presumably the fish move along the zone of convergence because such zones in the ocean are generally nutrient rich (Chapter 5). This inferred general pattern of migration in the albacore tuna is further supported by tagging data and seasonal changes in the catch rate.

Trapping

If an organism is trapped during the course of migration, it is possible to acquire much useful information about it. Measurements of size, body dimensions such as wing length, and amount of fat available for fuel plus the noting of sex, age, and reproductive condition all provide indication of the properties of migrants which can then be compared with the characteristics of other members of the population or of other species. If they are trapped live, individuals can also be marked and released (see next section) with the opportunity to collect data on future behavior, including migration pathways, and on life span. The ideal trap would sample with maximum efficiency with no damage to the organisms being caught. In practice the use of traps requires various trade-offs, the nature of which will depend on the nature of the questions being asked of the data collected and the nature of the organisms sampled.

The small size and usually large numbers of insects migrating minimizes the need to capture individuals alive and allows the design of traps that need monitoring only at periodic intervals. The traps themselves are of various designs (Gregg and Wilson 1991). Among the most common are light traps, which take advantage of the fact that many nocturnally flying insects will respond to light, especially if it contains ultraviolet wavelengths. The light can be combined with a fan to suck insects into a container of preservative or killing agent, allowing the trap to be emptied only at intervals. Similarly, traps can be baited with pheromone or other substances. These traps can be very effective, but a drawback is that they attract insects from an unknown distance and are not specific to migrants. They may in fact be biased toward nonmigrants because migrants may be unresponsive to the stimuli that enhance trap effectiveness (see Chapter 2). One way to avoid this difficulty is to use only *interception* traps that do not have an attracting agent, although such traps are likely to collect fewer insects. Devices that collect both migrants and nonmigrants can still be used to monitor population levels and various behavioral tendencies from which migration can then be inferred.

Another way to avoid problems is to use a method specific to migrants. Some aphids, for example, respond to yellow wavelengths after a period of migration (Kennedy et al. 1961) and can be efficiently trapped in pans of water placed out over yellow cards or by mounting yellow "sticky traps" above the ground. An excellent example of the effective use of periodically monitored traps is the Rothamsted Insect Survey (Taylor 1979; 1986a). Traps were established and monitored for several years throughout Great Britain providing a vast amount of data on both the spatial and temporal fluctuations on an island-wide scale (Chapter 15). Another example of the use of large-scale catch data to determine migratory patterns is the extensive use of information available from commercial fishing (Harden Jones 1968).

At the other extreme from traps that provide mass catches of usually dead insects (or dead fishes) are various devices for trapping individual animals alive. Small mammals, for example, may be caught in Longworth traps or other devices that trap them alive when they enter boxlike containers in response to a bait. Aquatic organisms such as fishes and large crustaceans can also be induced to enter baited traps, and indeed this form of trapping with "lobster pots" is the basis for the commercial lobster industry in the North Atlantic. Predatory birds can also be caught with baited traps (Kerlinger 1989), but most birds are caught in "mist nets" made of fine thread with a receiving fold in the bottom. The birds either hit the net and become entangled on contact or drop into the fold and become entangled there. All these traps have the advantage of capture with minimum risk to the animals caught but are not necessarily specific to migrants (although they can be made more specific with appropriate placement) and are labor intensive because they must be monitored at frequent intervals. They also catch relatively few individuals.

This last problem can be obviated to some extent by the use of devices to channel organisms into a trap or traps. A good example of the use of such a device was the study of Gill (1978a,b) on the migrations of newts (*Notophthalmus viridescens*) between ponds in the Appalachian Mountains of the eastern United States. Gill used drift fences to channel newts into pitfall traps where they could be classified as to size and sex and individually marked by toe clipping. In this way they could be identified with respect to breeding pond and whether they returned to their natal pond or bred elsewhere. These studies were the basis for Gill's "Metapopulation Model" of a species as a large population subdivided into smaller populations with limited interchange via migration to non-natal breeding sites. Drift fences are particularly useful for organisms such as amphibians, small mammals, or ground beetles whose travel is along the ground surface.

Walker (1985a,b; 1991) also used funnel traps to extend his studies, already introduced above, of butterflies migrating through peninsular Florida. For 10 years Walker operated flight traps near Gainesville, Florida, that intercepted migrating butterflies flying within approximately 3 meters of the ground. In addition to the cloudless sulphur, the gulf fritillary, and the long-tailed skipper already mentioned, the traps also caught large numbers of the buckeye (*Precis coenia*). The traps were operated from early March to early December each year from 1979–1988 to capture butterflies migrating during that period. The traps were constructed along the ENE-WSW line perpendicular to the axis of Florida (Chapter 8 and above). These migrants encountered the traps at a central barrier of wire mesh ("hardware cloth"), which directed them over the barrier and hence into a duct that led them through valves into holding cages. Butterflies encountering the barrier from the north (338 ± 90°) and from the south (158 ± 90°) were directed into separate holding cages. The difference in the numbers between the holding cages indicated the net direction of movement perpendicular to the trap. As an example of the results obtained from these traps, the seasonal patterns of migration in the gulf fritillary and the buckeye are illustrated in Figure 4–1.

The traps so far discussed suffer two major drawbacks in addition to the various disadvantages already mentioned. These are, first, that they require the organism to come to the trap, and secondly, that with the terrestrial traps sampling is either on the ground surface or only a few meters above it. For insects and other small organisms traveling in the atmosphere, these problems have been dealt with

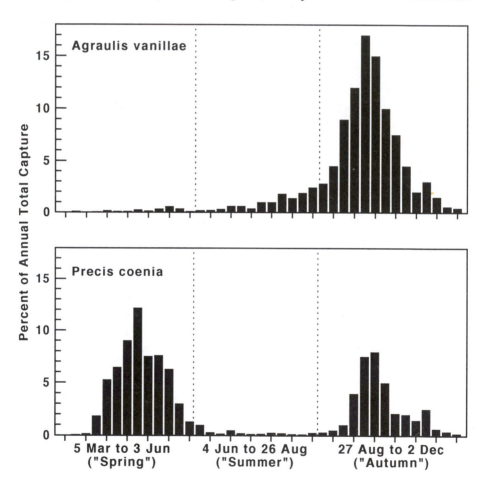

Figure 4–1. Seasonal patterns of migration through peninsular Florida in two butterflies, the gulf fritillary (*Agraulis vanillae*) and the buckeye (*Precis coenia*), as revealed by traps placed perpendicular to the ENE-WSW axis of Florida. (Data from Walker 1991.)

by attaching nets to some sort of vehicle. To sample large volumes of air close to ground level for *Heliothis* moths, as a case in point, the division of Entomology at the Commonwealth Scientific and Industrial Research Organization (CSIRO) in Australia has developed a large net attached to a frame and mounted on a vehicle so that moths can be sampled as the vehicle (in this case a small truck) moves slowly along (Drake 1991 and Fig. 4–2). Sampling at altitudes up to 200 m has been accomplished with kiteborne nets that intercept insects and other organisms being carried downwind (Farrow and Dowse 1984).

The obvious solution to sampling above ground level is to tow a net with an aircraft, a method pioneered by Perry Glick in the 1930s (Glick 1939). Glick used nets attached to a small single-engine airplane flown over the southeastern United States and demonstrated the presence of many species of airborne insects and spiders up to altitudes of thousands of meters, although with a rapid falloff in numbers above a few hundred meters. The problems faced in attaching nets to aircraft are cost, complexity of the device, sampling rate, and the necessity of flying slowly enough to avoid damage to the catch (or at least avoiding the use of

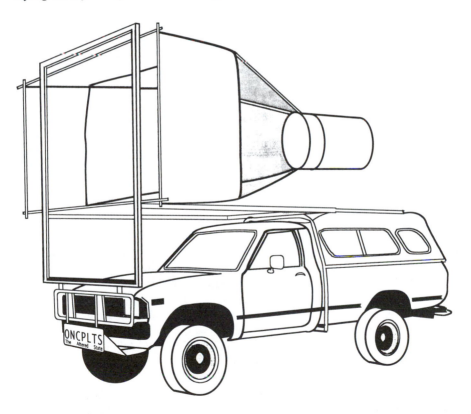

Figure 4–2. Truck-mounted net for sampling insects flying at 2–4 meters above the ground developed by CSIRO, Australia. (From Drake 1991.)

complex mechanisms such as expansion chambers to prevent catch damage). Helicopters can fly slowly and, indeed, effective large-mouthed helicopter-borne samplers have been employed (Hollinger et al. 1991), but helicopters are difficult to pilot and are expensive to purchase, maintain, and fly. Greenstone and colleagues (1991) have solved these problems by designing a net with relatively simple attachment mechanisms that can be towed by a slow-moving fixed-wing aircraft or carried by an automobile. The aircraft is a single-engine 1941 Piper Cub, which can fly at the very slow airspeeds of 65–72 km·hr^{-1}, speeds at which not even soft-bodied insects are damaged. Long-range fuel tanks have also been fitted so that the aircraft can remain aloft for 8 hours or more. At about 160 m above ground on a summer day, 44 arthropods were collected in 30,000 m^{-3} of air, which gives an idea of the sorts of aerial densities encountered. Over 3000 were sampled from a similar volume of air when the net was automobile mounted.

Mark and Recapture Methods

More precise information on migratory origins, destinations, and routes than is provided by simple observation or by natural markers (usually) can be obtained by marking or tagging individual migrants. Ornithologists have been banding (ringing) birds for decades and have gleaned an enormous amount of information from the recoveries of the banded birds. The method involves clamping a numbered lightweight metal band around the leg of a bird on which is an address to which the band should be sent if the bird is found dead or to which the band number and information on location, date, and condition can be sent if it is captured and released (the U.S. Fish and Wildlife Service in North America). Mammals can be marked with metal ear tags, which can also be imprinted with information; or, in some circumstances, mammals can be individually identified by having their fur bleached or dyed. Individual identification tags have also been used in the case of migrating butterflies. The best-known example is the monarch, *Danaus plexippus*; thousands of individuals were marked in eastern North America by Urquhart (1960) using a tag folded over the leading edge of the wing and glued. These tags included the address of the Ontario Museum, and enough recoveries were obtained to determine with certainty the long-distance autumn migration of the species. Tags attached to fins or other parts of the body are the most common means of identifying fishes (Harden Jones 1968), but coded beads stitched through the back or "tattoos" of various sorts have also been used.

Often it is not possible or practical to mark animals with individually identifiable tags, and in this case mass marking techniques may be useful, especially where large numbers of organisms of similar status (e.g., with regard to age or sex), can be marked at a single location. Mass marking methods have been particularly useful in fishes where many individuals can be netted (Harden Jones 1968). Studies are often carried out in connection with important fisheries, among which are salmon; juveniles can be netted and marked as they exit their natal rivers and recaptured, often by commercial fishermen, when they assemble to re-enter rivers at a later date. Birds like the red-billed quelea are also amenable to mass marking because they occur in large colonies where, in one case, they were sprayed with droplets of paint (Bruggers and Elliot 1989). Paints, dusts, and dyes, in many cases with fluorescent properties, and various dietary inclusions have all been used to mark insects (Drake 1991). Insects, in particular, can be reared en masse in a laboratory before mark and release, although this always raises the question of whether they would behave as they normally would in the field because they were raised under different food and crowding regimes from those they would experience in nature. If released from a crowded container, for example, recovery distances may indicate "escape" and dispersal but not migration. Migratory behavior may also have been selected against in the laboratory.

There are two major constraints on mark and recapture methods. The first, which applies to any system involving capture, is that whatever the trapping technique, it may be a biased sample of the population (Kerlinger 1989 summarizes nicely). Juveniles, say, may be more subject to capture than adults, or trapped individuals may be less healthy. If older individuals happen to migrate farther and are more difficult to catch, migration distances could be underestimated.

A second more serious constraint is that in order to be most effective, mark and recapture methods require very large sample sizes of marked individuals and a large contingent of observers who are likely to make and report recaptures. Bird banding has been very effective because both criteria have been met. A good example is the European lapwing (*Vanellus vanellus*); almost half a million birds have been banded and some 9500 recovered (Imboden 1974; Alerstam 1990). Even though the recoveries represent only about 2% of the birds originally tagged, a remarkable amount of information has been accumulated on the routes and wintering ranges of lapwings banded in different breeding areas across northern Europe. Similarly, some 47,000 bluefin tuna (*Thunnus maccoyi*) were tagged in Australian waters and nearly 7,000 recovered (Harden Jones 1984). At the opposite extreme, however, is the situation in terrestrial Australia, an area about the same size as the lower 48 states of the United States but with only about 17 million people (in 1992). Most of the population is concentrated in a coastal band in the Southeast, running from Brisbane to Adelaide, that includes six of seven of the continent's largest cities (Perth is the exception). As a result most birds are banded and recovered near population centers and vast areas of the continent, where many migrants apparently winter, report no recoveries (Fullagar et al. 1988). Even a large-scale banding effort on the yellow-faced honeyeater, an abundant migrant, failed to produce much in the way of results (Purchase 1985). There is also suggestive evidence of quite a bit of north-south movement in several "nomadic" Australian birds, but few data and little chance of obtaining more by banding recoveries (Ford 1989).

The problem is even more acute in insects. A herculean effort in China lasting 3 years and involving the mark and release of hundreds of thousands of oriental armyworm moths and the cooperation of hundreds of research stations monitoring traps netted less than 10 recaptured moths, although the recaptures did indicate the broad outline of migratory routes in eastern China (Li et al. 1964). More recent work using other methods has been far more effective in discerning the migratory cycle (Chen et al. 1989). In spite of the constraints, however, mark and recapture methods can be effective under the right circumstances.

Tracking Migrants

The best way to acquire an understanding of migratory pathways would obviously be to follow migrants over their entire journeys rather than simply noting times of departure, arrival, or passage at points en route. For a number of reasons, not least of which are the distances traveled and the difficulty of actually observing migrating individuals, this has only rarely been done. Nevertheless both direct and indirect methods have sometimes been quite effectively used to follow migrants. The sometimes surprising success in determining tracks and the migratory behavior occurring along them is a tribute to the determination and ingenuity of many investigators.

The most frequently used indirect method of determining tracks is to construct a backtrack from the place of arrival of migrants to their known or presumed place of origin. For aquatic or aerial migrants this requires a knowledge of the movement characteristics such as speed and altitude or depth at which the migration takes place, and a knowledge of the wind or current speed over the period of the migration. An example of how to construct a downwind trajectory is illustrated in Figure 4–3. In panel A it is assumed that over a given interval, usually 3 hours with standard meteorological data, the wind is at constant speed and direction as indicated by the two streamlines that we can imagine as having been taken from a wind-field map. The line XY represents the distance moved by our migrating organism at a given constant windspeed without change in wind direction under the assumption that its airspeed is zero. In panel B the wind changes direction over the interval as indicated by the dashed streamlines and, let us also say, increases in velocity. A reasonable estimate of the distance traveled in this case would be the line X'Y'. In this way a three-hourly (or some other appropriate time interval) plot of the movement of migrants can be constructed by adding the various segments. The whole process can also be computer automated. In panel C is an example of a hypothetical backtrack estimated in this way. Complications arise because wind speeds may vary with altitude, so that it is often useful to estimate more than one trajectory, and because near storms or major surface features like coastlines and mountains, wind fields can be erratic and rapidly changing (Scott and Achtemeier 1987). The method has, nevertheless, yielded enormous amounts of information on the tracks of migrating organisms (Pedgley 1982).

Backtracks will always be unsatisfactory to a certain degree because they involve assumptions and estimates. These problems can be solved only by making direct observations, and there have been various attempts to do this. The simplest

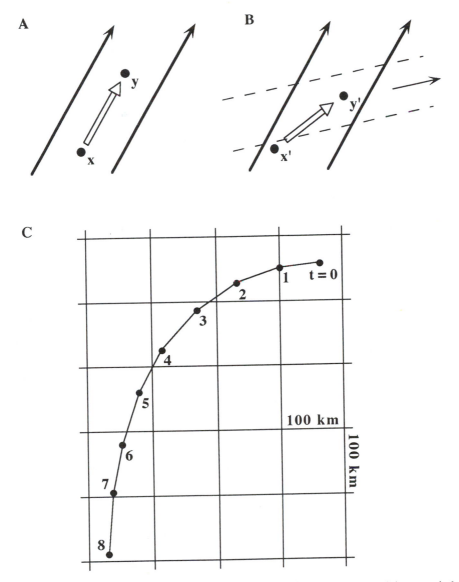

Figure 4–3. General method for constructing a backtrack for migrants arriving on winds. *A.* The presumptive path of a migrant between two points during a time interval given wind of a certain direction and speed (*solid arrows*). *B.* The presumed path taken when the wind changes direction during the time interval as shown and path is for averaged wind speed and direction. *C.* Schematic backtrack plot for eight time intervals beginning with present (t=0). (Modified from Scott and Achtemeier 1987.)

involves following a migrant by eye, or with the aid of binoculars or telescope, until it is no longer visible. We have seen an example in the case of the vanishing points of Walker's butterflies in peninsular Florida, and vanishing points are frequently used by ornithologists studying orientation in birds (see Chapter 8). Under the right circumstances and with large migrants, individuals can be followed for long enough periods and much can be learned about their behavior while en route. Pennycuick and colleagues (1979), for example, were able to follow cranes (*Grus grus*) for about 300 km with a small aircraft to observe soaring behavior, as will be detailed in Chapter 5. The information gained from following, however, can be greatly enhanced with the aid of various devices to aid in tracking.

An extremely effective method for increasing the ability of an observer to follow and track the movements of a migrant or other organism is radiotelemetry. This involves attaching a small radio transmitter and power supply (usually a small battery) to the organism to be tracked. The transmitter can be programmed to produce a unique signal which is then picked up by a receiver tuned to the signal in question. The technology of radiotelemetry is quite sophisticated and devices can be attached to relatively small organisms such as songbirds or bats (but so far not insects) with minimal influence on their performance with respect to added weight or increased drag (Obrecht et al. 1988). The success of tracking with the aid of telemetry is nicely illustrated by Cochran's (1987) study of a Swainson's thrush (*Catharus ustulata*) migrating north in the spring. The bird involved weighed only 39.8 grams at the time of capture in Illinois, but the transmitter at 1.3 g was a mere 3% of the mass of the thrush. Following capture in mid-May the bird was followed by car for seven nights (it is a nocturnal migrant) and over 1500 km until contact was lost in southwestern Manitoba. On some nights the thrush flew over 8 hours, but this was reduced to 3–5 hours when the bird encountered a cold front. Gradual changes in heading along the route correlated with slight clockwise changes in magnetic north and in the position of the sun's azimuth at civil twilight (see Chapter 8). Headings were not affected by complete overcast. Thus a great deal was learned about flight performance and orientation in this single bird because it could be followed continuously over a long distance.

The effectiveness of telemetry is limited by signal range, which is in turn influenced by the size of the power source and the sensitivity of the receiver. These constraints render it less useful for very wide ranging animals because the transmitters, especially, must have large power sources to transmit for long distances. To a major extent these difficulties can be overcome, at least for larger organisms, by the use of satellite tracking. Using satellite telemetry, Jouventin and Weimerskirch (1990) tracked wandering albatrosses (*Diomedea exulans*) as they foraged during incubation shifts. Six 10–12-kg males breeding on the Crozet Islands in the southwestern Indian Ocean were equipped with 180-g transmitters. The output of the transmitters was relayed via satellite to France, where it was collected and then transmitted to the observers on the Crozet Islands. The data revealed that the birds covered 3600 to 15,000 km on a single foraging trip lasting several days and flew at speeds over 80 km an hour for distances up to 900 km in a single day. Both the daily distances and the flight speeds are the highest estimates for any bird, whether migrating or not, and are presumably possible because albatrosses make optimal use of the wind by dynamic soaring (see Chapter 5). Long-distance movements occurred only during daytime hours, but the birds did continue to fly at night, a fact not previously known. Satellite tracking is obviously limited by the costs, the availability of satellites, and the lower size limits of the organisms to be tracked; but if these limitations can be overcome, it is a very effective technique for obtaining information from species that move very long distances. Other examples of animals that have been tracked in this way include Old World vultures (Berthold et al. 1991), swans (Nowak et al. 1990), elephants (Lindeque and Lindeque 1991), and seals (McConnell et al. 1992).

At the opposite end of the spectrum from megascale tracking by satellites are methods for tracking very small organisms. The problem here is that even the most miniaturized energy sources powerful enough to drive a transmitter are still too large to be carried by, say, an insect. The solution is to deliver the power to the transmitter from an external source that does not have to be carried by the migrant. This is the principle of the harmonic radar designed by Mascanzoni and Wallin (1986) to study the movements of insects. The device was originally developed by engineers in Sweden to locate avalanche victims. It consists of a microwave transmitter and receiver built together into a battery-operated portable gun-like unit. Glued to the back of an insect is a Schottky diode, which when illuminated by the microwaves re-radiates the signal at a harmonic of the original to minimize false reflections. To increase the reflecting efficiency, the diode can be equipped with a short length of wire to serve as an antenna. Because the diode is glued to the insect's back and so prevents flight, the harmonic radar is limited to species such as ground-dwelling beetles that move about primarily by walking. For these it has proved extremely effective, as indicated by its ability to successfully detect beetles even when they are beneath dense vegetation or burrowed into the soil.

Tracking Underwater Animals

Monitoring the movements of aquatic animals presents special problems. For the terrestrial observer, aquatic animals are invisible most of the time, and they are often inhibited in their behavior if the observer dons an air-breathing apparatus and tries to approach them in the underwater realm. One solution is to use remote sensing devices that can be kept at the surface; the two methods commonly used for tracking are radio and ultrasonic telemetry (Winter et al. 1978; Nelson 1987).

Each method has advantages and disadvantages. An important feature of radio tracking is that the animals can be located with an aerial and receiver that do not need to be in contact with the water; they can thus be carried by planes or trucks as well as boats. It is also relatively easy to program each transmitter to send on a different frequency so that several individuals can be identified simultaneously, a task that requires elaborate decoding techniques in the case of ultrasonic transmission. An important limitation of radio telemetry, however, is that radio signals attenuate with increasing conductivity and depth, a feature that limits their use to freshwater or to tracking marine animals that spend much of their time at the surface. Ultrasonic devices do not have this limitation and can also be made sensitive enough to discern brief and small-scale movements in addition to relatively large course changes. Costs of ultrasonic systems tend to be higher because of power demands and the need for underwater hydrophones. Both systems require the attachment of transmitters to the animals. In spite of increasingly sophisticated technologies, these devices are large enough to restrict their use to organisms that are relatively sizable (see above). Winters and colleagues (1978) describe several means of attaching transmitters to fishes.

The accuracy of tracking and the amount of information received can be enhanced in various ways, limited primarily by cost (Nelson 1987). One problem is accuracy of the bearing (compass point) when one is using only a single beam. The solution is to use two well-separated directional receivers yielding two simultaneous bearings that provide a more accurate position point so long as they cross at a favorable angle. Ranging and measuring can be further improved by means of *transponding* or *timefix* systems. With transponding a pulse is sent out from the tracker to the unit on the animal, which then sends a reply pulse. Because times of transmission and reception are known, the receiving equipment can calculate the distance as a function of the round-trip time of the pulse. Two receivers again increase the accuracy. A simpler system is the timefix, which uses two synchronized clocks, one on the animal and one in the receiving unit. If the synchrony is precise, then the delay in the reception of a pulse from the organism is due to the time of travel through the water, again allowing calculation of distance. Maximum effectiveness of tracking can be achieved with a fixed array of widely spaced unmanned receivers attached to buoys or other anchored structures and linked to a computer based X-Y plotting system to allow continuous monitoring of animal movements. So far most tracked movements have been of relatively short distances, with some notable exceptions such as the north-south and inshore-offshore movements of white sharks (*Carcharodon carcharias*), off the coast of California (Klimley 1985). Improving technology is expanding the possibilities, but costs are more limiting than either technology or investigator ingenuity. We shall, however, see several examples of the successful telemetric tracking of aquatic migrants in this book (e.g., in Chapter 5).

Radar Observations

The tool that perhaps more than any other has advanced our knowledge of migratory movements over the past few decades is radar. Its history as a method to study migration (and other movements) goes back 50 years to 1941 when the radar units developed in Britain for wartime use in detecting German aircraft also recorded the passage of unidentified targets or "angels" (Eastwood 1967). Among the scientists recruited to wartime service to work on the operational problems of using radar were the ornithologist David Lack and the entomologist G. C. Varley. Experiments performed by them and others in radar operations demonstrated that the radar angels in many cases were in fact, birds (Lack and Varley 1945). Lack, especially, realized the potential for the use of radar in ornithological work and together with Eric Eastwood and others developed the field of radar ornithology as summarized in Eastwood's (1967) now classic book, *Radar Ornithology*. Among the discoveries made by Lack and his colleagues was the vast amount of bird flight back and forth between Britain and continental Europe across the North Sea. One of the more interesting observations was that movement continued throughout the winter, often triggered by severe weather. Since those beginnings, radar has made many contributions to our understanding of bird migrations (Kerlinger 1989; Gauthreaux 1991).

The principles of radar (from *ra*dio *de*tection *a*nd *r*anging) are straightforward, although the accompanying electronics and hardware are necessarily quite sophisticated (Eastwood 1967; Schaefer 1976; Skolnik 1980; Riley 1978,1989). Radar is one form of so-called echo ranging systems, that is, systems that depend on the reflection of pulses back from objects to determine characteristics of those objects. A radar requires (1) a transmitter to produce pulses of high-powered radio waves and (2) a very sensitive receiver to detect the weak echoes of these pulses reflecting back from an object. Usually the transmitter and receiver are combined into (3) a highly directional antenna via a device that isolates the receiver from the aerial during the transmission of a pulse. The receiver echoes are then (4) displayed on a cathode ray tube (CRT) that consists of a screen and phosphor much like a television screen. Converting the time for the echoes to return into units of distance gives twice the distance from the radar unit to the object, and if the object is moving, its position on the CRT can be plotted and its rate of motion estimated.

The transmitted pulses of radio waves can be of various sorts. Wavelengths, for example, can be in the 3 cm (X-band), 10 cm (S-band), or 20–30 cm (L-band) ranges. The shorter wavelengths are better at detecting smaller or more distant objects, but suffer a disadvantage in that if, say, one wishes to detect aircraft, the radar will also detect raindrops so that it can be difficult to detect the aircraft in the presence of rain. For this reason surveillance radars are often of a compromise wavelength around 25 cm, although shipboard and airborne radars are often of the short wavelengths because these allow antennas to be much smaller. On the other hand, if one wishes to detect individual insects, they can be most easily detected at the shorter wavelengths of the X-band systems and for very small insects, 8 mm or K-band systems have been developed (e.g., Riley 1985, 1992; Riley et al. 1987). These latter are capable of detecting tiny insects like leafhoppers and planthoppers, but are of restricted use when it is raining. Because they do detect rain, radars of the shorter wavelengths can be very effective as weather radars used to locate fronts, storms, and weather systems in general.

The shape of the transmitted radio beam can also be varied depending on the task required of the radar. For airport surveillance radars (ASRs) where many single targets need to be detected and a large volume swept, "fan beams" are most appropriate (Fig. 4–4). The angle of the beam width in the vertical plane may be up to 40°. The azimuthal width is usually quite narrow (~1°), but a full 360° coverage is obtained by scanning with the antenna. A radar of this sort can determine the bearings of aircraft to 1°. Because of the vertical size of the beam, altitude cannot be determined (Gauthreaux 1991), and must be obtained from other sources. The fan beam is of little use for tracking individual targets. For the latter function a more focused "pencil beam" like that illustrated in Figure 4–5 is required. If the radar incorporates a steerable antenna, it is possible to "lock on" to a target and follow it for the range permitted by the power of the equipment. Pencil beams are also suitable if many targets are present, as is often the case with insects or birds, and are required if one is to obtain an accurate estimate of the altitude of a flock of birds or concentration of insects.

The most common form in which received echoes are displayed is by means of a plan position indicator or PPI. This is a circular CRT that shows echoes in map form. The center of the display represents the position of the radar and concentric range rings indicate the distance to the targets and the "scale" of the "map." A map outline of a particular geographic area can also be projected electronically onto the screen so that the position of objects detected can be more readily ascertained. The interpretation of echoes received from large numbers of small objects and displayed on the PPI is illustrated in Figure 4–5 for a radar with a pencil beam sweeping in a circle above the aerial. The beam sweeps a conical

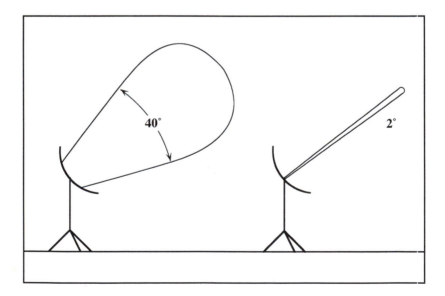

Figure 4–4. Two different sorts of radar beams, a broad (in the vertical plane) "fan beam" (*left*) and a narrow "pencil beam." Fan beams can track many targets; pencil beams are used to "lock on" to individual targets.

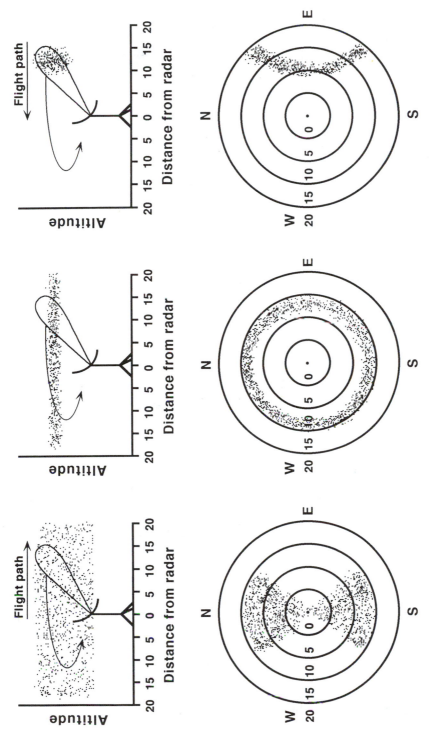

Figure 4–5. Schematic of the path of a radar beam through a group of insect migrants of varying shape and direction of movement (*top panels*) and the resultant image of the group on the radar screen or plan position indicator. Full explanation in text.

volume whose maximum extent is 15 units outward in all directions from the radar. In the left panels a large concentration of insects is uniformly distributed above the radar moving from west to east and also, for the sake of simplicity, the heading of every individual is due east. In this case the insects extend over an altitudinal range such that the beam is entirely occupied by insects. Insects observed side-view produce echoes 10 to 100 times stronger than those seen end-view (Schaefer 1976) with the result that the PPI displays a "dumbbell" pattern, as shown, whose axis of symmetry coincides with the head-tail direction of the insects. In the center panels the insects are in a narrow band whose lower limit is still some distance above the radar, and in this case we'll assume that the headings of individuals are completely random. The PPI display now shows a ring of echoes (because of course nothing is detected at lower altitudes), and the display consists of the top and bottom of the layer in the region swept by the pencil beam. Lastly, in the right panels a linear concentration of insects is moving west toward the radar with again random headings of individuals. The display appears as a "boomerang" because the two ends of the line are farther away from the radar than is the center. The displays shown in Figure 4–5 are simplified for clarity; in actuality any combination can appear on the PPI, depending on the configuration of the concentrations and individuals being detected.

In addition to raindrops and clouds, which can obscure short wavelength radars, echoes will also be received from stationary objects such as buildings or hills in the vicinity of the transmitter sometimes masking on the PPI the objects one wishes to detect. A bird flying in front of a hill, for example, would not be detected. A way around this problem is to use the principle of the Doppler effect because wavelengths emanating from a body approaching an observer (or from an object the observer is approaching) appear to be shorter and their frequency greater. Similarly, as a body recedes from an observer the apparent wavelengths are longer and the frequency less. This is the reason the rumble of a diesel locomotive goes from a higher to lower pitch as a train passes an automobile going in the opposite direction and is also the cause of the red shift used by astronomers to calculate the rate of movement of receding stars and galaxies. Similarly the radial movement of an object with respect to a radar causes a Doppler shift in the radio beam, which with the right equipment can be detected. This is the principle of the radars used for the enforcement of speed restrictions on vehicular traffic. A stationary target or one moving at very low radial velocity will produce no Doppler shift. With a Doppler frequency-sensitive radar, the echoes from such objects can be suppressed and only moving targets displayed. The most common type of Doppler radar, termed a moving target indicator or MTI, can be used to eliminate echoes arising from ground clutter or raindrops. The MTI is also useful in an additional way. If a target such as a flock of birds is moving tangentially to the radar, the MTI filter will create an apparent wedge across the display of the flock because the bird echoes are suppressed when their radial velocities are too slow to produce adequate Doppler shifts. The bisector of the wedge angle is perpendicular to the direction of movement of the flock and so indicates the direction of

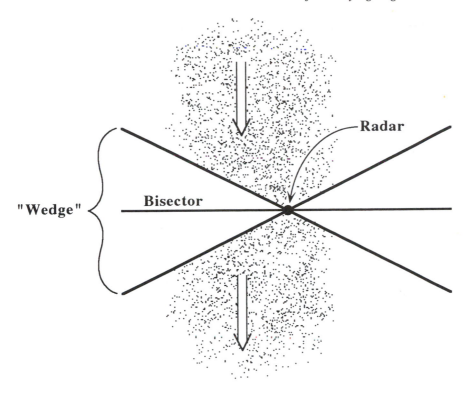

Figure 4–6. Wedge across a moving target (e.g., a flock of birds) produced by the filter of a Moving Target Indicator (a form of Doppler radar). The wedge forms where radial velocities are too slow to produce detectable Doppler shifts. The bisector of the wedge is perpendicular to the direction of movement of the target.

movement of the birds. The situation is illustrated in Figure 4–6, and the method was used, for example, by Drury and Nisbet (1964) to estimate flight directions of migrant birds over the northeastern United States.

Radar has been an enormously effective technique for studying the movements of insects and birds, and indeed much of our understanding of these movements, of the association of migrants with winds, for example (Chapter 5), would not have been obtainable without the use of radar. It is also extremely useful for determining movements at night. There are, however, constraints. Probably foremost is the fact that the units are expensive, although smaller marine radar units are certainly well within the means of most investigators with grant funding. The larger radars are a different matter, but many workers have been able to share time effectively on weather or airport surveillance systems. The latter, however, in addition to being designed for purposes other than the study of migration, are not mobile and so may not be located where most migration is occurring. The problem of mobility has been solved in various ways from self-contained mobile laboratories using marine radars carried in a recreational vehicle (Gauthreaux 1985b) to highly sophisticated airborne radars. The latter can be used very effectively to sample over wide areas and altitudes (Hobbs and Wolf 1989). Radar can sometimes be used to identify individual species by their "signatures" of wingbeat frequency (Schaefer 1976; Vaughn 1978), but usually it is necessary to identify the targets by other techniques, either visually or as in the case of insects by aerial sampling (e.g., Greenbank et al. 1980). In spite of these limitations, however, the sometimes extraordinary success of radar methods is attested by the tremendous expansion of our knowledge of migration gained through its use, and it is one of the most valuable tools available for studying both bird and insect movements.

Laboratory Methods

Many questions concerning migration are very difficult or impossible to answer using field studies alone, but rather require some sort of laboratory assessment of migratory performance. In particular, studies of physiology including behavioral physiology, orientation, genetics, and the relation of migration to life histories are all likely to depend on some combination of controlled environments, laboratory measurements, and experimental manipulations. Successful studies thus depend on the adequate indexing of migratory behavior. To this end a number of techniques have been tried, all with significant advantages and disadvantages.

Perhaps the organisms most frequently used for laboratory studies of migration have been insects because their small size allows large samples, and many species can be easily reared in growth chambers or other types of controlled conditions. With a few notable exceptions, insects migrate by flying, and so the usual analysis requires the indexing of migratory flight. The most commonly used method is the simple tethering of test insects in still air with the subsequent measurement of flight duration. An adumbrated list of insects tested in this way includes terrestrial (Dingle 1965) and aquatic (Fairbairn and Desranleau 1987) Heteroptera, Homoptera (Cockbain 1961), Orthoptera (McAnelly 1985; Fairbairn and Roff 1990), Lepidoptera (Rankin et al. 1986) and Coleoptera (Davis 1980; Rankin and Rankin 1980a,b).

Tethering usually involves attachment of the insect at the prothorax to a pin or stick with wax or glue. Release of tarsal contact by lifting the tethered insect from the substrate usually induces flight, and it is then a simple matter to record duration (Fig. 4-7). Longer flights are presumptively migratory. Sometimes flight can be prolonged by keeping the test individual in a wind stream. The method can also be modified by allowing the insect to hold an object, such as a styrofoam ball, which substitutes for the substrate. This can then be released at will and allows an estimation of flight threshold, which can be an important component of migration in addition to duration. There are several shortcomings with tethering, not least of which is that by keeping the insect stationary, most of the sensory inputs it would receive under natural conditions are eliminated. In spite of this, however, the

Figure 4–7. A migrant milkweed bug, *Oncopeltus fasciatus*, flying while tethered to a small stick. Duration of such tethered flight can be used as an index of migration. (Photo H. Dingle.)

method does seem to index migration in the species examined, because known migrants brought in from the field and tested in this manner consistently display flights that are of longer duration than those of unselected controls (Dingle et al. 1980; Rankin and Rankin 1980a,b; Davis 1981). The success of this crude method in assessing various aspects of migratory performance has been quite remarkable as we shall see throughout this book.

Tethering can be made more sophisticated in a number of ways. Among the most common is the use of flight mills, which have been used for insects as disparate in size as locusts and mosquitoes (Krogh and Weis-Fogh 1952; Rowley and Graham 1968). The disadvantage here is that a prescribed flight path is forced on the test individual, but because it moves, the insect does receive optomotor visual inputs and stimulation from wind. Other investigators have used sensing devices to measure free flight of relatively large insects in cages (e.g., Macaulay 1972), but here individuals are constrained by encounters with the container walls and often do themselves considerable damage by battering against them. Gatehouse and Hackett (1980) have developed a flight balance apparatus that not only allows a tethered armyworm moth (*Spodoptera exempta*) to control its own takeoff and landing but also provides freedom to yaw, bank, and fly forward. The moth is attached to a lever arm with a flag at the end opposite the moth; this flag then activates a photo switch when the moth takes off from or lands on a paper drum (which rotates allowing the moth to walk) thus permitting automatic recording of the durations of flights. The moth can also be affixed to the apparatus while still a pharate adult within the pupal case so that flight can be determined precisely from time of emergence. The spontaneous takeoff times of moths on the apparatus matched closely those of insects monitored in the field, suggesting that their flight behavior while tethered corresponded closely to their flight under natural conditions. With the addition of computer technology relatively large samples of moths can be flown simultaneously, flight variables automatically recorded, and various flight parameters derived.

The most sophisticated attempt to duplicate natural conditions is the use of a flight chamber to study the behavior of free-flying aphids (Kennedy and Booth 1963; Fig. 4–8). The device consists of a large box open on one side to permit the investigator to view the interior and painted flat black on the inside walls, floor, and ceiling. To one side of the box is a long lever arm with a clamp at the top to hold the landing platform for the aphid (usually a leaf) and a joint at the bottom so that it can be rotated to one side of the box out of the way. Aphids were placed in the chamber on the landing platform and were induced to fly by a bright light produced by fluorescent lamps at the top. The lever arm could also be twisted to jolt them off the landing platform. Wind came down through the top of the chamber, and its speed could be regulated by a butterfly valve to exactly match the rate of climb of the aphid. By recording the valve setting, the rate of climb could be continuously monitored. The butterfly valve was used to regulate wind speed because it gave much finer and more rapid control than trying to regulate the rate of fan revolution. The aphid could be induced to settle by presenting a leaf or card with the lever arm below it as it flew, and the arm could then be twisted to induce further flight if this was required. Aphids would fly in this chamber for periods lasting up to several hours. The use of this chamber for studying the subtleties of aphid migratory behavior and records of flight performance produced were discussed and presented in Chapter 2 (see Fig. 2–2).

Although an ingenious device for studying free flight, the chamber does have two significant drawbacks. First, the size of the insects to be studied is severely limited, although modifications have been incorporated to allow the study of small heteropterans larger than aphids (Laughlin 1974). Secondly, use of the device is extremely time consuming because it is virtually impossible to automate response to feedback from the flying insect to exactly balance the rate of climb by downward airflow, at least without considerable expense. Nevertheless, experiments with aphids flying in the chamber have provided insights into the control of migratory behavior that would have been possible in virtually no other way, and for that reason it is an extremely powerful means for studying migration by flight, as the earlier discussion has shown (Chapter 2).

The migratory biology of larger animals is much more difficult to study in the laboratory than is the case for insects, and adequate sample sizes are also more of a problem. Nevertheless, some aspects of migration have been investigated with considerable success. A good example is orientation in migratory birds. The device

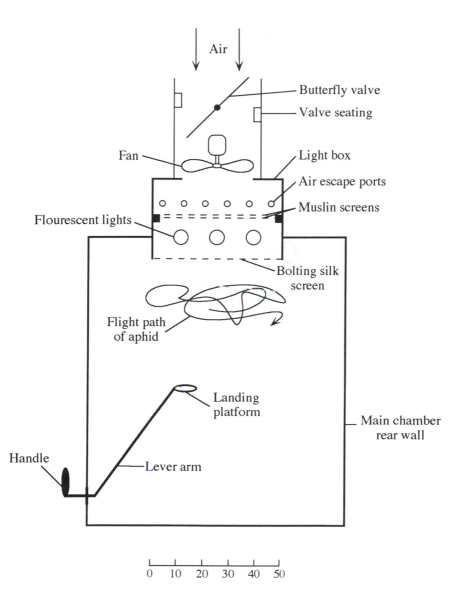

Figure 4–8. The Kennedy flight chamber used for studies of aphid migration. The lever arm can be twisted to shake the aphid off the platform. Free flight is maintained by wind from the top of the chamber, and the windspeed is varied with the butterfly valve whose setting thus indicates the wind speed balancing the aphid's rate of climb. (Redrawn from Kennedy and Booth 1963.)

that has made investigations of orientation relatively amenable to experimentation under controlled conditions is the Emlen funnel cage (Emlen and Emlen 1966 and Fig. 4–9). The device consists of a circular cage, the sides of which are constructed of white blotting paper rolled and stapled in the shape of a funnel with its constricted end downward. The funnel is mounted on a base that holds an ink pad. A bird is placed on the pad, and the entire cage is covered with a screen or clear glass or plastic to prevent the bird's escape. The bird can see only directly overhead, which means that its "sky" can be manipulated. Various experimenters have presented natural sky, a sky altered in a planetarium, or a sky of points of light ("stars") in an otherwise opaque sheet (see Chapter 8 for full discussion). When the bird hops onto the sloping paper funnel, it slides back to the base leaving

Figure 4–9. The Emlen funnel cage. A bird is placed in a circular cage with an ink pad on the bottom and sloping walls. The cage is covered with mesh screen and placed in a planetarium (as here) or exposed to the sky. As the bird attempts to fly the traces of its inked feet on the walls indicate its direction of orientation. (Modified from Emlen 1975b.)

a streak of inked footprint on the paper. The accumulation of inked footprints provides a record of bird activity and indicates in which direction the bird hopped most frequently. The fact that between hops the bird is confined to a small space at the bottom of the cage also avoids problems of parallax; in a large cage that allowed a bird to hop back and forth, the "sky" could appear to be at different latitudes or longitudes in different parts of the cage if under a planetarium or other artificial sky. Other sorts of paper such as typewriter correction paper (Munro and Wiltschko 1992) can also be used so long as it records a trace of the bird's feet.

Other sorts of apparatus must be used if the purpose of the experiments is to determine the performance of fully active animals. Usually this has involved measurements of muscular or metabolic activity. In the case of birds, training individuals to fly in wind tunnels has allowed the recording of data from actively flying individuals (Tucker 1968, 1972). Birds are trained to fly with a mask over the face that permits measures of gas exchange; the wind tunnel can even be tilted to induce ascending or descending as well as level flight. The obvious limitations are that it takes some time to train birds, birds can be flown only one at a time, and not all species adapt readily to the apparatus. Still, interesting information on general flight performance has been generated, although this has not been specific to migrants.

Similar sorts of devices have been used to measure swimming performance in fishes by the generation of water rather than air currents (Beamish 1978). There are two basic designs. The first is a circular chamber with a trough that is rotated, and the rotation can be at the swimming speed of the fish, which remains stationary relative to the observer. The second is a chamber with a current generated by a pump. Problems of the head at the pump and turbulent flow can be eased through modifications of the size and shape of the chamber or by the insertion of diffusion grids to smooth the flow pattern. Again these chambers are usually used to measure some aspect of performance without specific reference to migration. Snyder (1988) tried to address the question of migration with a tank in which sticklebacks were introduced in the center and could swim either upstream or downstream through a series of partitions allowing water flow through openings. Distance moved up- or downstream was then determined by the number of chambers transited by the fish, with the assumption being that migrants would move upstream while nonmigrants would either move downstream as the course of least resistance or show no preference. The most effective device for a given experiment will be determined by a number of factors, not the least being the size and swimming capabilities of the particular species of fish. The variety of the different sorts of apparatus is nicely summarized by Beamish (1978), who also illustrates several of the circular chambers and flumes used by ichthyologists.

Summing Up

To estimate the numbers of migrants, biologists often make observations of individuals or concentrations of migrating organisms passing a particular point.

Often this location can be a geographic feature that concentrates the migrants. Peninsulas that funnel migrants prior to an overwater crossing are favorite spots for these sorts of counts, and several locations are justly famous for migratory passages. Falsterbo, Sweden; Cape May, New Jersey; and Point Pelee, Ontario, are all well-known examples. The great concentrations of raptors passing over the Panama Canal on autumn migration indicate that isthmuses also can be excellent locations for tallying migrants during their journeys. One need not rely on concentrating geographic features, however. Migrants can also be counted when they cross the full moon, as in Lowery and Newman's (1966) classic study of birds over North America, or between two landmarks placed by the investigator, as in Walker's (1985a) study of butterflies proceeding southward through penisular Florida. All counts are limited by the ability of the observers to see the migrants and by the uncertainties inherent in making a point observation of a behavior that takes place over often considerable scales of time and space.

Estimations of the extent and path of migratory movements can also be assisted by trapping methods. A caveat that must always be stated, however, is that migrants may not enter traps for a number of reasons so it is well to have an independent assessment of trap bias. Nevertheless traps can be very effective. Some examples are the collection of amphibians returning to breeding ponds by drift fences (Gill 1978a,b), the extensive use of light and suction traps for insects (Drake 1991), the capture of birds with mist nets, and the funnel traps used to collect migrating butterflies (Walker 1991).

If migrants are captured alive, they can be marked and released, affording the possibility of recapture distant in time or space with possible elucidation of a migration route. The extensive use of bird banding has made the pathways of birds the best known of any class of migrants. Banding success, however, depends on rate of recovery; it thus works well where human population density is high, as in Europe, but less well in places like Australia, where few people live in some of the regions seasonally occupied by migrants. Various other tags have been used for organisms as diverse as butterflies and trout. Not all markers need be artificial. Natural markings on large animals like whales have been used to track identified individuals very successfully, and there are several cases where migratory populations of birds can be identified by specific plumage or song characteristics.

Mark and recapture methods do not tell us how migrants are behaving en route. For that, continuous monitoring is necessary, and for this many investigators have used radiotelemetry. Telemetering systems have become quite sophisticated and when combined with satellite tracking, migrants can be followed for long distances. Telemetry is mostly limited to vertebrate migrants but use of radar technology has permitted tracking of ground-dwelling beetles, at least while on foraging excursions (Mascanzoni and Wallin 1986). Fish present special problems because water, and especially seawater, attenuates radio transmissions. In this case ultrasonic rather than radio transmitters may be more effective.

Following movements of migrants flying at night or at high altitudes, when they are invisible to ground observers, has been considerably aided by the use of radar. Weather and airport surveillance radars have both been used effectively to follow the movements of birds and insects. Among the more interesting of the observations resulting from radar studies has been the fact that both birds and insects make extensive use of winds aloft during the course of their journeys. Although radar studies have proved extremely useful, they also are subject to constraints. Among these are that they are limited in mobility because of the elaborate electronics required and because they are relatively costly. As a consequence, they have been used by few investigators.

Finally, many aspects of the biology of migrants, especially with respect to physiology and life histories, require study under controlled conditions. This means that ways must be devised to study migration in the laboratory. Because of their small size, insects have been most successfully studied in this way. The migratory performance of several species has been measured using tethered flight, with other aspects of physiology and reproductive output analyzed simultaneously. In the case of aphids, classic studies were performed with free-flying aphids in a specially designed flight chamber (Kennedy and Booth 1963). Successful laboratory studies are not limited to insects, however. The orientation of songbirds has been analyzed repeatedly with the funnel-shaped Emlen cage (Emlen and Emlen 1966), and the energetics of both fish swimming and bird flight have been studied with current chambers and wind tunnels, respectively. Like the other examples discussed throughout the chapter, the ingenuity and perceptiveness of investigators has provided some remarkable means for overcoming difficult problems to provide insights into the behaviors of migratory organisms.

PART TWO

Proximate Factors
in Migration

The chapters in this section focus on the proximate mechanisms that influence the basic elements of migratory behavior and the pathways that migrants follow. Chapter 5 opens the discussion by outlining the ways in which migrants are influenced by perturbations in air and water. Because I suspect that most readers, like myself when I began to study migration, are only broadly familiar with the atmospheric and oceanic factors that generate winds and currents, I begin this chapter with a general survey of the planetary and other physical forces that generate the patterns we observe. In doing so I proceed from very large scale events to those occurring on such a minute level that they deserve to be called "nanoscale." Following this brief introduction to weather and to ocean circulation patterns, I give several examples of migrants that use winds or currents to aid them in their migratory journeys, again following an outline that leads from large to very small scale phenomena. Throughout I stress the interactions between the behavior of migrants and wind and current patterns. It is through these interactions that a marvelous array of observed migratory pathways has evolved.

In Chapter 6 I examine the physiological basis for migration. Because migration results in changes in habitat over (usually) large spatial scales, it is energetically demanding. It therefore is an advantage for migrants to anticipate environmental change by using cues such as photoperiod. I discuss the relation between such cues and migration as well as the endogenous rhythms used by migrants to synchronize their activities with the external environment. In the mobilization of physiology to support migration, hormones play an important if not always fully understood role. Frequently they are adapted from other functions and this hormone "capture" and other aspects of hormonal influences on migration are discussed at some length, especially for the best-studied insects and vertebrates. Finally, I examine fuels for migration, especially fat—the most efficient fuel by far in generating the energy necessary for migratory movement.

Migrants are subject to a number of constraints on their performance, including aspects of both morphology and energetics. In Chapter 7 I discuss some of these, beginning with an examination of the limits imposed on the size of ballooning spiders by the properties of their silk threads. I then examine the problems of optimizing fruit and seed size in plants relying on vertebrate transport of seeds, followed by an examination of swimming mechanisms in migrant fish. Birds are the best studied of migrants with respect to the analysis of constraints, and so I examine extensively various aspects of energy demands and wing shape and how these factors affect the performance of migrating birds of different sizes. Migration is, of course, not the only aspect of performance influenced by size and shape in birds, and I outline the trade-offs that have occurred as a result of the conflicting demands of migration and the need to be maneuverable in certain habitats. I conclude this chapter with some case studies of migrants that face the special problems presented by extreme environments such as the Sahara Desert crossing made by many Palearctic birds and the long overwintering period experienced by monarch butterflies in Mexico and California. All the examples illustrate the behavioral flexibility that has made migrants so successful.

In Chapter 8, the final chapter on proximate factors in migration, I address questions of orientation and navigation. In some migrant organisms, little orientation capability may be necessary, with required direction, if any, imposed by transporting currents or winds. Many migrants, however, possess sophisticated orientation or navigation mechanisms that rely on detection of both chemical and physical cues. The former have been extensively studied in salmon and that work I discuss at some length, followed by a brief treatment of the controversy surrounding odor navigation in birds. Visual cues turn out to be very important to migrants and both a sun and a star compass are used by organisms in diverse taxa. I discuss results from insects and several vertebrates. Other physical cues discussed include relatively simple ones like beach slope and the very complex magnetic field around the earth. The latter appears to be used by a surprising number of migrants. To conclude the chapter, I discuss the general conclusion that migrants use not just one cue, but rather a hierarchy of cues to coordinate navigation and orientation over the variety of migration routes they have evolved.

5

Migration, Winds, and Currents

To the insects, birds and other living beings which the air carries . . . the atmosphere is by no means the uniform and structureless medium which it may appear to the earthbound biologist.

R. C. RAINEY, Foreword to Pedgley, 1982

Our planet is a diverse place, and much of this diversity is a direct result of weather and climate. Because migrants move widely over the surface of the earth, they experience a range of climatic vagaries, and indeed it is the variation in climates and seasons that provides most of the ultimate basis for migratory journeys. To fully understand migration, it is therefore necessary to understand how migratory behavior may be influenced by conditions in the atmosphere and in the oceans. For migrants traveling through air or water, as most do, winds and currents can play major roles in determining pathways, timing, and distances traversed. This is true for large active swimmers and flyers as well as minute insects, spores, or plankton. The influences of weather and current patterns on migration extend from events occurring on a global scale down to minute changes in the microclimate.

In this chapter I first discuss briefly both worldwide and local atmospheric and oceanic patterns that are known to influence migrating organisms and the factors that cause them. I do this for both air and ocean by proceeding from large to small, beginning with *planetary scale* phenomena and working down to events at such a minute level that they may appropriately be called *nanoscale*. In between are phenomena ranging from *synoptic scale*, which includes the patterns depicted in the usual weather maps covering a portion of the earth's surface, through *mesoscale* to *microscale* patterns. These are summarized in Table 5–1, which also gives examples of atmospheric events that fall into each scalar category. For the information on which the sections on atmosphere and ocean are based, I have relied heavily on Wallington (1977), Pearce (1981), Barry and Chorley (1987), Drake and Farrow (1988), and the opening chapters in Alerstam (1990). Other references on more specific points are cited in the usual fashion in the text. Following the discussion of physical factors, I analyze migration both in the air and in the oceans with examples of both short- and long-term behaviors that allow organisms to adjust to winds and currents in some surprising ways. Here again, I proceed from large-scale to small-scale phenomena to illustrate the diversity of forms that migratory patterns can take.

The Meteorological and Oceanographic Background

Atmospheric Circulation

We live on a globe suspended in space and possessing characteristics having profound effects on the lives of organisms, including migrants. First, our planet orbits the sun whence it derives its energy, primarily as light and heat. Second, it spins on its axis, with the result that we have day and night *and* that objects on its surface travel at different rates, depending on where they are relative to the equator. At the equator the rate of travel is roughly 465 m sec^{-1}, but it declines to zero at the Poles, where the axis of rotation passes through the spinning earth. Finally, this axis of rotation is not vertical but is tilted at approximately 23.5°, a tilt euphoniously named the *obliquity of the ecliptic*. The spherical shape of the earth means that solar energy is differentially distributed from the equator toward the Poles, and the tilt means that this energy will undergo a regular fluctuation depending on where the earth is in its annual tour about the sun. The consequence is the seasons, with "winter" occurring in the half of the earth leaning away from the sun and "summer" occurring in the half leaning toward the sun. These seasonal changes are accentuated as one moves away from the equator, which is at the center of a band lying between the Tropics of Cancer and Capricorn (marking the position of the sun at the two solstices, the northern and southern reaches of its annual excursions) where there is little annual fluctuation in solar energy. The seasons in the "Temperate Zones" and Arctic regions outside the two Tropics bring with them major changes in weather and in the length of the days.

Weather occurs in a blanket of air called the atmosphere. Because of the pull of gravity the atmosphere is thickest near the surface, gradually thinning to a near vacuum as one proceeds outwards into space, and it is layered primarily on the basis of temperature. The lowest layer is the *troposphere*, containing about 75% of the atmospheric gases and extending upward to about 12,000 meters, varying

Table 5–1. Relative Sizes of Atmospheric and Oceanic Features Influencing Migrants

Size	Horizontal Dimension	Examples
Planetary scale	10,000 km	Jet streams, trade winds; Equatorial Currents
Synoptic scale	1000 km	Warm and cold fronts; coastal currents
Mesoscale	100 km	Sea breezes; tidal stream currents
Microscale	1 km	Thermals; estuary currents
Nanoscale	<10 m	Air or water moving over a surface

somewhat with season, latitude, and topography. On average, temperature in the troposphere declines ("lapses") with altitude at a rate of about $6.25\,^{\circ}$C km^{-1}. At the 12,000-meter altitude, more or less, ozone (O_3) begins to form. Ozone occurs primarily between 20 and 60 kilometers, but it causes temperatures to cease declining below this level and eventually to increase because of the absorption of the sun's ultraviolet (UV) radiation by the ozone. This region of change in temperature from a lapse to a *temperature inversion* is the *tropopause*. The inversion is located in the *stratosphere* and forms an effective cap, confining most weather events within the troposphere. Above the stratosphere are the *mesosphere* and finally the *thermosphere*, which need not concern us here.

The air within the troposphere is influenced in particular by the unequal distribution of the sun's heat between the equator and the Poles. In the Tropics, air is heated and rises, resulting in low pressure near the surface and high pressure aloft. In contrast, at the Poles air is cooled so that it sinks, with air pressures the reverse of those at the equator; that is, low aloft and high at the surface. All else being equal, the rising of air over the equator and sinking at the Poles would set up a general circulation pattern of air moving toward the Poles aloft beneath the tropopause "ceiling" and flowing toward the equator at the surface. However, because of the rotation of the earth, there is considerable modification of the pattern.

The full complexities of the circulation, and its tripartite nature are indicated in Figure 5–1. To begin with, the fact that the earth is spinning means that the *relative* speed of air moving toward the Poles will increase because a given point on the surface at the equator moves faster than points north or south. A parcel of air leaving the equator would thus increase in relative speed as it moved toward higher latitudes, and if one could view this parcel from space, it would tend to accelerate relative to the surface beneath it. Because the earth rotates from west to east, the air would drift to the east as it gained speed relative to the earth. In other words it would drift to the right in the Northern Hemisphere and to the left in the Southern. This tendency to drift to the right or left is termed the *Coriolis force*. Conversely a parcel of air moving from a Pole toward the equator would slow down relative to the earth's spin and drift west or, again, to the right in the north and to the left in the south as the Coriolis force took effect.

Outside the Tropics the Coriolis force is balanced against the pressure gradient in both hemispheres. The result is that when air moving poleward reaches mid-latitudes between about 40° and 60° north or south, it is deflected sufficiently to flow on average from west to east, forming the prevailing westerlies of the Temperate Zones. The temperature and pressure extremes are also greatest at these latitudes so that for the most part the strongest winds blow there, producing, for example, the notorious "roaring forties" of the Southern Hemisphere where winds are largely unimpeded by continental land masses. It is largely because of the constant high winds of the roaring forties that large seabirds such as albatrosses, which depend on the wind for soaring, are most abundant in the Southern Hemisphere (Walker and Venables, 1990). It is also in the mid-latitudes that winds

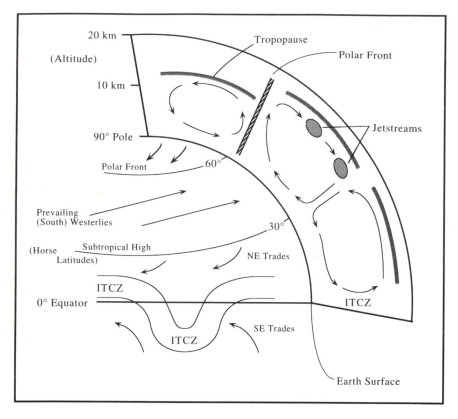

Figure 5–1. Major atmospheric wind patterns and convergence zones over the surface of the earth, shown here for the Northern Hemisphere. See text for full explanation. ITCZ = Inter-tropical Convergence Zone. (Modified from Alerstam 1990.)

are channeled into high level *jet streams*. The exact processes that cause this channeling are not fully understood, but the rate of air movement in these tubes of air that come and go can be an impressive 150–250 km hr^{-1} and sometimes even higher. Generally there are two and sometimes three jet streams in each hemisphere, with the polar front jet stream and the subtropical jet stream, which moves along at about 30° latitude, being the usual ones. The high-speed west winds and jets of the middle latitudes have another important consequence. High-level air moving from the equator toward the Poles meets the subtropical jet at about 30° latitude and is deflected downward to form a subtropical high pressure band around the earth. This air then moves along the surface northward and southward to become, under the influence of the Coriolis force, northeast and southeast trade winds in latitudes below 30° and surface westerlies above 30°. These latter tend to be from the southwest in the Northern and northwest in the Southern Hemisphere because of some slowing down from friction at the earth's surface.

The movement of air downward at 30°, and then its bifurcation to flow north and south, also has further effects. In the higher latitudes, the airflow meets air flowing down from the Poles, and the meeting of these air masses results in rising air along a *polar front*, where stormy conditions often arise. At the equatorial end, warm air masses from both north and south of the equator converge, meet, mix, and rise along the *Inter-Tropical Convergence Zone* (ITCZ). As the warm air rises, it cools and loses its moisture, creating precipitation as a rainy season that can be of considerable magnitude if the rising air has collected moisture following passage over an ocean. The ITCZ moves north and south with the passage of the sun, so that some tropical areas, especially near the equator, may have two wet seasons as the ITCZ passes over them first in one direction and then the other. Except for the sometimes intense winds of local storms, the tropical convergence zone is characterized by weak variable winds and extended periods of calm, the classic doldrums of the sailing ship era. The ITCZ can be of great importance to migrants able to reach areas where the rains bring flushes of new plant growth, especially in regions with lengthy dry seasons.

Finally, in the area of subtropical high pressure near 30° latitude, there is neither much wind nor rain. Here the descending air is dry and warms during compression, thus absorbing moisture instead of giving it up. The result is continuous warm clear weather, and it is here that the great deserts of the world are found in North and Southwest Africa, the Middle East, the southwestern United States and western Mexico, along the central west coast of South America, and in Australia.

There are some important regional modifications of this general hemispheric pattern of circulation. One of these is the *monsoon* of Southeast Asia. As the sun moves northward following the vernal equinox, it heats up the great land mass of central Asia. This warmed air rises and sucks in warm moist air originating over the Indian Ocean; when this moisture-laden air rises and cools over the land, it deposits the monsoonal rains. In the winter the land mass cools the air, forming a large area of high pressure; moist air is no longer drawn over Asia, and a winter dry season results. A second deviation from general circulation patterns is a trough of low pressure that tends to form over eastern North America in both summer and winter (Barry and Chorley 1987). The general effect of this trough is that low-pressure cyclonic patterns tend to move southeastward over the Midwest bringing colder polar air, while the lows farther east move northeastward up the Atlantic Coast. In the spring, especially, this pattern results in flows of warm moist air out of the Gulf of Mexico up the Mississippi Valley and over the eastern states (and causes tornadoes when it collides with cold polar air). As we shall see, these winds often offer important assistance to northward migrants.

The weather at any given time and place is largely determined by the pattern of passing low pressure (depressions or cyclones) and high pressure cells (highs or anticyclones). Generally speaking, lows and highs succeed each other from west to east, especially in the mid-latitudes. (Along the equator, rising air at the ITCZ creates a succession of weak lows with only slight high-pressure ridges between them.) Air tends to flow into a low, making it unstable so that it brings storminess, but the tracks of lows are generally stable, with the notable exception of hurricanes and typhoons, and they often move in a more or less continuous path. Highs, on the other hand, are stable internally, but they move fitfully, often ceasing to move at all and so blocking out lows bringing storms. The summer dry season on the lower west coast of North America, for example, is largely due to a blocking Pacific high, and long-duration Bermuda and Azores highs form over the Atlantic in midsummer.

High and low pressure cells also have characteristic patterns of wind direction associated with them. Air moves down the pressure gradient from a high to a low, but once again this fundamental characteristic of airflow is influenced by the Coriolis force. This is illustrated in Figure 5–2. Here the "expected" flow from high to low pressure is indicated by the heavy arrow. As the air moves outward from the high, it is deflected to the right (as shown here for the Northern Hemisphere) and flows clockwise around the center of the high pressure. Close to the ground the airflow is slowed by friction with the result that the wind direction is somewhat across the *isobars* or lines of equal pressure, as shown by the dashed arrows. With increasing altitude, friction becomes an ever decreasing factor and at about 500 m, the balance between the pressure gradient and the Coriolis force maintains wind flow parallel to the isobars as a *geostrophic* wind, with speed and direction a function of the local atmospheric pressure gradient. This tendency to change wind direction with altitude from somewhat across to parallel to the isobars is known as the *Ekman spiral,* and the change in wind speed is known as *wind shear.* Among other influences wind shear produces turbulence or gustiness in the winds.

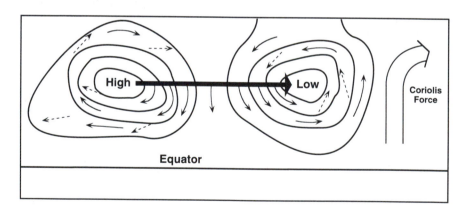

Figure 5–2. Characteristic patterns of wind direction in high and low pressure cells and shown here hypothetically for the Northern Hemisphere. "Expected" wind direction down the pressure gradient is shown by the heavy arrow. Deflections occur as a result of the Coriolus force as shown by their arrows. Friction slows winds closer to the ground, causing deviations indicated by dashed arrows.

As the airflow moves closer to the center of the low pressure cell, its tendency to deflect to the right is countered more and more by its acceleration along the steepening pressure gradient. The result is an airflow counterclockwise around the low. Once again this tends to be somewhat across the isobars at ground level because of friction and parallel to the isobars in the geostrophic winds above about 500 m, as indicated by the dashed and solid arrows in Figure 5-2. South of the equator the Coriolis force deflects winds to the left so that airflows around cyclones and anticyclones are opposite to those in the Northern Hemisphere—in this case, counterclockwise around a high pressure center and clockwise around a low.

The changes of wind speed and direction with height above the ground take place within the *planetary boundary layer* (PBL), the lowest 1–2 km of the troposphere, defined as the zone in which the frictional effects of the earth's surface have a direct influence on the atmosphere. The PBL can be in either a convective form, with vertical mixing via thermally rising air, or in a stratified form with little vertical mixing. The stable PBL occurs when the earth's surface is cooler than the air above, causing a rapid increase in temperature with height. This often results in a *temperature inversion* confining the zone in which the Ekman spiral operates and geostrophic winds occur to the lower 100–300 m or so of the atmosphere.

The PBL becomes convective when the ground or sea surface is warmer than the air above. Warmed air rises, usually until it encounters a capping inversion. Cooler air is displaced and descends to complete the circulation pattern. Large "bubbles" of convectively rising air known as *thermals* frequently form over land areas and can be important factors for migrant birds (Kerlinger 1989) or butterflies (Gibo and Pallett 1979) in gaining altitude by soaring (see below). Convectively rising thermals may also produce clouds, if they cool sufficiently for moisture condensation, and these clouds may rise out of the PBL to become the classic thunderstorm, producing cumulus with "anvils" spreading beneath the tropopause. If storms develop, convergent inflows may concentrate migrant insects to be washed out and deposited by rainfall (Dickison et al. 1983). Convective air movement in the form of sea breezes and land breezes may also form in the PBL under suitable conditions. Typically during the day, land areas heat up, drawing in cooler air as an onshore wind that can extend up to 500 or even 1000 m with a return flow out to sea at higher levels. The reverse situation occurs at night, producing an offshore land breeze. Sea breezes, especially, are known to transport migrant insects, sometimes for considerable distances inland (Drake 1982 and below).

The weather phenomena I have outlined so far take place over a large scale relative to most organisms, but for many small creatures important events occur very close to the ground or to plant surfaces. Air velocity gradients within thin boundary layers may change markedly over even minute distances (see Fig. 5–10) and can influence the takeoff behavior of migrants. Turbulence induced by vegetation and other objects can affect distances traveled, especially in the case of migrants that are not active flyers, such as parachuting seeds or ballooning spiders

(Chapter 3). Following a suggestion of Roger Farrow, I am calling these small-scale events *nanoscale* phenomena (Table 5–1). We shall see their importance for migrants at several places in this book.

Oceanic Circulation

The global system of oceanic currents is driven by the low-level planetary winds and the Coriolis force. The prevailing westerlies of the temperate latitudes and the subtropical oceanic high-pressure wind flows are of particular importance. Currents flow swiftest near the ocean surface where they are directly influenced by wind, are deflected to the right (Northern Hemisphere) or left (Southern Hemisphere), and decrease in velocity with depth resulting in a downward Ekman spiral.

On either side of the equator the prevailing trade winds drive surface waters westward as the North and South Equatorial Currents. At the western rims of the ocean basins, these waters pile up against the margins of the continents. Between the two currents, along the equator itself, there is essentially no Coriolis effect and only weak winds to influence water movement. In this region water thus flows down the hydrographic gradient in the opposite direction (eastward) to the Equatorial Currents as the Equatorial Countercurrent.

North and south of the Equatorial Currents the accumulated water of the continental margins swings poleward and is increasingly deflected by the Coriolis force. The result is clockwise circulation around the ocean basins in the Northern Hemisphere and counterclockwise circulation in the Southern Hemisphere forming large *gyres* on each side of the equator. Water starting out on these circuits is warm so that warm currents characterize the western edges of the basins, but it is then cooled as it passes the polar regions, and cold currents move along the east sides of the oceans. The cold currents passing along the western edges of Australia, South America, and Africa contribute to desertification in those regions, because air moving eastward over this cold water collects little moisture. A particularly interesting situation occurs in the western Atlantic. Here the Caribbean Basin and the Gulf of Mexico act as a constricted bowl, enhancing the pileup of water from the North Equatorial Current. This water is released through the Florida Straits as the narrow and swift-flowing Gulf Stream, which carries warm waters all the way to northern Europe producing a latitudinally uncharacteristic mild climate. Even in the winter the effects of the Gulf Stream are felt far inland, making western Europe a major overwintering area for migrant birds (Alerstam 1990). The general pattern of current flow in the Atlantic is shown in Figure 5–3.

Currents also affect migrants at a smaller scale. Inshore and along the continental shelves, the influences of local winds, waves, and tides can produce currents that predominate over the large-scale water movements of the ocean basin (Harden Jones 1984). Closest to shore, shoaling and breaking waves are the main current generators. Breakers arriving obliquely, for example, generate along shore currents in the surf zone that move away from the angle of incidence of the waves. These currents can be persistent and predictable in regions where winds and waves

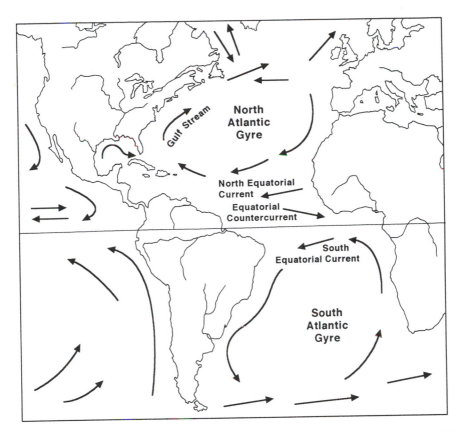

Figure 5–3. The overall pattern of current flow in the Atlantic Ocean. Superimposed on this general circulation pattern is considerable local variation that is not shown.

prevail from a consistent direction. More local are rip currents, which move outward below breaking waves.

The rise and fall of sea level due to tides will also generate currents. In the open ocean these rotate (the Coriolis force again) with changes in sea level, but with variation in coastlines and seabed topography, a great deal of local variation in tidal currents results. In some regions tides may produce simple reversing currents with inshore moving water on the rising tide and an offshore movement with the ebb. For estuarine organisms these tidal streams may provide an important means of transport for migratory movements within the estuary.

Winds also have an important role in current generation along the continental shelf. These currents are about 2–3% of wind speed and flow at about 45° to the wind direction (left in the Southern Hemisphere, right in the Northern due to the Coriolis effect). Because of friction an Ekman spiral occurs with depth, until some tens of meters below the surface, currents are only about 4% of the surface speed and the direction is nearly opposite to the wind direction. Under certain conditions this Ekman transport system is responsible for coastal upwelling. In the Southern Hemisphere a wind blowing with the coast on its right, say a south wind on the west coast of South America, will cause an offshore movement of water with a compensating onshore movement along the ocean bottom. The resultant upwelling of cooler water brings nutrients to the surface, increasing biological productivity. Shear processes at the bottom of the Ekman layer can also cause onshore veering of deeper water to provide another source of upwelling. Cold water rising from various causes can combine to create important areas of productivity, and transient shifts in the positions of these areas may cause changes in the pathways of migrating animals that need to feed en route. A major concentration of blue whales, for example, occurred in Monterey Bay, California, during migration in the autumn of 1986 when local upwellings produced an abundance of krill.

Thermohaline gradients in water density can also cause currents. In an upwelling, for example, colder denser water is brought to the surface against a coastline. This means that sea level inshore is lowered relative to the open ocean, generating an onshore flow down the gradient. On the east coast of a Southern Hemisphere continent such as Australia, this would generate a southward flow. The opposite situation occurs where a large river enters the sea. Freshwater is less dense and creates a raised sea level inshore and an offshore-flowing current. On the west coast of North America winter rainfall increases river outflows and generates just such a situation to produce the northward-flowing Davidson Current (offshore flow steering right).

Finally, where water masses of different types come together, "fronts" may form. Often these result from the convergence of currents, but they can also arise from other causes, such as when a large mass of low-salinity water from a river outflow meets the high salinity of the ocean. Temperature fronts are likely to form at the boundary of an eddy, caused by the sheer stresses of currents flowing past each other, or under some conditions of tidal mixing. Fronts create concentrations of nutrients, plankton, and fish that are major features of the distribution of pelagic organisms including migrants (Hamner 1988). Ocean currents can thus provide both transport for migrants and, in the form of fronts and upwellings, ecological conditions that influence movement patterns, distributions, and survival.

Migration in the Atmosphere

> We notice also that when a favoring wind springs up, whether by day or by night, migrating birds generally hasten to take advantage of it and even neglect food and sleep for this important purpose.
>
> FREDERICK II OF HOHENSTAUFEN
> *The Art of Falconry,* Chapter XX, circa 1250

In this remarkably astute observation, the Emperor Frederick foresaw what has become increasingly apparent in studies of migrating organisms, namely that migratory behavior and migratory pathways are often influenced by meteorological events and the patterns of atmospheric circulation. It was some 700 years, however, before the pioneering work of Wellington (1945) and Rainey (1951), especially, brought the importance of winds and rain to the forefront in understanding insect migrations, an importance that was emphasized by Johnson (1969) in his book on insect migration and dispersal. Johnson summarized a large body of evidence showing that even in tiny forms such as aphids, travel on winds was not an accidental consequence of passive transport, but was rather a programmed event in migratory life cycles. This viewpoint was updated and expanded by Pedgley (1982),who described the windborne movements not only of insects, but also of numerous other pest and disease organisms of both plants and animals. A still more recent update of the impact of atmospheric circulation on insects is provided by Drake and Farrow (1988). This impact can involve not only large-scale weather systems, but also events occurring at the microclimatic level where the atmosphere meets the surfaces on which organisms live.

Parallel to studies of weather and insect migration, it is becoming increasingly apparent that winds and weather fronts are perhaps equally important to the migrations of birds, even large and powerful flyers like raptors or seabirds. The classic radar studies of David Lack (1963), growing out of radar observations in World War II, first showed that bird movements were predictable from atmospheric conditions. This is a rapidly developing area of ornithological research that has been summarized recently by Kerlinger and Moore (1989) and Alerstam (1990). These summaries show that events from planetary-scale wind systems to microscale development of thermally heated rising air masses can influence the timing, the pathways, and ultimately the success of avian migrations. The examples that follow are chosen to illustrate the many-faceted interactions between migrants, from large powerfully flying birds to tiny drifting insects, and the movements of the air above them.

Planetary-Scale Bird Migration over the Western North Atlantic

Studies over the last several years have revealed that many land birds and shorebirds undertake often astonishing transoceanic journeys over most of the major oceans of the world in the course of their migrations from breeding to wintering grounds (Williams and Williams 1990). The best information on these overwater migrations comes from the western North Atlantic. Since the now classic radar studies of Drury and Nisbet (1964; Drury and Keith 1962; Nisbet and Drury 1967), evidence has accumulated that at least some small land birds and shorebirds migrate over the Atlantic from the northeast coast of North America directly to the Caribbean islands and South America, a distance of some 3000 to 4000 km. Examples include the American golden plover (*Pluvialis*), semipalmated sandpiper (*Calidris pusilla*) and blackpoll warbler (*Dendroica striata*) (although for an alternative view of the blackpoll warbler, see Murray 1989). Other equally impressive migrations take place over the central Pacific from Alaska to Hawaii, including that of the Pacific golden plover (*P. fulva*), and there is evidence that in the western Pacific, shorebirds such as the ruddy turnstone, *Arenaria interpres*, and the grey tailed tattler, *Tringa brevipes*, embark on a 5000 km nonstop journey from the northwest coast of Australia to the coasts of China (Lane and Jessop 1985). In the western North Atlantic it takes about 80–90 hours to reach South America and about 65–70 hours to reach the Caribbean (Williams et al. 1978). There is some dispute about the energy requirements for the journey, revolving around whether or not the birds actually need tailwinds, but whatever the energy storage capacity of the birds, it is clearly an adaptive advantage if tailwinds assist the journey. Simulation studies suggest that after takeoff following northwest winds are essential if small slow-flying passerines are to complete the migration; winds later in the journey are less critical (Stoddard et al. 1983).

In the mid-latitudes of both hemispheres, high- and low-pressure areas tend to succeed each other in a west to east progression (see above). Because air flows clockwise around a high and counterclockwise around a low, in the Northern Hemisphere winds will be out of the south when a high is east of a low and out of the north when a low precedes a high (Fig. 5–4). If birds are to take advantage of transporting wind systems, they should migrate following the passage of a high in the spring and of a low in the autumn, and the available evidence suggests that by and large they do this, although there is much variability that can be related to preferred direction or to local conditions (Richardson 1990). In the case of autumn migration over the western North Atlantic, birds should embark on their journeys after a low-pressure cell passes and on the winds behind the cold front at the leading edge of the ensuing high (Fig. 5–4); indeed there is evidence for increased fall migration as air pressure increases (Richardson 1990).

Radar studies of both shorebirds (Richardson 1979) and passerines (Richardson 1980) indicate that major departures of migratory flights occur under the expected wind conditions. Shorebirds, for example, tended to depart in a broad front generally southeastward from New Brunswick and Nova Scotia, following a heading that would take them out over the North Atlantic (Richardson 1979). These flights were particularly likely to occur on the relatively cold northwest winds in the high immediately following the passage of a low pressure. Interestingly, in embarking on these tailwinds, the birds must sometimes catch up with the cold front and pass through any storms that are generated. That the offshore flights were indeed migratory was demonstrated by the fact that large numbers of birds were involved, rapidly ascending to altitudes of 2–4 km above ground level.

Obviously if the birds maintained a southeasterly course, they would make landfall on the coast of Africa, if they survived the journey at all. As they do not end up in Africa, but instead in the Caribbean or South America, they must make a course change. Radar studies and other observations suggest that the birds fly first southeast and then curve south over or to the east of Bermuda, thus journeying around the "Bermuda High" that usually occupies the western North Atlantic at this time of the year. Once they do this, they enter the easterly "trade winds" generated by the planetary circulation, and the turn toward the southwest back toward the

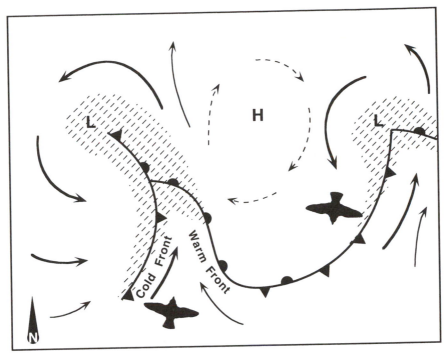

Figure 5–4. Wind patterns when high and low pressure cells succeed each other in the Northern Hemisphere. To take advantage of winds, birds migrating north in the spring should depart after passage of a high, and when migrating south in the autumn, they should depart following passage of a low, as indicated by the silhouettes. (Redrawn from Richardson 1990.)

Caribbean islands or the South American mainland again generates a situation with considerable tailwind assistance (Richardson 1985, 1990; Williams and Williams 1990). Simulations (Stoddard et al. 1983) indicate that for passerines there is no need to postulate a deliberate course change; a constant heading of greater than 150° will result in landfall on the coast of South America in less than 100 hours because of the effects of wind drift. For the more rapidly flying shorebirds leaving Nova Scotia at a heading of less than 150° (Richardson 1979), the simulations suggest that a reorientation is necessary en route. The migration system is illustrated in Figure 5–5.

Not all migrant birds from northeastern North America take this overwater Atlantic route, but a good many of them do and so gain the considerable energy savings from the prevailing tailwinds. Evidently the various weather systems involved are reliable enough so that this passage is favored by natural selection and has persisted, presumably, at least from the last glaciation. The alternative route is down the southeast coast of the United States with a short overwater flight from

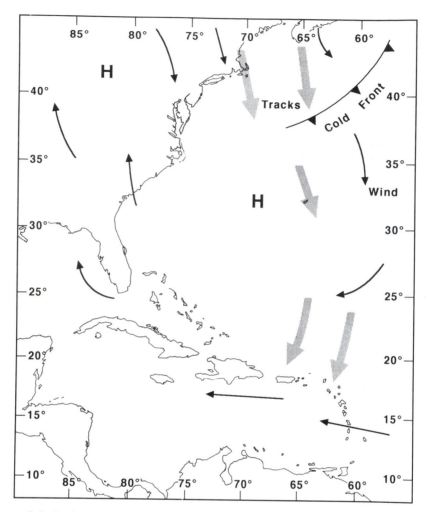

Figure 5–5. Typical weather pattern and tracks of migrant birds (*shaded arrows*) over the western Atlantic during the period of autumn bird migration from southeastern Canada toward the West Indies. Solid arrows show primary winds. (Redrawn from Richardson 1985.)

Florida to the Caribbean islands and, for some, on across to Central and South America. Many birds appear to do this, but it involves a much longer flight with its attendant risks. The advantage is continual movement over land with frequent opportunities to stop and refuel. Except for a few obvious cases such as swallows, which feed continuously on migration, we have much to learn about the factors favoring one route over the other. In the spring the landward route is favored by the prevailing winds (see below), and a majority of the birds taking the Atlantic route in the autumn do in fact take the predominantly overland route northward.

Synoptic-Scale Events

Winds moving northward up the Mississippi Valley of the United States and inland from the Gulf of Mexico are a potentially important means of travel for migrants, and they are used by a variety of insects, fungal spores, and other organisms (Johnson 1969; Rainey 1978). The arrival of corn-leaf aphids, *Rhopalosiphum maidis*, has been studied in east central Illinois following transport on these southerly winds (Achtemeier et al. 1987; Irwin and Thresh 1988). The study is especially illuminating because several methods were used to observe the flight of the aphids and to determine their probable origin. These included radar observations, trap captures using a specially equipped helicopter, computation of flight durations from estimated fuel consumption, and backtracking based on synoptic weather charts. The migration of these aphids is of considerable practical significance because they transmit virus diseases of crops, including barley yellow dwarf and maize dwarf mosaic viruses. Therefore, the ability to predict aphid movements and invasions is an important objective, which was the basis for the Illinois "Pests and Weather" project, a multidisciplinary approach to monitoring aphid movements (Achtemeier et al. 1987).

In conjunction with this project, radar observations in the early morning hours (0537 Central Standard Time) of 9 August 1984 detected three layers of small flying insects. Preceding this observation, there had been a moderately strong SSW flow of warm, moist air into Illinois as part of a prefrontal pattern. Although the layers weakened somewhat after midday, they could still be detected several hours later at 1300. The radar also indicated directional orientation toward the northeast, especially in the uppermost layer.

To determine what insects contributed to each layer, a helicopter with mounted collectors sampled flying insects between 50 m and 1000 m above the ground between 0655 and 0800 hours. The collections showed that the upper layer of insects, estimated from radar to be from 900 to 1200 m in altitude, consisted almost entirely of *R. maidis* at a density of around 4800 individuals per 10 cubic meters. The middle layer (500 to 650 m) contained a mixture of many different insect species. Temperature profiles indicated an inversion above 600 m and an isothermal layer above 1100 m. The uppermost layer, almost exclusively of aphids, was just below the isothermal layer in the zone of highest wind speeds and hence the greatest potential for long-distance transport. The middle layer of aphids, flies, and wasps coincided approximately with the temperature inversion.

Mean lipid content of aphids from the different layers was then determined and compared with laboratory data on aphids flying for different durations of tethered flight (Cockbain 1961; Liquido and Irwin 1986). From these data the probable flight duration of aphids from the middle layer was estimated at from 2–5.5 hours while that from the upper layer was from 5 to more than 24 hours. Because ground-level suction traps had failed to reveal any local aphid takeoffs during the previous 12 hours, the flying aphids had evidently come from some distance away, and the estimates of flight times could also be used to compute approximate distances traveled based on the wind velocities obtained from synoptic weather summaries. From the estimated flight durations, it was assumed that the middle layer had taken off near dawn about an hour before being first detected. The upper layer, on the other hand, likely consisted of aphids that took off during daylight the previous day and had been flying for 11.5 to almost 26 hours.

These estimates of flight times were then used, along with synoptic records of windspeed and direction at the observed altitudes, to determine a backtrack to a probable place of origin. In the middle layer at 600 m the wind patterns suggested that the aphids had come from an area in Illinois about 50–80 km westward of the observation site. The upper layer of aphids must have come from much farther away, and their origin was placed at somewhere between 400 and 1100 km to the southwest in a broad band lying between northeastern Texas and central Missouri. Thus by flying at altitudes with effective wind transport, even relatively minute insects like aphids can move great distances during a migratory episode. The fact that layers were associated with particular temperatures also suggests that aphids (and probably other insects) select certain wind conditions. Certainly they do not fly when temperatures are too low (circa 17°C, based on data from Kansas in Berry and Taylor 1968). To this extent aphids control their flight and are not simply passively transported by whatever chance winds are encountered aloft. Note also that they are not necessarily dispersed by wind streams, but may be concentrated along trajectories or even in quite confined local areas if suitable conditions are encountered.

Northward transport in the spring by southerly airflows, similar to that of the corn aphid, has been documented for a number of insects of agricultural importance in the Northern Hemisphere. These include moths such as the fall armyworm (*Spodoptera frugiperda*) in the southeastern United States, the Oriental armyworm, *Mythimna separata*, in China (Sparks 1979; Han and Gatehouse 1991), and several warm-climate moths transported into Europe as far as Finland (Mikkola 1967, 1986). The planthoppers *Nilaparvata lugens* and *Sogatella furcifera*, which are major pests of rice, not only invade northern China, but are annually carried by synoptic weather conditions to Japan. Because they cannot diapause and overwinter there, few survive, and Japanese rice fields are re-invaded every year from the Asian mainland (Kisimoto 1976; Noda and Kiritani 1989). The populations of *N. lugens* in northern China, on the other hand, return southward in the autumn on winds blowing toward the equator (Riley et al. 1991).

Not all out-of-range transport is necessarily catastrophic, as displacement can lead to colonization events. Perhaps the most interesting is the apparent colonization of the New World by the Old World desert locust, *Schistocerca gregaria* (Kevan 1989; Ritchie and Pedgley 1989). This locust under outbreak conditions can form large swarms as will be discussed in Chapter 11. There is now a well-substantiated record of the transport of *S. gregaria* from Africa across the Atlantic to the West Indies and northern South America on the easterly winds of a developing tropical storm (Rainey 1989; Ritchie and Pedgley 1989). The evidence from taxonomic comparisons also indicates that the several species of *Schistocerca* in the New

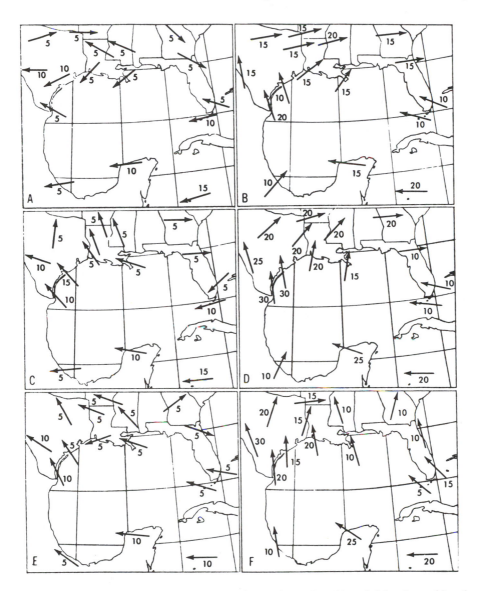

Figure 5-6. Wind speeds (in knots) and directions at the surface (*three left-hand panels*) and at approximately 900 meters (*right-hand panels*) averaged from 16 stations around the Gulf of Mexico over the years 1965–1967. Winds are for March (*top, A* and *B*), April (*middle, C* and *D*), and May (*bottom, E* and *F*). (From Gauthreaux 1991.)

World are all descended from *S. gregaria,* possibly as a consequence of several colonization events each followed by some adaptive radiation and speciation. Other apparently wind-assisted colonizations are those of the Australian grey teal (*Anas gibberifrons*) in New Zealand (Frith 1982) and of the fieldfare (*Turdus pilaris*) in Greenland (Williamson 1975).

As the grey teal and fieldfare illustrate, the use of synoptic wind systems for transport is by no means limited to insects; they are also used effectively by birds. Radar has been an important tool for studying the movements of migratory birds, and beginning with the early studies of Lack (1963), it has revealed that migration is most likely to occur in the presence of following winds. This makes sense because there is little point in a small bird embarking on a long flight in the face of a strong headwind. Even a bird's remarkable powers of fuel storage (Chapter 6) would be of little avail under such circumstances. In an impressive long-term study of bird migration in the southeastern United States, Sidney Gauthreaux (1991) has shown that birds can be quite selective in choosing suitable wind conditions for migratory journeys.

Since 1964 Gauthreaux has studied the flights of migratory birds in both fall and spring in Louisiana, Georgia, and South Carolina using a combination of radar and various types of visual observations to verify where possible the sources of echoes on the radar screen. Meteorological information obtained from the National Weather Service indicated that at the times of the year the birds were migrating, suitable winds were available for transport, especially at 300 m or more above ground level. An example is shown in Figure 5-6, which indicates wind directions and speeds during the spring at the surface and at about 900 m altitude (the 900 millibar level) around the Gulf of Mexico, as averaged for the years 1965–67. Surface winds are somewhat erratic and often not suitable in direction for migrant birds. At 900 m, on the other hand, winds generally blow toward the north or northeast during April and May, the directions in which migrant birds will be traveling at that time. The question then becomes one of whether birds actually fly at altitudes where they can make use of these wind streams.

This question was addressed using weather surveillance radar data on the altitudes of bird movements taken in the years 1964–1975. The data are given in Table 5-2 and show clearly that most migration took place at around 305 m above ground level with a scattering of observations up to about 2300 m. Gauthreaux first plotted the wind directions at 305 m, obtained from National Weather Service balloons, for all 79 of the spring and fall observations of birds flying at that

Table 5-2. Altitude of Greatest Concentration of Nocturnal Migrants

Altitude in Meters	Number of Observations	
	Spring	Fall
150	1	—
305	57	22
450–1000	11	7
1200–1500	6	4
1600–2300	4	6

SOURCE: From Gauthreaux 1991.

altitude. For 47 of the 57 cases of spring flight the mean direction in which the winds were blowing at 305 m was 37.6° or toward the northeast with little variation. In the remaining 10 cases winds were not so favorable, but to reach winds that were generally only slightly more favorable, the birds faced a further climb of 1000 m or more. This suggests a trade-off, accepting somewhat less favorable winds to avoid the cost of a long climb. Similarly in the fall, winds at 305 m generally blew to the southwest and on those few occasions when they did not, more favorable winds were at much higher levels. The birds, then, seemed to prefer flying at an altitude of around 305 m. Usually winds here were in a favorable direction, but even if they were not, birds seemed willing to settle for this level if it meant avoiding a long climb to reach more suitable wind streams.

Several of the observations in Table 5–2, however, indicate that birds sometimes fly well above 305 m. These higher altitude flights occurred 21 times during spring migration and 17 times in the fall. For these 38 occasions Gauthreaux plotted flight altitude as the dependent variable against the altitude of the most favorable winds aloft. The most favorable wind was defined a priori as a wind blowing toward the north-northeast in the spring and toward the south-southwest in the fall and occurring at the lowest possible altitude. The results are shown in Figure 5–7 and indicate a quite remarkable correlation between the level of bird flight and the altitude of winds blowing in the most suitable direction to assist in transport, with some 93% of the variance explained by the regression line. This is even more remarkable when one considers the errors inherent in making the measurements.

To verify these results Gauthreaux used a completely different data set of measurements made with an airport surveillance radar (ASR) rather than the weather surveillance system used previously. The ASR has a wide vertical beam (it is, after all, used to detect numerous airplanes) and therefore is less accurate at measuring altitudes, but these could be estimated from range measurements. In this case 38 of a total of 197 measurements between 1971 and 1983 showed birds migrating above 500 m, and these measurements were plotted once more against the altitude of most favorable winds with the result shown in Figure 5–8. Here the correlation is not as great, with only about 32% of the variance explained ($r=0.57$), but it is nevertheless statistically significant ($p < .0002$), and reduced accuracy is expected because of measurement errors with the ASR radar. Thus data from various sources all strongly suggest that birds select following winds during their migratory passage.

The behavior of migrant birds during passages of fronts also revealed their choice of suitable winds. Behavior was monitored under three conditions: a deep strong cold front in spring, a weak shallow cold front in spring, and a deep strong cold front in fall. The response to a strong cold front in spring is shown in Figures 5–9A and B, which depict composite drawings of combined data from a number of radar observations. Before the passage of the front a major migration was under way, with many echoes indicating that birds were moving with a tailwind toward the northeast. With the passage of the front, the birds climb along the leading edge

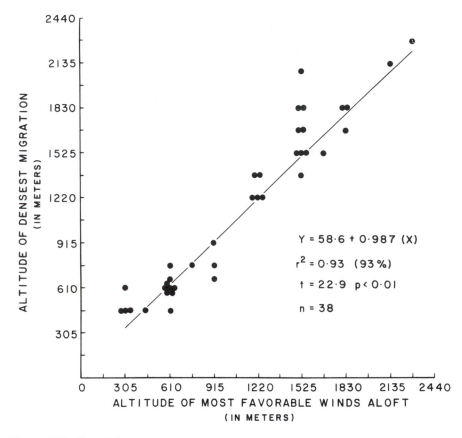

Figure 5–7. Correlation between the altitude of the densest bird migration at night as determined by weather surveillance radar and the altitude of winds blowing in the most favorable direction for spring and autumn migration near the Gulf of Mexico. (From Gauthreaux 1991.)

and shift direction somewhat more to the east as the front brings winds blowing toward the southeast. Except for a few large birds such as waterfowl, virtually all movement ceases behind the front.

Weak spring cold fronts and strong fall cold fronts produced rather different results from strong spring fronts. With a weak spring front a good migration is usually in progress prior to the arrival of cooler air. As the front passes, the birds tend to rise along its leading edge to eventually fly above it (Fig. 5–10). Rather than ceasing to fly, the migrants change their altitude to maintain contact with suitable following winds. In the fall passage of a strong cold front actually increases the number of individuals migrating (Fig. 5–9C,D). As the cold front moves past the radar, it brings with it winds from out of the northwest, and the density of echoes from migrants increases dramatically as birds take off and make

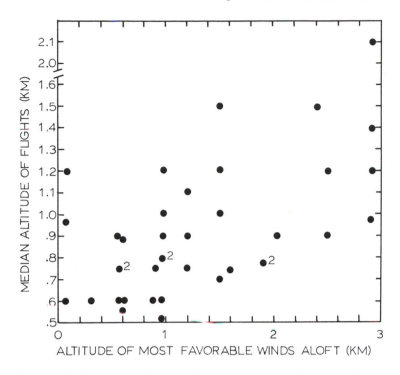

Figure 5–8. Correlation between the mean altitude of nocturnal migration as detected by an airport surveillance radar and the altitude of most favorable winds for birds flying above 500 m. (From Gauthreaux 1991.)

use of the favorable wind fields to move southward. It is thus clear that birds alter their behavior according to both season and prevailing weather conditions to pick winds most suitable for assisting travel on the appropriate migratory route.

Mesoscale Movements: Sea and Lake Breezes

Sea breezes and the breezes that can arise in association with large lakes are good examples of mesoscale events that can affect migration. The effects have been especially noted in the case of insect movements. A clear instance of insects moving on a sea breeze front was recorded in Canberra, Australia, on the evening of 22 December 1980 by Drake (1982). Sea breezes are common in the austral summer in Canberra, which is located on a tableland of hilly terrain at about 600 m above sea level and about 120 km to the west of the eastern coast of Australia. The evening sea breeze arises when air over the tableland is heated by afternoon temperatures, allowing penetration of cooler air from coastal regions to the east. On the day in question Canberra was experiencing light southwesterly surface winds. Aloft there were also light winds but from the north-northwest under the

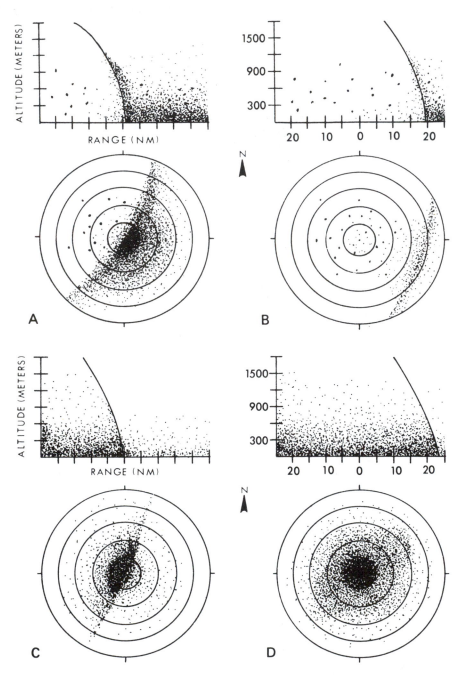

Figure 5-9. Composite drawings based on radar displays showing the responses of migrating birds to strong cold fronts in spring (*A* and *B*) and fall (*C* and *D*). In the spring, birds climb along the leading edge of the front to fly above it. In the fall the passage of the front brings with it winds out of the northwest, and the number of migrants increases after the front passes. (From Gauthreaux 1991.)

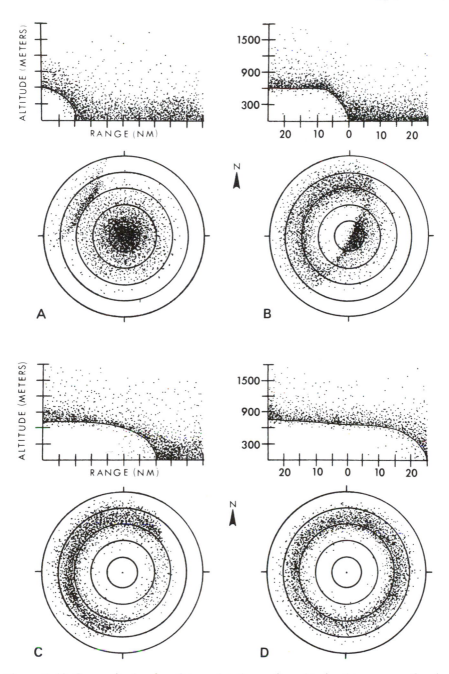

Figure 5–10. Composite drawings from radar observations showing the response of spring migrant birds to the passage of a weak, shallow cold front. The migrants increase altitude to maintain contact with following winds causing a "donut" to develop over the course of time on the radar display (*A–D*). (From Gauthreaux 1991.)

influence of a northerly airflow circulating around a high-pressure system lying offshore in the Tasman Sea. Because these conditions resulted in considerable late-afternoon warming over the tableland, they generated conditions favorable for the development of a sea breeze front.

A Commonwealth Scientific and Industrial Research Organization (CSIRO) entomological radar was being operated in an agricultural area just outside Canberra, and this radar picked up the movement of insects on the sea breeze front. Successive photographs of the plan position indicator (PPI) display (see Chapter 4), allowed the position of the insects to be plotted as the sea breeze moved inland. The passage of the front across Canberra was also recorded at local meteorological stations independently of the movement of insects noted on the radar, confirming the association between the insects and the sea breeze.

The radar image of the insects on the sea breeze front is shown in Figure 5–11. In this photograph, taken at 1927 hours (9 minutes after sunset), the line of insects is apparent to the east-northeast of the position of the radar at the center of the image. The line-echo of insects appears somewhat bowed because the portions of the front farther from the radar are detected at a greater height than the portions closer to the radar. Also apparent on the screen is some "clutter" around the radar itself and some blotchy areas ahead of the frontal mass of insects (moving from right to left) and to the west of the radar. These echoes are concentrations of insects in thermal convection cells and had been even more densely distributed on the PPI earlier in the evening before cooling reduced the amount of heat-driven convection. These thermally induced concentrations of insects are absent behind the sea breeze front. What this pattern of pre- and post-front concentrations suggests is that as the sea breeze progressed inland, it swept up concentrations of day-flying insects occurring in the convection cells ahead of it. Radar images later in the evening, some 2 hours after the passage of the front, picked up the flight of insects taking off after sunset. The movement of the front was plotted and followed for the next 45 minutes as it moved some 20 km to the south-southwest away from the radar set.

Transport on a sea breeze front such as this one is potentially important for migratory insects, especially in moving them away from coastal areas where movement in the opposite direction would take them out to sea. In southeastern Australia, for example, major insect migrations occur on warm anticyclonic northwest winds that carry insects from inland areas toward the coast and sometimes out to sea (Farrow 1975a; Drake et al. 1981, and below). The sea breeze is a good mechanism for bringing insects back inland at least part way. Many insect migrants depart at dusk and so in coastal and adjacent areas are likely to be transported on sea breezes, if they fly at the proper altitudes, and carried inland rather than offshore. The potential thus exists for an adaptive mechanism of takeoff and migration, but solid evidence for such adaptations is difficult to demonstrate and caution with adaptive explanations is certainly in order (Drake and Farrow 1989).

One instance of potentially adaptive use of winds analogous to sea breezes, but

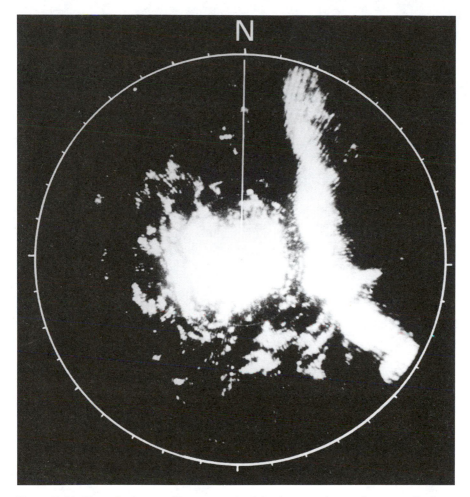

Figure 5–11. The radar image of insects moving inland on a sea-breeze front near Canberra, Australia, at 1927 hours, just after sunset, on a December (summer) evening. The line of insects appears as a somewhat bowed echo east-northeast of the position of the radar at the center of the circle; it moved westward some 20 km over the next 45 minutes. (Photo courtesy of Alistair Drake.)

in this case arising as lake breezes, occurs in the migratory locust, *Locusta migratoria*, in the Niger River Delta of West Africa. The movement on lake breezes is part of the complex annual cycle in these locusts, a cycle which is closely integrated with the northward and then southward movement of the Inter-Tropical Convergence Zone (ITCZ) (Farrow 1975b, 1990). The so-called Niger River Delta lies between the Niger and Bani Rivers southwest of the city of Timbuktu in the Republic of Mali (Fig. 5–12). As the ITCZ moves north early in

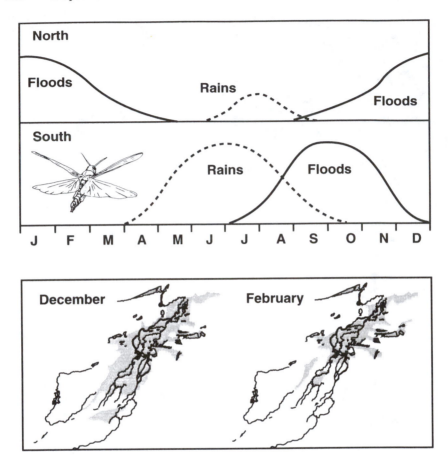

Figure 5–12. The pattern of rains and floods in the Niger River Delta of West Africa. As the Inter-Tropical Convergence Zone (ITCZ) moves north it brings first long rains in southern parts of the region beginning in April and then short rains further north beginning in July (*top panel*). These rains are followed by flooding of the Delta with the floods waxing and waning over the season (*bottom panel*). (After Farrow 1975b.)

the year, it brings monsoonal rains first to the southern part of the region, where the rains may continue for 5–6 months, and finally to the north, where they are much briefer (top panel in Fig. 5–12). As the rains continue, the Niger and the Bani fill and then overflow their banks in the large region known as the Delta (lower panel, Fig. 5–12). As the ITCZ moves back southward bringing the northeasterly cooler "harmattan" winds, the rain stops, and the floods recede through the early months of the year. The cycle then repeats.

The locusts breed first at the start of the rains as the vegetation greens on both sides and between the rivers. Here they can complete up to three generations.

There is a tendency for newly fledged migrating adults to move northward on the southwesterly airflows to the areas newly greening after rain and to move out of the area between the rivers as it becomes flooded. Because population densities tend not to reach outbreak levels, most of the locusts are in the solitary phase (see Chapter 11) and migrate at night. This nocturnal movement comes into play in an interesting way when the rains cease and the floods recede. The very large inundation lake arising from the floodwaters can cause a lake breeze airflow pattern. Thermal warming of the land during the day causes a lake breeze to blow over dry areas, and the residual heat of the warmer water at night causes a land breeze to blow toward the flooded areas. The previously flooded regions, as they are exposed, are again suitable for locust breeding, and the nocturnal wind flows apparently assist the night-flying migrants in reaching these suitable habitats. The situation is not simple, however, because the northerly harmattan winds prevalent at this time can also exert an influence. Locusts from the more southern portions of the Delta migrate *against* this prevailing wind to reach the line of the retreating floods, movements that have been confirmed by radar observations (Riley and Reynolds 1979).

Two points are worth making as a result of these observations on West African locusts. First, the use of winds for transport by migrating organisms may not be straightforward. In even the most simple of weather systems, several types of wind and directions of wind flow are likely to be present in different regions or altitudes or at various times. More than one of these, one of these, or none of these may be used, depending on circumstances. Second, some organisms, especially if they are strong flyers like locusts (or birds), are quite capable of moving across or against the wind, at least for limited periods. Wind systems can provide tremendous aids to migrants, but we should not assume that they necessarily do so.

Thermals and Other Microscale Phenomena

The bubbles of warm rising air known as thermals can allow migrants to attain relatively high altitudes, so that in using them there is considerable conservation of energy over that consumed in using powered flight to attain the same heights. It is thus not surprising that migrants have been observed using thermals, especially under conditions in which following winds aid movement in the preferred direction. Two cases of the use of thermals by migrants have been particularly well documented involving, respectively, the monarch butterfly, *Danaus plexippus*, in North America and the crane, *Grus grus*, in Europe. The large size disparity between these two species makes the comparison of their migratory flight tactics an interesting one.

Soaring in thermals and other energy-efficient flight tactics of monarch butterflies were observed in Ontario, Canada, by Gibo and Pallett (1979). The monarchs were studied while on their southward migration in early September and detailed observations were made on 358 butterflies, with the behavior of many others noted briefly.

When conditions were favorable for the development of thermals, some 90% of the butterflies observed were using soaring as their primary means of flight, with only brief bursts of flapping by some individuals, presumably to remain in regions of lift. Individuals in lift would soar in circles, often changing direction as they gained altitude, and pairs or small groups frequently soared within a meter of each other in formation, circling and ascending as a group. Sometimes as many as 30 individuals soared together in loose assemblages. These instances of "team flying" were common and appeared to increase the ability of the members of the group to remain in stronger sections of the thermal. The behavior of these migrants seemed to be finely tuned to the optimal use of ascending air to gain altitude with a minimum of energy expenditure. Schmidt-Koenig (1985) reported similar observations of monarchs using thermals as part of an overall flight "strategy" while on fall migration along the Atlantic Coast of North America.

Ground observers are limited in their ability to see high-flying butterflies and hence to estimate how high they might ascend. Gibo and Pallett (1979) observed butterflies ascending to approximately 300 m but were unable to estimate to higher levels. This prompted Gibo (1981) to obtain observations of southward migrating monarchs from glider pilots, who noted individuals up to 1200 m above the ground. Some of these butterflies were recorded as soaring on thermals. Attainment of these altitudes adds considerably to the ability of the migrants to traverse long distances with little energy expenditure (see also Chapter 6). The glide ratio, or ratio of horizontal travel to sinking, in these butterflies is about 4:1. On this basis Gibo estimated that a butterfly reaching 1200 m in a thermal could glide in still air 4×1200 m or almost 5 km before it had to resume powered flight. At this altitude the butterflies would also enter the higher wind velocities characteristic of geostrophic winds, so that under tailwind conditions they could glide for potentially very long distances with very little energy expenditure. Because of energy savings, thermal soaring apparently constitutes an important element of the flight tactics of migrating monarchs.

Energy conservation via soaring seems to be equally important for the common crane, a much larger and more powerful flyer, which migrates between wintering grounds in the western Mediterranean to breeding areas in northern Scandinavia. Cranes have a glide ratio of about 16:1 and so gain a tremendous advantage in energy conservation if they can reach altitudes that allow long-distance glides. This is especially true because these large birds use a great deal of energy in flapping flight (Chapter 7). The soaring behavior of European cranes was studied by following a flock moving north over Sweden in the spring with a small aircraft (Pennycuick et al. 1979; Alerstam 1990). Over the Baltic Sea, where there are no thermals, cranes must use flapping flight, but they switch to thermal soaring and gliding when they again reach land. A radar operating near the southern tip of Sweden directed the aircraft toward flocks of cranes coming over the coast after their flights across the Baltic from Germany.

One particular flock was followed for nearly 150 km from Lake Vombsjön north to Lake Vidöstern, a traverse that took the cranes about 3 hours. This represents a migration speed of close to 50 km hr^{-1}, a speed that the birds maintained even if it meant they had to resort to flapping flight in regions of weak thermals (see Chapter 7 for a discussion of the trade-offs involved in maintaining various migration speeds). The reasons for maintaining this rate of passage are unknown. Suffice it to say, however, the cranes effectively used thermal soaring. At the outset, when first spotted by the plane, soaring conditions were excellent with sun and scattered cumulus clouds, and the birds moved northward rapidly with essentially pure soaring and gliding flight over a 500 to 1500 m range in altitude. They then entered a region of cloud cover and weak thermals where they steadily lost height down to about 300 m. At this point good thermal soaring conditions returned, and they soared with short breaks of gliding back up to near 1500 m. At this point they entered a region of hazy conditions indicative of stratified air and, although they tried to make use of weak temporary updrafts, they again lost height and had to resort increasingly to flapping flight. They were in continuous flapping flight when the airplane pursuit was broken off.

It is not entirely clear how the cranes find thermals. It is possible that they can read cloud formations, although there is no evidence one way or the other. They do seem to cue on other soaring birds, and do gain an advantage from formation flying both aerodynamically (see Chapter 7) and because they can monitor the behavior of other birds. Some birds may also monitor other species; various soaring raptors apparently do this (Kerlinger 1989). By whatever means the thermals are located, the cranes, like monarch butterflies, use them very effectively during their migratory travels at a tremendous savings in the energy required for the journey (see Chapter 6 for detailed discussion of fueling and energy expenditure).

In contrast to cranes and monarch butterflies, oceanic birds soar not on thermals, which are generally not present over the oceans, but by a technique known as dynamic soaring (Walker and Venables 1990). Birds such as albatrosses, fulmars, and shearwaters (the "tube noses") have evolved long narrow wings that provide a great deal of lift (see Chapter 7). Dynamic soaring depends on gradients of wind speed near wave crests and on the increase in wind speed occurring from the sea surface up to about 100 m altitude. These birds gain air speed and height by passing from low on the lee side of a wave (sometimes almost touching the water surface with a wing tip) where the air is moving slowly to the more rapidly moving air coming over the crest (Scorer 1978). This allows the birds to rise 2–3 m. To climb further, the birds must head into the wind and move upward as they gain air speed. At some point the ascent can no longer be sustained, and the bird turns and glides downwind, gaining energy for the turn into the wind at the ocean surface for a repeat of the process. One of the joys of birding at sea is watching the great wheeling movements of albatrosses, the masters of dynamic soaring.

Seabirds of both Northern and Southern Hemispheres practice dynamic soaring to migrate long distances. Wind conditions must be suitable for such soaring, and the necessary winds occur most often in the temperate latitudes, accounting in part for the relative paucity of seabirds in the Tropics. (Nutrients and productivity are also higher in colder waters.) Optimal wind conditions occur in the Southern Hemisphere, where between 40° and 65° latitude some 98% of the earth's surface is ocean. This is where open ocean birds are in their element, and where they are most numerous. Albatrosses are the largest of these and are characteristic of the Southern Hemisphere, where some also take advantage of the steady wind flow of the prevailing westerlies of the "roaring forties" to undertake regular circumpolar migrations (Alerstam 1990).

Boundary Layer "Nanoscale" Gradients

Air moving over a surface is influenced by friction even at the most minute levels such as the surface of a plant leaf. The gradients in wind velocity established, even though they may occur over distances measured in millimeters or less, can still affect the takeoff of tiny organisms attempting to launch into the air for transport. The organisms so affected include spores, seeds, and microarthropods such as mites, spiderlings, and the early instars of lepidopterans. Many of these use silk threads to aid transport, but others do not, instead launching themselves bodily from the surface. A case of the latter involving specialized takeoff behavior to take advantage of "nanoscale" wind velocity gradients was studied in detail in the minute (circa 0.4 mm) first-instar crawlers of the ice plant scale insect, *Pulvinariella mesembryanthemi*, by Washburn and Washburn (1984).

In wind tunnel experiments the Washburns estimated the velocity profiles in the laminar boundary layer for air moving over a surface, say a leaf, at 1.8 and 4.0 m · $\sec^{-1}$ (Fig. 5–13). In each case velocity was estimated to increase from near zero at the surface to a maximum at a height of about 0.4 mm. These velocity profiles reveal that the gradient is greatest near the surface and over the range that is almost precisely matched by a crawler rearing up for takeoff. The takeoff behavior involved first raising the antennae above body height and rotating to face downwind. Crawlers then lifted the prothoracic legs (forelegs) and the anterior portion of the body off the substrate and arched backwards, supporting the body with the remaining four legs. The axis of the body was between 45° and 90° to the substrate, and the crawlers held this posture for up to 50 seconds or until they were lifted off and carried downwind. The effect of rearing up was to place the anterior part of the body into the near-maximum wind velocity above the surface. This created maximum drag on the body and the necessary force to engender liftoff, overcoming the forces created by the tarsi clinging to the surface. When crawlers walked with all six legs and with the body parallel to the surface, they were sufficiently low in the wind gradient that liftoff did not occur. The tendency to exhibit takeoff behavior increased as the crawlers became older, presumably because an initial period of feeding on the parent host plant is desirable to acquire energy stores for the journey and to seek feeding sites on a new host.

Once airborne, crawlers arch backward while extending the legs and antennae, a posture that produces greater surface area and so enhances lift. The average

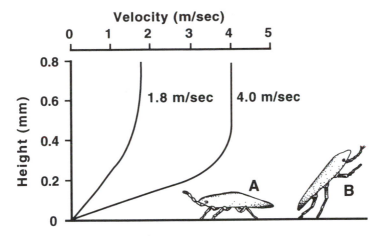

Figure 5–13. Gradients of wind velocity near a surface for winds moving at 1.8 and 4.0 m/sec and the takeoff behavior of ice plant scales. The scale insects first rotate to face downwind (*A,B*) and then raise the anterior end of the surface over a height matching the gradient in wind velocity (*B*). (Reprinted with permission from Washburn and Washburn 1984. Copyright 1984, American Association for the Advancement of Science.)

terminal velocity of live crawlers, measured when dropped through a hollow tube, was 26.2 cm sec^{-1} compared to 32.7 cm · sec^{-1} for dead crawlers, which tended to fall like simple spheres. Live crawlers thus enhance their chances of being carried into the planetary boundary layer and being transported for considerable distances downwind. The specialized behavior that ensures wind transport in these minute migrating insects thus includes both postures leading to takeoff and postures increasing the chances of continued "flight." These behaviors are part of an overall syndrome that prior to takeoff also involves a positive phototaxis and negative geotaxis that promote arrival at upper leaf surfaces where launching activities are most effective. Like much larger migrants, these minute crawling insects have evolved behaviors allowing them to take advantage of the wind conditions that enhance their journeys.

Migration in the Ocean

Like their terrestrial counterparts, the creatures of the oceans are subject to the dynamics of the medium in which they live. In none is this more true than in migrants, for in moving from place to place, whether in estuaries or in the open ocean, these travelers depend on the currents to assure that they will reach their destinations. Much of the pioneering work has been summarized by Thorson (1964) for marine larvae and by Harden Jones (1968) for fishes, and more recent research has confirmed and refined the earlier studies. Some of the most remarkable of all migratory journeys, such as those of Atlantic eels and the western rock lobster of the Indian Ocean, have evolved in close synchrony with current patterns over large areas of the ocean basins. Many estuary and inshore migrants, on the other hand, more than make up for what they may lack in drama with precise behavioral refinements allowing the adaptive migratory patterns that are the keys to their survival. Thus, as with atmospheric circulation, the movements of ocean waters at all scales influence in profound ways the migrations of the seas' inhabitants.

Planetary-Scale Currents and Atlantic Eels

There are some 15 species of eels (*Anguilla* spp.) in the world, all but two of which occur in the Indo-Pacific (McDowall 1988). So far as is known, all species are catadromous, with breeding in the ocean basins and a period of maturation in freshwater, but no breeding sites have ever been found, not even for the two most-studied species, the European (*A. anguilla*) and American (*A. rostrata*) eels of the North Atlantic. However, since the early work of Johannes Schmidt, who accumulated records of some 12,000 larvae of both species, it is thought that Atlantic eels breed in the Sargasso Sea (Schmidt 1912, 1922). The complete life cycles of the two species are similar and make extensive use of the currents in the North Atlantic Basin (McCleave and Kleckner 1985). The cycle of the American eel is outlined in Figure 5–14.

Based on what is known of life cycles and the current patterns in the North Atlantic, Power and McCleave (1983) generated a simulation model for the movements of larval eels. These larvae are called leptocephali after the generic name, *Leptocephalus*, assigned to these transparent forms before it was realized that they were in fact young eels. The model was based on two aspects of eel biology: the fact that leptocephali rise into the top 300 meters of the ocean after hatching and that they then begin a movement northward. It also included two assumptions: (1) that larvae would be concentrated in regions of net water convergence (downwelling) and (2) that they would be dispersed by net water divergence (upwelling). The model indicated, first, that the larval eels initially drift slowly to the northwest along the eastern edge of the Bahamas and accumulate to the northeast of these islands. There is some recent oceanographic evidence to suggest that there is an anticyclonic circulation cell in the waters in this area of the Atlantic that would cause the young eels to accumulate (Olson et al. 1984). Secondly, from this accumulation leptocephali would be introduced into the Gulf Stream over several months, roughly from March to November. Historical samples taken off eastern North America were consistent with this temporal sequence. Direct transect sampling across the Gulf Stream and the area of predicted larval accumulation generally supported the model, but the particulars of the distribution of larvae differed somewhat from predictions. For example, there was no accumulation northeast of the Bahamas, possibly because of disparities between actual and assumed currents (McCleave and Kleckner 1987). Especially interesting,

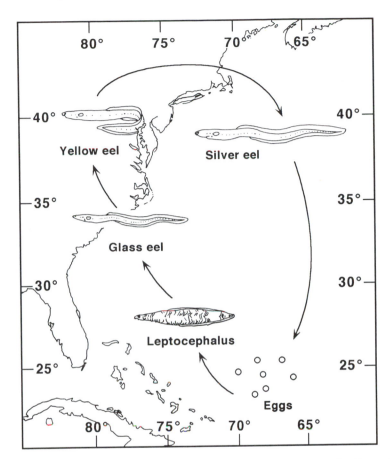

Figure 5–14. The life cycle of the American eel in the North Atlantic. Reproduction apparently takes place in the region of the Sargasso Sea and is followed by movement north in currents, mainly the Gulf Stream. Following metamorphosis eels mature in fresh or brackish water before returning to the ocean to breed. (Redrawn from McCleave and Kleckner 1985.)

however, was the fact that the simulations, supported by the transect data, could account for the previously puzzling observation that leptocephali of American eels of approximately equal age appear more or less simultaneously up and down the North American coast just prior to metamorphosis and migration into fresh water. The initial gradual release of leptocephali into the Gulf Stream, with similar-aged larvae then carried varying distances, could readily account for this.

Leptocephali of American eels drift for up to a year in the Gulf Stream and those of European eels for up to 3 years. Young of American eels are probably released onto the continental shelf by variable eddies and filaments, and they may even be able to select certain of these via diel vertical movements (commuting by "vertical migration"), which they are known to perform (McCleave 1987). Just before entering freshwater the larvae metamorphose into glass eels, which assume the adult eel form but are, as their name implies, semitransparent. In the American form, these glass eels, aided by selective tidal transport (McCleave 1987), enter brackish and fresh water from the Gulf of Mexico to Labrador. They swim upstream and gradually mature into yellow eels. This nonmigratory form spends several years in freshwater and then undergoes a second metamorphosis into a silver eel for a return to the breeding areas in the Sargasso Sea. The morphology of silver eels suggests that they migrate in the upper regions of the ocean, because they are countershaded dark above and silvery below, characteristics that would only be useful in the upper photic zone; have large eyes suitable for vision in dim blue light; and possess a swim bladder that probably is unable to function below about 150 m depth. Unfortunately there is no direct confirmation, as no oceanic adult eels have ever been captured. The European eel has a life cycle virtually identical to its American counterpart but taking place in freshwaters from Iceland to North Africa.

Because no silver eel has been captured in the ocean, the course and timing of the migration route back to the breeding areas is unknown. Again, however, simulations suggest what might be happening (McCleave and Kleckner 1985). First, the simulations indicate that because of the general pattern of current drift, the orientation mechanism need not be very precise. For American eels a generally southward heading, with perhaps some deviation, toward the southeast or southwest, depending on point of departure from the continent, would be adequate. Studies in the North Sea of European eel headings during current drift suggest that orientation is actually better than this. It is not known how eels recognize the breeding area (imprinting to odors in the leptocephali?), but there seem to be plenty of cues available. These could be provided by a lengthy east-west frontal zone across the Sargasso Sea that has been revealed by satellite imagery. A transition from waters north of the front to those south should be readily detectable using known mechanisms such as odor imprinting, so that no new sensory response is required to enable eels to recognize the breeding areas (McCleave 1987). As in Pacific salmon (Chapter 3), the eels apparently die after a single semelparous breeding episode.

Because of the absence of data with respect to many key points, much about migration pathways, behavior, and even breeding areas of Atlantic eels remains speculative. It is apparent, however, that oceanic and inshore currents are important in the life cycle. The planetary scale Gulf Stream, in providing a vehicle for drifting larvae and, probably, for modulating the return pathways of adults, is the most important in this regard, but other currents and eddies matter as well. Although it appears the eel life cycle has evolved under selection from oceanic current patterns, we have much to learn about what the details are and the exact nature of the behaviors that have evolved in concert with them.

Synoptic-Scale Currents and the Western Rock Lobster

The western rock lobster, *Panulirus cygnus* (Palinuridae), is limited in its distribution to the western coast of Australia between North West Cape (22°S) and Cape Naturaliste (34°S). Its life cycle in relation to the current morphology of the eastern portion of the Indian Ocean and the Australian continental shelf has been analyzed by Phillips (1981), using data on currents and larval distribution gathered by the International Indian Ocean Expedition of 1959–1965, which involved some 40 research vessels from 20 nations. It is obvious from these data that the current systems in question are not particularly well developed and are subject to many vagaries. This poses problems for an organism that depends on currents for the movement and redistribution of its planktonic young. The specialized migratory behavior of the larval stages of *P. cygnus* has apparently resulted in adaptive adjustments to the variable current systems encountered and in the successful recruitment of young in the region of its distribution.

Like other members of its genus, *P. cygnus* initiates post-hatching development as a transparent phyllosoma larva released during the summer by the bottom-dwelling females. There are estimated to be 15 phyllosomal instars before molt to the next stage of development, the puerulus, a stage that looks more or less like a miniature adult but retains large pleopods, making it a strong swimmer. The puerulus in turn molts to become a settled juvenile. The behavior of the phyllosoma and puerulus stages is attuned to the movement of currents occurring within the range of *P. cygnus*.

As with many crustacean larvae, the early phyllosomas are strongly photopositive, with the consequence that when released by the female they rise from the bottom to enter the plankton. The positive phototaxis is to low light intensities, and these early instars also display diel vertical commuting movements rising to within 30 m of the surface at night and descending to depths of 100 m or more during the day. As the phyllosomas mature they show progressively less positive response to low light intensities and hence reduced vertical movement; in these later instars they are found near the surface only on the darkest nights with little or no moonlight (Rimmer and Phillips 1979).

The vertical movements of early instar phyllosoma larvae appear to explain an apparent conundrum in their pattern of distribution. Although there are much less extensive eastern and western continental boundaries to the Indian Ocean than to the other major ocean basins, the pattern of current flow is similar to those in the Atlantic and the Pacific, although much less well defined. Thus at the eastern margin in the vicinity and latitude of Australia, there is a generally easterly flow of oceanic water because of the location of the continent south of the equator. This represents the return flow of water, around the Indian Ocean gyre, from the south equatorial current proceeding across to and south along the coast of Africa. The problem is that the phyllosoma larvae move generally westward off the coast of Australia to concentrate in areas 1000 km or more offshore, apparently in the face of the easterly currents. There is, however, a westward-moving surface current, caused by offshore winds, overflowing the generally eastward water flow, and the hatching time of the lobsters coincides quite well with the maximal offshore surface flow, which exceeds that of deeper onshore flow. The vertical movements of the phyllosoma thus expose them to wind-driven ocean surface transport at night and to a more or less steady drift of around 5.25 km · day^{-1} away from the continental shelf where they were born. Because of the somewhat erratic current flow, they become dispersed over a broad area of ocean, although with higher concentrations between about 500 and 800 km offshore from the center of the range of the lobsters. The area of concentration offshore seems to be roughly that reached by the larvae before their positively phototactic responses cease, and they no longer display the nightly rise to the surface into the offshore currents. The position of the larvae far offshore also allows them to escape a southward flowing current, the Leeuwin Current, that develops in March–May as high-temperature and low-salinity tropical waters move down along the Australian coast.

The transport of the larvae back to the coast of Australia appears to be complicated and not well understood in its details. The tendency of the older larvae to remain deeper in the water column will, however, result in their remaining for most of the time within water masses that are moving eastward. There may also be some generally eastward moving eddy systems. Suffice it to say, the result will be net movement of the larvae back toward Australia and the sites where they will settle out to begin benthic existence.

As they approach the continental shelf, the larvae face another problem. Marker buoy studies show that the water circulation patterns of oceanic and shelf components are essentially independent, and that the eastward-flowing oceanic currents bend southward rather than transporting water (or larvae) up onto the continental margins. Any larvae in these oceanic currents would thus be transported away from the areas where they must settle to survive. Here is where the behavior of the puerulus enters in. Lobsters at this stage are strong swimmers apparently evolved for swimming across the shelf, and escaping south-flowing currents, to settle in the shallow inshore reef area. Currents on the shelf itself are generally northward flowing at the time of settlement, which is in early summer some 10–11 months after hatching, so that the pueruli display oriented swimming movements across these currents (Phillips and Olson 1975). These oriented movements are apparently of particular importance because some puerulus settlement occurs throughout the year, during which time currents along the shelf may flow in different directions.

The migratory movements of the western rock lobster can be summarized as follows. There is, first, westward movement of recently hatched phyllosoma larvae in wind-driven surface currents, facilitated by upward movement of the larvae at night. Second, as the nocturnal upward movement tendency declines, there is net eastward movement back toward the Australian coast on the prevailing deeper current flow. Finally, the puerulus stage completes the planktonic cycle by swimming back over the continental shelf to shallow settlement areas regardless of current direction. The planktonic life cycle of 9–11 months is a long one compared to most marine crustaceans, including other *Panulirus*, and arises presumably both because of the need to time transport in suitable currents and because of the scarce food supply in the Indian Ocean where most growth and development takes place. Other panulirids (and other Crustacea) tend to release larvae when food is plentiful, for example, at the time of upwellings, and to rely on more predictable current regimes. The long pre-settlement period and the specialized behaviors of the phyllosoma and puerulus of *P. cygnus* have apparently been adjusted by natural selection to the low nutrients and less reliable currents met by these animals during the migratory portion of their life cycles.

Micro- and Mesoscale Continental Shelf and Estuary Currents

Along the margins of continents, current systems are dominated by local water movements driven by tide or wind rather than by the forces determining the major circulation patterns of the open ocean. Many migratory organisms have become quite proficient at using these currents. One of the primary mechanisms is to use vertical movements in the water column to effect *modulated drift* by *selective tidal stream transport* in tidally generated currents (Arnold 1981). In the case of many larvae of inshore organisms this modulated drift is effectively used to move away from the parents but to remain within an estuary. Even when larvae are transported out of the estuary, long shore and counter currents just offshore will usually keep them within 100 km or so of the estuary mouth. The commercial blue crab, *Callinectes sapidus*, population of the Chesapeake Bay is apparently maintained in this way (McConaugha 1992). Models of the movements of larvae indicate a selective advantage to remaining in the home estuary rather than traveling long distances (Strathman 1982).

A case of selective tidal stream transport by a crustacean was described by Cronin and Forward (1979) for the estuarine crab *Rhithropanopeus harrisii* from on the southeastern coast of the United States. Field-caught larvae of this crab display a tidal rhythm of vertical movement that involves moving upward in the water column during flood tide and downward during the ebb. The estuaries in which *R. harrisii* occurs show a classic estuarine pattern of stratified currents with a high salinity current of net landward flow underlying upper more brackish net seaward flowing water. The result of selective tidal transport is, then, a minimization of seaward movement by the larvae during the ebb and a maximization of movement up the estuary during the flood, and the larvae remain in the estuary where they must eventually settle. The specialized migratory behavior serves to keep the migrating larvae concentrated within the estuary while still moving away from the parents and presumably dispersing over favorable habitats. Interestingly, lab-reared larvae fail to show tidal vertical movements, implying that the behavior is cued by changing conditions in the estuary during the tidal cycle. The mechanisms are unknown.

Selective tidal stream transport is also practiced by fishes and has been particularly well studied in the North Sea (Greer Walker et al. 1978; Arnold 1981; Arnold and Cook 1984). Once they leave rivers, eels (*Anguilla anguilla*) evidently use tidal currents to move from the continental shelf to the depths of the open ocean for a return migration to the Sargasso Sea spawning grounds. Plaice (*Pleuronectes platessa*) evidently practice tidal stream transport to make relatively rapid migrations between spawning, feeding, and nursery grounds. In plaice the basic pattern of vertical movement has been shown to be diurnal; that is, one circuit every 24 hours. The tidal cycle in the North Sea is, however, semidiurnal with two high and two low tides each day. Migrating plaice adopt a diurnal pattern of vertical movements, modified by the timing of the tide, to "lock on" to the appropriate currents, but just what cues and mechanisms are involved remains a mystery. Suffice it to say these migrants use the tidal stream system in the North Sea very effectively.

The movement patterns of individual plaice and in one case those of an eel have been studied using tagging and telemetry. The track chart of a silver eel, captured in its downstream journey in the Humber River system in Yorkshire and released in the North Sea, is shown in Figure 5–15 along with the track of a plaice caught with an otter trawl in the North Sea and released south of where it was caught. The fish were fitted with transponding acoustical tags and were tracked with sonar. Both fish (as well as seven other plaice) showed a consistent pattern of upward movement into the northward-flowing tidal stream at flood tide followed

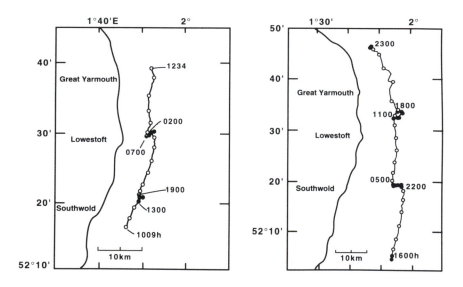

Figure 5–15. The tracks of a plaice (*left panel*) and a silver eel (*right panel*) in the North Sea illustrating tidal stream transport. Open circles indicate north-flowing tide and filled circles indicate either slack water or south-flowing tide. The two species move over the bottom during north moving tides. (Modified from Arnold 1981 and Arnold and Cook 1984.)

by a descent during slack water to remain on the bottom during ebb tides. By moving on the tidal streams both fish reached speeds over the ground of over 1 m · sec^{-1} and the 97-cm-long eel even reached speeds of almost 2 m · sec^{-1} for a brief period. Both covered impressive distances northward during the day to day and a half they were tracked, demonstrating the effectiveness of the tidal transport system. The system is energy efficient and minimizes the necessity for elaborate orientation mechanisms because the ability to select the current based on diurnally timed movements allows the fish to move in the "correct" direction (Leggett 1984).

Inshore currents along continental margins may also be wind driven and therefore subject to change when winds shift. The larvae of the capelin, *Mallotus villosus*, a fish that spawns in shallow gravel beds along the coasts of Newfoundland and Labrador, use current shifts to program emergence from the gravel and entry into the plankton (Leggett 1985). In the areas where capelin spawn offshore westerly winds prevail, driving warmer waters offshore and permitting cold upwelling currents to dominate the beach areas close inshore where the capelin deposit eggs. At irregular intervals the winds shift and blow onshore driving warm water up onto the beaches and suppressing the cold upwellings. The cycle is illustrated in Figure 5–16. During these warm-water intrusions, the capelin larvae initiate their migration and swim upwards out of the gravel in which they hatched and into the plankton. They accumulate in the upper 20 cm or so of the gravel after hatching but do not emerge into the water column until immersed in the warmer water brought by onshore easterly winds.

This behavior of the larvae was further demonstrated in laboratory experiments in which eggs in gravel were placed in five separate laboratory-maintained aquaria, all supplied with water pumped from local inshore areas. When a wind reversal brought warm water onshore, the larvae in all five tanks emerged simultaneously. The behavior is apparently selectively advantageous because the planktonic organisms on which the capelin larvae feed are more abundant in the warmer waters, while the jellyfish and chaetognaths that are the major predators of the larvae are more numerous in the colder upwellings. The emergence into the warmer water also begins a migratory period in which the larvae undergo extensive transport by ocean currents.

Summing Up

The shape of the earth, its tilt, and its spin around a central axis all interact to generate the major weather and current systems of the atmosphere and oceans. Because of the way the seasons change and the way the Coriolus force and other factors direct winds and currents, many of these atmospheric and oceanic streams are sufficiently predictable to be used by migrating organisms for assistance in transport. Thus birds flying northward in the spring in eastern North America can depend with fair certainty on southerly winds to aid them in flight (Gauthreaux 1991), and the oceangoing segment of the migratory life history of Atlantic eels

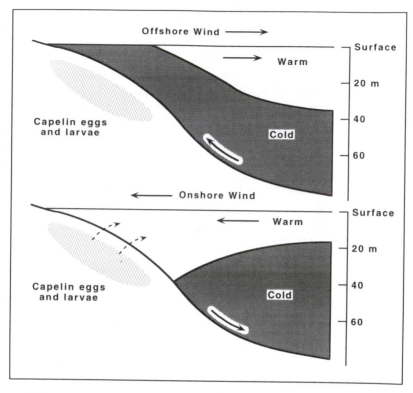

Figure 5–16. Currents and the emergence of capelin larvae. Offshore winds drive warm water away from shore allowing cold water intrusions that keep hatched larvae in the gravel. When onshore winds bring warm water, the larvae leave the gravel (*dashed arrows*) and enter the plankton. (Redrawn and modified from Leggett 1985.)

depends on the persistent northward and then eastward flow of the Gulf Stream around the margins of the North Atlantic Basin to deliver the leptocephalus larvae to estuaries and river mouths (McCleave 1987). The certainties of the daily tidal cycles both program and assist the within-estuary migrations of the larvae of crabs and other demersal organisms (McConaugha 1992).

Across a range of temporal and spatial scales the restlessness of the atmosphere and the oceans influences the timing, course, and distance of migratory journeys. Proximately, a thermal, a wind stream, or a current may provide a period of assistance to a migrant on its travels; ultimately, the patterns generated by air or water movements may provide the basis in selection for the evolution of specific migratory journeys. Size and degree of locomotor efficiency are not necessarily predictors of wind or current transport. Large active swimmers and flyers and minute "passive" travelers are all likely to take advantage of the oceanic or atmospheric patterns of structure and motion they encounter. It also needs to be stressed, however, that the more robust migrants may sometimes move across or into winds and currents should the need arise. Behavioral flexibility interacts with the movements of wind and water to produce migratory pathways.

Migratory journeys may be over only short distances or they may be very long. In either case the organisms undertaking them are often characterized by a seemingly uncanny ability to choose a vehicle traveling in the required direction. Thus the diurnal vertical movements of the western rock lobster permit transport offshore and into the Indian Ocean by wind-driven surface currents (Phillips 1981), and selective tidal stream transport assists European eels in their migrations to the open sea to reach their spawning ground (Arnold 1981). North American bird migrants fly at the altitude of most favorable winds when moving north from the Gulf of Mexico in the spring, and autumn migrants that fly out over the Atlantic key their departure to the passage of low pressure cells that signal the coming of tailwinds (Gauthreaux 1991; Richardson 1985, 1990). At the level of nanoscale events, the ice plant scale takes off only when wind velocity in the laminar boundary layer at leaf surface is within a favorable range (Washburn and Washburn 1984). At intermediate scales, monarch butterflies and common cranes identify and use the lift provided by rising thermals (Gibo and Pallett 1979; Alerstam 1990), West African locusts use lake breezes (Farrow 1975, 1990) and capelin emerge only in wind-driven onshore currents (Legget 1985). When necessary, migrants like locusts and rock lobsters also move across or against current or wind.

During migrations transport by winds and currents may take place over very brief or very long periods of time. For western rock lobster and Atlantic eel larvae the period of current-borne transport may last for weeks or months, and in the case of the European eel even for years. Terrestrial migrants move at much greater speeds, and their journeys are far less long. Even birds moving across the equator between hemispheres with stops to restore fuel supplies make the journey in a month or so, and the monarch butterfly takes about the same amount of time to fly from Canada to Mexico. Individual legs are far briefer. The corn aphid covers the 1000 km between northern Texas and Illinois in about 24 hours (Irwin and Thresh 1988), and songbirds flying over the North Atlantic from New Brunswick to the Caribbean islands cover that distance in only about 65–70 hours (Williams et al. 1978). The month-long journey of the monarch butterfly must include multiple brief periods, measured in minutes or hours, spent in rising thermals followed by downwind or descending flight (Gibo and Pallett 1979; Gibo 1981). The movements of eels, plaice, and crab larvae on tidal streams are also hourly in duration. Like the migratory journeys themselves, the durations of assisted transport and the behaviors that assure it are fine-tuned by natural selection to meet the demands facing the particular organism.

The examples in this chapter demonstrate that natural selection has designed a broad array of migratory programs that take advantage of the motions of air and water. These programs depend on the interactions between migrants and their physical environment. As we have seen, migrating organisms, even tiny and seemingly helpless ones, are not simply passively carried by winds and currents.

Rather, as pointed out so emphatically by J. S. Kennedy (1985), they are active participants in the process, launching when it is advantageous on transporting vehicles but staying put or moving under their own power when it is not. Thus, biology is as important as meteorology or oceanography to migrants, and together physical and biotic factors are integrated to produce the wonderful array of migration systems we observe.

6

The Physiology of Migration

Introduction: Migration as Enhanced Locomotory Behavior

There are two major aspects to the physiology of migration. The first is the central nervous system (CNS) control of migratory behavior by means of interactions between migration and "settling" responses. This aspect was discussed in Chapter 2, where migration is defined. The nervous control of migratory behavior by the CNS was analyzed for the aphid, *Aphis fabae,* by J. S. Kennedy, as we have seen, and it involves not only the enhancement of locomotory behavior but also the reciprocal inhibition of "vegetative" responses such as feeding and reproduction, even in the face of stimuli that would usually trigger those responses. Although there is strong suggestion of reciprocal interaction between migratory and vegetative behaviors in other animals, many songbirds, for example, migrate at night when they would otherwise be sleeping (a vegetative response) and other examples occur throughout this book, Kennedy's aphid study is virtually the only one in which the action of the CNS has been explicitly considered. Analagous reciprocal mechanisms occur in the migratory physiology of plants, but they, too, have been little studied from a perspective of promoting migration (see Chapter 2). Even though CNS control is so fundamental to migratory and indeed all behavior in animals, the lack of understanding of the CNS function and role in migration, and the almost total lack of studies investigating that role, is undoubtedly the single biggest lacuna in our entire conception of the migration phenomenon. Unfortunately there is little one can do at this point except plead that the conceptual foundation established by Kennedy for aphids be followed up in studies of migratory behavior in other organisms.

The second aspect of migratory physiology, the mobilization of hormonal and metabolic pathways to stimulate migration and support its high energetic demands, is much better studied. This chapter deals primarily with this component of migratory activity. This is hardly to imply that this aspect of migratory physiology is by any means completely understood, but there is a tradition of studies, especially with some of the better-known migratory birds, fish, and insects. Not surprisingly these studies reveal complex systems of regulation and mobilization. These include the response to exogenous inputs, neurosecretory and hormonal action, and the organization of energy reserves and suitable fuels. We can begin our discussion of physiological mechanisms and responses with a brief examination of how organisms synchronize their migrations to environmental fluctuations.

Environmental Stimuli

Organisms must be able to respond to changes in the environment if they are to survive and reproduce. In many, if not most cases, however, they must alter their physiologies to respond appropriately, and it is, therefore, greatly to their advantage if they can *anticipate* changes that will occur (Follett 1982). This is certainly true of migration, which at the very least requires sequestration of energy for the journey and in some cases, as with wing polymorphic insects or polymorphic seeds (Chapter 13), may require the reorganization of the developmental program to produce the migratory apparatus. In these cases anticipation is a must, or the organism will be caught short without migratory capability. The most common and frequent environmental changes are those associated with the seasons, and indeed seasonal migrations are probably the best known migratory phenomena (Chapter 9). Outside the Tropics the most reliable predictor of seasonal change is the annual change in day length, and it is therefore not surprising that a broad array of organisms, including migrants, respond to changes in the photoperiod. Responses range from diapause in insects to flowering in plants, and studies of these phenological reactions have spawned an extensive literature (reviews in Vince-Prue 1975; Beck 1980; Follett and Follett 1981; Tauber et al. 1986). In migrants the photoperiod may serve not only to synchronize migratory movements with the seasons, but also with reproductive and other cycles such as the molt in birds and smoltification in juvenile salmonid fish.

In birds and insects the original discoveries of photoperiodic responses occurred in association with studies of migration. Insect photoperiodism was discovered by Marcovitch (1923, 1924), who demonstrated that the winged sexual autumn migrant females of the rosy apple aphid, *Dysaphis plantaginea*, were produced in short-day conditions. Marcovitch succeeded in producing these sexual migrants in June, rather than autumn, by keeping them in a greenhouse for several weeks under an artificial day length of 7.5 hours. He actually thought the effect came through the host plant because his studies followed on the earlier pioneering work of Garner and Allard (1920, 1923) on plant photoperiodism. It wasn't until the extensive later studies of Shull (summarized in Beck 1980) that a direct response to photoperiod, independent of the plant host, was shown in aphids.

In the case of bird migrants, the association between day length and migration was demonstrated by Rowan, who showed first with winter-captured juncos, *Junco hyemalis*, in Alberta (1925) and later with American crows, *Corvus brachyrhynchos* (1932), that birds kept in artificially lengthened light periods moved north when released, while controls moved south. He was also the first to produce evidence linking hormones to migration: whereas intact birds moved north after exposure to long day, castrates persisted in southeastward movement like the controls. This suggested that spring migration depended on gonadal condition, but fall migration did not (see below). The connection between day length, annual reproductive and other cycles, and migration is now well established in birds (reviewed in Farner and Follett 1979), fishes (Hoar 1988), and other migrants and will be a general theme pervading discussions of migratory physiology in this chapter.

In many insects, migration is associated with an additional photoperiodic response, adult reproductive diapause (Dingle 1985). Two good examples of this are tropical milkweed-feeding species that have successfully invaded the Temperate Zone in the New World—the monarch butterfly, *Danaus plexippus*, and the large milkweed bug, *Oncepeltus fasciatus* (Dingle 1978b; Brower and Malcolm 1991; and Chapters 3 and 9). In both these species, short days in the autumn result in diapause in the adults, characterized by the shutting down of the reproductive system (Dingle 1974; Barker and Herman 1976). This reproductive shutdown has the consequence of permitting long flight, because reproduction inhibits flight, and of making energy otherwise channeled to reproduction available for migration. Both species leave their northern breeding areas to overwinter farther south, the monarch in spectacularly large aggregations in the mountains of Mexico (Chapter 7).

The two-generation spring migration in the monarch begins at the end of the diapause period when the butterflies are refractory to photoperiod. In this generation, migration is over a much shorter distance and so without the necessity of a diapause to permit long flights. Diapause, of course, does not occur in the lengthening days of late spring. In the large milkweed bug, the spring flight also is two generations in the making. The diapausing overwintering generation, after emerging from diapause, breeds in the early spring across the southern tier of the eastern United States where they overwintered. Their offspring are also largely refractory to the short days of early spring, a response induced maternally from their parents. The offspring of diapausing mothers, in other words, don't diapause (Groeters and Dingle 1987). Relatives of the monarch butterfly and the large milkweed bug do not display a photoperiodically induced diapause and are confined to the Tropics and subtropics. The evolution of diapause, allowing the migrants to escape south in the autumn, is evidently what allowed the monarch and the large milkweed bug to successfully invade the Temperate Zone in the summer to take advantage of the large milkweed crop growing there (Dingle 1978b and Chapter 9).

In many insects, including the monarch butterfly and the large milkweed bug, the diapause response can be overridden by high temperature. The milkweed bug, for example, responds to short day lengths with diapause at 23°C, but not at 27° (Dingle 1974). At 27° this bug also fails to display long-duration tethered flights, even in short days, probably because such temperatures indicate favorable breeding conditions under all photoperiods (Dingle 1968). Birds are endotherms and so are free of temperature effects, except as their physiologies respond to maintain internal homeostasis. They may, however, delay or advance the timing of migratory journeys depending on local climatic conditions, including temperature. Ectothermic vertebrates such as salmonid fishes, on the other hand, do show temperature modifications of photoperiodic responses. Salmonids are cold-water fish with maximum responses to photoperiod around 12°C; higher or lower temperatures inhibit reactions to photoperiod (Hoar 1988).

The rainfall regime may also modify the response to photoperiod. In the Australian noctuid moth *Mythimna convecta*, for example, long days delay oocyte maturation from 3 to about 10 days after adult eclosion and thus allow a longer period of migration in the females (McDonald and Cole 1991). Crowding has the same effect. Crowding and long days are what the moths experience in southeastern Australia during the summer dry season. In order to give their offspring a reasonable chance of survival, adults maturing during the summer dry period must migrate to the nearest rainfall areas close to the coast. The delayed egg maturation induced under long days or crowded conditions allows them to reach these regions and so to breed successfully.

Endogenous Factors

In addition to environmental factors, endogenous timing mechanisms may play an important role in migration. Often they interact with external stimuli to synchronize physiological responses with seasonal or other environmental events. There are various kinds of endogenously rhythmic cycles in organisms, with the two most important for migrants being daily and annual rhythms. Tidal rhythms may also be important for some marine species, especially as they serve to time entry into appropriate currents, as we have seen in Chapter 5. When organisms are kept under constant conditions, their daily and annual endogenous rhythms will "free run" independently of external inputs, although with periods that only approximate a day or a year. Hence they are referred to as *circadian* and *circannial* rhythms, meaning approximately (Latin *circa*) a day or a year in duration, respectively. In natural populations the rhythms are precisely a day or a year in period length because they are synchronized by environmental *zeitgebers* or "time givers," in most cases deriving from the photoperiod. For photoperiodically controlled circadian rhythms the zeitgeber is either dawn or dusk, and for circannual cycles the zeitgeber is some form of the annual photoperiodic regime. The subject of rhythms is more fully reviewed by Aschoff (1960), Pittendrigh (1960), Brady (1982), Saunders (1982), and Gwinner (1986a,b).

A good example of the role of a circadian rhythm in synchronizing migratory behavior occurs in the large milkweed bug (Caldwell and Rankin 1974). In this insect there is a daily pattern of activity with mating and feeding peaking at the end of the daily light period and oviposition and migratory flight peaking in the middle, 6–8 hours after "dawn" (Caldwell and Dingle 1967; Dingle 1968). On the face of it, migration and oviposition would seem to be in conflict because they peak at the same time of day, but in fact this is not the case because they are separated by ontogeny. When insects first eclose to the adult stage, there is a period during which the cuticle hardens (the teneral period; from the Latin *tener*, soft); in many insect Orders including Hemiptera, it also thickens. During this period milkweed bugs direct energy toward cuticular growth rather than reproduction. Once the cuticle hardens, energy is redirected to reproduction, but with the cuticle now hard, sustained flight is possible. Reproductive activity shuts down migration, but before oviposition begins, there is a "window" of time during which migration can take place. Under long days this window is short in milkweed bugs, and there is just a brief burst of migratory flight; in short days, the bugs enter reproductive diapause, and there is a long period of migration (Dingle 1978b and see above). In either case, because of their ontogenetic separation, there is no midday clash of flight and oviposition activities. Neither activity conflicts with feeding and mating because these occur at the end of the day. The relationships among the various activities are illustrated in Figure 6–1.

These relationships among flight, reproduction, and feeding in milkweed bugs are consistent with Kennedy's (1961, 1985) model in which migration suppresses nonmigratory behavior and vice versa (Caldwell and Rankin 1974 and Chapter 2). This is evident in two ways. First, migratory flights are limited to a specific stage in the life history, after the cuticle has hardened but before reproduction begins. This ontogenetic relation between migration and reproduction in insects has been called the oogenesis-flight syndrome by Johnson (1969) and will be discussed further below. Secondly, migratory flight is segregated by time of day from various nonmigratory activities involving foraging and station-keeping behavior. Thus, circadian cycles and ontogeny interact to program the life cycle of the large milkweed bug in line with the Kennedy model, albeit on a longer time scale than the reciprocal interactions occurring during aphid flight.

Juvenile American eels display a circatidal rhythm enabling them to use tidal stream transport for upstream movement during migration (Wippelhauser and McCleave 1988). "Glass eels" were captured during migration through the Penobscot River estuary in Maine and tested for rhythmic activity in the laboratory. When maintained under constant conditions, eels collected from tidal water swam up into the water column at 10.5- to 11.2-hour intervals, or in other words, with a circatidal periodicity. These peaks could be entrained to a period of 12.1 hours if the eels experienced a current reversal every 6.2 hours, that is, with a tidal cycle. Eels captured in the river above the last point where it was influenced by tides did not show a circatidal rhythm. These results are consistent with the hypothesis that an endogenous clock times the semidiurnal migratory activities of these eels to synchronize with the flood tide during their passage from the Atlantic into freshwater rivers.

A relation between circannual rhythms and migration has now been documented for many species of bird (Berthold 1984a,b; Gwinner 1990a). Once it was realized that seasonal photoperiod influenced the annual cycles of avian migrants, a potential problem arose regarding those species migrating across the equator. For example, birds migrating south of the equator in the northern autumn would encounter the lengthening days of the Southern Hemisphere spring and would in theory run the risk of premature regrowth of the gonads and return migration

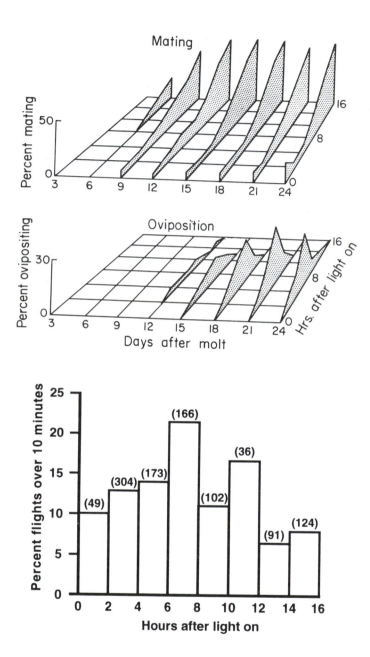

Figure 6–1. Circadian and developmental control of mating, oviposition, and flight in the large milkweed bug, *Oncopeltus fasciatus*. Top two panels show daily mating and oviposition levels as a function of age after the molt to adulthood. The bottom panel shows the daily cycle of flight (sample sizes in brackets), most of which occurs before the onset of oviposition. (Redrawn from Dingle 1968 and Caldwell and Rankin 1974.)

northward. That they do not respond in this way implies the existence of some endogenous mechanism that synchronizes the migratory, molting, and reproductive programs of the birds with the ambient conditions all along the migratory routes. This mechanism has now been shown to be a circannual rhythm.

The role of a circannual rhythm in bird migration has been well documented in the garden warbler, *Sylvia borin* (Gwinner 1989,1990a and Fig. 6–2). This species breeds throughout the western Palearctic roughly west of the Urals and north of the Pyrenees, Alps, and Caucasus. Wintering occurs in the southern half of Africa from a few degrees north of the equator south to the northern edge of the Kalahari and Namib Deserts. Captive birds kept in a constant 11-hour day display a circannual cycle of testicular growth, molting, and migratory restlessness that free runs with a period of just less than a year (Fig. 6–2). Birds kept in a sinusoidally changing photoperiod of 12 months' duration match the annual cycle of wild birds, while those kept under a cycle with a 6-month period also show biannual cycles in each of the measured variables, demonstrating that the annual variation in photoperiod is the zeitgeber in this case.

The role of photoperiod in synchronized physiology and behavior over the geographic range of the migratory cycle is illustrated in Figure 6–3. Birds wintering at the equator experience a declining day length in the autumn, which then levels off at the equator; birds flying beyond the equator to 20°S experience first shortening and then, after they cross the equator, lengthening days. The birds wintering at 20°S must fly farther when returning north, and this requires an earlier start. These birds begin the molt period as early as January, show increased gonadal growth by February, and display migratory restlessness by the middle of

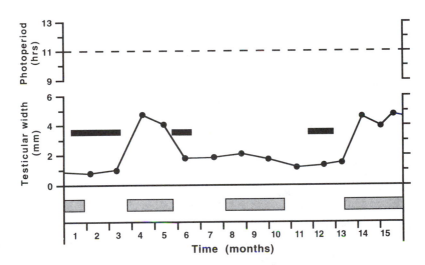

Figure 6–2. Changes in testicular width (*line*), the occurrence of molt (*solid bars*), and the occurrence of migratory restlessness (*shaded bars*) in garden warblers (*Sylvia borin*) held on a constant 11-hour photoperiod. (Adapted from Gwinner 1989.)

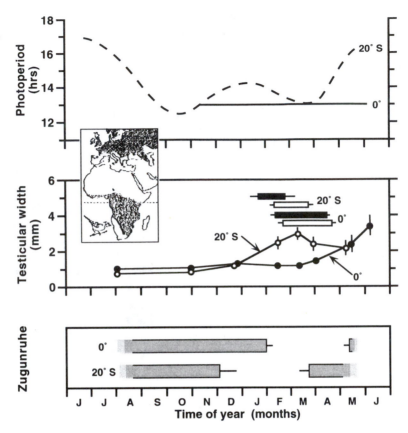

Figure 6–3. Changes in testicular width (*lines*), the occurrence of molt (*solid and hollow bars*), and the occurrence of migratory restlessness (Zugenruhe) in garden warblers experiencing the photoperiods of their respective wintering grounds at 0° and 20°S, as shown in the top panel, in two groups of garden warblers. Birds at 20°S initiate physiological processes leading to spring migration earlier than their 0° counterparts. (Redrawn from Gwinner 1989.)

March. In contrast, birds wintering at the equator do not molt until February, delay gonadal growth until late April or May, and display migratory restlessness only at the beginning of May, almost 2 months later than their counterparts from farther south. The circannual rhythms thus interact with the ambient photoperiodic regime to produce the behavioral and physiological response that matches the migratory route followed. Similar results derive from comparisons between species. The pied flycatcher, *Ficedula hypoleuca*, for example, winters exclusively north of the equator in Africa, while the closely related collared flycatcher, *F. albicollis*, winters from approximately 10°N to 20°S. When exposed to various annual photoperiodic regimes, the pied flycatcher shows no response to photoperiods corresponding to 0° and 20°S while responding to that at 10°N, whereas the collared flycatcher shows appropriate gonadal, molt, and behavioral responses under all three regimes, as would be predicted from its wintering range (Gwinner 1989, 1990a).

Likewise, circannual rhythms have been implicated in the migratory cycles of fish. In salmonids the smoltification transformation from the cryptically colored parr to the silvery smolt (Chapter 3 and below) is apparently controlled by an interaction between a circannual rhythm and the photoperiodic regime. Ericksson and Lundquist (1982) kept Baltic Sea salmon (*Salmo salar*) under a constant LD12:12 light cycle at 11°C for 14 months. These fish proceeded through smoltification and then about 10 months later reverted to parr (a reversion also sometimes seen in the wild). Like the circannual rhythms of birds, this one thus "free runs" on a less than annual cycle, and also like those of birds, it is synchronized to 12 months by the photoperiod actually experienced by the fish. Other species of *Salmo* and species of Pacific salmon (*Oncorhynchus*) also show evidence of circannual rhythms under constant conditions (Hoar 1988).

Endocrine Influences on Migration: Insects

Migration and reproduction tend to be alternative elements in the life cycles of migratory organisms. The reproductive cycles are under the influence of a battery of hormones, and some of these hormones also turn out to be important in migration through either direct or indirect effects on migratory behavior. Hormones not only stimulate migratory activity, they also influence interactions of migration with reproduction, including energy mobilization, timing, and the partitioning of available energy between the two activities.

In many insects migration seems to involve a trade-off with reproduction. The frequently observed ontogenetic separation of migratory behavior from reproduction has been called an oogenesis-flight syndrome (above and Johnson 1969), and the assumption underlying the concept is that migration and reproduction are alternate physiological states. This idea is reinforced by the observation that the conditions that produce delayed reproduction and adult diapause, such as short days, may also induce migration, and that migrants are often similar to diapausing adults in having immature ovaries and hypertrophied fat bodies. Thus, both physiologically and ecologically migration is an alternative to immediate reproduction (Rankin and Burchsted 1992). Once migratory flight has occurred, reproduction can take place in a new location, and indeed such flight may actually accelerate subsequent reproductive activity. Laboratory flight tests have shown that acceleration occurs in milkweed bugs and the lesser migratory grasshopper (Slansky 1980; McAnelly and Rankin 1986) and in other insects (Rankin 1989). Many migrant insects are highly evolved with respect to the balance between reproduction and migration, a balance that is mediated by the neuroendocrine system (Rankin et al. 1986; Rankin 1989). Because of the highly evolved balance, costs in terms of trade-offs between migration and reproduction appear to be minimized (Rankin et al. 1986; Rankin and Burchsted 1992).

As originally conceived, the notion of an oogenesis-flight syndrome implied that migration and reproduction were essentially mutually exclusive processes. Recent comparative studies both across and within species, however, reveal that the relations between migration and reproduction occur in many combinations and permutations. In some cases, mating may occur prior to or during migration, as in the monarch butterfly in the spring. In other cases, mating terminates migration. A further complication may be male quality; females mating with low-quality males may continue to move, perhaps because males are deficient in transferring accessory resources with their sperm (McNeil et al. 1994). The situation is thus more complex than first proposed in the oogenesis-flight syndrome, so the model should not be applied uncritically (Rankin et al. 1986; Sappington and Showers 1992; McNeil et al. 1994). Some of the complexities inherent in the physiological integration of migration and reproduction will be evident in the examples that follow.

The relation between migration and reproduction has been explicitly examined for three species of insect, the ladybird beetle, *Hippodamia convergens*, the grasshopper, *Melanoplus sanguinipes*, and the monarch butterfly, by Rankin and her colleagues (Rankin et al. 1986). In all cases migratory flight, as measured on tethered insects in the laboratory, preceded full development of the female reproductive tract (Fig. 6–4). The relationship could be modified by factors such as season and previous diapause (e.g., Fig. 6–4A for *H. convergens*), but the basic pattern holds for all three species. There was no obvious relation between migratory flight and measures of male reproductive tracts, and it is likely that migratory activity in males is terminated by encounters with sexually receptive females. This is especially likely in cases of extensive mate guarding or when males must respond to a pheromone released by females.

The obvious relation between flight and reproduction in migratory insects made it logical to examine reproductive hormones with respect to an influence on migratory behavior. In insects the primary hormone influencing oogenesis is the juvenile hormone (JH). This is synthesized in the corpora allata (sing. corpus allatum), which are paired neurosecretory organs that, along with the corpora cardiaca, are located along a short nerve pathway behind the insect brain (Fig. 6–5). In juvenile insects, JH prevents the individual nymph or larva from molting into the adult form while promoting growth; in the adult female, JH stimulates the ovary and along with other hormones promotes vitellogenesis, leading to incorporation of yolk into the developing oocyte. The JH is eventually broken down by the enzyme juvenile hormone esterase. A summary of the influence of hormones on insect development and reproduction can be found in Happ (1984).

In addition to its role in development and reproduction, JH has a direct effect on insect migration (Rankin et al. 1986; Rankin 1989, 1991). Rankin and her colleagues have demonstrated this effect with three types of experiments. In the first of these, JH or one of its analogues is applied topically to an insect's cuticle

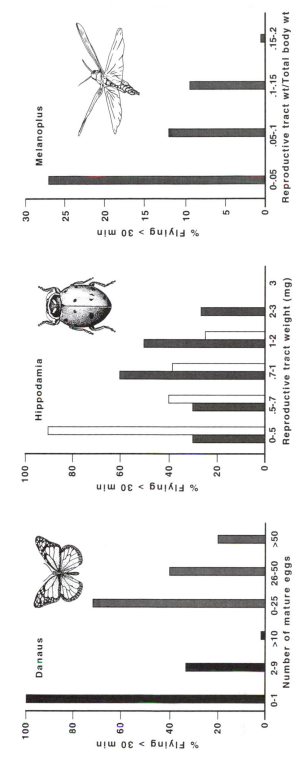

Figure 6-4. The oogensis-flight syndrome in three insects, the monarch butterfly (*Danaus plexippus*), convergent ladybird beetle (*Hippodamia convergens*), and lesser migratory grasshopper (*Melanoplus sanguinipes*). Panel for *Danaus* shows percent of females flying on a tether as a function of number of mature eggs in the oviduct for lab-reared (*solid bars*) and field-caught (*shaded bars*) females. In *Hippodamia*, flight is shown for females captured in March (*shaded bars*) and midsummer (*open bars*) as a function of reproductive tract weight. In *Melanoplus*, flight in lab-reared females is shown as a function of reproductive tract weight to total body weight. (After Rankin and Rankin 1980a,b), Rankin et al. 1986, and McAnelly and Rankin 1986.)

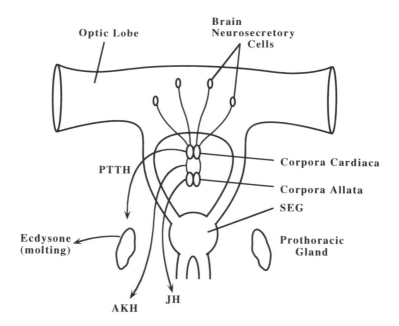

Figure 6–5. The insect brain and endocrine system showing sources of hormones influencing migration. AKH = adipokinetic hormone; JH = juvenile hormone; PTTH = prothoracico-trophic hormone; SEG = subesophageal ganglion. (Drawing by Mitchell Baker.)

from whence it is absorbed into the hemolymph. The results of such applications on the large milkweed bug, the convergent ladybird beetle, and the monarch butterfly are illustrated in Figure 6–6; in all three cases application of JH resulted in an increased proportion of individuals flying for long periods in a tethered-flight situation, relative to controls that were untreated or treated with acetone carrier alone. In the second type of experiment, the source of juvenile hormone, namely the corpora allata (CA), is removed, and the experimental insects are then given replacement therapy either in the form of CA implants or topical application of JH. An example using CA implants (without prior CA removal in this case) in the large milkweed bug is shown in Figure 6–6 and clearly indicates that implantation of these neuroendocrine organs in the large milkweed bug stimulates individuals of this species to long-duration migratory flights. The third type of experiment involved the application of precocene, a substance that suppresses the action of allatotropin, the hormone stimulating the CA to produce JH (Unnithan et al. 1977). Precocene application thus has approximately the same effect as removing the CA surgically. The results of precocene application to the convergent ladybird beetle are also shown in Figure 6–6 and indicate that, in contrast to the action of JH, the antiallatotropin suppresses migratory flight. In addition to the three insects illustrated in Figure 6–6, JH has a positive influence on migratory flight in the

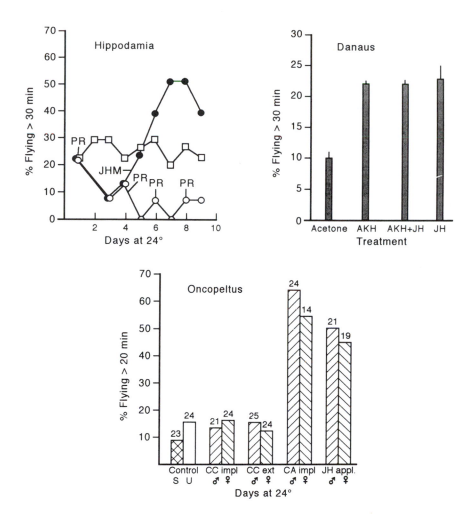

Figure 6–6. Influence of juvenile hormone and other hormones on the tethered flight behavior of the convergent ladybird beetle (*Hippodamia convergens*), the monarch butterfly (*Danaus plexippus*), and the large milkweed bug (*Oncopeltus fasciatus*). For *Hippodamia* there were three treatments: 10 μg application of precocene II (a juvenile-hormone inhibitor) every 2 days, precocene II followed by juvenile hormone mimic (JHM), and a control. Precocene II suppresses flight while JHM stimulates it. In *Danaus* either JH or adipokinetic hormone (AKH) increases flight over acetone controls. In *Oncopeltus* implants or extracts of corpora cardiaca (CC) do not increase flight over sham (S) or unoperated (U) controls, while implants of corpora allata (CA) and JH applications have a marked effect increasing flight. Numbers above bars indicate sample sizes. (Adapted from Rankin et al. 1986.)

cotton boll weevil, *Anthonomus grandis*, and experiments using the JH inhibitor fluoromevalonate (FM) have shown stimulation of migratory flight by JH and its suppression by FM in the western corn rootworm beetle, *Diabrotica virgifera*, a major pest of corn in much of North America (Coats et al. 1987).

Because JH stimulates both migratory flight and reproduction, the question arises as to how the trade-off between these two activities is regulated. The answer turns out to be a function of the titers of JH in the hemolymph. At low titers neither activity is stimulated, at intermediate titers migratory flight occurs, and at high titers reproductive activities and oogenesis are induced, which in turn suppress migration. Rankin (1978) demonstrated the relationship nicely for the large milkweed bug by taking advantage of the technique of artificial selection (Chapter 14). Normally migration in these bugs takes place 12–15 days after adult molt. Rankin, however, selected for delayed flight so that migration did not occur until some 30 days after bugs eclosed to adulthood. The selected population was then compared to the unselected control. There were two correlated responses to selection for delayed flight. First, oviposition was also delayed, and began after a decline in migratory activity. Second, the JH titer in the blood remained at relatively low levels for a much longer period before rising to a peak that followed maximum flight levels (Fig. 6–7). The data indicate that when titers are at intermediate levels, flight is stimulated, but that once they reach high levels, they induce oogenesis and reproductive behavior, so ending the migratory period. The decline in JH occurring just as oviposition starts indicates its removal from the hemolymph during vitellogenesis.

These results can also be related to diapause in milkweed bugs. During long days and at high temperatures, JH titers rise rapidly, following cuticle hardening, to stimulate reproduction, which in turn suppresses migration. With short days and low temperatures, JH titers are not sufficient to induce oviposition, but they do stimulate migratory flight. The bugs are then capable of migration throughout the long period necessary to move south to suitable overwintering sites (Dingle 1985).

Another interesting case of the relation between JH and the timing of migratory and reproductive behaviors occurs in the true armyworm moth, *Pseudaletia unipuncta* (Cusson and McNeil 1989). In this and several other species of migratory moth, females attract males by assuming a characteristic behavioral posture termed "calling" and emitting a pheromone. The occurrence of this pre-mating behavior suppresses migration. The age at first calling, which determines the duration of the migratory period, is probably determined by the brain via release of the hormones allatotropin and allatostatin, which stimulate or inhibit, respectively, JH production by the corpora allata; JH in turn stimulates calling and pheromone production. Under the short-day, cool-temperature conditions of spring and fall when these moths migrate, JH titers probably remain lower for a longer period than they do in summer. Hence calling and pheromone release do not occur during migration. Once JH titers reach high levels, calling occurs and migration ceases. The genetics of similar relations between reproduction and behavior have been studied in other moths, by Gatehouse and his students (Han and Gatehouse

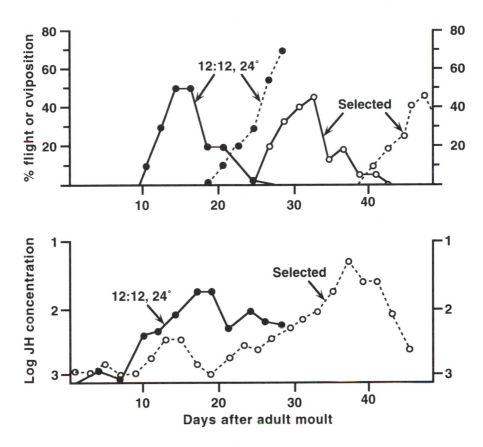

Figure 6–7. Relations between migration and oviposition in the large milkweed bug, *Oncopeltus fasciatus*. Top panel: Tethered flight of more than 30 minutes (*solid lines*) or oviposition (*dotted lines*) in unselected bugs versus bugs selected for delayed onset of flight. Bottom panel: Juvenile hormone (JH) titers in the selected and unselected lines. Note that flight occurs at intermediate JH titers whereas oviposition follows titers at maximum levels. (Redrawn from Dingle 1985 after Rankin 1978.)

1991; Hill and Gatehouse 1992; Wilson and Gatehouse 1993) and are discussed in Chapter 14.

In addition to displaying behavioral variation in migration, many insects are wing polymorphic; different populations or individuals within populations possess wing forms that vary from long (macroptery), to short (brachyptery), to absent (aptery). Frequently the long-winged morphs serve as migrants, while brachypterous and apterous forms display higher reproductive capacity (Roff 1990). The ecology and evolution of the phenomenon is discussed in Chapter 13. Because the wingless morph of at least hemimetabolous insects resembles the nymphal stages, Kennedy and Stroyan (1959) and Southwood (1961) suggested some time ago that juvenile hormone might be responsible for producing the apterous form. It is only recently, however, that this action of JH has been confirmed. Hardie (1980) demonstrated apterization by JH applications in the aphid, *Aphis fabae*, and Iwanaga and Tojo (1986) showed that high JH and low population densities contributed to brachypterization in the brown planthopper, *Nilaparvata lugens*, an important pest of rice. The most extensive studies of the relation between wing form and JH, however, have been done by Anthony Zera, Kristina Tiebel, and their colleagues on the wing polymorphic ground cricket, *Gryllus rubens* (Zera and Tiebel 1988, 1989; Zera et al. 1989; Zera and Tobe 1990).

Zera and Tiebel (1988) first showed that JH application either in the penultimate or early last nymphal instar would induce short-winged adults. Their results were particularly interesting because JH induction of brachyptery occurred both in lines selected in the lab for long wings or in populations raised under crowded conditions to produce long-winged adults. In other words, regardless of whether the "long-winged" populations were genetically or environmentally determined, JH application resulted in short wings. Titers of JH, however, differ little between nymphs destined to develop into long- or short-winged adults (Zera et al. 1989). The key factor turns out to be juvenile hormone esterase (JHE), the enzyme primarily responsible for degrading JH, especially at the end of larval development prior to metamorphosis (Hammock 1985). In last-instar nymphs of long-winged crickets, JHE activity is considerably higher in the middle of the instar than it is in mid-instar nymphs destined to be short winged (Zera and Tiebel 1989 and Fig. 6-8), indicating that it is the removal of JH by JHE that permits metamorphosis into a macropter.

A further interesting aspect of this JH control system is that the critical modulation of JH titer occurs *after* the major decline in JH titer that takes place early in the instar. This early decline is what permits metamorphosis into the adult rather than another juvenile stage. During the first 3 days of the last-instar JH titers, rates of JH biosynthesis, and JHE activities in both wing morphs are the same; it is thus subtle elevation of JH in the middle of the instar when JHE is low that produces brachypters (Zera and Tobe 1990). There is also evidence that this subtle rise in JH titer acts in concert with a reduced titer of ecdysteroids, the so-called molting hormones from the prothoracic glands (Zera et al. 1989 and Fig. 6-5). Although there is much still to be learned about the hormonal control of wing morph, it is quite evident that there is a finely tuned and complex interaction between ontogeny and hormones.

Juvenile hormone is also involved in insects in which the flight muscles histolyze following migration. In the cotton stainer bug, *Dysdercus fulvoniger* (Davis 1975), and bark beetles of the genus *Ips* (Borden and Slater 1969; Unnithan and Nair 1977) the protein from histolized wing muscles is used for oocyte development; JH applications result in muscle degeneration. In the Colorado potato beetle, *Leptinotarsa decemlineata*, the short days of fall induce movement to a

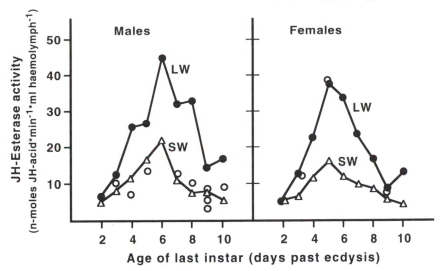

Figure 6–8. Relation between median juvenile hormone esterase activity and the occurrence of long wings (LW) or short wings (SW) in the ground cricket, *Gryllus rubens*. Data are from LW and SW genetic stocks measured during the last nymphal instar. Open circles are measurements on individuals of the LW line that molted into SW adults. (Redrawn from Zera and Tiebel 1989.)

diapause site, low JH titers, and muscle degeneration. In the spring, long days trigger increasing JH production, movement out of diapause sites, muscle regeneration, and finally reproduction (de Kort 1969; de Kort et al. 1982; Rankin 1989). Thus JH can apparently have opposite effects on flight muscle degeneration in different insects, depending on the ecophysiologies involved.

In addition to influencing the flight system directly, hormones are involved in the mobilization and use of flight fuels. The best studied of these are locust adipokinetic hormones (AKH), neurosecretory products of the glandular lobes of the corpora cardiaca (Goldsworthy 1983; Wheeler 1989; Fig. 6–5). In addition, AKH may have a direct influence on flight speed (Rankin 1989), at least in locusts, although the relation between AKH and specifically migratory flight is not always clear. The reevaluation of phase relationships in locusts, with the gregarious-phase swarms now seen as an adaptation for extended foraging, with migration occurring mostly in the solitary phase (Farrow 1990 and Chapter 10), also will affect how the role of AKH in locusts is interpreted. In the monarch butterfly, however, preliminary data do suggest a role for AKH in migratory flight (Rankin et al. 1986 and Fig. 6–6A). In this species, injections of AKH either alone or in combination with JH increase long-duration tethered flight in the laboratory. Both hormones appear equally effective, and their effects are not additive. The exact role of AKH in this situation remains to be determined, as measurements of hemolymph lipid or muscle metabolism were not made in these experiments.

Endocrine Influences on Migration: Vertebrates

There are many parallels between endocrine influences on migration in insects and vertebrates. Chief among these is the "capture" of hormones evolved for other functions to modulate migratory physiology and behavior. In insects such hormone capture is evident in the case of juvenile hormone. In vertebrates, hormones involved in a number of processes, ranging from energy mobilization to reproduction, also play a role in migration. Such cross-functioning hormones are evident, first, in the migratory life cycles of fishes.

The vast majority of fish migrations take place within the oceans (Harden Jones 1968) or within freshwaters, in the latter case especially in the Tropics (Lowe-McConnell 1975). The best known and best studied, however, occur in Temperate Zone diadromous species, both because these are the most accessible to concentrations of fish biologists and because they include major commercial species such as salmon. Relative to other migrant species these have been studied to a disproportionate degree, and this is especially reflected in the analysis of hormones. Keep in mind, however, that our knowledge of fish migratory physiology is heavily biased by the very few well-studied species, and although general trends are likely similar, details may or may not reflect what is going on in the majority of migrant fishes.

Two prime examples of diadromous commercially important fishes are the catadromous North Atlantic eels (*Anguilla anguilla* and *A. rostrata*) and the anadromous salmon from both the Atlantic and the Pacific. Dramatic behavioral and physiological events accompany movements of the various species from the ocean to freshwater (eels) or the reverse (salmon). These events characterize specialized migratory behavior with enhancement of locomotory activity of the sort I describe throughout this book. Both sets of species undergo a metamorphosis. In eels this involves transformation from a leptocephalous larva to a transparent glass eel. The new elvers show an increasing preference for freshwater when given choices in experimental tanks, and a decreasing photonegativity in their response to light (Deelder 1958; Fontaine 1975). Behavioral changes can be even more dramatic during smoltification of young salmon. Fontaine (1975) describes the behavior of Atlantic salmon smolts from the Adour basin of France as follows: "The smolt leaves the gravel, rises midway toward the surface, swimming vigorously, even leaping frequently out of the water, and this is how the spring floods carry it down to the sea. . . . it no longer finds the relatively calm zones where, hiding under stones, it could shelter from the current . . . of spring floods."

Accompanying these changes in behavior are equally dramatic changes in the smolt's physiology (Hoar 1988). First, smolt transformation involves changes in purine nitrogen metabolism. Synthesized crystals of two purines, guanine and hypoxanthine, are deposited in two distinct skin layers to give the silvery appearance of smolts. This silvering is apparently of survival value in the pelagic marine environment, as it is characteristic of most species living there. Second, the rate of oxygen consumption increases with heightened catabolism of carbohydrate, fat, and protein to provide energy for the migratory journey. Accompanying these changes is increased absorption efficiency of the intestine, including increased proline flux (Collie 1985). Third, there is an increase in the complexity of blood hemoglobins that preadapts the smolt to the lower oxygen tensions of the ocean environment. Finally there is an increase in the activity of Na^+, K^+- ATPase in the gills to prepare the smolt for the passage to the osmotically very different seawater medium. The gills are an important component of the capacity to hypoosmoregulate. They excrete salts in the hyperosmotic marine environment to keep plasma electrolytes at about one-third seawater concentration. Changes in the gill are accompanied by reduced glomerular filtration in the kidney and by transport of monovalent ions across the hindgut wall. Water follows these ions into the tissues with the excess ions then excreted by the gills. These and other more subtle changes transform the salmonids from a freshwater-adapted parr to a fish that is well adapted to living in the sea. The smolt stage is brief, however, and if the fishes do not enter the sea during this time, they revert to the parr condition. It should also be noted that the different genera and species of salmonids show variations on the theme of smoltification, depending on habitats, life cycles, and possibly phylogenies.

These physiological changes between parr and smolt are also accompanied by changes in the levels of various hormones (Fig. 6–9) which have been observed in both hatchery-reared and wild-caught fishes (Whitesel 1992; Snyder et al. 1995). Given the major metabolic mobilization that occurs in migratory smolts, it is not surprising that increased levels of thyroid hormones are among these, a conse-

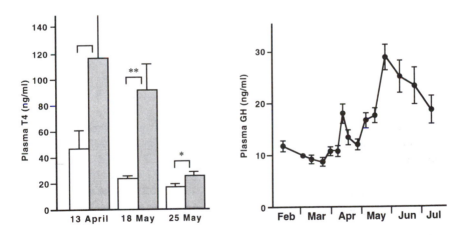

Figure 6–9. Changes in plasma T4 and plasma growth hormone (GH) during transformation from freshwater parr to the migratory smolt in salmon. In the left panel measurements are for coho pair caught in a lake in British Columbia (*open bars*) and for smolts caught at a fence placed downstream from the lake (*p < .05; **p < .01; data provided by Randal Snyder). The right panel shows changes in growth hormone during parr-smolt transformation in fish reared in freshwater. (Redrawn from Prunet et al. 1989.)

quence of a dramatic increase in thyroid activity (Hoar 1988). In particular this increase involves thyroxin (tetraiodothyronine or T_4), whereas levels of triiodothyronine (T_3) remain constant. This makes biological sense as T_3 is usually in the higher concentrations in situations where iodine is in short supply, such as in the freshwater from which the smolts are about to depart. The rise in T_4 levels has been shown to coincide with the new moon in some studies (Grau et al. 1981), but in others no relation to lunar phases is apparent (Hoar 1988; Snyder et al. 1993). The evidence indicates that thyroid hormones do not actually trigger smolting. Rather they intensify the physiological and behavioral changes that accompany smolt transformation. These actions of thyroid hormones include stimulation of gill Na^+, K^+- ATPase activity (Prunet et al. 1989), stimulation of lipid mobilization (Sheridan 1986), induction of purine deposition and increased hemoglobin complexity (Hoar 1988), and general enhancement of a more active metabolism (Hoar 1988). Increased thyroid activity has also been found in other migrating fishes, including sticklebacks, *Gasterosteus aculeatus* (Baggerman 1962) and juvenile American eels (Castonguay et al. 1990).

Three other hormones with significant influences during the parr-smolt transformation are prolactin, cortisol, and growth hormone (Sheridan 1986; Hoar 1988; Young et al. 1989; Prunet et al. 1989). Calcitonin, on the other hand, appears not to be involved, which seems counterintuitive given the relatively much higher calcium ion content of seawater (Björnsson et al. 1989). Of the three active hormones, prolactin actually decreases during smoltification. This hormone has an inhibitory effect on gill enzyme activity, and its decrease probably allows gill Na^+, K^+- ATPase to increase as part of the overall preadaptation for movement into the sea. Injection of prolactin into two species of stickleback shifted salinity preference toward freshwater (Audet et al. 1985), further supporting the notion that it promotes physiological responses appropriate to a hypoosmotic environment. Prolactin also apparently has a positive role in migration in that along with cortisol, growth hormone, and T_4, it stimulates lipid mobilization through enhancement of lipolysis in the parr just prior to smoltification (Sheridan 1986). Cortisol, in addition to its role in lipolysis, also appears to stimulate gill Na^+, K^+- ATPase activity (Richmond and Zaugg 1987). Finally, growth hormone promotes hypoosmoregulatory ability, as well as stimulating growth and lipid catabolism (Young et al. 1989).

There are interesting parallels between the involvement of hormones in migratory diadromous fish and the migrations of urodele amphibians between terrestrial and aquatic habitats (Rankin 1991). In the latter case the two primary hormonal effects come from thyroxin (T_4) and prolactin. Thyroxin promotes metamorphosis and also movement away from water, requiring physiological responses, such as water conservation, similar to those required when fish enter the sea. Prolactin antagonizes the action of T_4 in metamorphosis and is reduced during that process; this suggests the possibility that the absence of prolactin as much as the presence of T_4 may produce the "land drive." The return to water for breeding is promoted by prolactin. This hormone also seems to promote water uptake, an increase in specific gravity, and an increase in the osmotic pressure of the blood, all of which result in a tendency for the animals to sink (Moriya and Dent 1986). Thyroxin apparently has opposite effects, causing the animals to float and facilitating movement to land. These effects led Rankin (1991) to propose that the osmoregulatory functions of these hormones were probably important in the evolution of their influences on migration (hormone "capture," see above), and indeed this seems likely for both diadromous fish and urodeles. The case for anuran amphibians is much less clear. Thyroxin is a prime mover in metamorphosis, and some correlative evidence suggests association with movement away from water. Experimental data, however, are lacking, and experiments with prolactin have produced negative results (Dent 1985). More field and experimental data are clearly needed to clarify the situation.

Analyses of the contributions of hormones to the migratory physiology and behavior of birds have again involved only a few species, most notably the North American white-crowned and white-throated sparrows (*Zonotrichia leucophrys* and *Z. albicollis*) studied by Donald Farner and his students. A few other species such as the Asian red-headed bunting (*Emberiza bruniceps*) have also received intensive study. As with fishes, this concentration on a few species constrains the revelation of generalization, but the results so far do at least hint at some common threads.

Repeating the pattern seen in other taxa, migration in birds involves several physiological functions. As outlined by Wingfield and colleagues (1990), some of these functions are (1) deposition of fat, (2) integration of enzyme systems for energy storage and mobilization, (3) increased hematocrit (number of red blood cells) for enhanced oxygen transport, (4) hypertrophy of flight muscles, and (5) development and synchronization of migratory behavior, including organization of diel patterns. Not surprisingly hormonal regulation of these processes is complex and because of this complexity is not clear in many of its details. In the case of muscle hypertrophy and hematocrit, essentially nothing is known of hormonal relationships. With respect to the other functions, the knowledge we do have once more suggests that hormones evolved for a number of different actions have been incorporated in support of physiological mobilization for migration.

Because of the relation between migration and breeding in Temperate Zone migrants, attention has been directed at the possible role of gonadal hormones on vernal migration. In his early experiments with photoperiod, Rowan (1925, 1932) postulated such a role because intact juncos and crows exposed to midwinter long days moved north, while castrates and nonphotostimulated birds moved south. Later experiments on *Zonotrichia* sparrows showed that the removal of the gonads during shortening days of early winter eliminated subsequent vernal premigratory flattening and considerably reduced *Zugenruhe* or migratory restlessness (Wingfield et al. 1990). A detailed study of white-crowned sparrows by Mattocks (in Wingfield et al. 1990) showed that castration before the winter solstice eliminated hyperphagia and premigratory fattening, and reduced zugenruhe. There was no effect of castration, however, on autumn migrants, suggesting that migration in the fall has a different hormonal basis from spring (Fig. 6–10). This is not surprising,

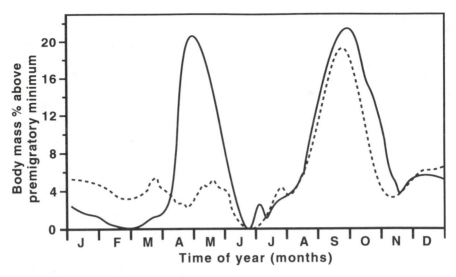

Figure 6–10. Seasonal changes in body mass in intact (*solid line*) and castrated (*dashed line*) male white-crowned sparrows, *Zonotrichia leucophrys gambeli*. (After Mattocks 1976.)

considering that spring and fall migrations occur in very different physiological states. Replacement therapy with small amounts of testosterone restored spring zugenruhe and fattening in castrates of both sexes, indicating that it is important for regulation of vernal migration in females as well as males. Schwabl and colleagues (1988) obtained similar results for the same species, and Schwabl and Farner (1989) likewise reinstated migratory responses with estradiol implants in ovariectomized females. Blockers of testosterone metabolism prevent effective therapy. There may thus be a synergism of metabolites of testosterone active at the target organ level (Wingfield et al. 1990). Other evidence suggests testosterone may also act by increasing the release of prolactin. Although their mode of action is not completely clear, a role for androgens in spring migration seems quite well established. In contrast, the endocrine basis for fall migration remains to be elucidated.

The influence of testosterone or testosterone-prolactin on spring migration is apparently under central control by the median eminence and infundibular nucleus regions of the hypothalamus (they are part of the hypothalamus-pituitary axis). Lesions in the median eminence reduced testicular growth, vernal premigratory fattening, and zugenruhe in white-crowned sparrows. Administration of testosterone, prolactin, or a combination to the median eminence of the lesioned birds induced fattening, but not zugenuhe (Yokoyama 1976). The hypothalamus is the area of the brain sensitive to photostimulation, so the assumption of endocrine regulation via this region is a reasonable one (Yokoyama and Farner 1978). Various other lesioning experiments (summarized in Wingfield et al. 1990) also indicate central control of the influence of hormones on bird migration. There is some evidence from ring doves and white-crowned sparrows that this central control may be regulated by specific prolactin receptors in the hypothalamic region (Buntin and Ruzycki 1987; Schwabl et al. 1988; Rankin 1991).

Thyroid hormones are also involved in bird migration and seem to interact with photoperiod, prolactin, testosterone, and possibly growth hormone. The best evidence comes from the red-headed bunting (Pathak and Chandala 1982a,b; Thapliyal et al. 1983; Thapliyal and Lal 1984; Lal 1988). In this species, circulating T_3/T_4 ratios increased prior to spring migration, although interestingly, not prior to migration in the autumn. Thyroidectomy inhibits nocturnal premigratory restlessness and long-day–induced fattening and increases in body mass. Injections of T_3 and T_4 increased locomotor activity and restored body mass to pretreatment levels. After 4 months under long days the buntings become photorefractive, no longer responding to long days behaviorally or physiologically. At this time T_4 treatment had no effect; however, prolactin injections alone or in combination with T_4 or testosterone increased body weight. Prolactin had no effect in thyroidectomized birds, and likewise testosterone in the absence of thyroid hormones failed to elicit a response. All three hormones may therefore be necessary for the complete spring migratory response. Results from a scattering of other species further indicate that these three hormones, plus thyroid-stimulating hormone (TSH) and perhaps growth hormone, acting more or less in concert facilitate migratory preparedness in vernal birds (Wingfield et al. 1990). There seems to be no clear pattern of thyroid hormone involvement in autumn migration in any bird.

Perhaps the most intriguing model of hormonal action in migratory birds derives from the work of Albert Meier and his colleagues on the white-throated sparrow (Meier and Fivizanni 1980; Meier and Wilson 1985). Meier proposes that the phase relationship of two circadian rhythms, those of plasma levels of corticosterone and prolactin, varies seasonally as a function of day length and/or a circannual rhyhm. The phase angle between the rhythms then regulates premigratory hyperphagia, fattening, and even the direction of migration, whether south in the fall or north in the spring. For example, in birds made photorefractory by long exposure to long days, daily injections of corticosterone followed by injections of prolactin either 4 or 12 hours later resulted in fattening and increased migratory restlessness. Even more interesting, the orientation of birds under the night sky (Chapter 8) was south with the 4-hour treatment interval and north with the 12-hour interval (Martin and Meier 1973). Analyses of plasma adrenal steroids and pituitary prolactin throughout the year were alleged to show a 5–9-hour interval between peaks of the two hormones in August ("nonstimulatory") and a 10–14-hour interval in May ("stimulatory"), thus supporting the experimental results.

There are problems with the phase angle model, however. First, evidence of temporal synergism has not been found in other migratory species, including the closely related white-crowned sparrow (Farner and Gwinner 1980; Vleck et al. 1980). In addition, a careful analysis by Rankin (1991) indicates inconsistencies between the data on relative levels of prolactin and corticosteroid levels and the model. In spite of assertions to the contrary, the observed annual circadian changes in hormone levels in at least some cases appear to be opposite to those predicted by the model. Finally, many of the experiments were not done with sensitive techniques such as radioimmunoassays, and data were obtained from pooling individuals. These and other problems make the experiments vulnerable to several sources of error and results difficult to interpret. This is not to say that circadian rhythms are not important. The well-established effects of prolactin on fattening (see above) do appear to depend on the time of day of administration (Meier and Fivizanni 1980; Vleck et al. 1980). Experimental evidence also suggests a role for corticosteroids in fattening and migratory behavior, although the data are sometimes equivocal (Wingfield et al. 1990). Cause and effect relationships may also be difficult to sort out because corticosteroids are associated with stress responses. Migrating individuals from a partially migratory population of willow tits (*Parus montanus*), for example, have higher baseline levels of corticosterone than residents in territorial flocks, but this may also indicate stress due to poorer body condition or lower social rank or both (Schwabl and Silverin 1990). In contrast, plasma corticosterone levels in 32 individuals of seven passerine species migrating through an Alpine pass were generally low, suggesting that these birds were not stressed (Gwinner et al. 1992).

Turning now to mammals, we find that data regarding hormones and migration are extremely sparse; there seems, in fact, to be only one major study. Noting that males of small mammals are more likely to migrate than females, Holekamp and colleagues (1985) tested the possibility that postnatal emigration of Belding's ground squirrels (*Spermophilus beldingi*) was influenced by testosterone. Their first finding was that castrated males were briefly delayed in making their exodus; their emigration rate was otherwise similar to intact males. Neither intact nor castrated females showed much evidence of emigration, although there were only seven of the latter. In a second experiment, females were injected with testosterone propionate within 36 hours of birth and 8 of 12 of these treated females emigrated, as revealed by recoveries after 60 days. Unfortunately no control-injected (oil only) females were recovered, raising the possibility that they emigrated as well, although indirect evidence implied they did not. Holekamp and colleagues proposed that testosterone has an organizing effect on the connections in the central nervous system, a phenomenon well known in the literature on hormonal influence on the mammalian brain (e.g., Maclusky and Naftolin 1981), and that this masculinization prompted the postnatal exodus of treated females. These results are interesting, but because of small sample sizes and the aforementioned difficulty with controls, they must be considered preliminary. There is also the problem that because the postnatal movements of small mammals are continually referred to as "dispersal" without a behavioral test (see Chapter 2), it is not entirely clear whether these ground squirrels are ranging or migrating. There is obviously much to be done concerning hormones and mammalian movements.

A summary of the actions of vertebrate hormones on migration is diagrammed in Figure 6–11. Two things are obvious from the figure and the preceding

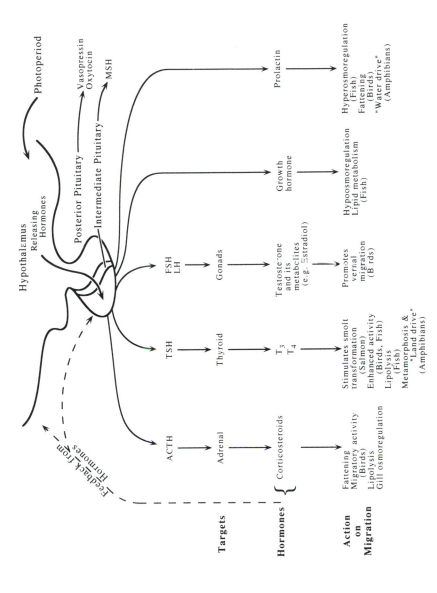

Figure 6–11. Summary of the action of various vertebrate hormones on the physiology of migration. ACTH = adrenocorticotrophic hormone; TSH = thyroid stimulating hormone; FSH = follicle stimulating hormone; LH = luteinizing hormone; MSH = melanocyte stimulating hormone. (Modified from a drawing provided by Eugene Spaziani.)

discussion. First, there is commonality of function across taxa. Prolactin, for example, promotes hyperosmoregulation in both fishes and amphibians; it is reduced when salmon leave freshwater for the sea and increases when amphibians return to freshwater to breed. Similarly, thyroid hormones promote enhanced activity and metabolism in both birds and fishes embarking on migratory journeys. Secondly, to reinforce the notion introduced at the beginning of this section, hormones evolved for other functions have been incorporated into the physiology of migration. Thus the reproductive hormone testosterone promotes avian vernal migration, and growth hormone promotes hypoosmoregulation in fish migrating to the sea. Finally, it is worth noting that there is a great deal of ambiguity and often contradiction in the results of hormonal experiments and analyses in vertebrates, although the broad outlines seem evident. I suspect that much of the inconsistency derives from the varied developmental histories of experimental animals. Here insects offer a real advantage because many can be reared for many generations under controlled conditions. Nevertheless, I urge vertebrate biologists, and especially physiologists, to pay more attention to the genetic and developmental histories of their organisms than has been customary so far.

Fueling Migration

The overwhelming majority of migrants rely primarily on fat to fuel their migratory journeys. Relative to other potential fuel sources, fat has great advantages (Blem 1980, 1990; Alerstam 1990; Ramenofsky 1990). When oxidized, a gram of fat yields about 9 kilocalories of energy, whereas the yields of both carbohydrates and proteins are only about half this value. Furthermore, each gram of carbohydrate, stored in the form of glycogen, requires in addition 3 grams of water. In contrast, fat is stored without water or protein, thereby considerably reducing the bulk per unit of energy produced. A summation across these values shows that it takes 8 grams of carbohydrate to generate the energy available from 1 gram of fat. It is hardly any wonder that fat predominates as the fuel for migration.

Before setting off on their journeys, migrants can store enormous amounts of fat. A monarch butterfly, for example, under favorable conditions can store up to 125% or more of its lean dry weight as fat (Beall 1948; Brower 1985), and even tiny insects such as the aphid, *Aphis fabae*, may store fat proportions of around 30% total dry weight (Cockbain 1961). Birds that fly long distances, such as the North American scarlet tanager (*Piranga olivacea*) or ruby-throated hummingbird (*Archilochus colubris*), both of which winter in South America, virtually double their body weights by laying on premigratory fat (Blem 1980). This much fat in turn incurs a cost because it, too, must be transported. A bird with 50% body fat consumes about 40% more energy than a bird covering the same distance with 10% fat (Alerstam 1990). There are also other trade-offs involved (see Chapter 7).

Neverthless, for most animal migrants there is a positive relationship between the amount of fat storage and the distance traveled. An example is shown for fish in Table 6–1. Various species of Caspian Sea shad migrate up the Volga River to spawn, and their travel distances match well the amount of fat they store. In plants, the relation between fat storage in seeds and distance traveled is indirectly influenced by travel mode, as is seed size (Table 6–1). Seeds transported by wind or animals, and therefore traveling longer and farther, have increased proportions of fat. The smaller wind-transported seeds are presumably selected for lower wing loading (the ratio of weight to surface area; see Chapter 7), and animal-transported species should be selected for tough, protected seeds and less seedy fruits. Energy levels in both cases could be conserved by increasing the proportion of fat. The tremendous advantages of fat are also shown in some calculations by Alerstam (1990). Using Pennycuick's (1975) models of bird flight (Chapter 7), he estimates that a fat load of 10% body weight will carry a bird up to 20 hours and 750 km. Raising the fat load to 50% allows 3 to 4 days of flight and distances of 4000 km before fuel exhaustion, comparable to flights actually observed in some long-distance migrants.

Even with the high energy available from fat, migratory journeys often stretch capacities to the limit. Sockeye salmon migrating 1200 km upstream to spawning sites may consume 70% of their lipids and even up to 40% of their protein content (Idler and Bitners 1958, 1959). Less than 10% of this cost goes to gonadal development, even in the egg-producing females, indicating the high demands of migration.

Body proteins may also be used by birds, especially when making long transits of inhospitable terrain (Bairlein and Totzke 1992), and migrant butterflies may need carbohydrates from nectar sources en route (DeVries and Dudley 1990; Gibo and

Table 6–1. Percent Fat Content as a Function of Distance Traveled in Caspian Shad and Mode of Transport in Large Samples of Seeds

Species	Percent Fat	Migration Distance (km)
	Caspian Shad	
Caspialosa kessleri	16.0	1000
C. volgensis	8.7	500
C. caspia	7.5	100
C. saposhnikovi	5.6	0

Transport Mode	Percent Fat	Seed Size (mg)
	770 Seed Species	
Passive[*]	$10.1 \pm .7$[†]	49.7 ± 9.1[†]
Wind	$25.5 \pm .8$	24.8 ± 5.4
Animal	25.0 ± 1	265 ± 90

[*]No special transport mechanism other than being dropped by the plant
[†]Values ± standard errors

SOURCE: From Fontaine 1975 and Lokesha et al. 1992.

McCurdy 1993). Given this high cost, it is not surprising that migrants regulate the balance between fat and other fuels. This was first demonstrated by Krogh and Weis-Fogh (1951) with tethered desert locusts, *Schistocerca gregaria*. They measured the respiratory quotient (RQ) or the ratio of carbon dioxide produced to oxygen consumed and found that this was about 1.0, indicating carbohydrate metabolism, during short flights, but gradually shifted toward 0.7, the value for lipids, as flights became longer. Similar transitions occur in other migratory insects like the moth *Anticarsia gemmatalis* (Hammond and Fescemyer 1987). Migratory hummingbirds face the special problem of accumulating large fat deposits for migration, while still fueling their energetically costly foraging flights, resulting from high metabolic rates and hovering behavior. Measurements of RQ in trained rufous hummingbirds (*Selasphorus rufus*) showed they burned fat while resting but quickly switched to carbohydrates while foraging for nectar (Suarez et al. 1990). Burning carbohydrates derived from nectar during foraging spares the fat stores and also avoids the necessity of synthesizing fatty acid from glucose and then burning the fat. Fat is used during migration because of its relatively high energy yield per unit weight, as discussed above.

In birds the accumulation of extra fat for migration seems to occur in three ways. First, premigrants greatly increase their appetites. This hyperphagia involves more rapid rates of feeding rather than increases in meal size (Bairlein 1990; Ramenofsky 1990). Second, accompanying enhanced appetite are an increase in the synthesis of lipids (hyperlipogenesis) and rapid deposition of fat in subcutaneous and visceral "depots." These fat deposits are visible through the skin if the feathers are spread, allowing visual assessment of the condition of a bird. The hyperlipogenesis depends not only on the increased dietary intake, but also on increased rate of fatty acid synthesis in the liver, implying that metabolic adjustments occur in addition to those involved in feeding.

The third contributing factor to increased fat accumulation is diet shift (Bairlein 1990). Observations on a number of species in both the Old and New Worlds have noted that autumn migrants feed heavily on fruits. This appears to be a significant adaptation because fruits are often locally abundant and easily procured, and they provide a rich carbohydrate source to support fat deposition. In addition, fruits of some regions used for migratory stopovers, notably around the Mediterranean, generally contain relatively high concentrations of lipids (Herrera 1987). At least part of the diet shift is the result of deliberate choices made by migrants, as confirmed by choice experiments in the laboratory. Garden warblers, when offered choices among diets varying in fats or proteins independent of caloric levels, consistently take foods higher in these two nutrients (Bairlein 1990). North American catbirds (*Dumatella carolinensis*) and mockingbirds (*Mimus polyglottos*) also preferred fruits higher in fat content, but the predominantly seed eating white-throated sparrow did not (Borowicz 1988). Even migratory shorebirds may switch; at an inland stopover site in Texas, seeds comprised 37% of the diet, and up to 100% of the diet of some individuals (Baldassare and Fischer 1984). It would be interesting to know what the relative fat contents were of seeds versus the more usual animal food. Where high-energy rewards justify it, some migrants may switch to specific insects. Premigratory sedge warblers (*Acrocephalus schoenobaenus*) in northern Europe will specialize in feeding on the explosive populations of aphids in their reed-bed habitat, consuming an average of some 40 aphids a minute. The high sugar content of the aphids is probably what allows these warblers to build up their fat levels to 50% of body mass in only 3 weeks (Bibby and Green 1981). Diet shifts to foods rich in energy thus seem to play an important part in premigratory fattening and so are also an important element of migratory syndromes.

Summing Up

Specialized migratory behavior is accompanied by a syndrome of physiological activity that functions at several levels, involving hormones, metabolic pathways, and the actions of the nervous system. Far more is known, however, about metabolic and hormonal functions accompanying migration than about the role of the nervous system in regulating it. The physiological syndrome is synchronized with the external world by both circadian and circannual rhythms that allow precision responses of great complexity. Migrant birds, for example, respond to seasonal changes in day length by altering metabolic pathways to favor fat storage and determine molt cycles; they can respond to asymmetric changes in external time givers when they cross the equator on migratory transits so that the timing of their migrations corresponds to seasons in the opposite hemisphere. Insects use daily rhythms to incorporate diapause and migration into their life cycles, and fishes can time their swimming to correspond with the incoming tide.

Metabolic and other physiological functions of migratory organisms are coordinated by a system of hormone actions. Primary among these is the development of energy reserves, most often in the form of fat deposits, to sustain the migratory journey. This often involves profound reorganization of metabolic pathways to facilitate fat storage. The efficiency of the process can be seen in the fact that the monarch butterfly may accumulate 125% of its lean dry weight in body fat prior to migration and many birds virtually double their weight, sometimes in just a few days. Other hormonal actions include muscle hypertrophy, enhanced oxygen transport, modification of osmoregulatory function, and, as demonstrated with insect juvenile hormone, direct stimulation of migratory behavior. Other behavioral changes engendered by hormones include diet shifts, suppression of reproductive activities, responses to ion concentrations, and the timing and magnitude of locomotory patterns. The precise nature of hormonal action is a function of phylogeny, environment, and mode of life. In spite of major differences in all three of these factors, however, migratory physiologies across a broad taxonomic range are often strikingly similar.

Meier and Fivizanni (1980) noted the similarities and further stressed the point that whereas behaviors may be highly variable, physiological mechanisms tend to be conservative. Certainly this is the case for migratory physiology. We can illustrate the point with hormones. In both insects and vertebrates, for example, hormones evolved for a variety of different functions are "captured" for use in migratory syndromes. Juvenile hormone in insects serves to maintain larvae or nymphs in a juvenile state, to stimulate reproduction, to stimulate pheromone release, and to stimulate migratory flight. Similar hormonal actions occur across a range of divergent insect taxa. Thyroid hormones in vertebrates enhance metabolic activities in the service of a number of adaptive functions, of which migration is one; prolactin influences fat metabolism and osmoregulatory changes between fresh and salt water or between water and land. Other hormones play similar multiple roles. The evolution of migratory behavior in all its often spectacular variety has been accompanied by basic physiological mechanisms that have diverged but little. In spite or perhaps because of their conservatism, hormones are an important intermediary step in the translation of genetic information into behavioral and life history outputs (Rankin 1991; Ketterson and Nolan 1992). We have seen two direct demonstrations from insect migration in this chapter. In the first, the expression of wing length in a wing polymorphic cricket is determined by the interaction between juvenile hormone and juvenile hormone esterase (Zera and Tiebel 1989 and Fig. 6–8). At a more complex level, selection for delayed flight in milkweed bugs produces correlated responses in juvenile hormone titers and in reproduction (Rankin 1978; see also Chapter 14). These examples demonstrate the importance of studying the proximate hormonal causes of gene action and of genetic correlations involving behavior and life history syndromes. It is perhaps surprising, therefore, that such studies are so scarce. Migratory syndromes should offer ample opportunities to manipulate both genes and hormones to further our understanding of how genomes relate to life histories.

In spite of the elucidation of some interesting generalities, studies of migration physiology are still hampered at two levels. First, there is a lack of a common conceptual basis for migratory behavior, with the result that it is often not clear what behaviors are in fact being compared. We have seen the problems raised by calling a number of behaviors "dispersal" in Chapter 2, and these problems sometimes surface in studies of migratory physiology. Thus, are the ambiguities of hormonal involvement in the movements of Belding's ground squirrels (Holekamp et al. 1985) the result of experimental difficulties or of the fact that the animals are perhaps ranging for vacant homesites rather than migrating? Similarly, much of the work on fat metabolism in locusts (*Locusta* and *Schistocerca*) assumes the gregarious form is the migrant, when application of specific behavioral criteria indicate that it is the solitary form that migrates (Farrow 1990 and Chapter 11). Second, there are still very few species that have been studied in any taxon, a point also made by Rankin (1991). The adumbration of apparent generalities is encouraging, but with so few species considered, it is best to interpret them cautiously. Finally, it is worth repeating that one of the biggest gaps in our knowledge concerns the direct actions of the central nervous system on migratory behavior and their mediation. The criteria that define migration as specialized behavior (Kennedy 1985 and Chapter 2) should set some guidelines and serve as a challenge to physiologists to uncover the underlying nervous mechanisms.

7

Biomechanical and Bioenergetic Constraints on Migration

In order for an organism to perform any movement it must have the apparatus to do so and the energy to drive the apparatus. So much is obvious. The situation becomes complicated, however, if one considers the fact that different types of movement require different types of structures for optimal performance. Among hawks, for example, the short rounded wings and relatively long tail of accipiters allow high maneuverability among the trees of a forest but are not very efficient for soaring. The bird watcher's field mark of several flaps and a short glide characteristic of accipiters flying in the open is the result. We have already seen in Chapter 5 how the long pointed wings of albatrosses are highly adapted to dynamic soaring on the steady winds and waves provided by the "Roaring Forties" in the southern oceans. Any long-distance movement, as frequently occurs during migration, also requires an augmented supply of energy. Frequently this is in the form of fat, as we have seen in Chapter 6, but the storage of large amounts of fat can change the body mass of a migrant dramatically with consequences for flight performance. In this chapter, I consider various biomechanical and bioenergetic costs and trade-offs that constrain locomotion as it relates to migration. These are not the only trade-offs, of course; energetic, genetic, and life-history constraints also enter into the migration equation. These latter are considered in Chapters 6 and 14. The term "constraint" has been somewhat loosely used at times, and there have been calls for a moratorium on its use (Antonovics and van Tiederen 1991). But the notion of constraints is central to evolutionary biology (Perrin and Travis 1992) because "evolutionary constraints are restrictions or limitations on the course or outcome of evolution" (Arnold 1992). It is these sorts of constraints that I deal with in this book.

Not all forms of locomotion are equally efficient. Flight is a far more effective means of transport than is either swimming or any form of terrestrial locomotion (Schmidt-Nielsen 1975). In terms of the energy required per unit body mass or unit distance, it takes far less energy to fly than to swim or walk. Swimming, although not as efficient as flight, still takes relatively less energy than terrestrial locomotion. With flight and swimming, organisms can also further increase their efficiency by using winds or currents (Chapter 5). All else being equal, we would expect more migration, or at least more long-distance migration, among flyers or swimmers than we would among those groups constrained to move over the land surface. In general this does seem to be the case, although some large animals like wildebeest in Africa or caribou in the Arctic move impressive distances. Within flying and swimming taxa, nevertheless, there is still great variation in all aspects of performance, including performance relating to migration. A further aim of this chapter, then, is to explore migration-related constraints within those groups. Before doing so, however, I shall consider the rather special case of migration that involves devices, using as examples the silken parachutes used by spiders and some lepidopteran larvae and the seeds carried in vertebrate guts.

The Constraints of Silken Threads

Two major groups of terrestrial arthropods have evolved transport mechanisms using silk threads, the spiders and several families of Lepidoptera. Both groups use silk for other purposes as well, spiders for webs and lepidopteran larvae for webs or cocoons, so its use in travel may be a consequence of its "capture" for an additional use. In lepidopterans, ballooning with silk is largely confined to the early-instar larvae of forest-dwelling species. In some of these cases, ballooning is a secondary mode of travel because the adults possess wings and are quite capable of travel themselves. Transport on silk threads is the sole means of movement over all but the most local distances in many other species of lepidopterans whose adult females are flightless. These, especially, are likely to be denizens of forests where the structure of the environment, notably the presence of large trees, which are physical barriers to movement and reduce wind currents, is apt to prevent larvae from moving very far on their parachutes. Simulation modeling based on empirical relationships incorporating wind, size of the organisms, and so forth indicate that less than 2% or so of the airborne larvae would be carried more than a few tens or hundreds of meters (Fosberg and Peterson 1986; Taylor and Reling 1986a; Roff 1991). This is still an important means of transport, if the alternative is virtually no movement at all. It is also possible that ballooning in lepidopteran larvae functions for true dispersal (via ranging?), in the sense of increasing the mean distance among individuals (Southwood 1981 and Chapter 3). These early-instar larvae feed mostly on the newly emergent leaf flush in the spring, and dispersal over the resource could be an effective means of reducing sibling competition in particular and intraspecific competition in general. The question of the adaptive value of ballooning in lepidopteran larvae could stand more investigation from this perspective.

In spiders as well as lepidopterans the effectiveness of ballooning will be determined by the presence of suitable takeoff behavior (Chapter 2), the mass of the individual, the diameter and length of the silk filament from which the animal is suspended, and the wind velocity and turbulence (Humphrey 1987; Roff 1991). The tensile strength of the silk sets an upper limit to the mass that can be transported, and Humphrey's analysis indicates that this limit is in the range of 8 to 32 mg. Using the mass/length regression for spiders given by Rogers and colleagues (1977), this translates into approximately 5.9 to 9.4 mm total length. In general, then, we would not expect aerial transport of spiders above 9.4 mm total length or 32 mg mass. Below this upper bound, the mass that can be transported by the frictional drag of the air will be proportional to wind speed and (approximately) to filament length (Humphrey 1987), and because of the proportionality between filament length and spider size, the required length increases rapidly as spiders become larger. The nature of the relationships among spider mass, filament length, and wind speed is illustrated in Figure 7–1.

Filament length is likely to be further restricted by the tendency for long filaments to become tangled in the vegetation, and Roff (1991) therefore concludes that a length of 2 m is likely to be an upper limit. Filament length may also be restricted by the energetic cost of producing silk, a cost that can influence social behavior (Jakob 1991). If 2-m thread length is a constraint, an upper limit to body size for effective transport of spiders will be on the order of 5 mm total length or

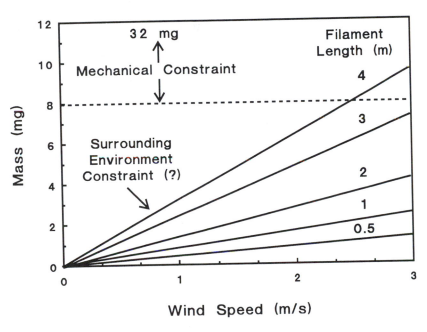

Figure 7–1. Relations between mechanical factors and the ballooning ability of spiders. The maximum wind speed shown is approximately that at which spiders cease to balloon under natural conditions. At about 2.5 m sec^{-1}, wind vectors shift from predominantly vertical to predominantly horizontal alignment. Because of possible entanglement with vegetation, a filament length approaching 4 m becomes an environmental constraint, although there is some question about this. The mechanical constraint refers to the maximal mass that can be supported by a filament, ranging from 8 to 32 mg. (Redrawn from Roff 1991 after Humphrey 1987.)

a mass of about 5 mg with an optimum considerably below this. Roff then compared the mean sizes of migrant and nonmigrant British spiders and found the former averaging 2.07 mm total length and the latter 4.89 mm, a difference that was highly statistically significant. Some larger spiders such as those in the genus *Pardosa* average 5 to 8 mm as adults, but balloon as immatures when their body lengths are around 2 mm. The figure of 2 mm may be an optimum size for spider ballooning. If the size distribution of all British spiders is considered, the modal length is around 1.75 to 2.25 mm (Fig. 7–2) indicating that natural selection has favored this size; ballooning constraints have probably been an important selective factor.

This latter notion is also supported by data on the size of aeronauts themselves (Fig. 7–3). Data for spiders captured with aerial tow nets in eastern Australia and sticky traps in Missouri show that the modal size of aeronauts is around 0.5 mg (or approximately 2 mm total length from Fig. 7–1) in both samples (Greenstone et al. 1987). The mass-frequency distributions of spiders collected from the vegetation at both sites also reflected a preponderance of small individuals, but did not show nearly the same degree of skew. Furthermore, spiders larger than 26 mg were common in the areas sampled, which they are not among the aeronauts. The largest aeronauts were 25.5 mg in Missouri and 19.1 mg in Australia, both well below the

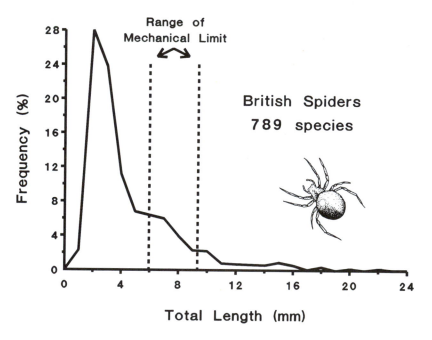

Figure 7–2. The size-frequency distribution of British spiders. The range of mechanical limit indicates a region above which the tensile strength of silk will no longer support the weight of the spider. (Redrawn from Roff 1991.)

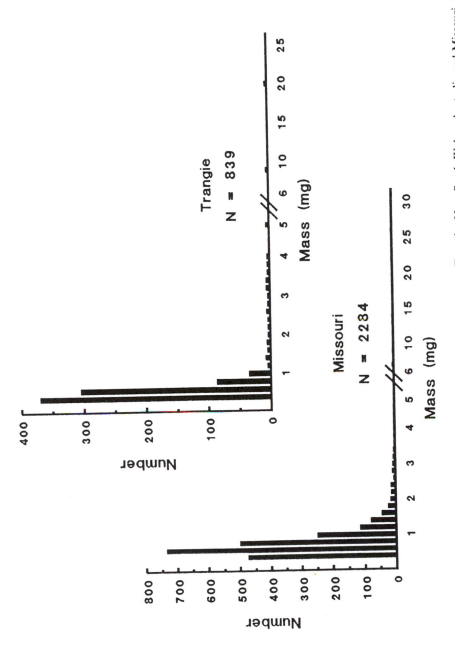

Figure 7–3. Mass-frequency distributions for spiders captured in aerial tow nets over Trangie, New South Wales, Australia and Missouri, U.S.A. (Redrawn from Greenstone et al. 1987.)

upper limit of 32 mg predicted by Humphrey (1987) on the basis of the tensile strength of silk. Theoretical considerations and the evidence available thus both indicate a biomechanical constraint on spider migration, with migration confined either to species whose adults are small (circa 2 mm total length) or to the juvenile stages of species that are large.

Seed Size and Fruits

Insects and spiders are not the only organisms that migrate with the aid of special devices that involve constraints. There are two good examples from plants. The first is the occurrence in many species of a parachute-like pappus that enhances transport by the wind, much like silken threads. Two familiar examples are the "parachutes" of milkweeds and common dandelion. As with ballooning spiders, seeds that rely on a pappus will also be size limited. The migrations of seeds with pappi are discussed in Chapters 12 and 13. The second example of devices to aid plant transport is the widespread presence of fleshy fruit, often rich in fats or sugars; these attract various vertebrates, which then carry the seeds away from the parent plant in their guts. We saw in Chapter 6 how many migrant birds may switch to energy-rich fruits during the period of premigratory fattening. An interesting case of constraints involving seed size and the behavior of fruit-eating birds is that of the sandalwood tree (*Santalum album*) of the Indian subcontinent (Hegde et al. 1991).

Sandalwood is a small deciduous tree well known for the high-priced aromatic oil derived from its heartwood. It bears single-seeded fleshy fruits (10–15 mm in diameter) that are dark purple when fully ripe and are fed on by the common koel (*Eudynamys scolopaceae*), a jay-sized bird of the cuckoo family. These birds swallow the fruits whole and either regurgitate the seeds or pass them through the gut. The seeds vary in size, and it turns out this size variability imposes a trade-off. Using total leaf area of seedlings as their measure, Hegde and colleagues found that large seeds were more successful at germinating and establishing than were small seeds. Large seeds, however, are characteristic of fruits that are less preferred by koels, and when they are consumed, the seeds are regurgitated not far from the parent tree. Fruits with smaller seeds are more preferred by the birds, and in this case the seeds are more frequently passed through the gut. The longer time spent in the gut also means the seeds are transported farther. From the perspective of transport efficiency, therefore, the advantage lies with the smaller seeds. The trade-off is imposed because the small seeds are less successful at germination and establishment. The optimal seed is presumably of some intermediate size. Fruit size itself seemed to have little influence on migration but may have an effect in other bird-transported fruits (Howe and Wesley 1988). Trade-offs in seed size between transport efficiency and seedling establishment also occur in wind-transported seeds (Ganeshaiah and Uma Shaanker 1991).

Constraints in Swimming

Swimming requires sustained motion through a relatively dense and slightly viscous fluid, a fluid that is denser than air. The generation of motion thus places relatively high energy demands on the swimming organism, which must produce thrust to create the required motion. In fishes this thrust is generated by oscillations of the entire body in the case of "anguilliform" or eel-shaped fishes or of the caudal fin in the case of "carangiform" fishes; that is, fishes with a distinct tail fin. In general, swimming efficiency is increased with a fusiform or torpedo-shaped body, which reduces the drag due to turbulence on the body and so reduces the parasite power (Table 7–1) and hence the energy necessary to overcome drag. In general, migratory fishes that travel long distances or against currents—for example, herring, tuna, or salmon—are fusiform in shape whereas nonmigratory fishes such as bass, sunfish, and many coral reef fishes assume other shapes, especially lateral flattening. Such generalizations, however, mask a great deal of variation and numerous exceptions.

Studies of the biomechanics and bioenergetics of fish migration are still very much in their infancy, but a few additional conclusions can be drawn from work in this still nascent field (Weihs 1984). Comparisons of body shapes show that the drag on the body of an anguilliform fish is quite high when it is swimming, being about four times as high as when the straightened body is simply towed through the water (Lighthill 1971). This suggests that selection in migrant Atlantic eels, for example, must maintain this shape in the face of relatively slow and inefficient swimming. Probably this selection comes from the need to maintain the ability to exploit a variety of habitats, and even to move across necks of dry land, during the freshwater phase of the life cycle, but the details, especially as they relate to the constraints of body shape and means of propulsion, remain to be analyzed. In contrast to anguilliforms, the drag increment in active swimming carangiform fishes is only about one and a half times that of the gliding drag, but because a smaller part of the body—the tail—is used for thrust, some sort of efficient propulsor is

Table 7–1. Definitions of Some Terms Used in the Biomechanics of Performance

Term	Definition
Induced power	Kinetic energy per unit time imparted to the medium to produce momentum
Parasite power	Power required to overcome drag forces on the body
Profile power	Power needed to overcome drag on propelling structures (wings, fish tail, etc.)
Wing (fin) chord	Width of the wings (fins) measured along the direction of flight (swimming) (variable along the span)
Wing (fin) span	Distance from tip to tip when wings (fin) are spread to the fullest extent
Aspect ratio	Ratio between span and average chord
Wing loading	Weight (not mass) per unit wing area (weight = mass + gravity)

SOURCE: From various sources, especially Pennycuick 1989 and Ellington 1991.

required. One such propulsor is the "lunate" or crescent moon-shaped tail. This form of the caudal fin has a large aspect ratio (Table 7–1), which increases the ratio of lift to drag up to a maximum value of around 5 to 10 and creates a very effective mechanism of propulsion. Its effectiveness is further increased by stiffening the fin so that it doesn't bend as it is oscillated back and forth. Again, many migrant fishes have highly evolved lunate tails, notably various species of tuna. Aspect ratio will be further discussed when I consider bird wings below.

Considerations other than migration also determine fish shape so that migrants do not necessarily assume optimum shape, as we have already noted in the case of Atlantic eels. In part to overcome constraints on shape and in part to reduce energy costs of swimming and migration in general, migratory fishes have a number of behavioral options available. One of these is to swim at the optimal speed, or the speed at which energy expenditure is least per unit of distance traveled. This value scales to fish size roughly according to the formula $S_O = kL^{0.43}$ (Weihs 1984) where S_O is the optimal swimming speed, k is a constant that has a value of approximately 0.5 if speed is measured in meters per second, and L is the fork length of the fish, measured from the nose to the point at which the tail forks. As can be seen, the optimal speed will be greater with larger fishes, but the gain increment declines at larger sizes. Because of the scaling of optimal speed with size, migration should be favored in larger fishes because with higher speeds they can travel farther with relatively less energy expenditure.

The situation becomes more complicated, however, if fecundity trade-offs are factored in. Roff (1991) defines a migration distance corresponding to the energy equivalent of the gonads as "fecundity equivalent distance" (FED, which for the average fish is given by FED [km] $= 52 \, GL^{1.03}$ where G is the ratio of gonad weight to somatic weight and L is again fork length). Like optimal speed, FED scales to size such that the potential trade-off between fecundity and migration distance decreases in larger fishes. Thus, for reasons of both optimal swimming speeds and fecundity, selection should favor larger migrant fishes. Indeed empirical data indicate this is the case at several taxonomic levels. Within species, for example, migrant stickleback (*Gasterosteus aculeatus*) populations are larger in mean size than nonmigrant populations (Snyder and Dingle 1989) and among the Order Gadiformes migratory species are on average significantly larger than nonmigrants (Roff 1991). What are needed now are the kinds of detailed analyses that will sort from among the not mutually exclusive theories predicting a positive relation between migration and body size.

Aside from optimal swimming speed, there are other possible advantageous behaviors. We saw some of these in Chapter 5 in the examples of the use of currents and selective tidal stream transport. A measure of the gain realized from using tidal stream transport is seen in plaice, which are flatfish essentially without adaptations for efficiently powered swimming. In the North Sea, however, they make an annual spawning migration of 280 kilometers. Swimming the whole distance would require about 10 grams of fat, but this is reduced to about 3.5 g by selective tidal stream transport (Harden Jones 1981). Another means of reducing energy costs is by schooling. As a fish swims it induces a velocity in the water immediately behind it, opposite to the direction of swimming, so that another fish behind it would be swimming "upcurrent." Diagonally behind it, however, a forward velocity is induced in the water, and a fish in the diagonal position would be "riding" this forward velocity and require less effort to swim than a solitary individual (Weihs 1984). It is therefore not surprising that migratory fishes and indeed fishes in general move in schools. Schooling can also have other advantages such as protection of the individual against predation, as predicted by so-called "selfish herd" models (e.g., Krebs and Davies 1984). Finally, migrating fishes frequently swim close to the bottom where the bottom boundary layer of water moves at reduced speed because of friction at the interface, a factor that can be of considerable importance in migrations up high-velocity streams and rivers. Body shape and propulsion devices are important considerations in the analysis of the evolution of migration, but they are not ultimately limiting. Rather, they are but two of the terms that must be entered into any trade-off equation for migratory performance. Other terms include a variety of ecological and behavioral variables relating to all aspects of the lifestyle. These sorts of factors have been considered in far greater detail in birds than in fishes or other aquatic organisms, as will be discussed in the next sections.

Constraints on Flight

Migration in birds, bats, and most insects depends on flight, and flight performance is subject to constraints. These have been examined in several excellent reviews (Rayner 1988, 1990; Kerlinger 1989; Pennycuick 1989; Alerstam 1990; Ellington 1991; Welham 1994), and I have drawn heavily from these in the following account. A good place to start a discussion of flight and the constraints placed upon it by migration and other factors is a consideration of what it takes to become and remain airborne. The force required to raise a body in a fluid and overcome gravity is known as *lift*, and a body of suitable design will be influenced by a lifting force if placed in a current. A design that accomplishes this is one that forces the fluid to pass more rapidly over the upper surface than over the lower surface, at the same time directing the current downward slightly or in other words, a wing or airfoil. Bernoulli's law states that the sum of kinetic pressure and static pressure around a body is constant. When the velocity of the current and so its kinetic pressure increases, the static pressure drops and vice versa. Air moving over the upper surface of a wing moves faster than over the lower surface (Fig. 7–4) and decreases the static air pressure above relative to the underside of the airfoil, generating lift. The lower pressure above the wing contributes about twice the lift of the higher pressure below so that a plane or a bird is actually suspended in the air rather than resting on it. Note that lift does not have to be directed upward, but occurs at right angles to the incident flow (cf. fish tails above).

Another way to assess lift is by describing the localized rotational air

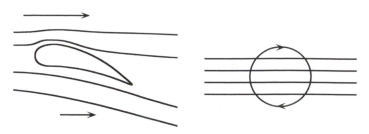

Figure 7–4. Relation between wing shape and lift. A wing develops lift because the downward deflected air under the wing is slowed down relative to air moving over the wing. This can also be represented as a bound vortex attached to the wing and superimposed on the undisturbed flow (*right diagram*). (After Pennycuick 1989.)

movements or *vortices* created by the airfoil as it moves through the medium. This vortex system forces air to move downward behind the wing, and the reaction experienced by the wing to this downward momentum is lift. The vortex around the wing is known as a *bound vortex,* and the development of lift by this vortex can be expressed diagrammatically as the right panel in Figure 7–4. Behind the wing there is always a wake of trailing vortices with the same circulation pattern as the bound vortex; in the right conditions these form the "vapor trails" of high-flying aircraft. Jeremy Rayner and his coworkers have developed ingenious methods, using helium-filled soap bubbles and high-speed cinematography, to visualize the wakes of flying birds. In general the vortices generated in slow flight are a series of large circular rings or "donuts," while those of fast or cruising flight undulate in a concertina fashion, although this oversimplified statement masks some very complex processes. The concertina wake is generated by shortening the flapping wing during the upstroke, and it reduces the energy cost of flight because no transverse vortices (which require energy) are formed. This concertina type of wake formation may be partly responsible for strong performance in long-distance migration, but with the trade-off that energetically costly ring vortices must be formed during slow flight.

In order to move through a medium an organism must generate power (= energy consumption per unit time). A flying bird must, first, provide power to accelerate the air downwards. This *induced power* (see Table 7–1 for definitions) increases with weight—obviously it takes more power to move heavier bodies—but decreases with flight speed, wing span, and the density of the air. But just moving through the air is not the end of it for bird flight. Movement generates friction on the bird and turbulence around it, and both of these factors cause drag. The power used to overcome drag on the body is called *parasite power* (because the body is only being carried along) and that to overcome the drag on the wings is the *profile power.* Both of these tend to increase as flight speed increases. The sum of the three power outputs is the total power needed to propel the bird through the air, a value which of course also varies with weight, speed, and air density. The relationships among the three kinds of power and the total are illustrated in Figure 7–5, which shows the estimated flight performance of a barn swallow (*Hirundo rustica*) according to a model generated by Rayner. In fact the model is likely not to be very accurate at low speeds because of differences in wingbeat kinematics at these low speeds. Those few instances where metabolic power has been measured also suggest flattening of the curve as speeds decline, rather than a rise contributing to a U-shape. These metabolic data and other considerations are discussed in Ellington's (1991) excellent critique of the various models.

On the total power curve the lowest point, at which flight speed requires the minimum power, is designated V_{mp} (velocity at minimum power output). At first glance this might seem the optimum speed for migration because it conserves power and hence fuel, a prime consideration in long-distance flights. The problem, however, is that flight speed at V_{mp} is likely to mean that a bird takes a long time to reach its destination, so using more fuel, not less. By drawing a line from the origin tangential to the total power curve, we obtain the speed at which the ratio of power output to distance flown or V_{mr} (mr = maximum range) is minimum, and this may be a better predictor of flight speed during migration (Fig. 7–5). Radar

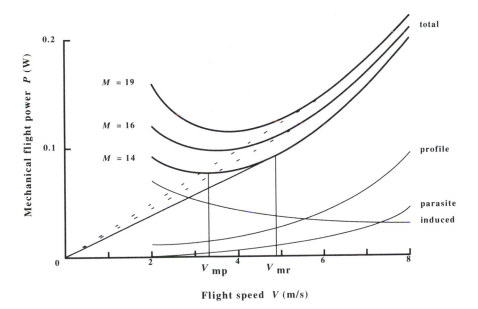

Figure 7–5. The relation between mechanical flight power, P(W), and flight speed for birds weighing 14, 16, and 19 grams and having wingspan and area approximating that of a barn swallow (*Hirundo rustica*) showing also profile, parasite, and induced drag components. The tangent of the lines from the origin to the curves indicate the maximum range speed, V_{mr}, which is also the minimum cost of transport. V_{mp} is the speed at minimum power output. Note that V_{mr} rises with increased mass, but so does the cost in terms of P(W). (After Rayner 1990.)

studies of flight speeds (Alerstam 1990) suggest that V_{mr} is the speed chosen, at least in the spring when there is selective advantage to returning as early as possible to the breeding grounds to establish territories and begin breeding. Smaller birds have a surplus of power available for flight (due to a large difference between V_{mp} and V_{max}, the maximum flight speed) and many use this "extra" power to fly faster during migration, probably to minimize time in transit (Lindström and Alerstam 1992; Welham 1994). Autumn flights may be slower, unless the birds are crossing major barriers like deserts or open water.

Scaling relationships also influence the generation of power in large birds (Rayner 1988, 1990; Ellington 1991). Mechanical flight power increases with mass as $M^{7/6}$, a more rapid increase than resting or feeding metabolism, which scale at $M^{3/4}$, so that large birds face tight constraints on the amount of energy they can devote to flight. In general they have responded to these constraints by evolving relatively longer wings, by spending less time in flight (and in the case of rattites and some island rails perhaps becoming flightless partly for this reason), or by spending high proportions of their flight time gliding or soaring. It is worth noting in this context that the majority of migrants are small birds. Also because mass-specific power scales negatively with size, very large birds may not be able to generate the power to reach V_{mr}; they may thus be confined to slower speeds as diagrammed in Figure 7-6. Radar tracking of migrating whooper swans (*Cygnus cygnus*) shows, for example, that they do not fly at their predicted V_{mr}, quite possibly because for them this flight speed is outside the range of their maximum power output (Alerstam 1990).

As indicated, one way for large birds to "defeat" the constraints imposed by size on power is by soaring. We discussed the gains made by gliding and soaring cranes in Chapter 5 and noted that they had a glide ratio, the ratio of gliding speed

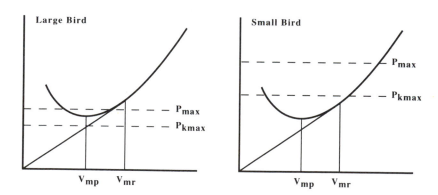

Figure 7-6. Relations between mechanical flight power (ordinate) and flight speed (abscissa) for a large bird versus a small bird to indicate consequences of scaling. P_{max} the power available for bursts of flight, including anaerobic flight, and P_{kmax} indicates the power available for continuous or cruising flight which is aerobic. Note that a large bird may be restricted to speeds below V_{mr} or maximum range speed.

to speed of descent, of about 16:1 so that they gained considerably in energy conservation by soaring on and gliding between thermals. The best gliding properties are associated with long, narrow wings of high aspect ratio. Not only do these sorts of wings provide a great deal of lift, but the pointed tips are able to shed tip vortices that create drag because they must be pulled along by the bird (Kerlinger 1989). As a result profile power is reduced, making these more efficient wings as well. The quintessential soaring and gliding birds are albatrosses, which attain the highest glide ratios of any animal at around 24:1.

Small birds do not glide, or at least not very much, and they do not routinely glide during migration. The reason is shown elegantly in Pennycuick's (1989) calculations comparing flapping and gliding performances of a 5.5-kg crane with a 17-g ovenbird (*Seiurus aurocapillus*). If it soared in thermals with air rising at 3 ms^{-1}, an ovenbird's cross-country speed during soaring and gliding would be about 60% of its V_{mr}, about the same percentage as for a crane although the actual speeds are less than half those for cranes. The problem for the ovenbird is that it requires 63% as much energy per kilometer when soaring in thermals with this rate of rising air as when in flapping flight at V_{mr}, as contrasted with only 9% as much for a crane. With a thermal of 1 ms^{-1} the ovenbird actually requires 6% *more* energy to soar relative to flapping flight because basal metabolism is so much higher a fraction of total energy requirements in a small as opposed to a large bird. The reduced speed that is the consequence of soaring and gliding means the ovenbird spends up to three times as long covering each kilometer, and the extra time simply uses up too much fuel in basal metabolism. Any fuel saved by gliding is negated.

The best design for gliding is one that combines high aspect ratio with low wing loading. In contrast more rapid gliding and flying is possible with higher wing loadings, but the latter reduce range. Birds that fly far and/or spend a great deal of time soaring indeed do have high aspect ratios, and they tend to be large (Figs. 7–7 and 7–8). There will be much less selection for high aspect and low loading in small birds; compare, for example, the relatively large body outline of the thrush in Figure 7–7 with the large soaring birds (the tern is a somewhat special case, to be discussed below). One group of birds, the waterfowl, is characterized by high aspect ratio and short wings with high loading. This group is clearly specialized for migration, and Rayner (1988) suggests that the short, relatively narrow wings represent a compromise between migration economy and flight speed, especially in arctic ducks, which have particularly short wings and may need fast flight for movements in adverse weather and high winds. For whatever reason, there is a significant positive interspecific correlation between relative wing loading and breeding latitude in Arctic ducks.

A particularly interesting adaptation to migration occurs in small birds that put on large amounts of fat to sustain the migratory journey (Chapter 6). In some cases, a migrant will double its lean mass. As the mass of the bird changes over the course of the migratory journey so will the balance between the power output by the flapping wings and the power input by the flight muscles. A wing is a fixed

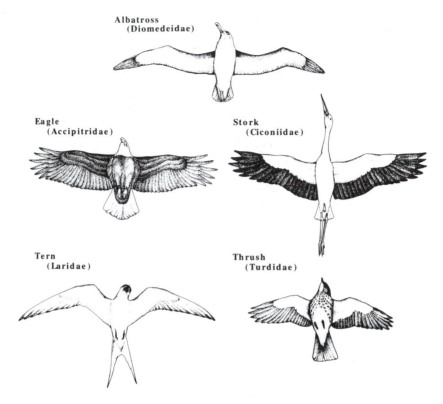

Figure 7–7. The profiles of some birds in flight as viewed from below. All are drawn to the same scale. Compare especially the high aspect ratio wings of a soaring bird like the albatross with the rounded wings of a thrush, which have been selected to promote maneuverability. (Drawings by Mitchell Baker.)

structure optimized to output a given level of mechanical power that, because of the changing mass, cannot be at the best balanced level throughout the long flight. One way to solve this would be to evolve a set of muscles that could provide a range of power. Although not impossible, this would be difficult, because in essence it would be asking natural selection to solve an equation with a number of variable terms representing annual and shorter term variation in the amount of fat deposited, vicissitudes of the winds along the route, and so forth. It would also require flight muscle of several fiber types.

Instead, the solution, it turns out, is behavioral (Rayner 1988, 1990). The birds insert pauses between bouts of flapping so that the mean power output averaged between flapping and pausing is at the required level, and the birds travel with a characteristic undulating or "bounding" flight. The flight muscles are all of a single fiber type and deliver a constant quantity of energy with each contraction. For

Figure 7–8. A Cape gannet (*Sula capensis*) approaching a landing at a nesting colony near Capetown, South Africa. Note the long wings of high aspect ratio typical of pelagic seabirds. (Photograph by Mitchell Baker.)

reasons of scaling the behavior is restricted to smaller birds, and it is confined to species with wings of low aspect ratio. It is the typical flight pattern of many small passerines and also turns up in some relatively small nonpasserines such as hummingbirds, the smaller kingfishers and owls, and some woodpeckers. The selection driving the evolution of bounding flight may have been from migration or as a way of flexible habitat exploitation (or both), but in any event, the behavior seems to be all but essential to cope with at least the early stages of migration when there is a full fat load. Those species such as swallows, swifts, and woodswallows, which hawk on the wing during migration, do not show the very high mass increases due to fat deposition (e.g., Fig. 7–4) and do not display bounding flight. Although the total energy costs of migration are not reduced by bounding, the behavior permits a greater fuel capacity with a consequent greater range than would be possible with steady flapping flight.

A behavioral response that can increase energy savings during flight is flocking. Because air is driven downward by the wings in flight, there is a compensatory movement of upward-moving air off the end of a bird's wing (cf. also fish schooling above). This will be strongest in the vortices that form just off the wingtips (Alerstam 1990), and it means that birds flying close to each other can in effect gain some free lift (almost 40% even in just three birds flying wing tip to wing tip) by flying in the upward airflow. On the other hand, they would lose lift if they flew either directly behind, above, or below other birds in regions where the air movement is downward. These considerations largely determine flock formations in the absence of other factors such as predation. The optimal flock shape from the perspective of energy savings is a horizontal "V" or bow shape. These formations, with birds flying just off the wing tip but just behind or in front of their neighbors, allow flight without actually touching wing tips (which would have unhappy consequences), but with their near overlap to take maximum advantage of the updrafts just ahead and behind the distal edge. The more drawn out horizontally the flock, the greater the energy saving. There will be a slight trade-off because the optimal migration speed (V_{mr}) will be reduced by the reduction in the induced power requirement. In any case many migrants, especially the larger waterfowl, do travel in V- or bow-shaped flocks, suggesting benefits to the energy-saving behavior. (The "leader" of the flock does not gain, which may be why leadership frequently changes.)

Turning now to insects, the relation between body size and migration has been examined by Roff (1991). The limit to the distance an insect can fly on a given fuel load is given by the equation

$$\text{distance} = \frac{\text{flight speed} \times \text{stored calories}}{\text{metabolic rate during flight}}$$

Flight speed increases with body mass raised to the power 0.38, and metabolic rate increases as mass to the power 0.84 on average for those insects on which there are measurements. For reasons of geometry the stored calories can be assumed proportional to body mass. Combining these various relationships gives the equation

$$\text{distance} \propto (\text{body weight})^{0.54}$$

which says that the distance an insect can fly, all else being equal, increases as a function of body size. Insects, as we have seen in Chapter 5, are masters at taking advantage of winds to facilitate migration, so size will place no absolute limits on distance of travel. Still, it should in general be the case that migrants on average will be larger than nonmigrants, especially in those groups that rely heavily on powered flight. For very small insects, where wind transport assumes a more important role, there is unlikely to be a relation between size and migration. Also the prediction is that migrants will be larger, not that large insects will be migrants.

The prediction that (large) migrants will on average be larger than nonmigrants was tested by Roff (1991) by comparisons both among populations and species and within populations. For British butterflies, for example, the wingspan of migrants averaged 15 mm wider than for nonmigrants. The larger migrants also tended to grow faster, as would be expected of species in more ephemeral habitats. Migratory milkweed bugs, *Oncopeltus fasciatus*, are larger than nonmigratory species confined to the Tropics and within *O. fasciatus* the migratory Iowa bugs are larger on average than the nonmigratory population from Puerto Rico (Dingle et al. 1980). Furthermore, selection for increased wing length, which also resulted in larger size, increased the proportion of migratory individuals in the Iowa but not the Puerto Rico bugs (Palmer and Dingle 1986; Dingle and Evans 1987; and Chapter 15). Comparisons within populations, however, are equivocal. Roff cited examples where the relation between size and migration seemed to hold, as it did in the selection experiments in milkweed bugs, but Rankin and Burchsted (1992) point out that some of the studies in question may not have been dealing with migrant insects. The latter authors also failed to find a relation between size and migration within populations of four species of insect they tested. Although several reasons are possible for differences in results concerning within-population correlations between flight and size in insects (differences among populations, differences in methods, etc.), the issue is at present without resolution.

Ecology and Wing Morphology

Another aspect of the notion that migration is but one of many factors influencing a morphological adaptation such as body size is that a particular adaptation cannot favor all aspects of morphology and behavior simultaneously. Nowhere is this more true than for wing morphology and flight performance (Rayner 1988), so that compromises among various aspects of shape and size will be the inevitable result. General trends, however, are still likely to be apparent, and these can be seen in the shapes of the birds illustrated in Figure 7-7. First, some general principles should be noted. Low cost of transport, for example, is associated with high aspect ratio, especially if combined with low wing loading to reduce required power. Reducing aspect ratio and increasing wing loading increase power requirements, but also increase flight speed (cf. water birds above). Speed can be increased behaviorally by flexing the wings as seen in the "power glides" of peregrines and other hunting raptors. Maneuverability is enhanced by short rounded wings and a long broad tail that can be fanned to give added lift or twisted to provide ruddering. The latter characters are common in small birds that live in dense cover such as reedy marshes or thick scrub.

With these general principles in mind, we can ask what determines the shapes of the birds in Figure 7-7. It is apparent immediately that the two marine birds, the tern and the albatross, are characterized by long pointed wings of high aspect ratio; the same is true of the gannet in Figure 7-8. These are undoubtedly adaptations for continual soaring by these birds, which can fly for sustained periods of weeks, months, or even years in the case of the long prebreeding period in albatrosses. Even much smaller birds that spend a great deal of time on the wing, such as swifts and swallows, have this characteristic pointed wing shape. Both terns and albatrosses make long-distance migration flights with the prevailing winds, and because they can obtain food more or less at any time while migrating, they are under little pressure to evolve high-speed flight (Rayner 1988). Even terns that breed in inland areas do so on littoral marshes, and so like their marine cousins and the albatrosses (and other marine or littoral birds like gulls), they experience no environmental constraints on wingspan and wing shape. Because biomechanical considerations probably predominate, long, pointed, aerodynamically efficient wings are the result.

A very different set of considerations applies to the thrush. Thrushes spend much of their time on or near the ground, often in relatively thick cover or in semi-open areas near cover, the sort of habitat found in forests and woodlands where they are common. Under these conditions selection will favor maneuverability, and the relatively short rounded wings and comparatively large tail are the sorts of shapes that should be optimized. Many thrushes are migrants or partial migrants, but because they are small birds there would be no selection for soaring efficiency with accompanying long wings of high aspect ratio. The balance of selection is therefore in the direction of traits facilitating maneuverability in cover.

The situation regarding storks and eagles, the final two birds in Figure 7–7, is less clear. On theoretical grounds a high aspect ratio wing should be preferable for terrestrial soaring as it is for marine soaring (Pennycuick 1972a,b). Yet terrestrial birds, such as hawks, eagles, storks, vultures, and pelicans, consistently have broad wings with square tips and separated primary ("slotted") feathers, rather than the high-aspect pointed wings of marine species. Although there is more speculation than resolution regarding the difference, some ecological and behavioral factors distinguishing marine and terrestrial species are obvious. Terns and albatrosses, for example, do not have to maneuver around and land on cliff faces and castle walls or peaked roofs the way eagles and storks, respectively, do. Both terrestrial species must also on occasion move among trees. Eagles pursue prey in flight; storks do not, and this may account for the difference in tail shape. The very short tail of the stork provides little in the way of steering ability. Variation in wing shape is also related to roosting habits and takeoff patterns, but the exact mechanisms are not very clear (Pennycuick 1983; Rayner 1988). The slotting of the wing tips seems to help with the shedding of vortices and thus aerodynamic performance, but there is some controversy about this (Kerlinger 1989). In any event both storks and eagles are very capable at thermal soaring, and those species that migrate, such as the white stork and several species of eagle, can and do take advantage of this ability (Kerlinger 1989).

The discussion of eagles and storks makes obvious the fact that the wings of migrant birds are used not only for flight during migration, but must be used during other periods of a bird's life to meet other demands of the lifestyle. Birds fly to forage, to establish territories, and in some species to perform aerial displays. Migration and these other behaviors may well place constraints each on the other, and the nature of such constraints has been examined in considerable detail by Bernd Leisler and Hans Winkler (Leisler 1990, 1992; Leisler and Winkler 1985, 1991; Winkler and Leisler 1985, 1992). They have examined small passerines in particular because aerodynamic constraints on small birds are relatively weak (Rayner 1988), and ecological influences on wing shape and other aspects of morphology are likely to be obvious. They used multivariate analyses to examine the relationships among migration distance, habitat characteristics, and 32 morphological characters that included measurements on wings, tail, bill, legs, and feet.

Results from multiple regression analysis using both Old World (Sylviidae) and New World (Parulinae) warblers are shown in Figure 7–9 (Winkler and Leisler 1992). Stepwise regression was used to search for combinations of characters that provided parsimonious predictions of migration distance. For the Old World sylviids the best set with fewest traits was a combination of the number of slotted primary feathers in the wing and the length of the carpometacarpus, the bone at the distal end of the wing. The regression is shown in Figure 7–9, and some 83% of the variance in distance is explained by the two morphological traits in question. Longer distal portions of the wings (the carpometacarpus) are associated with greater migration length, but slotted primaries entered with a negative coefficient so that longer migration was associated with less slotting.

To see if the results from sylviids could be extended to other groups, stepwise regressions were run with a set of five variables that formed a common subset among the sylviids. These variables were tail length and bill width plus three skeletal characters, femur, ulna, and carpometacarpal lengths. Of these the skeletal characters explained 81% of the variance in sylviid migration with 76.4% explained by the two wing bones alone. The regression equation associated increasing migration distance with a lengthening distal portion of the wing and a shortening proximal portion represented by the ulna. Applying the sylviid regression equation to parulines revealed a good correlation between predictions generated and the actual migration distances recorded; however, many predicted distances were negative, which means that the predictions can only be applied to within-group rankings. Performing a multiple regression using these three traits within the parulids did not improve the regression but did change the intercept (Fig. 7–9). These three characters explained 54% of the variance in migration distance in parulids. This suggests convergent evolution with respect to migration distance between Old and New World warblers.

Winkler and Leisler then extended the analysis further to address directly the question of constraints. To be a candidate for the action of a constraint, a single character or a multivariate measure of a suite of characters had to be of functional significance in at least two ecological contexts, in this case migration distance and mean vegetation height of the breeding habitat. A constraining feature was defined operationally as one that showed simultaneous correlations with both ecological variables. Of the 32 characters examined, two turned out to be constrained in sylviids according to these fairly stringent criteria. These two were aspect ratio, for which migration distance and vegetation height explained 88% of the variance (migration distance contributed somewhat more), and femur length, in which 62% of the variance was explained (vegetation height was slightly more important). What this says is that migration sets an upper bound—that is, constrains—femur

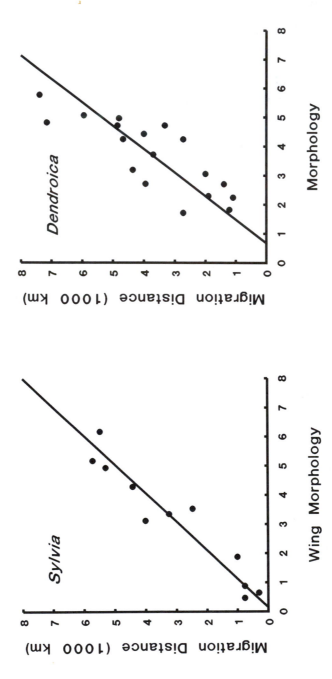

Figure 7–9. Migration distance as a function of wing morphology in Old World (*Sylvia*) and New World (*Dendroica*) warblers. The abscissa indicates a gradient of shorter, more rounded, more slotted wings near the origin to longer narrower wings. (Data from Winkler and Leisler 1992.)

length. Birds that need to cling strongly to perches tend to have shorter femora (Leisler et al. 1989) and require more force and hence muscle mass for clinging. Migrants have longer femora with less muscle mass and should be, therefore, less likely to exhibit this sort of behavior; and this in fact has been observed to be true.

An interesting picture also emerges when multivariate associations among characters for migrants and nonmigrants are examined with respect to migration and habitat usage. For example, Leisler and Winkler (1991) used principle component analysis (PCA) to look at 12 wing and tail characters in 11 species of Old World warblers in the genus *Sylvia*. These birds generally occupy habitats that are shrubby in scrub, thickets, or forest edges with dense low trees and brushes. Under these conditions maneuverability is at a premium, and selection should favor shorter, more rounded, and slotted wings and longer tails, traits that allow movement through a cluttered habitat. Migration, on the other hand, with its high demands for more efficient flight, should select for longer, narrower wings without slotting and shorter tails. The plot of the first two components of the PCA (Fig. 7–10) shows the relationship among these traits in the 11 sylviids. At the lower right of the plot are two mostly resident species, the Dartford and Sardinian warblers, that show in extreme form the morphotype of short, round, slotted wings and long tails, so much so that they are often referred to as having a "magpie type" morphology because they exhibit a characteristic magpie-like body shape (Leisler, personal communication). The Sardinian warbler is a resident of the islands of the western Mediterranean and the Spanish east coast; the Dartford warbler is mainly resident in Spain and Italy, with some northward extension of its range in the west into northwestern France and the very southern coast of Britain. Both species are characteristic of particularly dense habitats of scrub, maquis (chaparral), and for the Dartford warbler in the northern part of its range, heather. At the upper left of the plot are four species, all of which are long-distance migrants from Africa to Europe; some western populations of the blackcap are also migratory within Europe (Chapter 12 and 15). These species are characterized among this group of warblers by their longer, narrower (short tertials) wings of high aspect ratio and relatively high wing loading (which increases speed) along with relatively short tails. All these characteristics would be expected of migrants. The other five species in the plot are more or less intermediate in both migration and morphology between the more extreme groups.

The behavioral use of habitat by migrants and nonmigrants was examined in a group of eight species of chat, which are open-country birds belonging to the thrush family (Leisler 1992). Leisler compared the behavior and morphology of species that migrate between Eurasia and Africa with those that are residents within Africa (Fig. 7–10). All these birds exploit similar types of habitat where they were studied in Kenya, but the migrants were somewhat more opportunistic in their feeding, and they foraged differently. These differences show up in the PCA of 15 behavioral and 25 morphological traits. At the lower right in Figure 7–10 are three especially long distance migrants from Africa to Eurasia that are characterized by longer, thinner wings of higher aspect ratio, but long tails and bills. These latter

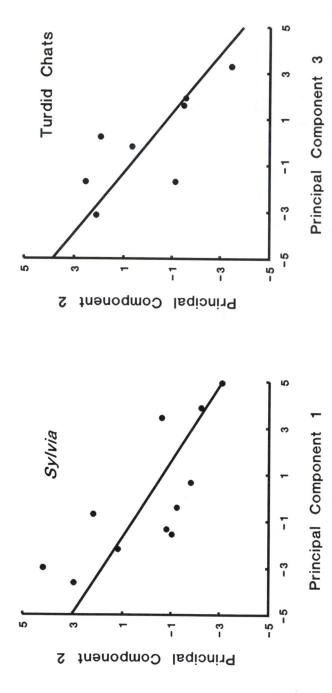

Figure 7-10. Principal component analyses of morphology in two groups of birds, Old World *Sylvia* warblers and Turdid or thrush-like chats wintering in Africa. The *Sylvia* graph plots a component indicating shorter rounder wings with many notched primaries and longer graduated tails (*abscissa*) against a component indicating unslotted wings of high aspect ratio and wing loading, but shorter tails (*ordinate*). Points to the upper left represent more migratory species. In the chats, PC 2 includes behavior with increasing values indicating slower, less sallying, more sedentary behavior while PC 3 is a measure of morphology with increasing values indicating more pointed, longer wings and a longer tail. The three left-most points are African residents, while the rest are migrants. (Data from Winkler and Leisler 1991 and Leisler 1992.)

two characters reflect the fact that these species forage more by sallying flights (longer wings), which also requires maneuvering (longer tails). The primarily resident species have shorter wings, as expected, and this also translates into less sallying and fewer flights with as a result less demand on maneuverability in flight. It thus appears that migration and foraging method interact strongly to shape the ecomorphology of these species. An interesting exception seems to be the rock thrush, which is a long-distance migrant but falls in the upper left of the figure. This species on its breeding grounds is a bird of rocky, dissected high-mountain slopes where it moves and forages in and out of crevices. It may be that in this kind of habitat sallying and pouncing are less adaptive behaviors than in more open areas, and this takes selective precedence over adaptiveness for migration. Finally, the more flexible feeding of migrants is also associated with less aggression. In interspecific encounters, residents without exception dominated migrants even when the residents were smaller; evidently a resident's more restricted foraging strategy also requires the ability to hold one's territory in the face of challenge from migrants.

As with size and migration in insects, within-population comparisons may fail to reveal a correlation between migration and morphology. Such is the case with the North American dark-eyed junco, *Junco hyemalis* (Mulvihill and Chandler 1990). In this species females migrate farther than males (Ketterson and Nolan 1983 and Chapter 12), yet it is males that are larger and have longer wings, a difference that persisted even when corrected for the size difference between the sexes. Males also had more pointed wings, again contra expectation based on migration distance. Considerations other than migration routes apparently dominate in selection for wing length in juncos, and in fact there may not be selection for longer migrations in female birds per se, but rather intraspecific competition may be forcing the females to adopt a "best of a bad job" strategy (discussed at length in Chapter 11).

In flight, migratory or otherwise, the wings must of course be driven by muscles that supply the power. In birds the thoracic musculature is dominated by two muscles, the pectoralis and the supracoracoideus. The former is the larger and provides the power to depress the wings. The supracoracoideus has a unique arrangement in birds; it is located beneath the pectoralis, and its tendon passes over a pulley-like arrangement and through the foramen triosseum (the canal formed at the meeting point of scapula, clavicle, and coracoid) to insert on the *dorsal* side of the humerous (Pennycuick 1972b). This muscle was long thought to raise the wing, but recent work on aerodynamics indicates things may not be so simple, with a reassessment needed (Rayner 1988). In any case flight muscle mass may be the best indicator of flight capability. Comparative studies of the ratio of flight muscle mass to unladen body mass suggest this may indeed be the case (Marden 1987; Ellington 1991). When takeoff ability was marginal, this ratio was around 16–18%, just enough for a standing takeoff. Interestingly this was true not only for birds, but also for bats and insects. In birds this low ratio was found in weak-flying aquatic or semi-aquatic birds such as rails (which also tend to be flightless on islands). The highest values were found in birds that make quick takeoffs requiring a burst of power, such as pigeons, grouse, or sandpipers. There seems to be no obvious correlation with migration, however. Many rails are proficient migrants and many grouse are completely resident. Strong flight per se thus does not seem to be a requirement for migration; too little is known of the biochemistry and physiology of flight muscle to understand adaptations for endurance (Rayner 1988).

In bats the arrangement of the flight muscles differs in that the elevators of the wing have their origins dorsally on the scapula which is large relative to birds. Bats also have only a small sternal keel. This arrangement is quite as efficient as that in birds (Rayner 1988; Ellington 1991), so that bats seem to be just as good flyers. Relatively fewer bats are migrants than are birds, and it has been argued that this is in part because (1) the flight muscles are apparently well adapted for slow, maneuverable flight, but not the high-speed flight more suitable for migration, and (2) the wings are thin airfoils of high camber, producing high lift at low speeds but creating drag at high speeds (see Dingle 1980 for a more complete summary of the arguments). Bats do have somewhat lower muscle-mass–specific induced power than comparably sized birds (Ellington 1991), but in view of the lack of correlation between migration and relative muscle mass in birds, as noted above, it is hard to see how this could limit migration capability in bats. The reasons for few migrants among bats must therefore be sought elsewhere, probably in their ecology. It is perhaps worth noting in this regard that among potoos, nightjars, and frogmouths— birds that like most bats capture relatively large insects at night—there are also few migrants (I found 9 out of 96 species).

We can summarize this section on ecomorphology and migration by noting once again that adaptation for migration must compete against other selective trends. Evolutionary compromises will clearly occur and in many cases morphological adaptations to feeding mode or other aspects of the ecology may supersede adaptation for migration. Migration does not in fact seem to impose the extreme selection on structure that it was once thought to do. At least insofar as flight is concerned, once efficient flying ability evolved, migration seems to be well within performance capability, with further morphological modification simply enhancing that capability relative to other demands on flight.

Special Problems Faced by Migrants

Migrants must often deal with situations that place particularly high demands on them in addition to the already severe demands of long-distance travel. Many European bird migrants to Africa, for example, must cross the Alps, the Mediterranean, or the Sahara with different routes crossing one, two, or all three of the barriers. On the other side of the Atlantic, some migrant birds in the autumn migrate out over the ocean from the northeastern coast of North America around and over Bermuda to first landfall in the West Indies or even the northern coast of South America (Chapter 5). In both spring and fall, migrants make an overwater crossing of the Gulf of Mexico. Favorable winds may assist these crossings, but considerable energy resources are still required to successfully complete them. Insects migrating to diapause sites must face long periods without feeding and, in northern latitudes, often extreme cold. If they emerge prematurely in a warm early spring, they run the risk of death by freezing if there is a subsequent cold snap. Thus, the diapause phase of a migration-diapause cycle can also act as a "barrier" in a migration system. In both barrier crossings and diapause periods (and other special problems) we need to ask what sorts of adjustments migrants might make to the stresses imposed.

This question has been addressed by Frank Moore (1991; Loria and Moore 1990) with respect to small migrant birds transiting the Gulf of Mexico in the spring. Three species were studied: two thrushes, the wood thrush and veery, and the red-eyed vireo. All three expend a great deal of energy in crossing the Gulf, as indicated by the fact that about half the birds captured on the coast of Louisiana after completing the crossing had almost no externally visible signs of fat. By capturing and weighing birds, and then recapturing and weighing them again after they had spent up to several days at the site, Moore demonstrated that there was a negative correlation between mass at first capture and subsequent rate of gain in mass. Although only 6% and 19% of the variance in gain was explained for wood thrushes and veeries, respectively, the correlation was statistically significant and indicated that lighter, leaner birds were feeding more and putting on more mass, presumably in the form of fat for the continuing journey. It was also the case that in general leaner birds tended to remain longer, while fatter birds continued the migratory journey with little stopover time.

Feeding was studied by direct observations in red-eyed vireos. Loria and Moore (1990) predicted that lean vireos would adjust their foraging behavior to compensate for increased energy demand by being more flexible as measured by (1) moving faster, (2) turning more following attempted feeding, (3) broadening use of microhabitat, and (4) expanding their feeding repertoire. This prediction was based on earlier ideas of Morse (1971), who suggested that contingencies arising during migration would place a premium on plasticity, and of Real (1980), who in considering environmental uncertainty concluded on theoretical grounds that diversification of behavior would minimize uncertainty of outcomes. Uncertainty enters into the equation with lean migrants because these birds experience both increased energy demands and increased uncertainty that the demands will be met. In this case the behavior of lean vireos was significantly more diverse with respect to both maneuvers and substrates used. While maneuvering, lean birds hawked and hovered extensively in addition to the usual vireo foraging method of leaf gleaning, whereas fat birds mostly gleaned. In use of substrates, lean birds used bark extensively, and also the ground and air, in addition to using foliage, whereas fat birds used foliage 85% of the time with only 15% of the feeding efforts on bark and none in the air or on the ground. The prediction of greater diversification of feeding in lean vireos was thus realized. Morphological constraints and increased exposure to predation probably restrict the increased behavioral flexibility to stressful and uncertain conditions.

To all appearances at least, the Sahara Desert may present an even more formidable barrier to small migrant birds than an overwater crossing such as the Gulf of Mexico. For many years the scenario for trans-Saharan migration was derived from the studies of Moreau (1961, 1972). The wide expanse of desert was seen as an essentially totally inhospitable place that birds, especially in the autumn, had to cross in a single nonstop flight. This flight would have to traverse at minimum 2200–2500 km, requiring some 40–60 hours of continuous flight. In the spring it was assumed that a lack of suitable stopover sites on the coast of North Africa meant the nonstop flight also had to include a Mediterranean crossing. This hypothesis of a nonstop migratory flight seemed to be supported by only very few records of birds on the ground in the desert when, given the huge numbers of birds making the journey, many observations might be expected if they made the transit in shorter hops. Furthermore, birds were observed to put on large amounts of premigratory fat, as would be expected if they were about to undertake a long journey. More recently, however, conflicting evidence is appearing, and one model, based on considerations of both energetics and water balance, suggests that a nonstop flight across the Sahara in the autumn would be impossible. Water loss at high daytime temperatures would restrict migrants to flying at night and hence to intermittent 10–12 hour flights (Carmi et al. 1992).

The picture of trans-Saharan migration that is emerging is largely due to the work of two German ornithologists, Franz Bairlein and Herbert Biebach, who study migration in the western and eastern Sahara, respectively (Bairlein 1987, 1988, 1992a,b; Biebach 1990, 1991, 1992; Biebach et al. 1986). Bairlein and Biebach considered three basic but not mutually exclusive hypotheses for how small passerines might manage the Sahara crossing. First, the birds might stop over in oases or coastal areas (Red Sea or Atlantic) where they could rest or feed. Second, they could stop either in oases or elsewhere without feeding. Finally, they could make nonstop flights to regions south of the Sahara where food was again available.

Considering nonstop flights first, the distance to be traveled, especially in autumn, mandates that birds store high fat loads. For a bird such as the garden warbler, which weighs less than 20 g lean mass before migration (Kaiser 1992), it would take at least 8–9 g of fat to complete the journey, and captures reveal that many birds take off on migration with less than this amount. In the spring there is a chance of refueling on the north coast of Africa with its growth of vegetation following winter rains, but at this time the birds are also much more likely to encounter headwinds. In fact it is probably to the advantage of spring migrants to fly east and north or west and north, rather than in a broad front directly across the Sahara as they seem to do in the autumn with more favorable winds. The presence of far more migrants flying north through the rift valley of Israel in the spring than in the fall supports this notion (Carmi et al. 1992). Although the model of Carmi and colleagues (1992) indicates a nonstop flight would not be possible, Biebach (1990) concludes that such a flight would be possible if birds flew in air of 10°C to minimize water loss. The problem here is that Biebach assumes, based on older

Air Ministry Meteorological Office data, that this temperature will occur at night around 1000 meters altitude where autumn winds are also apparently most favorable. It is more likely, however, that an air temperature of 10° will occur closer to 3000 m both day and night, even though a temperature inversion will cause night temperatures to be cooler than day near the ground (V.A. Drake, personal communication). Suffice it to say, there is no compelling evidence for nonstop flights, and there is some evidence against them, or at least suggesting they would be very difficult.

Because of the difficulties associated with nonstop flights, intermittent crossings of the Sahara, either with or without feeding during stopover, should be the tactic of choice for small passerines. Is there any evidence that such crossings might take place? To test for this possibility, a number of trapping stations were established in the Sahara both at oases and at sites where feeding would be essentially impossible. One striking result from the trapping data is that many grounded migrants are not exhausted or depleted of water or fat, but are quite ready to resume migration. Their stopover times even when at oases are very brief, as judged from retraps, and in cages these healthy birds show migratory restlessness at night. Observations of the birds in the field revealed that often they are simply resting in the shade during the heat of the day when it is physiologically disadvantageous for them to continue flying because of water loss. These fat, resting birds were a much higher proportion of the trapped birds in the open desert than in oases.

In contrast to the fat birds, lean birds were a much higher proportion of trapped migrants at the oases. This suggests that the choice of a landing site during the migratory journey is strongly influenced by the amount of fat present. These lean birds also spent a much longer time at stopover sites and frequently showed a significant weight gain and increased fat score before they again departed. Lean birds also did not show nocturnal migratory restlessness the way the fat birds did. All the evidence available points to lean birds choosing oases as landing sites and then using these as staging areas to store fat before proceeding with the next stage of their migratory journeys. Intermittent flights thus do seem to be the predominant mode of travel across the Sahara, both with and without feeding at major oases to refuel. Apparently the reason so few grounded birds had been noted previously, leading Moreau (1961, 1972) to propose a nonstop crossing, was that although many were present, they were resting inactive in the shade and so went unnoticed.

Even though birds make rest stops, it is still likely that favorable winds are needed for the crossing. The reason for this is illustrated in Figure 7–11, using the example of the garden warbler. Here are plotted the body mass distributions of garden warblers trapped while grounded at two sites in the Sahara—Hadjar, Algeria, in the western part and a site in the Libyan desert of Egypt in the east. Biebach (1992) then calculated the flight range of birds of different body mass at three different wind speeds. Also plotted are the distances from two sites to the Sahel region at the southern edge of the Sahara, which is where the birds in the autumn would experience the first globally suitable habitat after leaving Europe.

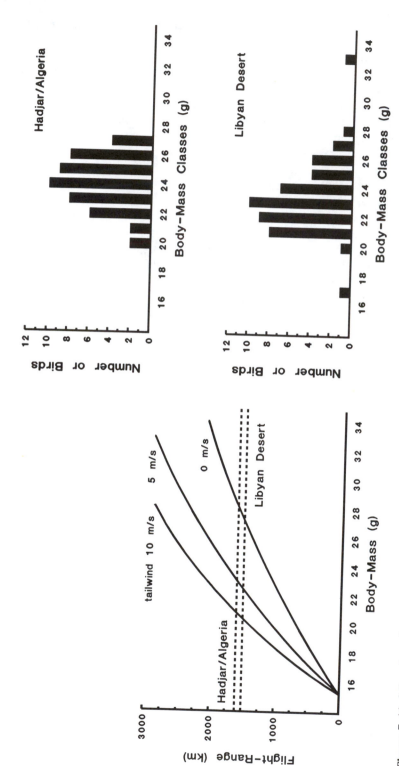

Figure 7-11. Migration of the garden warbler, *Sylvia borin*, over the Sahara. The left panel shows flight ranges at tailwinds of different speeds calculated for birds over a range of body masses. The dashed lines indicate distances from Hadjar, Alergia, and a Libyan Desert stopover site to the first suitable overwintering habitat in the Sahel. The right two panels plot the size frequency distributions for birds captured in Algeria (Bairlein 1987) and the Libyan Desert (Biebach et al. 1986). (Redrawn from Biebach 1992.)

As can be seen, with no wind virtually no birds could fly the required distance. For most birds a tailwind of about 5–8 m sec^{-1} is necessary for a successful journey. As indicated above, for reasons of water balance the birds should fly at around 2000–3000 m. Some radar data (Schafer, cited in Moreau 1972) suggest birds fly around 2000 m over the Sahara, but lack of knowledge of flight altitudes is still one of the major gaps in the understanding of trans-Saharan bird migration.

Another formidable barrier to migrants is formed by the mountains and deserts of central Asia and Kazakhstan between the Caspian Sea on the west and the Himalayan massif and Tibetan plateau on the east. There migrants of several species have been studied extensively by Dolnik (1990). These birds migrate between Siberian breeding areas and wintering areas either in Africa or in the Indian subcontinent. In spite of the fact that they are arid, cold, or both, the deserts and the Eburz, Hindu Kush, Pamir, and Tian Shan ranges apparently provide plenty of stopover sites suitable for refueling. The migrants thus adopt an intermittent flight tactic, but the stops are for refueling, not simply resting as often occurs in Sahara migrants. Birds captured before departure do not have enough fat to provide fuel for the entire journey. In the spring, deserts provide the most suitable stopover sites, often in irrigated areas of lowlands and foothills, and migrants pass between the eastern Eburz mountains and the western Hindu Kush, Pamirs, and Tian Shans, making use of stopover sites in the Kara Kum desert areas. In the autumn mountainous regions are more favorable for stopovers, with the exception of the high cold Pamirs, which are crossed without stopping in both spring and fall. Dolnik estimates that a day of feeding provides fuel for about 1.1 hours of nocturnal flight in the spring and about 0.6 hours in the autumn. Birds average about 4.5 hours of flight during the night in both spring and fall, suggesting that they probably spend several days at each stopover. Observations do in fact indicate that the rate of travel is at a moderate rate constrained by the limited opportunities for refueling, again in contrast to the rapid transit of the Sahara for most individuals migrating to tropical Africa.

For insects migrating to diapause sites, the basic problem is the same as that for birds migrating across barriers, namely, to transit a stressful period on a limited fuel supply. The problem for diapausing insects is to burn enough fuel (or provide enough antifreeze) to keep from freezing to death but not so much as to run out of reserves. The antifreeze option is used almost exclusively by nonmigrant insects, which often produce substances such as glycol to prevent freezing (see Danks 1987 for an excellent review). Migrants usually move to and from diapause in sheltered places, as is the case with the European milkweed bug, *Lygaeus equestris*, which overwinters in old buildings and rock crevices (Chapter 3). Diapausing insects also possess mechanisms to prevent emergence and reproduction in late-winter warm spells, although the exact nature of these is very poorly understood (Chapter 3 and Tauber et al. 1986).

The outstanding example of diapause tactics designed to survive active overwintering is the monarch butterfly, whose eastern populations migrate to Mexico and western populations migrate to the coast of California (Chapter 3). The problem for the butterflies is to avoid freezing, desiccation, heat stress, and predation while drawing on nonrenewable lipid reserves for at least 90 days (Masters et al. 1988). In Mexico large (up to 15 million individuals) colonies occur on southwestern-facing sunny slopes in high-altitude (3000 m) mesic forests of oyamel fir (*Abies religiosa*) (Brower et al. 1977, Calvert and Brower 1981, 1986). The combination of high altitude and tropical latitude provides a generally stable daily and seasonal thermal regime during the winter with daily shade temperatures below flight threshold for the butterflies. In California winter aggregations are of the order of tens of thousands rather than millions of individuals and occur along the coast from just north of San Francisco southward in stands of eucalypts (primarily *Eucalyptus globulus*) and Monterey pine (*Pinus radiata*). Butterflies cluster where there is filtered sunlight, shelter from wind, and access to moisture (Leong 1990). In both regions butterflies form clusters year after year in the same locations, suggesting that they are choosing rather specific site characteristics.

Just what these specific characteristics might be was investigated at California sites by Leong and colleagues (1991) using multivariate analysis of microclimatic variables. The results indicated that butterflies select groves of trees with less direct light (≤ 0.08 cal cm^{-2}min^{-1}), minimal wind movement (≤ 0.85 m $\cdot$ sec^{-1}), and moist conditions (vapor pressure deficit ≤ 0.2 mm Hg). In all groves with monarchs present, recorded values of these variables were less than the values given, and in all groves where monarchs were absent, but which otherwise appeared suitable, recorded values were higher. Within groves butterflies moved around as conditions changed so that clusters centered on the most favorable areas, and the density and shape altered somewhat as monarchs dispersed or concentrated. It was also apparent that the microclimate of cluster trees was influenced by the trees immediately surrounding the roosting areas within groves.

Masters and colleagues (1988) show that in Mexican overwintering clusters not only are the site characteristics important but so too is the behavior of the butterflies in promoting survival on the available fat reserves. They note that tens of thousands of butterflies bask in the sun on clear or partly cloudy days, that many fly up to 1 km to drink, that there is much shivering and crawling or flying up off the ground, that there are hovering flights above the trees when basking individuals are suddenly shaded, and that flights above the colony are common on clear cold winter days. Given that the costs to shiver or fly at the flight threshold of 15.5°C are 25 and 28 times energetically as expensive, respectively, as resting, why is there so much activity going on in the colony? It turns out that there are different answers for the different behaviors. Basking behavior raises the thoracic temperature very quickly without the butterfly itself using any energy. Once above flight threshold temperature, the monarchs takeoff, whereby they can either fly to a more shaded area or cool off with flight or, especially, gliding. Remaining in direct sun and raising body temperature, and hence metabolic rate, would cause too rapid consumption of lipid reserves. The butterflies must therefore be "sun minimizers," and the choice of colony site plus the relation between basking and flying are means toward that end.

The clouds of flying butterflies on clear days reflect the behavioral thermoregulation that is going on. The flight response to sudden shading is apparently a response to avoid being thermally trapped where they might be exposed to freezing or predation. Body temperature below −2°C begins killing monarchs so they must avoid situations, more likely at ground level, where freezing may occur. Shaded individuals in the colony fly up and hover for 3 minutes or so before reaggregating while those away from the colony, for example, drinking, instantly fly back. In the latter case the risk also involves cooling to below flight threshold so that they can fly only 1–2 km before becoming too cold to fly. It therefore behooves them to fly back at once before the possibility of a temperature drop from the shading prevents them from reaching their destination. Shivering allows monarchs to raise body temperature enough to crawl or fly off the ground onto vertical surfaces and avoid freezing or mouse predation. The results of shivering and crawling can be seen after winter storms knock butterflies to the ground, when huge numbers are gathered on the lower reaches of tree trunks just above ground level. The choice of cool sites and the low body temperatures resulting from thermoregulatory behavior also maintain diapause and so prevent premature reproductive activity (Herman 1985), but the conservation of lipid reserves is also a major function and thus a key element in the maintenance of the monarch migratory cycle.

Summing Up

Perhaps the first and most obvious conclusion to come from the survey in this chapter is that with respect to migration constraints do matter. Adult tarantula spiders do not migrate by ballooning on silk threads; the length and strength of the thread necessary to support a tarantula are simply too great for natural selection to have acted to produce ballooning as a viable means of transport. Ovenbirds and other small songbirds do migrate and often for very long distances. They do not, however, do so by soaring because any energy gains are negated by the metabolic costs of being small accumulating over the extra time that soaring takes. Size in vertebrate-transported seeds is apt to be a compromise between increasing the likelihood of long-distance transport (small seeds) and increasing the likelihood of successful germination and establishment (large seeds).

At the same time that migrants have been influenced by constraints, they have been superb innovators at working around those constraints. Biomechanically, plaice and eels are relatively inefficient swimmers poorly designed for migration. Yet migrate they do, with the Atlantic eels in many ways the quintessential of long-distance oceanic travelers. The solutions to the limitations imposed by body plan or muscle power lie in behavior. Plaice, as we saw in Chapter 5, are masters at using tidal streams, and eel larvae use oceanic currents to reach their destinations in estuaries and freshwater rivers on both sides of the Atlantic. Songbirds facing the extremes of daytime temperatures in the Sahara land and rest in the shade before proceeding on their journeys. Monarch butterflies seek out and congregate in sheltered, cool locations and behaviorally thermoregulate while there to conserve the energy reserves necessary to see them over the winter diapause portion of their migratory cycles. Modulation of behavior thus allows migrants to deal both with their own limitations and with the environmental exigencies they face.

Looking now at the other side of the coin, migration is one of many factors influencing the body plans of organisms. Long, narrow wings or lunate tails are optimal designs for flying and swimming migrants, respectively, but other factors also enter in. Migrant birds may also need to maneuver in dense thickets and fishes to maneuver in reed beds where short, rounded wings and laterally compressed bodies with broad tails are better designs. The design of the organism is perforce a compromise between these conflicting demands. By understanding the nature of these compromises and the way natural selection organizes them, the effects of selection arising from the demands of migration become apparent. Within a group of short-winged birds of brushy habitats, the migrants tend to be longer winged and shorter tailed; the fact that leg length can also be influenced indicates that it is the multivariate nature of the whole organism that migration, or any other imposed selection, influences. In a sense it is because organisms, migrant or otherwise, cannot optimize everything at once that they have become the behavioral innovators they are. Migration is one of the innovations.

When it comes to coping with environmental hazards, migrants are often astonishing in their abilities to overcome inhospitable habitats. Thus small birds successfully transit the Sahara or the cold high mountains of central Asia by employing tactics suitable for each region. In the Sahara many birds rest in the shade even well outside oases and in central Asia they slow their journeys and exploit the resources of even the smallest valleys. Among migrant insects milkweed bugs can shelter from the most severe of Baltic winters and monarch butterflies can spend an entire winter in Mexico without feeding by a combination of physiological and behavioral adjustments. Once again innovations of behavior allow the minimization of constraints even under the most stringent conditions faced by migrants.

8

Orientation and Navigation

There seems to be an inherent bias against considering simplicity in migratory mechanisms, the magnitude of the bias being proportional to the scale of the migration.

J. D. McCLEAVE, 1987, p. 110

The ability to orient to the surroundings is a fundamental property of life, and probably all organisms orient at some point in their life cycles. At even the simplest of organizational levels, there exists the ability to process information, and that information is frequently used to orient in space. When it comes to migration, the ability to control one's position in space can become very important indeed. The degree of importance that orientation assumes, however, and the level of sophistication necessary in orienting are very much functions of the necessities imposed by the migration system. It is quite possible to imagine a migratory episode in which no orientation is necessary, for example if all surrounding habitats to which migration may occur are ecologically equivalent. This sort of situation is no doubt an extreme case, but it is still worth considering as a sort of "null model" for orientation behavior. Minimum orientation may also be the case if wind or currents provide the vehicle to a suitable habitat (Chapter 5). There are also situations where orientation may occur not during migration but after migration is complete, say in foraging or ranging activities to find suitable food or breeding sites. Such orientation may not be very complex, or, on the other hand, it may involve sophisticated use of landmarks or other cues if a home range or territory is involved.

Many migrants, however, home to very precise locations to breed, and in some cases, to spend the nonbreeding period. In these organisms, the ability to orient and navigate may become a necessity, and if so, it behooves them to use environmental cues that are both reliable and available (Able 1991a). If migration covers long distances, the reliability and availability must hold true on a global scale. It may also be necessary to rely on more than one cue because cues may vary depending on ambient conditions. Clouds may cover the sun, for example, or landmarks may be obscured by darkness or fog. The organism therefore must be imbued with sufficient plasticity in its orienting mechanisms to allow adjustments to changing conditions.

Complex orientation behavior and guidance systems are often divided into three general categories (Able 1980; Terrill 1991). The first of these is *piloting* or the ability to use fixed known reference points to orient or navigate. These are commonly referred to as landmarks and may occur either more or less as points on the landscape such as a hill or a tree or as "leading lines" such as a line of hills, a shoreline, or a river that may guide movement over considerable distances. The second category is *compass orientation* or orientation in a given direction without reference to landmarks and without reference to particular places of origin or destination. The organism, in other words, does not necessarily know where it is, but only in what direction it is facing or moving. Finally, there is *true navigation* or the ability to move toward a particular goal in completely unfamiliar territory in the absence of any sensory contact with that goal. The ability to navigate in this sense also implies that the organism possesses a "map" (often called a "cognitive map" after Tolman 1948) or internal representation of the geometric relations among points in the environment (Wehner and Menzel 1990; Downhower 1992). The above categories do not necessarily reflect increasing levels of complexity nor the order in which the capabilities evolved.

Migrants face a variety of orientation problems depending on their sensory and cognitive capabilities, the distances traveled, and the specificity of goals. There seem, however, to be only a limited number of ways in which these problems can be solved, as indicated by a number of common themes that seem to apply across migration systems and taxa (Able 1991a). The ultimate causal basis for the similarities is at present unclear. Their presence in phylogenetically distant groups may represent convergence, but there are very few, if any, data against which the hypothesis of convergence can be tested. The same can be said for the notion that common themes may exist because they represent characters that are phylogenetically ancient, and long conserved and honed during the course of evolution. Whatever their basis, they are clearly present, as we shall see. It is also clear that much work and thought must yet go into understanding the reasons for that presence.

Are Signposts Necessary?

McCleave (1987) notes what seems to be an inherent bias against considering simplicity in orientational and navigational mechanisms, with the magnitude of the bias proportional to the scale of the migration (see the quote at the beginning of the chapter). The point is well taken, for there is a tendency to be impressed with large-scale migrations and the apparent precision with which organisms arrive at goals at the conclusion of migratory movement. It is appropriate to ask, however, if precise guidance systems are really necessary to account for the observations, or whether much simpler processes will result in the arrival of migrants at their destinations. It should be obvious that the answers will depend on the particular migrations, so it is useful to consider the interactions between migration and guidance systems beginning with those in which quite simple mechanisms for the latter may suffice.

In Chapter 5 I discussed a number of cases where migratory organisms used winds or currents to facilitate transport during migration. In eels (*Anguilla*), for example, passive horizontal drift with modulation by active diurnal vertical movements, coupled with the characteristics of oceanic currents such as the Gulf Stream, is apparently sufficient to bring enough leptocephali near to continental shelves for subsequent successful invasion of river systems (McCleave 1987). The passage into rivers is then accomplished to a large degree by selective tidal stream transport. It is not necessary to postulate more sophisticated guidance systems because even with a large chance mortality, there are enough surviving leptocephali to assure success of the migration system. Selection has apparently acted to produce the highly fecund life history rather than sophisticated navigation. With appropriate adjustments for the particular pattern of oceanic currents it faces, the same can probably be said for the western rock lobster in the Indian Ocean off Australia (Phillips 1981 and Chapter 5).

It is also not necessary to postulate complex guidance systems for the migrations of the milkweed bug, *Oncopeltus fasciatus* (described in Chapter 9). In the spring and early summer a continuous series of northward-blowing winds bring many species of insect into the northeastern and upper midwestern United States (Chapter 5 and reviewed in Johnson 1969), and the phenology of the milkweed bug is consistent with a wind transport scenario. The bug, like many other insects, is photopositive, and its tendency to be more frequent in milkweed patches along water courses suggests the possibility of a simple orientation to brighter patches or streaks caused by the reflection of the sun off water (Dingle 1991b). This is not, however, a necessary mechanism for successful migration. In the autumn, wind directions are not as consistent as they are in spring or summer, although as the season advances there is an increasing frequency of winds blowing toward the south (Barry and Chorley 1987). The short days of autumn trigger an adult reproductive diapause in *O. fasciatus* (Dingle 1974 and Chapter 9), and this may be the key to the successful migration system because it allows a long period for successful movement to overwintering sites farther south.

Gabriel (1977) modeled the southward migration of *O. fasciatus* under several assumptions regarding wind transport. The first assumption was that autumnal movement of the bug was completely random with no directional effect produced by wind systems. Needless to say the bugs were not particularly good at getting south, but even though they represented only a small proportion of the population, a large number of bugs still did so. These bugs are tropical in origin and are unable to survive the winter in the northern summer breeding area; with *any* return to the south in the autumn, selection would favor migration. The abundance of milkweed in the north, with the resultant production of large populations, undoubtedly assures that enough bugs reach favorable wintering areas to restart the northward migration cycle the following spring. The next step in the model was to assume that bugs flew at an altitude of approximately 300 meters in the period late September to early October and were carried by winds at that altitude. Under these circumstances most of the bugs would have arrived in eastern Canada to start the winter, clearly not a favorable choice of overwintering site, but even so some bugs reached the Gulf Coast. Once again the net result of migration would have been sufficient numbers in the south to begin the next season's migratory cycle. Any tendency for directional winds to increase the number of bugs reaching the south, as winds at times and altitudes not considered in the model could do, would markedly improve the success of the migratory life cycle; so would any orientational capability. The point is, however, that migration can be a successful strategy even in the absence of any orientation mechanism that would aid the bugs in proceeding south.

There is, of course, overwhelming evidence that migratory organisms travel on winds and currents (Chapter 5). This being the case, it is logical to ask whether in any sense they "choose" a transporting vehicle that would carry them in the "right" direction without the use of specific orientational or navigational cues. We have seen one such case in Chapter 5 in the larvae of the crab *Rhithropanopeus harrisii*, which swim upward in the water column on rising tides and descend when the tide falls, with the result that they remain in estuaries while moving to new habitats (Cronin and Forward 1979; Forward 1987). Fishes using tidal stream transport also display what appear to be programmed vertical movements. Insects usually fly upwind while in visual (optomotor) contact with the ground, but in the absence of such contact above the species-specific "boundary layer," they turn and fly downwind (Taylor 1958). They may, however, fly at altitudes much above their boundary layers, and it is a frequent observation that different species at the same time may be flying at different altitudes in winds moving in different directions (e.g., Schaefer 1976). Are different species selecting different winds, and if so, how do they do it? At present we don't know, although possible cues are available. The different winds, for example, usually represent air masses of different temperatures. Another possible cue is air pressure, which changes with altitude. Among birds, pigeons are able to detect and react to changes in air pressure (Alerstam 1990), but we really have no knowledge of how they might use this information, nor whether insects use it. There is evidence that birds have some (unexplained) ability to sense winds aloft (Richardson 1991).

In addition to apparent selection of transporting vehicle, there is now abundant evidence (in what often seems an exponentially increasing literature) that organisms orient and navigate. The kind of evidence is nicely exemplified in Walker's (1980) study of butterflies migrating through peninsular Florida (see also Chapter 4). Walker observed three species of butterfly, the cloudless sulphur, the gulf fritillary, and the buckeye during autumn migration. These species are day flying and fly within their boundary layers; as a result, they are readily visible to ground observers. The directions of the flights of these butterflies were estimated by first observing them flying across an open field and then proceeding to the point where they were first sighted. They were then followed by eye until they disappeared from view, and the bearings of the vanishing points were read from a sighting compass. The directions of flight of all three species were little different from 141° or, in other words, from the axis of peninsular Florida (Fig. 8–1). The presence of the sun was not necessary for this course to be maintained, for the vanishing directions were essentially the same on both cloudy and sunny days. Because these butterflies fly so consistently in a quite precise direction, they must be at least capable of orienting. Can they also navigate or pilot? If so, what cues are they using, given that they can maintain course under cloudy skies? For these butterflies, and indeed for most organisms, we don't know the answers. Work over the last several years has, however, produced some very interesting insights into how at least a few well-studied species find their way about with what to a human observer are often astonishing capabilities.

The Ability to Orient and Navigate

The environment of any organism abounds with cues giving information not only on the position of the organism itself, but also on the condition and location of a host of biotic and abiotic factors of potential importance. These are in the form of an array of potential chemical and physical inputs. In order to be used, however, they must be detectable. We are familiar ourselves with those cues that we can detect through our five senses, but there are other highly reliable sources of information that we are apparently unable to detect, at least consciously (although there is some controversy about this), such as the earth's magnetic field or

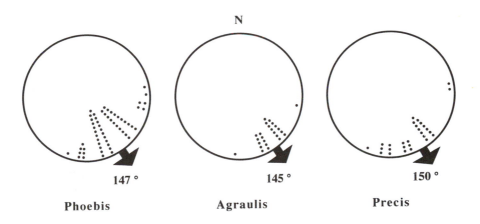

Figure 8–1. Orientation of three butterflies, the cloudless sulphur (*Phoebis sennae*), the gulf fritillary (*Agraulis vanillae*), and the buckeye (*Precis coenia*), during migration through peninsular Florida. Each dot represents one individual, and arrows indicate mean direction whose value is given. (Redrawn from Dingle 1985 after Walker 1980.)

polarized light. As we shall see, however, our own sensory abilities are not necessarily a reliable guide to the sensory world of many migrants. In the following sections I briefly survey some of the cues that migrants use to find their way to breeding sites or to places where they spend a nonbreeding period.

Chemical Cues

The enormous array of volatile chemicals that can be emitted into air or water provide highly specific information about their sources to those organisms with the ability to detect them. Organic molecules in particular, because they can be formed by increasing chain lengths or changing configurations of carbon atoms, provide an essentially unlimited number of chemical signatures, even more so when combined with either other organic or even some inorganic molecules to form blends (e.g., Wilson and Bossert 1963). The species and sometimes even individual specific pheromones that numerous taxa use for chemical communication are good examples of how useful and adaptable chemical cues can be. They are also used by migrants, especially, to locate a particular "home" area or site. Perhaps the best known and best studied use of chemical cues by migrants is the homing to stream odors by salmon, studied for many years by Arthur Hasler and his students and associates at the University of Wisconsin (Hasler and Wisby 1951; Hasler 1966; Hasler et al. 1978; Hasler and Scholz 1983).

Years of data based mostly on tagging and recapture of salmon when returning to freshwater to breed had indicated that by and large the fishes returned to their natal streams (Harden Jones 1968). Hasler and his colleagues developed and tested the odor hypothesis of stream homing based on three primary assumptions: (1) that each stream used for breeding has a unique odor because of the specific soil and chemicals occurring in the stream basin, (2) that juvenile salmon became chemically imprinted on the home stream, and (3) that adult salmon then used the imprinted odor to return to the home stream. Initially Hasler and Wisby (1951) used conditioning experiments to demonstrate that coho salmon (*O. kisutsch*) could be trained to distinguish between stream odors. The trained fish lost this ability when they were made anosmic or when the organic fraction was removed from the water. Streams were consistent, and therefore the odor cue reliable, over years, a necessary condition because salmon do not return to their natal streams until 2–5 years after hatching there. Additional evidence consisted of the fact that salmon in neutral water became more active when home stream water was added to their tanks, but not when other water was added, and that the electroencephalogram (EEG) of the olfactory bulb increased when the olfactory epithelium was bathed in home stream water. The odor hypothesis was field tested by removing adult salmon from above the branching point of a Y-shaped stream, marking them, and returning them to the stream below the point where it bifurcated. The nasal area was plugged in half of these fish, and 28 of the 70 recaptured returned to the branch opposite the one in which they were caught; in contrast, only 8 of 73 controls failed to return to the original branch of capture. In a series of transplant experiments, fish transplanted from their natal stream to another before the smolt stage returned later to the stream into which they had been transferred, but those transplanted after becoming smolts returned largely to the natal stream. Imprinting on stream odor thus occurred at a specific stage of the life history.

Perhaps the most conclusive proof of chemical imprinting to home stream odor came with experiments designed to induce coho to migrate into a stream where they had never been before. In the Great Lakes region of central North America, coho are introduced into streams where they do not naturally occur. They then migrate into one of the Great Lakes to mature before returning to the streams where they were released. In the experiments hatchery-raised coho were conditioned to the chemical morpholine, an organic molecule that does not occur naturally in the habitat of the salmon, prior to release into Lake Michigan. Just prior to the start of the return migration of coho to streams, morpholine was released into Oak Creek, a tributary stream flowing into the lake. Of the coho captured in Oak Creek during the upstream migration, 218 fish were from the morpholine-treated group; only 28 were from a control group not receiving any contact with morpholine. Furthermore, when coho in the lake were tracked with ultrasound after introduction of morpholine-exposed fish, they paused and milled about for a considerable period in a small test area where morpholine was directly introduced into the lake waters. When there were no morpholine-imprinted fish in the lake, this milling behavior was not observed. Taken together all these experiments make a strong case for imprinting to home stream odor during the juvenile (immediately pre-smolt) period.

A question raised by the demonstration that salmon home to natal stream odor is what is the nature of the behavioral mechanisms allowing them to do it? This question has been addressed by Johnsen (1987). The observations consisted of monitoring the movements of individual fish that had been imprinted to synthetic compounds when distributions of the odor stimuli were controlled in experimental channels. When an imprinted odor was distributed evenly across the width of a stream, the fish displayed a straight positive rheotaxis moving upstream against the current. In the absence of the odor, the fish showed negative rheotaxis and moved actively downstream until they again detected the scent. The presence of odor thus changed swimming behavior from negative (downstream swimming) to positive (upstream swimming) rheotaxis that would take the fish toward the source of the odor. In real streams with turbulence, eddying, and other factors, odor distributions will be uneven. This brings in another aspect of swimming behavior while orienting. When the odor was confined to one edge of an experimental stream, the fish used the interface between the scented and unscented water masses more than the odor stream itself. This they did by zigzagging along the edge of the odor corridor, making steady progress upstream. Apparently the triggering of switching between positive and negative rheotaxis and zigzagging by the odor present in ambient water is sufficient to ensure that the salmon reach the home stream.

As Johnsen also points out, these behaviors are closely analogous to the behaviors exhibited by insects in an airborne odor corridor (Kennedy 1983). The presence of a stimulating odor such as a pheromone induces an insect to fly upwind. As long as it is in contact with odor it zigzags steadily in an upwind direction toward the odor source. When it loses contact, it begins to "cast" in wide sweeping flights back and forth across the wind stream until contact is regained. The orienting behaviors of both salmon and insects are effective means of locating an odor source because in the turbulence of moving air or running water, the scent will be produced not as a steady plume, but rather as a series of bursts or puffs of varying odor intensity with which the organism will be alternately losing and gaining contact. The behaviors allow both the maintenance of contact in this unsteady stream (zigzagging) and the re-establishment of contact when the odor is undetected for intervals of more than a few seconds (negative rheotaxis in fishes and casting in insects).

Salmon may also use the odor structure of water masses to move from the ocean into river mouths (Westerberg 1984; Doving et al. 1985; Johnsen 1987). The odor profile in the ocean from a large-scale diffuse source like a river will consist of layers of water with odor at different concentrations (Fig. 8-2). The vertical distribution of the odor will be related to variables such as differential velocities of different layers, temperatures, and salinities. The result is considerable vertical fine structure as well as a horizontal quasi-mosaic pattern to the odor. Assuming that salmon can detect differences in odor concentration, they could orient to the river mouth by moving against the current shear in the fine-structured gradient layers (Westerberg 1984 and see Chapter 5 for a discussion of shear). The probability of local shear being in the direction of a river mouth is great enough that the mean direction of travel of fishes, were they using this information, would be in the correct direction even if by a somewhat irregular course. Because the structure is vertical as well as horizontal, fishes using this information source should sample by making vertical as well as horizontal movements. Atlantic salmon (*Salmo salar*) tracked by transmitter on their homing migration through a Norwegian fjord indeed showed such movements (Doving et al. 1985 and Fig. 8-2) and also displayed a preference for a particular water layer. The observations are at least consistent with the hypothesis that salmon can use olfactory inputs to discriminate among stratified water layers and use this information to locate river mouths.

Although in salmon the use of odors in orientation is quite clear, the question of odor orientation in birds has aroused considerable controversy and even acrimony (Schmidt-Koenig 1987; Waldvogel 1989; Papi 1990,1991; Wallraff 1991). The major issues are perhaps best summarized by Waldvogel (1989) in his attempt to assess the problem in a critical but unbiased fashion. Earlier workers had largely ignored the possibility of odor orientation in birds, mostly because the relatively small avian olfactory bulbs suggested that the sense of smell was poorly developed, and most observations of behavior suggested it was little used. More recently, however, there have been demonstrations that at least some birds do make use of odors. Turkey vultures (*Cathartes aura*), for example, possess a large olfactory system and use olfactory cues to locate carrion. Procellariiform seabirds (albatrosses, shearwaters, petrels, etc.) use odor to locate nesting colonies and food items on the surface of the sea (Hutchison and Wenzel 1980), interestingly by using

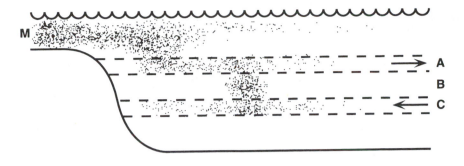

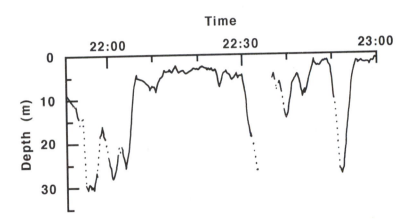

Figure 8–2. *Top panel*: Schematic representation of odor layer formation in near shore areas. Odor enters the ocean from a river mouth, M, and spreads at the surface. At a region of higher concentration some "leaks" into a current, A, which causes more rapid spread. In turn there is leakage into slower moving water, B, and again into a current, C. The result is a pattern of odor concentration varying horizontally and vertically. The horizontal scale is compressed. *Bottom panel*: Swimming record of a salmon during one hour showing pattern of vertical movements. Dotted lines indicate loss of data due to high signal to noise ratio and other factors. (Redrawn from Johnsen 1987, after Døving et al. 1985.)

a zigzag upwind flight much like that of insects or the upcurrent swimming of salmon. In one of the more interesting examples of odor use, starlings (*Sturnus vulgaris*) apparently use plant chemicals to identify nesting material that has parasite-repelling properties (Clark and Mason 1985, 1987). These observations do not, of course, address the question of whether birds use odor for long-distance orientation or even navigation, in the latter case with an olfactory "map."

Such a map, in which a bird is able to locate itself relative to home on the basis of a more or less mosaic pattern of atmospheric trace compounds, has been claimed for homing pigeons by Floriano Papi and his associates (Papi 1990, 1991) and Hans Wallraff (1991). The issues and experiments on which the claim is based are far too complex to cover in detail here. Briefly, however, the argument is that pigeons can locate home from as far as 700 kilometers away based on an atmospheric odor profile in some ways similar to the one possibly used by salmon to find natal rivers from the ocean (Fig. 8–2). This claim is based on experiments in which pigeons are unable to find their home lofts when rendered anosmic with local anaesthesia on the olfactory epithelium or surgical severing of olfactory nerves but can do so when they can still smell. Manipulations of the odors of or around lofts also seem to cause variation in homing ability. Problems arise either because other workers cannot repeat the results or differ in the interpretation of the experiments. Again, oversimplifying, other major counterarguments suggest that making pigeons anosmic has behavioral effects beyond simply disrupting smell and that atmospheric odor profiles will be far too unstable to provide reliable map information except very close to the home loft. Part of the contributing factors to differences between experimental results may also be differences between pigeons and/or between experimental situations. Wiltschko and colleagues (1989) found, for example, that they could demonstrate odor use in pigeons kept in a roof loft and exposed to wind, but not in pigeons kept on the ground where they were more sheltered. Suffice it to say that counterclaims to the olfactory map hypothesis are vigorously disputed. On balance, however, whereas pigeons may use odor (and certainly use other cues) to find the home loft once they are nearby, the claim of an olfactory map is far less convincing. I agree with Waldvogel (1989 and see also Berthold 1991d) that for a number of reasons, not least of which is atmospheric instability (Chapter 5), the use of odors by pigeons for long-distance navigation is extremely unlikely. If pigeons do use the very low concentrations of odors that would be present over distances, they would need to use a rather elaborate verson of zigzagging, and most likely casting as well, to maintain contact with the various odor puffs and streams. Anything resembling such behavior has yet to be demonstrated. Finally, as far as migrant birds are concerned, the whole issue is probably irrelevant, as not even the strongest proponents of the olfactory model claim an olfactory map would operate over the distances traversed by even the least traveled of migrants.

Visual Cues

Migrants travel both by day and by night, and the sky provides potential sources of guiding information at both times. During the day the sun is the dominant stimulus, and so it is not surprising to find it a prominent cue used by migrants to maintain and determine routes. There is also other information provided by the sun, including differential brightness of the sky, patterns resulting from the plane of polarization of light, and the position of the rising or setting sun. All these cues are evidently used by migrants to some degree. At night the stars and the moon are available, although the latter presents special problems because for part of its monthly cycle, it is not visible the whole night, and it is phase shifted by 50 minutes a day. The stars, in addition to providing an overall pattern, also rotate during the course of the night (around the North or Pole Star in the Northern Hemisphere), and this also turns out to be an important consideration in the use of stellar sources of information. In this discussion I begin with daytime-oriented movements.

Perhaps the simplest sort of information available as a consequence of the sun's light is contrasting brightness patterns in the sky, and this is used by newly hatched sea turtles of several species both in reaching the ocean and in continuing to move offshore (Mrosovsky and Kingsmill 1985; Salmon and Wyneken 1987; Wyneken and Salmon 1992). The first problem faced by hatchling sea turtles is finding the ocean from their nests, which are situated up on the beach. At the land-water interface the daytime sky is considerably brighter over the water, which reflects light from the sun, than over land, which tends to absorb sunlight. The hatchlings move immediately toward the brighter sky when they emerge from the buried nests and so find the ocean. Seaward movement is also facilitated by orientation away from dark objects such as trees, bushes, or dunes and by movement down slope (Salmon et al. 1992). Once in the ocean the turtles begin 24 hours of continuous swimming, known as the frenzy period, when all other activities such as feeding and resting are suppressed. It is thus another example of specialized migratory behavior (Salmon and Wyneken 1987) of the sort discussed in Chapter 2. This frenzy continues to be oriented toward the brighter seaward horizon for some 5 km offshore; after that the turtles must use other cues to continue seaward. Wyneken and Salmon (1992) postulate that because the frenzy period is the same in all three species they studied—the loggerhead (*Caretta caretta*), the green (*Chelonia mydas*), and the leatherback (*Dermochelys coriacea*) turtles—it probably serves to escape shallow water where predator risk is high. Under these circumstances the continued orientation to brightness is probably quite sufficient, with more complex orientation coming into play only later.

Orientation to brightness is the simplest use of visual orientation. At a more complex level is the use of the sun itself, which can be done in two ways. First and simplest, the organism can simply move toward the sun without correcting for its movement during the day. Under these circumstances orientation cannot be very precise because as the sun moves across the sky during the course of a day the direction of movement will also change. To some extent this can be minimized by flying only at a certain time of day, say for a few hours after warming up in the morning if one is a butterfly, but even so directional accuracy will not be great. The second way to use the sun, which vastly improves accuracy, is to compensate for the sun's movements and change the angle of orientation with respect to the sun so that a constant direction is maintained. Such a bicoordinate system with two reference points, the sun and the time of day, requires some sort of internal timing method or biological "clock." Such a time-compensated "sun compass" has now been demonstrated in a great variety of organisms. The first demonstration of the use of the sun in orientation was by Felix Santschi (1911) studying the homing of several genera of desert ants in Tunisia. In a now famous experiment, he altered the apparent position of the sun, as seen by the ant, by using mirrors and noted that the insect would change direction at the same angle as the displacement of the sun. In a charming historical essay, Rüdiger Wehner (1990) surveys Santschi's work (and his life in North Africa) and shows that Santschi was well aware that the mirror experiment worked better in some ant genera than in others, suggesting that the sun compass was part of an integrated navigational system specific to each ant and could play different roles depending on the ecology of the species. Santschi (1913) also suggested that animals might possess an internal representation of time, although he made no attempt to experimentally demonstrate it.

One other point about the sun is that, at least in theory, an animal could use it to determine latitude by knowing season and time of day. This is because the path or arc of the sun through the sky is dependent on latitude. At the equator the path of the sun travels directly overhead, but as one proceeds poleward the altitude of the arc is progressively reduced; it also changes with the season, from a high at the summer solstice to a low at the winter solstice. There is no indication, however, that animals use this information. Rather they use the position of the sun relative to the horizon, or its azimuth, an imaginary line traversed by the projection of the sun on the horizon (Fig. 8–3).

The monarch butterfly, *Danaus plexippus*, apparently orients to the sun's azimuth without any time compensation (Kanz 1977). There were, however, no clock-shift experiments done to demonstrate lack of time compensation (see below), so its absence is still not certain (cf. Schmidt-Koenig 1985). In any case summer nonmigrant individuals placed in large outdoor opaque flight cages would orient to the sun as it moved along its azimuth, but oriented more or less randomly under overcast skies or when the cage was transparent so that the surrounding terrain and vegetation were visible. Fall migrants, on the other hand, oriented to the sun's azimuth in both opaque and transparent cages, but again displayed no directionality under overcast skies. Because the butterflies do not start flying until they have warmed up in the morning, the general tendency in following the sun's path in the autumn will be more southwest. Gibo (1986), however, notes a preferred SW direction apparently independent of azimuth. In any case a SW preference will be in the direction of Mexico where the monarchs spend the winter. Schmidt-Koenig (1985), however, has observed what appears to be much more precise southwestward orientation during the fall migration, even under overcast skies, and suggests his data are consistent with magnetic orientation (discussed below).

The modern era of orientation studies of migrants and other organisms was ushered in with the publication of the simultaneous discoveries of a sun compass in honeybees (von Frisch 1950) and migratory starlings (Kramer 1950). Kramer took advantage of the migratory restlessness (Zugenruhe) of caged starlings and showed that this was directed rather than random. He used a circular "Kramer cage" (see Chapter 4) to show that starlings oriented at a specific angle to the position of the sun, and by using mirrors as Santschi had done with ants, to demonstrate that if the apparent position of the sun was changed, the birds oriented

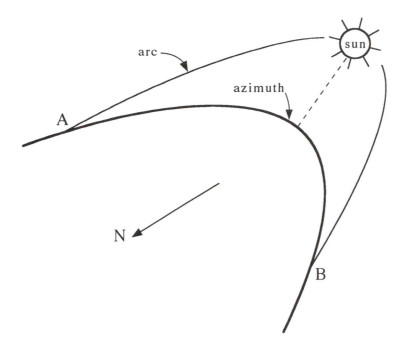

Figure 8–3. The sun arc, or path of the sun through the sky, and the sun azimuth, or path of the sun projected onto the horizon, diagrammed here for the Northern Hemisphere. The sun rises at A and sets at B.

at the same angle with respect to the apparent sun (Fig. 8–4). Later experiments used birds trained to search for food in given compass directions, making them independent of Zugenruhe.

Subsequent experiments showed that the starlings could find the food at one time of day even if trained at another, indicating that they compensated for the sun's daily movement across the sky, which, of course, required a time sense and an internal chronometer. This observation led to a series of experiments by Hoffmann (1954) in which starlings were clock shifted by keeping them in a day-night cycle that was out of phase with the naturally occurring cycle. When starlings were clock shifted, they displaced their orientation by a factor of 15°X, where X is the hours of clock shifting and 15° is the distance moved by the sun along the azimuth in one hour. In the original experiment starlings were shifted 6 hours counterclockwise, and the birds then displaced 90° in the same direction. This result demonstrated that the starlings were using a bicoordinate system, incorporating both the sun's position on the azimuth and the time of day, to orient. Extensive use of clock-shift experiments has allowed quite detailed analysis of the sun compass (reviewed in Schmidt-Koenig et al. 1991). Subsequent experiments with foraging honeybees revealed that they, too, can allow for the changing position of

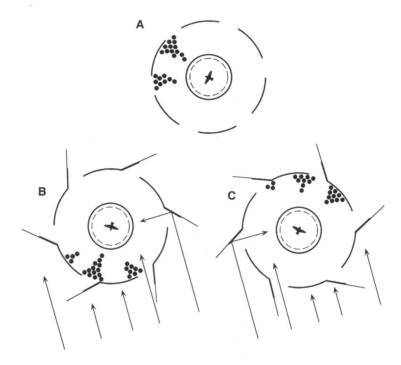

Figure 8-4. Kramer's experiments demonstrating a sun compass in birds. *A*. In a circular cage a starling orients in a particular direction. *B* and *C*. Mirrors are added that deflect the incoming sunlight to change the apparent position of the sun. The bird maintains a constant angle of orientation to the perceived sun. (After Kramer 1951.)

the sun (reviewed in von Frisch 1967) and can even compensate for the daily changes in rate of sun movement by maintaining a running average of azimuthal change based on the experience of the previous 40 minutes, approximately (Dyer and Gould 1983).

Sun compass orientation has since been found in a wide array of animals. It occurs, for example, in crustaceans as taxonomically distinct as decapods and amphipods (Herrnkind 1972), in ants (Wehner 1983), in fish (Leggett 1977), in assorted amphibians and reptiles (Adler and Phillips 1985), in mammals (Lüters and Birukow 1963), and in many birds (Able 1980 and for a general review). A general use of the sun compass in amphibians (and beach-living crustaceans) is orientation perpendicular to the shoreline or y-axis orientation (Landreth and Ferguson 1967). The direction of movement is determined by stage of the life cycle. In toads, for example, adults move away from the shoreline toward land whereas tadpoles move in the opposite direction toward open water (as do fish). Newly metamorphosed frogs and toads switch the direction of orientation when they emerge from ponds and begin their migration to terrestrial sites. The animals evidently learn the shoreline because they can be trained to a new axis of orientation if the x-axis of the shoreline is shifted. Clock-shift experiments demonstrate that this orientation can be by sun compass, although other cues such as polarized light and a magnetic compass are also involved, as discussed below.

In birds the sun compass is used not only during migration and homing, but also for other functions requiring orientation. In North American scrub jays (*Aphelocoma coerulescens*), for example, the sun compass is used to locate stores of pinyon pine seeds that have been cached (Wiltschko and Balda 1989). In an octagonal aviary, jays cached their seeds in sand cups and later probed the appropriate sector of the cage to recover seeds. When the jays were clock shifted, the sector of probing deviated by the amount expected from the extent of clock shift. It is interesting that the sun compass is used (because of reliability?) in spite of the fact that familiar landmarks were available. The presence of a sun compass in so many animals and its functioning under a variety of circumstances imply that it did not have to be especially evolved for migration but rather could be rapidly incorporated as necessary as part of migratory syndromes. A question that has not been answered, however, is how birds migrating to equatorial latitudes, and especially transequatorial migrants, adjust for the changes in the sun's path and azimuth that occur with latitude (Schmidt-Koenig et al. 1991).

In addition to information provided by its position, the sun also provides information via the plane of polarization of light or the e-vector. The e-vectors of the polarized light of the sky create a pattern in the celestial hemisphere, with a band of maximal polarization 90° from the sun's position, as illustrated in Figure 8–5. The band of maximal polarization is visible through the twilight period for up to 45 minutes before sunrise or after sunset. Thus considerable information is available from the polarization pattern if an organism can detect it. The mechanism of detection has been well established in ants and bees (Wehner 1989). A specialized portion of the retina located at the uppermost margin of the eye and constituting only 2.5% (honeybee) or 6.6% (desert ants, *Cataglyphes* spp.) of all photoreceptors is the only part of the eye sensitive to the e-vector, as revealed by experiments in which various parts of the eye were covered by light-excluding paint. These polarization receptors function in the ultraviolet range and are arranged morphologically in a way that matches more or less the distribution of the e-vector pattern in the sky; they act as "crossed polarization analyzers." The insect sweeps this "matched polarization filter" across the sky and translates the e-vector information into relatively simple changes in the temporal firing of receptor interneurones; that is, the neurones speed up or slow down depending on the degree of polarization (Fig 8–5). Ants (*C. bombycina*) crossing the desert stop at intervals and perform what Wehner and Wehner (1990) describe as "a graceful little minuet," which may function as the actual scanning behavior.

Central-place foragers like bees and ants move out to a feeding site and then must return to the central place, the hive or nest. The information used to do this can come from either a sun compass or from the polarization pattern of the sky, which can still be used if the sun is invisible (von Frisch 1967; Wehner and

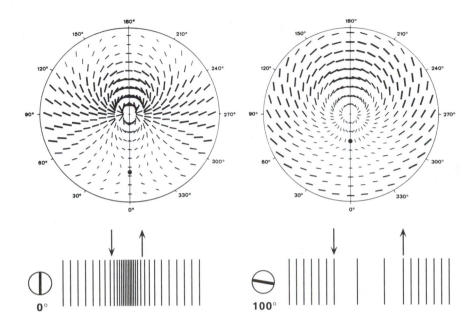

Figure 8–5. *Top panel*: Two representations of the e-vector pattern in the sky with the sun at 24° (*left*) and 60° (*right*) elevation above the horizon. The sun is indicated by the black dot in the central axis. The orientation of each black bar gives the e-vector direction, and the width of each bar indicates degree of polarization. *Bottom panel*: Diagrammatic representation of the response of an interneurone in the optic system of a cricket to changes in the polarization of incoming light. Arrows indicate onset and end of stimulus. (Modified from Wehner 1989.)

Wehner 1990). In a delightful analogy the Wehners liken the problem to the one faced by Ariadne and Theseus in Greek mythology. Theseus had to enter the labyrinth at Knossos on the island of Crete, kill the minotaur, and then escape so that he could sail with Ariadne to the island of Dia. He was given a ball of wool by Ariadne, which he unwound on the way into the labyrinth so that he could retrace his steps and return to his lover. If he had found a map with the minotaur, Theseus could have returned by a much more direct route. The Wehners ask whether a searching and foraging desert ant must use "Ariadne's thread" to return to its home nest or whether it possesses the analogue of a map, allowing return by a more direct pathway. It turns out that the photoreceptor system described above serves as a preprogrammed "hard-wired" representation of a celestial map, namely of the pattern of polarization in the overhead sky, and that the scanning of the sky allows the ant to use the map as a compass to indicate homeward direction without reference to the "thread" of its outward track. In addition the ants also more or less equalize right and left turns and make relatively small turns on average so that errors in interpreting the map are slight and tend to cancel each other out. Desert ants thus in effect integrate their outward movements away from the nest and by this process of path or route integration can then return home on a relatively straight-line course. Unlike Theseus they do not need Ariadne's thread.

The desert ant foragers studied by the Wehners are not migrants, but many migrants also can apparently detect and use e-vector information. We have already noted, for example, that amphibians can use a sun compass, and it turns out that they use polarized light as well. What is particularly interesting is that polarized light in these animals is apparently detected not by the eyes but only by an extraoptic receptor, which seems to be the pineal organ. In experiments on tiger salamanders, *Ambystoma tigrinum* (Adler and Taylor 1973) and on bullfrogs, *Rana catesbiana* (Auburn and Taylor 1979) y-axis orientation was examined in an arena where the e-vector alignment could be manipulated. The salamanders were first trained with an e-vector aligned with true compass bearing. They were then tested with the e-vector rotated 90°; both intact and blinded animals showed bidirectional orientation as expected (bidirectional because the e-vector itself has no polarity and so provides a bidirectional axis). When an opaque plastic disc was inserted under the skin of the head just above the pineal organ, the salamanders no longer oriented to plane of polarization even when optical vision was unimpeded, demonstrating that the polarized light can be used only via an extraoptic receptor. Similar tests were done with larval bullfrogs. Frogs have two potential extraocular photoreceptors, the pineal and its derivative, the frontal organ. Again, inserting opaque plastic under the skin over the pineal eliminated polaritaxis even when vision was unimpaired. The frontal organ may perhaps also be involved. Polarized light acting via extraoptic receptors thus provides orientational information to amphibians in addition to that available from the sun compass.

A characteristic of the celestial pattern of polarization is that at sunrise and sunset, but at no other time, both the e-vector and the band of maximum polarization are vertically aligned. The intersection of the vertical band of maximum polarization with the horizon provides a geographical reference independent of horizon height that can change depending on whether the sun sets or rises behind mountains, trees, or other vertical objects. Further, no time compensation is required to recognize the reference. There is much accumulating evidence that nocturnally migrating birds use the sunset to assist in setting course (Moore 1987), and the question thus arises as to exactly what information they use. This question was addressed by Moore and Phillips (1988) using the migrant North American yellow-rumped warbler (*Dendroica coronata*). After first determining that the birds could indeed detect patterns of polarized light, Moore and Phillips placed spring migrants in hexagonal enclosures and subjected them to three treatments; the control was an arrangement that allowed the birds to see the natural sunset and e-vector alignment. Under the natural conditions the birds oriented to 347° or approximately northwest as appropriate for spring migration. In the first treatment, the e-vector was rotated either clockwise or counterclockwise of north with the sunset still visible. The birds changed orientation in concordance with the new e-vector and, curiously, showed the bimodal directionality expected of polarized light in the absence of other cues. In the second treatment, the sunset was deflected with mirrors to south of its azimuth and the e-vector was again rotated. Once more the orientation of the birds was bimodal and aligned with the e-vector axis. In the final treatment, incoming light was depolarized so that the birds could not determine an e-vector and the sunset was again reflected in the south. In this experiment the birds did not orient biomodally, and their orientation was deflected from the real towards the reflected sunset. These results suggest that the birds orient preferentially with respect to the plane of polarization, but in its absence will use the position of the setting sun. Futher experiments (Phillips and Moore 1992) indicate that this position is itself calibrated by several days' exposure to polarized-light patterns. Why the e-vector orientation is bimodal when the sunset is visible and providing a directional cue is not so clear. Either the experimental apparatus rendered the sunset or other cues unusable or the natural pattern of polarization itself provides the necessary additional information. These alternatives could not be distinguished by the experimental design.

Nocturnally migrating birds sometimes continue to fly into the early morning hours in a direction that matches their nightly activity. Moore (1986) tested four species of North American warblers to see if they could use sunrise position and if polarized light was involved. Under clear skies orientation was bimodal, suggesting use of the vector, while under overcast there was no consistent direction of orientation. Rotating the e-vector resulted in bimodal orientation along the axis of rotation, again consistent with the use of information from polarized light. As with the sunset experiments, it is not clear why orientation should be bimodal or what cues, if any, are used under natural conditions to maintain orientation in the appropriate direction of migratory movement.

At this point, however, a note of caution must be injected into interpretations of polarized-light experiments with birds. Recent studies using pigeons have been unable to demonstrate any mechanism for polarized-light sensitivity in birds (Coemans et al. 1990; Martin 1991)—unlike insects, crustaceans, and fish, where mechanisms are apparent (e.g., Wehner 1989)—even though previous studies had purported to show polarization sensitivity in the pigeon eye (Kreithen and Keeton 1974). Furthermore, Coemans and colleagues were unable to demonstrate any behavioral response to polarized light in pigeons, again in conflict with the earlier results. The problem with polarized-light experiments is that it is extraordinarily difficult to eliminate differential brightness from a polarized stimulus even when using a matte black or other apparently uniform surface, and pigeons, at least, readily respond to such patterns. The fact that birds are influenced by the orientation of polarizing filters is not in question, but whether they use the e-vector axis per se or some other generated pattern is a question that must for the moment remain open, even allowing for the possibility that pigeons and migrants may differ in their visual capabilities.

A study of orientation at sunrise does not of course address the fundamental question of what cues nocturnally migrating birds use to find their way at night. As with diurnal orientation, Kramer (1950, 1951) was the first to demonstrate that the stars were necessary for orientation at night when he demonstrated that blackcaps (*Sylvia atricapilla*) and red-backed shrikes (*Lanius collurio*) displayed oriented nocturnal restlessness on clear nights but not when the sky was obscured by cloud. It was Franz Sauer (1957) and Stephen Emlen (1967), however, who demonstrated with planetarium experiments that nocturnal migrants could orient using stars alone, Sauer with European warblers (*Sylvia* spp.) and Emlen with the North American indigo bunting (*Passerina cyanea*).

The mechanisms of stellar orientation were extensively explored by Emlen (1967, 1970, 1975a,b) with indigo buntings tested in a planetarium. By shifting the artificial sky out of phase with the birds' subjective time and by selectively blocking out stars and different areas of sky, he demonstrated that the buntings use star patterns to orient, probably over a large area of sky but with the most important constellations being those closest to the polestar. Because of the fixed relationship among the stars, migrants can use them to orient without knowing time of night, season, or geographic location. Under spring skies birds in spring (reproductive) condition hopped northward in the Emlen funnel test apparatus, while under the same skies, birds in fall (nonreproductive) condition hopped southward. It was thus the condition of the bird that mattered and not the seasonal appearance of the night sky.

In further experiments Emlen also showed that the ability to use the star pattern depended on early visual experience. Nestlings reared in isolation from stars, or any other point sources of light, all displayed nocturnal activity in the test device under a stationary night sky, but none gave any indication of orienting when tested over several nights. The relevant experience necessary to use the star pattern turned out to be the axis of rotation of the stars about the firmament. This was demonstrated in two groups of experimental birds, one reared through early development under a normal sky with the axis of star rotation around the north star, and the other reared under a sky with the stars rotating around the large red star Betelgeuse in the constellation Orion. When the birds came into migratory condition in the autumn they were tested under a stationary planetarium sky. The two groups oriented quite differently. The birds reared under a normally rotating sky oriented south as they would have if migrating. The birds reared under the stars rotating around Betelgeuse, on the other hand, oriented in a direction that would have been south had Betelgeuse been the polestar (Fig. 8-6). In later experiments using European garden warblers (*Sylvia borin*), the Wiltschkos (Wiltschko and Wiltschko 1976; Wiltschko and Wiltschko 1978) showed that a natural star pattern is not necessary for orientation to develop. Any star pattern will do, even a purely arbitrary and very simple one, so long as the pattern rotates around a "star" that serves as a central reference point. This central star will then be treated by the birds as the polestar, with "north" and "south" orientation steered with respect to it. Like indigo buntings, garden warblers also acquire their star compass through early experience with a rotating stellar pattern (Wiltschko et al. 1987).

One final possible means of visual orientation that should be mentioned is the

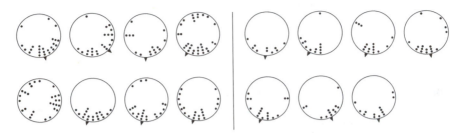

Figure 8–6. Celestial rotation and the development of migratory orientation in indigo buntings (*Passerina cyanea*). On the left are results from eight birds allowed to view a normal sky during development. These birds were tested under a stationary planetarium sky and oriented "south" when Polaris designates north as indicated by the vector arrows, which are statistically significant mean directions. On the right are seven birds that viewed a planetarium sky rotated around Betelgeuse as the pole star. They orient "south" when north is indicated by Betelgeuse. The designated "north" is at the top of each circle. (Redrawn from Emlen 1970.)

use of landmarks. At first sight, this might seem the simplest means of orienting, but this isn't necessarily so (Terrill 1991). Using landmarks to locate a specific goal requires assessing an often complex relationship among objects or patterns in the environment. Furthermore this relationship in many instances is not stable over time, and therefore may not be reliable. Among insects, landmarks may be used where goal orientation is not a factor. Thus migrating salt marsh butterflies, *Ascia monuste*, in coastal areas of Florida flew en masse on a straight course apparently using dunes, a railway line, and telephone poles as references. The insects deviated from these, however, when they no longer continued in the butterflies' preferred direction along the edges of coastal marshes (Nielsen 1961). This latter observation suggests that other cues had priority. The European cockchafer, *Melolontha melolontha*, develops in open fields, but adult beetles migrate to woodland edges after emergence, apparently cueing to dark silhouettes (Stengel 1974). Tinbergen (1953) in classic experiments with the bee-wolf wasp, *Ammophila campestris*, demonstrated, by manipulating a circle of pinecones around the nest entrance in the sand, that females returning from foraging trips use landmarks to locate the nest entrance. Manipulations have also demonstrated that bees use landmarks to return to their hives after foraging excursions (Lindauer 1976) and to identify food sources and hive entrances (Lehrer 1991). Among vertebrates, the use of landmarks is widespread among taxa, but the pattern of use parallels that in insects. Thus migrating raptors may use mountain ridges or river courses as "leading lines" and follow them for some distance, but there is no evidence that they are a primary source of information (Kerlinger 1989). Likewise a number of authors have noted that pigeons often seem to use local landmarks to complete their journey to a home loft, but that in more long distance movement, landmarks play little role, even when the pigeons repeatedly experience a particular route (R. Wiltschko 1991).

Piloting using landmarks, therefore, seems to play only a minor role in orientation behavior, and when landmarks are used, it is usually in connection with a familiar home area.

Physical Cues

In the spring on the sandy beaches of eastern North America, there is an annual migration of sometimes astonishing numbers of horseshoe crabs, *Limulus polyphemus*, which emerge at high tide to mate and lay eggs in the sand. The migratory cycle is driven by a tidal rhythm that assures that the *Limulus* move up the beach at the highest tide of the day regardless of whether it occurs by day or by night (Barlow et al. 1987). Males find mates visually, aided by a circadian rhythm that increases the sensitivity of the lateral eyes by as much as 100,000 times at night. Before mating, however, the animals must find their way up the beach, and after mating they must return down the beach to the sea. In both cases physical factors, rather than vision, are the primary cues. In getting up the beach, the orienting stimulus seems to be wave surge to which the crabs can orient when near the breeding beaches (Rudloe and Herrnkind 1976). How they get to this point from deeper water offshore, however, is still unknown. Because they emerge onto the beaches at the highest tide, they are well above the water line by the time mating is complete. In returning to the sea, they use the slope of the beach, orienting downslope until they reenter the water (Botton and Loveland 1987). Blinded individuals showed no impairment of ability to move downslope. On a flat beach both blinded and normal horseshoe crabs were disoriented, although the sighted animals less so, suggesting the possibility that the visual system might be used as a backup to the downslope orientation.

Beach slope is a simple cue that is obviously restricted in its availability only to those organisms living on beaches. A much more generally available physical cue is the earth's magnetic field—if, that is, migrants are able to detect it. The earth itself is a gigantic magnet with field lines exiting from its surface at the South Pole and extending around the globe to reenter it at the North Pole. As these field lines proceed around the earth they form continuously varying *angles of inclination* relative to the surface, being perpendicular at the Poles, parallel at the equator, and opposite in polarity on either side of the equator. At the equator the intensity of the geomagnetic field is also just about half that at the Poles. There are also local variations in the intensity of the field as a consequence of the properties of the rocks underlying the earth's surface; the most important variable is the concentration of coarse-grained magnetite (Fe_3O_4) in these rocks, which can cause "hills" and "holes" in an otherwise geomagnetically "flat" surface if it varies sufficiently (Kirschvink et al. 1986). At most places on the earth, magnetic north and "true" or geographic north do not coincide. The deviation between them, the magnetic *declination*, is large near and between the geographic poles, but is not great at lower latitudes. The general properties of the geomagnetic field are described in

Skiles (1985) and Wiltschko and Wiltschko (1988, 1991), and I have briefly summarized them in Figure 8–7.

After years of controversy and very mixed experimental results, there is now general agreement that animals can detect the geomagnetic field and use the information contained within it for orientation and perhaps also navigation. Many different workers using a range of methods have now demonstrated orientation to magnetic fields in a variety of organisms. At the most basic level, magnetic orientation occurs in magnetotactic bacteria (Blakemore 1975), implying that such responsiveness is phylogenetically very old. In addition to bacteria, response to changes in the magnetic field has been experimentally demonstrated in honeybees (Kirschvink and Kirschvink 1991), in beach-dwelling amphipod sandhoppers (Ugolini and Pardi 1992), in tuna fish (Walker 1984), in sockeye salmon fry and smolts (Quinn 1980; Quinn and Brannon 1982), in newts (Phillips 1986), in natterjack toads (Sinsch 1992), in the American alligator (Rodda 1984), and in several species of bird (summarized in Wiltschko and Wiltschko 1991). This list

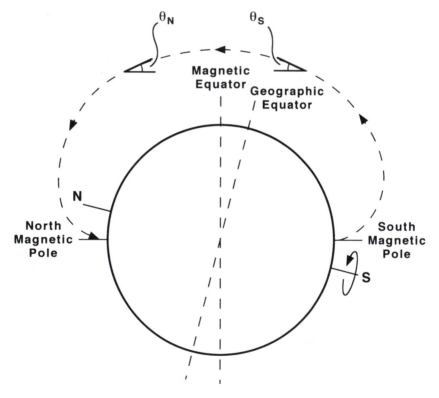

Figure 8–7. Diagram of the geomagnetic field of the earth relative to the true north-south axis. Field lines exit from the south magnetic pole and reenter at the north magnetic pole. They form continuously varying angles of inclination which are indicated north and south of the magnetic equator as θ_N and θ_S.

is by no means complete. Furthermore, a number of additional organisms have been found to have magnetite crystals present somewhere in their bodies, although—except in bacteria—no direct link between this biomagnetite and sensory transduction of the magnetic field has yet been demonstrated (reviewed and summarized in Kirschvink et al. 1985). Experiments with honeybees, however, including ones in which the orientation preference of bees was changed with pulse-remagnetization (subjecting the bee to a brief magnetic pulse), are at least consistent with a ferromagnetic process of sensory transduction (Kirschvink and Kirschvink 1991).

One of the most intriguing suggestions of magnetic orientation comes from the imaginative study of Kirschvink and colleagues (1986) in which the authors analyzed the location of beach strandings of a number of species of whales and dolphins on the east coast of North America. They first plotted satellite data on magnetic intensities of the seafloor just off the coast and noted the pattern of magnetic "ridges" and "valleys" representing regions of higher and lower magnetic field intensities, respectively. They then superimposed on this plot records of cetacean strandings catalogued by the Smithsonian Institution in Washington, D.C. These latter totaled a sample of 212 animals whose strandings fell along the coast within the boundaries of the geomagnetic survey. Statistical analyses of the data indicated highly significant tendencies for strandings to occur at coastal locations bordering local magnetic minima or valleys. Two species in particular, the pilot whales *Globicephala melaena* and *G. macrorhynchus*, had both a large number of strandings and a strong tendency to beach where magnetic fields were weak, but the general tendency was also present in the other 12 species with stranding records. One species, the dolphin *Delphinus delphis*, tended to beach more at magnetic maxima, but with a high variance associated with magnetic field strength. The data imply that these cetaceans might have a magnetic sensory system and that they can use the magnetic topography of the ocean floor to guide their long-distance migrations. They presumably do this by using a relatively straightforward tactic of following the paths provided by magnetic valleys.

Just how much animals can use changes in intensity in the earth's magnetic field either to orient or to navigate is a subject of debate. The problem is that these changes are usually very slight against the background of the total intensity of the field. This total intensity is around 80,000 nanotesla (nT; 100,000 nT equals the former 1 gauss) near the Poles and about 29,000 nT at the equator (Kirschvink et al. 1986), whereas most local intensity changes are on the order of less than 500 nT. Some animals such as pigeons (Beason and Semm 1991) may well have this sensitivity, but there is no evidence (yet) that it is used to orient using magnetic variation, although homing pigeons can be disoriented at magnetic anomalies (Wiltschko and Wiltschko 1988; Walcott 1991). What has been demonstrated, however, is that migratory birds can adapt to changes in field intensity as they move toward or away from the equator. European robins housed for 3 days at 16,000 nT, an intensity lower than any they would experience naturally, could orient at this intensity even though they could not do so initially (Wiltschko 1968).

There are two other major sources of information in the magnetic field, as indicated in Figure 8–7; namely, the polarity—the fact that the lines of force proceed from south to north—and the angle of inclination. Most of the evidence indicates that birds do not use polarity but rather possess an inclination compass (Wiltschko and Wiltschko 1988) distinguishing poleward and equatorward rather than north and south. The inclination compass was discovered in the European robin (*Erithacus rubra*) by Merkel and Wiltschko (1965) and has since been described for several other species of bird. These include at least four species of European warblers of the genus *Sylvia*; the pied flycatcher, *Ficedula hypoleuca*; the dunnock, *Prunella modularis*; and among North American species, the indigo bunting, the savannah sparrow, *Passerculus sandwichensis*, and the bobolink, *Dolichonyx oryzivorous*, all of which are nocturnal migrants. One possible exception is the Swainson's thrush, *Catharus ustulata*, followed with radiotelemetry during its spring migration between Illinois and Canada by Cochran (1987). In this bird the shifts in observed heading as this bird moved north were correlated with clockwise changes in magnetic north and not with the counterclockwise changes in magnetic dip. The experiments demonstrating the inclination compass are illustrated diagrammatically in Figure 8–8. Other animals that use an inclination compass are the sockeye salmon (Quinn 1980) and the North American red-spotted newt (Phillips 1986).

Two of the bird species shown to have a magnetic inclination compass, the bobolink and the garden warbler, *Sylvia borin*, cross the equator twice a year during the course of their annual migrations. This creates a special problem for them because for some distance north and south of the equator the magnetic field is horizontal and under these conditions birds become disoriented (Figure 8–8). Not only that, but the birds must migrate equatorward first, that is with the angle of inclination upward, and then poleward with the angle of inclination downward. Wiltschko and Wiltschko (1992) subjected an experimental group of garden warblers at the beginning of October, when they would be crossing the equator on migration, to a period when they experienced only horizontal field lines, simulating an equator crossing. When tested in the natural northern conditions of field lines directed downward toward the North Pole, these birds oriented to the north or poleward, while control birds that had not been subjected to horizontal fields continued to orient south. In other words the experimental birds switched to a preference, poleward, that in the Southern Hemisphere would have meant continuing to migrate south. In a similar experiment with bobolinks, Beason (1992) tested birds in a planetarium with fixed star patterns, but in a series of magnetic fields incremented each night from a northern angle of inclination through horizontal to a southern angle of inclination. The birds maintained a constant southward heading throughout the experiment, including the time they were experiencing horizontal fields. Evidently these birds can use the star pattern to maintain proper heading while crossing the belt of horizontal fields at the equator. As with the garden warbler, experience with the horizontal fields probably triggers a switch from equatorward to poleward orientation, although this experiment does

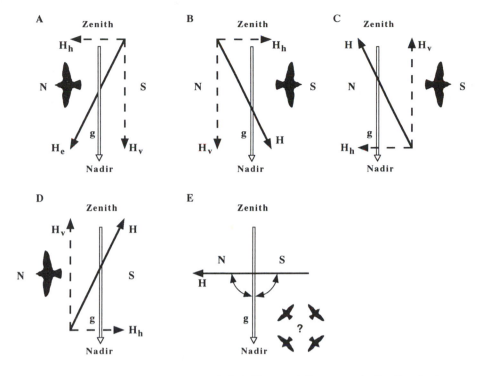

Figure 8–8. The inclination compass of birds illustrated diagrammatically. Panels show a vertical section through the magnetic field. *A*. Orientation in a local geomagnetic field with vector H_e; H_h and H_v are the horizontal and vertical components and g is the gravity vector. N and S indicate north and south. *B*. The horizontal component is reversed to yield vector H, and the bird orients south. *C*. The vertical component is reversed yielding the same result as *B*. *D*. The field is inverted, and the bird again orients north (vector at same angle as in H). *E*. In a horizontal magnetic field the birds would still receive information about the position of the North-south Axis but would no longer be able to decide which end was which. (Redrawn from Wiltschko and Wiltschko 1972.)

not rule out a fixed star pattern accomplishing the same thing. Suffice to say, a mechanism exists to cope with the equator crossing.

All the birds discussed so far with respect to magnetic orientation have been nocturnal migrants. A diurnal migrant that also shows magnetic compass orientation is the yellow-faced honeyeater, *Lichenostomus chrysops*, of eastern Australia (Munro and Wiltschko 1992, 1993; Munro et al. 1993). The largest population of this species breeds in southeastern Australia south of Brisbane and on the coastal side of the Great Dividing Range. In the autumn the birds migrate first northeastward, roughly following the Australian coastline to the latitude of Brisbane, and then northwestward, again roughly following the coastline to their northern overwintering areas. The route is reversed in the spring.

Munro and her colleagues tested yellow-faced honeyeaters trapped during autumn migration and examined their orientation in Emlen funnels both under the natural sky and in the laboratory with no celestial cues. Both sets of birds oriented to the northeast early in the season (March-April), but changed to a preferred northwesterly direction later (May to July); the birds thus possessed an internal seasonal clock (Chapter 6). The directional changes paralleled those observed in the field and also demonstrated that the birds could orient even when the sky was not visible. In the spring, birds oriented in southerly directions again with or without the sky visible. Laboratory experiments confirmed that these honeyeaters were capable of using magnetic fields to orient. If the angle of inclination of the magnetic field was reversed, the birds reversed direction, strongly suggesting that like their holarctic counterparts, they were using an inclination compass.

Whereas in migratory birds an inclination compass seems paramount, there is evidence that eastern red-spotted newts, *Notophthalmus viridescens*, have two magnetoreceptive pathways and use both magnetic inclination and north-south polarity (Phillips 1986). Newts exhibit both y-axis orientation to the shoreline, when they are in their breeding ponds, and homing ability that allows them to find their natal ponds (Gill 1978a). The animals switch from one type of behavior to the other when the water temperature is manipulated. A water temperature of 18°C and less than 5° fluctuation up or down results in y-axis orientation, while fluctuations from 3° to 5° at night up to 25° to 27° during the day causes orientation in the direction of the home pond. Reversing the angle of inclination of the magnetic field causes a reversal in y-axis orientation when the newts are orienting to the shoreline. After the fluctuating water temperature treatment, however, changes in the angle of inclination do not alter the ability to orient toward home. Evidently the newts have a map sense (probably acquired while residing in training tanks), and a response to the polarity of the magnetic field provides the compass component for navigation. The significance of temperature is probably that when it is stable it signals that newts are in the pond, but when highly fluctuating, as it would be on land in early spring, it is a signal to home to the breeding site (Phillips 1985, 1987).

Evidence for two magnetoreceptive systems in both amphibians and birds also comes from other experiments, along with data that suggest that at least one of these is associated with the visual system. Phillips and Borland (1992) tested newts showing y-axis orientation under full-spectrum light, and these continued to orient toward shore regardless of the position of the magnetic north pole relative to the shoreline. When tested under the same intensity of light limited to 450 nanometers wavelength, the behavior was unchanged. When the wavelength was 550 nm, however, the newts shifted their orientation approximately 90° counterclockwise. The fact that there was a shift in orientation rather than simply a disruption of behavior indicates that the wavelength of light was specifically influencing the directional information that the animals perceive via the magnetic field. Beason and Semm (1991) have examined the electrical activity of single units in the nervous system of pigeons. Earlier experiments had shown that pigeons transported in the dark to a release site behaved, when released, in the same way as those transported in a distorted magnetic field (Wiltschko and Wiltschko 1981). Evidence that retinal photoreceptors could be involved came from electrical recordings of cells in different parts of the visual projection area of the brain that responded to changes in the angle of inclination of the magnetic field. On the other hand, there were responses in ophthalmic nerve fibers coming from the upper beak to intensity changes in the magnetic field of as little as 200 nT, sensitive enough to detect magnetic anomalies and therefore possibly of use in a magnetic map sense. Although there was no direct evidence, Beason and Semm speculate that these latter responses may be associated with magnetite crystals in cells of the ethmoid region of pigeons. In any event, there appear to be two distinct magnetoreceptive systems in pigeons, one sensitive to inclination angle and one to intensity.

One other source of directional information potentially available to birds is infrasound, or sound generated by waves of very low frequency, below about 30 hertz, which are generally considered out of the acoustic range. In the pigeon, there are fibers in the ear that fire spontaneously and whose frequency of firing is modified by infrasound down to a frequency of as low as 1 Hz (Klinke 1991). Infrasounds emanate from sources such as distant thunderstorms or waves breaking on a beach. Under some circumstance these could provide directional information useful to a homing or migrating bird. There is, however, no direct behavioral evidence that infrasound is actually used.

The Coordination of Orientation Systems

One of the classic experiments on bird migration was that of the Dutch ornithologist A. C. Perdeck (1958). In its conception the experiment was simple enough, but it revealed a great deal about the behavioral organization of migration and orientation patterns. Perdeck captured starlings and chaffinches in Holland during their southwestward migration in the autumn, banded them, and transported them to Switzerland, where he released them to continue their journey. Some 10,000 starlings were banded and released, and the subsequent recoveries of the bands revealed an interesting picture of starling migration. During the ensuing winter, recoveries of first-year birds indicated that they kept going in a southwesterly direction and wintered in southeastern France and Spain about the same distance south of their usual wintering area as the displacement to Switzerland. Adult birds, on the other hand, corrected for the displacement and were recovered in their usual wintering areas in western France. After the initial wintering in the "wrong" area, however, the young birds subsequently returned to the breeding area where they had been raised. The experiment demonstrated that there were both endogenous (genetic) components to migration, such as the continued southwestern movement of the displaced young birds, and components obviously based on experience, such as the return of adults to their usual wintering area and possibly the return of the young to their natal area for breeding.

In the course of further experiments on endogenous or genetic components of migratory behavior, a navigational model has been developed called the "vector navigation hypothesis" by Schmidt-Koenig (1973; Berthold 1991a). The basic notion is that migration is composed of a time program based on a circannual rhythm (Chapter 6) coupled to a corresponding directional program. The time program determines the onset of migratory activity and departure based on fat deposition and the molt sequence. Once migration begins, the time program and a genetic directional component (Berthold 1991a; Helbig 1991a and Chapter 14) will determine where it goes and how long it takes to get there. In first-year birds the migration journey often takes considerably longer than in older birds, probably because they lack previous experience with the route. It is apparent that both genetic and environmental (experience) components (Chapter 14) are involved in the development and honing of vector navigation, and as we have seen in the above discussion, of orientation mechanisms as well. It is appropriate here to examine the relation between the ontogeny of cue use in orientation behavior and how the behavior is then integrated with migration syndromes and performance.

Over 20 years of studies on avian navigation have begun to create a picture of how genes and experience interact to produce the guidance system that a bird actually uses (Wiltschko et al. 1987; Able 1991a,b,; R. Wiltschko 1991; Wiltschko and Wiltschko 1991). These studies of several bird species have revealed, first, that the ability to orient using the inclination of the earth's magnetic field is genetically based. Hand-raised birds showed appropriate orientation in the absence of all other information. In nocturnal migrants, which have been the most intensively and extensively studied, the pattern of rotation of the stars and constellations around a fixed point is acquired by experience during the first several weeks of life. Under natural conditions in the Northern Hemisphere the fixed reference point is the North Star, but laboratory rearing has revealed that any rotating pattern of stars will do. Presumably a similar rotating pattern of stars is also used in Southern Hemisphere migrants, but has not to this point been tested. Experiments in which hand-raised birds were exposed to a magnetic field rotated away from N-S while they were simultaneously exposed to natural night *or* daytime skies showed that under either circumstances the birds behaved in subsequent tests as if the rotated magnetic field orientation was the correct one. In other words, under a natural sky and in a natural field, they rotated their direction of orientation in a direction predicted by the earlier rotation of the manipulated field. Thus even though the orientation of the migratory route based on the inclination compass is acquired genetically, the direction can be modified based on experience with celestial cues. Exactly how daytime skies serve to do this, if in fact they do, remains to be assessed. Experiments also show that the development of the compass mechanism that seems to be involved with migratory orientation at sunset involves experience gained during the first 3 months of life, even though the cues remain unclear. A conceptual model for the ontogeny of avian guidance systems is shown in Figure 8–9.

A role for experience in the ontogeny of guidance systems is also apparent in

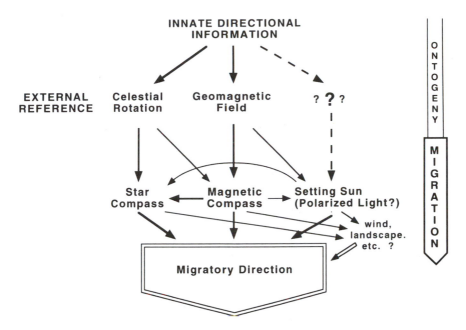

Figure 8–9. Wiltschko conceptual model for the ontogeny of avian guidance systems based on the interactions among stars, the earth's magnetic field, and sunset. (Redrawn and slightly modified from Wiltschko and Wiltschko 1988.)

other organisms. Eastern red-spotted newts, for example, evidently learn the location of their home ponds and can learn new "homes" when kept in laboratory training tanks (Phillips 1986). Frogs and toads also return to natal ponds for breeding, although in toads it is only males that do so, with displaced females orienting to the nearest breeding chorus of males (Sinsch 1992). Honeybees also learn the location of the home hive (Lindauer 1976) and American alligators display site-based navigation that improves with age (Rodda and Phillips 1992). Experience is thus an important component of acquiring a navigational "map."

Another general principle to emerge from studies of animal orientation is the similarity in the cues used even among distantly related taxa. All orienting or navigating organisms face similar kinds of problems with similar constraints in terms of the environmental cues available and the reliability of those cues. For this reason, it is not really surprising that there is much overlap in the approaches or strategies used (Rodda and Phillips 1992). If one scans the groups that have been studied, one observes that crustaceans, insects, fishes, amphibians, reptiles, and birds all use both sun and magnetic compasses. There are differences too, of course, often based on the sensory or cognitive capabilities of particular taxa. Fishes, insects, reptiles, and amphibians (Sinsch 1992) all seem to make prominent use of odor for orientation; the evidence for odor orientation to a home or natal area in birds is equivocal, and in any case does not have the conspicuous role across species or in migration that it does in other groups. In those groups that do use odor, the zigzag approach up-current to an odor source is remarkably similar where it has been studied. In contrast, birds may be alone in using a star compass, although there have been virtually no studies of possible stellar orientation outside birds. Orientation to the e-vector of polarized light has been clearly demonstrated in insects and fish and polarization analyzers have been described in the visual system. Although birds respond to polarization patterns, the evidence is not so clear on the use of e-vectors, and polarization analyzers have not been found. The diversity of the taxa with similar orientation mechanisms suggests that these mechanisms probably have a long evolutionary history (Terrill 1991) and have been adapted to migration rather than evolved especially for it.

The similarities among taxa extend a step further in that all use multiple cues and that the cues serve as backups to each other. Under cloudy skies, when celestrial cues are invisible, most animals still orient quite well by using the magnetic field, for example. Magnetic orientation may also be favored under certain special circumstances such as the orientation of salmon smolts under ice (Quinn and Brannon 1982). There is in all probability a cue hierarchy for each organism, but there is no consensus for any organism on exactly what the hierarchy might be. This is true even for the celestial versus magnetic compasses of nocturnally migrating birds and in the homing pigeon, the two subjects on which most of the work and thought have gone (Able 1991a). This state of affairs is again not really surprising, given the difficulty in ascertaining just how a bird makes a choice or for that matter, what cue it is actually using under natural conditions. Experimental conditions may also impose a response that would not occur in nature (Helbig 1990; Able 1991b,1993). A further complication arises from the transfer of information among orientation systems rather than the detection and use of each cue occurring in a separate channel (Able 1991a). It is probably safe to say that the complexity of the interacting systems serves as a way to deal with the spatial and temporal variability with which any organism, let alone a migrant, must deal in even the simplest orienting performance (Able 1993). Beyond that, however, the adaptive value of the complexity in any specific situation or for any given organism is not very clear. Most research, perhaps legitimately, has gone into the elucidation of mechanisms; greater attention to adaptive value, however, may in turn shed light on the mechanisms.

A final common theme to note is that orientation and navigation are not perfect. The variability that is described as "straying" in salmon that fail to return to their home streams or as "vagrancy" in birds that occur well outside their usual range is discussed in Chapter 12, where the possible evolutionary significance is also considered. It is worth noting in addition, however, that even in the most carefully controlled experiments, there is usually considerable scatter to the orientation of the subjects. In some experiments, especially those on magnetic orientation, the scatter can be considerable even when statistical tests indicate significant differences among test groups or between experimentals and controls.

This variation in performance is usually attributed to the unnatural conditions of the experiments, and this no doubt contributes a major share. Although recognized as a source of variance, individual differences are almost never examined in detail; the emphasis, rather, is on differences in means among groups. One wonders, though, if there aren't potential insights into the orientation process available in the scatter among individuals that is so prevalent. My own feeling is that they could stand more attention.

Summing Up

The past few decades have seen great progress in our understanding of the mechanisms migrants and nonmigrants alike use to find their way about. There have been some surprises. Thirty years ago, no biologist could have predicted the extensive use of magnetic information that we now recognize as common to many organisms from bacteria to birds, even though we still have not identified the sense organs or the transducing devices. The complexity of the interacting hierarchical systems could also not have been predicted, yet as a general principle of organization it is now well recognized (Able 1991a). In addition to magnetic fields, the stars, the sun, and planes of polarization of light are all used by various animals, often in combination. We also recognize that organisms capable of homing from some distance have some sort of "map" that combines recognition of a deviation from a reference direction and recognition of the distance to the home site or the wintering site. This is the case for both vertebrate and invertebrate migrants (Downhower 1992). To this point there has been more emphasis on the direction component than on the distance component, although the concordance of migratory restlessness with the time taken in the wild to migrate for species-specific distances (Berthold 1984a,b and Chapter 11) suggests a genetically based distance component (as incorporated into the vector navigation model). In a species such as the pigeon, or in breeding birds homing after long-distance displacement from their nests (R. Wiltschko 1992), the distance to home must in some way have been gained by experience. Downhower (1992) suggests that it would pay to give more attention to the distance component and its relation to the direction component than has so far been the practice.

Two other points seem worth stressing. The first is that we still have information concerning maps or compasses on very few species, and of these by far the most data come from a single species, the domestic pigeon. The Wiltschkos (1991) argue with some justification that mechanisms are similar across species and that as a consequence results from any one species can be generalized, but we really need more comparative data before the argument can be accepted with total confidence. The second point is that it is not clear whether differences observed among species are due to real biological differences or are rather the consequence of dissimilarities in experimental designs. There is a need for studies of a spectrum of species, employing similar experimental procedures to produce comparable results (Helbig 1990; Able 1991a, 1993). Experiments should also be interpreted carefully with the perspective of field observations (Moore 1987). Concurrently, we also need more data on the orientation of a few species under a variety of conditions to explore in greater detail the adaptive nature of guidance mechanisms so that we can place them in an evolutionary framework (Alerstam 1991). Because controlled conditions are so critical to the elucidation of orientation mechanisms, researchers must choose tractable systems. Rodda and Phillips (1992) suggest that much of the unexplained variability in, for example, pigeon homing may be the result of field conditions that are impossible to control. They suggest more effort be focused on amphibians and reptiles; an equally strong case can be made for insects, as the elegant experiments of the Wehners have demonstrated (e.g., Wehner and Wehner 1990). Both insects and amphibians have the advantage of rendering large sample sizes possible and allowing experiments under less confining conditions than cage experiments with birds. Finally, to return to a theme discussed at the beginning of this chapter, we need to ask continually how much complexity is really necessary. A healthy balance between skepticism and delight along with judicious applications of Occam's razor should guide both field and experimental studies.

PART THREE

Migratory Life Histories and Their Evolution

In Part II the focus was on the *how* questions of migration. In the chapters of Part III that follow, I turn to the *why* questions and consider data and theory that may help us understand the evolution of life histories that incorporate migration as one of their major features. Chapter 9 opens this section with a discussion of seasonal migrations, which are those likely to be most familiar. In the first part of the chapter I focus on particular migration pathways and outline what we know of the selective factors favoring their evolution. It is apparent from many studies that seasonal migration patterns are resource based, with an abundant, predictable, but temporally restricted food supply the most important source of selection. Examples from spiny lobsters to migrant land birds support this conclusion. The second part of the chapter directs attention more to the issue of why some species of a region or taxon migrate seasonally while others do not. Here the most important factor seems to be the extent of dependence on a particular set of resources, especially if that set is subject to high levels of spatial and temporal variation. Particularly noteworthy is the observation that in both birds and insects, long-distance migrants to the Temperate Zone belong to taxa that also display migration within the Tropics.

In Chapter 10 I discuss some pathways that involve matching the life histories of migrants with special requirements. Among these are refuges, with two examples of passages to refuge sites being the molt migrations of waterfowl and the flights of moths to high-altitude summer estivation sites. Other special requirements are restricted breeding locations. Thus most amphibians must return to ponds or other limited sources of water to lay their eggs and have evolved sometimes complex migration systems to accomplish this. Analogous to these amphibian migrations are the return of crustaceans to the sea to release eggs or young, and the return of pelagic birds or turtles to islands or beaches. At a microscale level, phoretic mites require matching of their migratory behavior to movements of the insects that transport them. I conclude the chapter with an extended discussion of diadromy in fishes, which includes many of the most interesting cases of the evolution of specialized migratory life-history syndromes.

If the persistence of habitats is highly variable through time, migration is favored as a strategy to escape from deteriorating habitats and to colonize new ones. A number of taxa inhabiting ephemeral habitats have incorporated migration into their life histories, and these are discussed in Chapter 11. Among the most variable of ephemeral conditions are those occurring in arid environments, and denizens of these arid habitats display some of the most highly evolved migratory life cycles. Two of these are the desert locust (and related locusts and grasshoppers) and the African armyworm, elaborated at some length in the chapter. They are particularly interesting because of their highly polyphenic morphology and behavior. I also discuss movements of dry-country birds, especially those of Australia. In concluding the chapter I consider the movements of some northern birds and conclude with a brief consideration of forest bark beetles, whose ephemeral resources are trees that must be in just the right conditions to support beetle reproduction and population growth.

Selection for migration varies in intensity and pattern, with the result that migratory behavior and life histories vary considerably both within and across species as discussed in Chapter 12. I first discuss the variation apparent among both populations and species of (especially) insects, birds, and fishes where these sources of variance have been relatively well studied. In these taxa, there can be major life-history differences between migrants and nonmigrants. I then proceed to examine the variation often present among individuals of the same population. Studies of this phenomenon have been considerably aided by the exploration of theoretical models, in particular that of the evolutionarily stable strategy or ESS. This has been used, for example, to analyze partial migration where some individuals migrate while others do not. Variation occurs also in the routes and distances of migration as well as in whether or not migration occurs. In birds migration distances can vary between the sexes, a difference also occurring in eels. Better studied are variations in migratory pathways, which are known for a number of organisms. One aspect of these variations is straying (fishes) or vagrancy (birds) in which a few individuals follow routes that can be quite different from those of the main population. From such deviations salmon colonize new breeding streams; in Europe, the blackcap warbler has evolved a new route to an overwintering area in southern Britain and Ireland. Rather than being simply "error variance," route and timing variants can have important evolutionary consequences.

The most extreme forms of migratory variations occur with polymorphisms and polyphenisms, and these are explored in Chapter 13. They are particularly prevalent in insects, where they take a variety of forms. These range from instances where wing muscles are histolyzed following migration through various environmentally induced combinations of winged and wingless forms (polyphenisms) to situations where wing form is almost entirely the result of genetic influences (polymorphisms). The different wing morphs are often associated with different life-history syndromes of varying complexity. Comparisons at different hierarchical taxonomic levels consistently suggest that pterygomorphisms are associated with degrees of habitat permanence, with total winglessness more likely in habitats that are long lasting. Polymorphisms and polyphenisms, especially the latter, are also found in the seeds of plants, with a partitioning between seeds germinating near the parent and those adapted for transport, allowing migration to more distant germination sites. Transport mechanisms include wind, water, and the fur of mammals.

In Chapter 14, the last in this section, I discuss the genetic basis for migratory life histories. Only if the variations observed in migration are genetically based can they be modified by natural selection, and the evidence suggests that indeed genetic variation in migratory behavior is widespread. The most often used method of demonstrating gene differences is the common garden experiment, which has revealed genetic influences in a diverse set of organisms that produce considerable variation among populations in migratory performance. In a few cases variation in migration is based in single Mendelian loci, but more often analyses reveal polygenic effects. Heritability estimates reveal that genetic variance for migration can be quite high. High heritabilities predict rapid response to selection, confirmed in laboratory selection on both birds and insects. Selection experiments also reveal the presence of migratory syndromes with sets of behavioral and life-history traits that share genes in common and that are evidently molded by natural selection. I conclude the chapter with a brief consideration of maternal effects that can also have major influences on migration, especially in certain insects. In all the instances discussed, migration is seen as a complex adaptation arising from multiple interactions between genes and the environment producing the great variety of migratory syndromes we observe.

9

Seasonal Migration

Seasonal Migration Patterns

Seasonal migrations involve movements among two or more areas that are occupied in different seasons during the annual cycle. Classically these are migrations to one location for breeding and to another for the remainder of the year. Annual migrations of birds like those of the arctic tern and the swift parrot outlined in Chapter 3 are fairly typical examples and illustrate the extremes of distances traveled. Such to-and-fro movements are not limited to birds, but occur also in organisms as diverse as whales and butterflies. It is also the case that seasonal movements can be among more than two places. The movements of wildebeest between three different parts of the Serengeti region of East Africa (Chapter 3), and the numerous cases of circuits to at least three areas of ocean—for breeding, nurturing of young, and adult feeding—in many marine fishes (Harden Jones, 1968; Dingle 1980) are good examples of multipartite migratory cycles. In these examples we see already some of the variability present in seasonal movements. This theme of variability and its significance will pervade the discussion in this chapter.

To indicate the variety present in seasonal migrations and to contrast these with other movements that may also be seasonal, I give some examples, using birds from three continents, in Table 9-1. Some points are worth noting. First of all is the fact that seasonality of movement is not restricted to migration. Both local wandering and local nomadism, rather arbitrarily separated from each other here by the distances moved, occur seasonally, especially outside the breeding season. Australian fairy wrens are more likely to wander in the winter or the dry season and European tits, blackbirds, and starlings do most of their wandering in the winter. Lorikeets are somewhat less seasonally restricted in their search for trees in blossom because the programming of blooming varies seasonally among the tree species visited. These wandering and nomadic movements are probably best considered extended foraging or ranging. The second point is that a species is not confined to a single type of movement. Starlings display extensive intra-continental migrations in Europe, for example, as well as nomadic movements, and the white-crowned sparrow is divided among populations that are sedentary, partially migratory, or wholly migratory. In the western United States the yellow-rumped

Table 9–1. Different Types of Seasonal Bird Movements, with Examples from Three Continents

Type of Movement	Australia[*]	Nearctic	Paleartic
Extended Foraging or Ranging			
Local wandering (< 1 km)	Fairy wrens (Maluridae)	Chickadees (Parus)	Tits (Parus)
Local nomadism (1–100 km	Lorikeets	Tricolored blackbird	Starling
Migration			
Partial migration	Silvereye	White-crowned sparrow	European robin
Altitudinal migration	Olive whistler	Yellow-eyed junco	Alpine accentor
Intra-continental migration	Yellow-faced honeyeater	Yellow-rumped warbler	Redwing
Irruption outside normal range	White-browed woodswallow	Bohemian waxwing	Common crossbill
Continental nomadism	Grey teal	Pintail (?)	Pallas's sandgrouse

[*]Australian examples from Ford 1989.

warbler is primarily an altitudinal migrant, while in the eastern United States it is an intra-continental migrant. Pallas's sandgrouse irrupts well beyond its usual range as well as wandering widely in western and central Asia, and pintails in North America are north-south migrants in addition to being nomadic.

With the exception of some species of Australian fairy wrens and lorikeets, which are not migrants, the movements listed in Table 9–1 occur within the Temperate Zone or between temperate and tropical regions. It would be misleading, however, not to discuss bird migrations within the Tropics because, contrary to the picture we usually get, they are actually quite common (Levey and Stiles 1992). The example of the quelea was discussed in Chapter 3; another good example from Africa is the grey-headed kingfisher, *Halcyon leucocephala*, studied in West Africa by Elgood and colleagues (1973). This kingfisher is primarily a savanna and open-woodland species and like others in its genus feeds terrestrially, mostly on insects. The pattern of its migration and breeding in Nigeria is shown in Figure 9–1 along with the general latitudinal distribution of the rainy season. The rains begin in March and last until November in the southern forest region, but last for only the three months of July through September north of about 12°. Grey-headed kingfishers begin breeding at the start of the rains in the south, but then a major part of the population moves north to commence breeding when the rainy season begins. There is then a postbreeding migration north into desert regions that are occupied during the local rains. The result is that *H. leucocephala* has nonoverlapping "winter" and "summer" ranges with migration between them. The related *H. senegalensis*, which is more of a closed-forest bird, is largely resident in the southern forested regions, but some do move northward to occupy desert regions during the middle of the year. It is thus a partial migrant, extending its range during the height of the wet season. Many other bird residents of Nigeria show intratropical migration patterns resembling those of these two kingfishers.

This pattern of north-south movement synchronized with the rains produced by the Inter-Tropical Convergence Zone (see Chapter 5) is also seen in locusts and bats of the Sahel and desert areas of West Africa. In locusts the breeding is

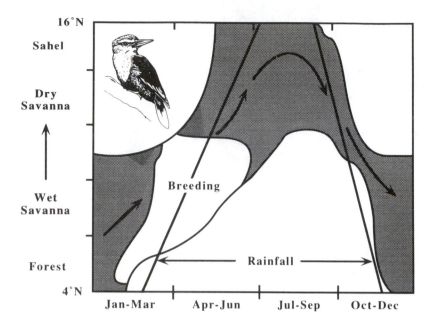

Figure 9-1. The migratory pattern of the grey-headed kingfisher (*Halcyon leucocephala*) in Nigeria. Breeding begins at the start of the rainy season in the forested south, but most of the population migrates north to breed during the rains in savanna regions. There is then a further postbreeding migration northward before a return south at the end of the rainy season. (After Elgood et al. 1973 and Sinclair 1983.)

associated with the inundation areas of the River Niger (Farrow, 1975b and Section 3 below). At least one species of bat, *Eidolon helvum*, migrates about 1500 kilometers annually to follow the rains into the Niger River basin (Thomas 1983), and other tropical African bats may also move some distance in response to rainfall (Fenton and Thomas 1985). Migration within the Tropics is thus a general phenomenon.

Given the considerable energetic costs of migration (Chapter 6) and the probable risks associated with hazards encountered en route, it is appropriate to ask at this point why organisms should migrate seasonally at all. Theories of seasonal migration focus on the abundance of resources available in spring and summer in the Temperate Zones and on the fact that winters are to a greater or lesser extent inhospitable. Older theories of bird migration, in particular, tended to have a thread of Lamarckism or group selection passing through them and assumed either that species evolved in the Temperate Zone needed to escape the winter (northern ancestral home) or that those evolved in the Tropics migrated to take advantage of resources in the north (southern ancestral home). As pointed out by Moreau (1951), however, the former ignored the extensive migrations occurring in the Tropics, and the mixture of zoogeographical elements in the bird faunas of wintering areas argues against these regions being ancestral to many migrants (Mayr 1957). Neither ancestral home theory was therefore very satisfactory. These and other older theories of bird migration are reviewed by Dingle (1980) and extensively by Gauthreaux (1982).

The major turning point between early and modern theories of bird migration was the publication of David Lack's (1954) book on the natural regulation of animal numbers, followed closely by Salomonsen's (1955) paper that specifically addressed the evolutionary significance of bird migration (Gauthreaux 1982; Dingle and Gauthreaux 1991). Lack in particular drew a clear distinction between ultimate evolutionary factors and proximate causes and outlined a cost-benefit approach that has dominated most thinking since 1954. In a nutshell Lack proposed that if the benefit-to-cost ratio of moving exceeds the ratio for staying, then migration will be favored by natural selection. That of course is true for all organisms, not just birds, and as a consequence Lack's contributions pervade all subsequent discussions of the evolution and ecology of migratory behavior, and attention has concentrated on trying to assess what the various costs and benefits might be.

Perhaps the first major attempt to evaluate the role of resources in the evolution of bird migration was that of MacArthur (1959). MacArthur used breeding bird censuses from North America to examine the proportion of Neotropical migrants in the faunas of three habitat categories: grassland, northeastern deciduous forest, and northern coniferous forest. As a result of his analyses he concluded that the density of breeding migrants was highest where the difference between winter and summer food supplies was greatest. This was more likely to be true for insectivores than seedeaters, and indeed there were more insectivorous than granivorous migrants. Insectivores tend to occur in forest, and forest had more migrants than grassland and deciduous forests more than coniferous forests. Willson (1976) reanalyzed MacArthur's data and added some additional censuses. She found no difference in proportion of migrants between the two forest types (contra MacArthur), but again both had a higher proportion than grassland. One reason for the lack of difference in proportion between the forest categories appeared to be that coniferous forests also had fewer resident individuals and species than deciduous so that a roughly equal number of migrants made a larger contribution to the former. Most of the migrants in conifers were parulid warblers, which seem to be specifically adapted to this habitat, although exactly what these adaptations are still needs study. The general conclusion in both the Willson and MacArthur studies is that seasonal differences in resources do influence the contributions migrants make to various habitats.

An underlying assumption of both papers is that northern areas do provide rich resources for Neotropical migrants to exploit. It is worth asking whether this is a justifiable assumption. On the face of it, at least, the Neotropical habitats, and especially the tropical forests that are left behind in favor of northern breeding areas, are lush and resource rich. From the perspective of breeding birds, however, appearances are evidently misleading. Many forest species are rare, and breeding involves many risks and difficulties (Karr 1990; Keast 1990). The rainforest, in particular, in Keast's words "does not have, at least year-round, the exuberant resources as commonly believed." One adaptation to cope with these less than exuberant resources would seem to be migration, with the great burst of resources in the northern spring and summer a potent selective mechanism favoring its evolution.

Neither MacArthur nor Willson could demonstrate much of an influence of latitude, as distinct from habitat type, on the proportion of migrant birds in North America. Herrera (1978) found, in contrast, a strong effect of latitude in Europe (Fig. 9–2). Herrera used multivariate regression models to examine the proportion of passerine migrants in various communities as a function of different combinations of geographic and climatic variables. The best single predictor was latitude followed quite closely by the temperature of the coldest month. There was no significant influence of habitat type per se. What best explains the proportion of migrants seems to be harshness of the winter coupled with total resources available during breeding. Hard winters coupled with productive breeding seasons produce the highest proportion of migrants. Herrera also speculates that the absence of a latitudinal effect in North America may have been in part because the censuses

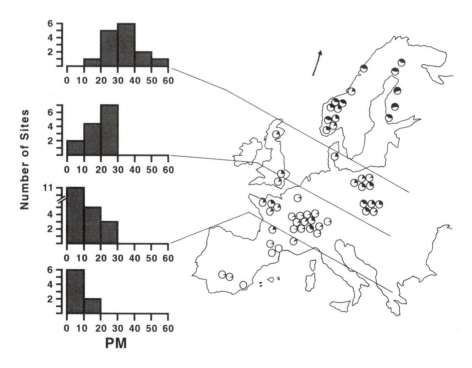

Figure 9–2. The proportion of passerine migrants (PM) in sites from different latitudes in Europe. The proportion increases at higher latitudes as shown both by the amount of shading in the circles on the map representing the sampled sites and by the frequency histograms at the left. (Redrawn from Herrera 1978.)

came from a somewhat limited range, with the more northerly parts of Canada underrepresented. In contrast the European samples covered a wide range of latitude.

The situation with respect to seasonal migration of birds in Australia is more complex than in either the Nearctic or Palearctic migration systems. Largely this is a consequence of the varied climatic regimes in Australia, encompassing as they do an arid northwest and interior, a tropical north and northeast, a Mediterranean southwest, and a temperate southeast with much altitudinal variation. Although the continent does serve as a wintering area for Palearctic breeding birds, most of these are shorebirds; the primary migrations of land birds and wetland species is within Australia or in some cases between Australia and New Guinea (which forms part of the same faunal province). The most intensive attempt to analyze the ecological factors influencing bird migration in Australia is that of Nix (1976). Nix based his analyses on the climatic factors controlling the growth and development of plants, on the premise that the food of birds depends directly or indirectly on plant primary productivity, and that the course of plant dry matter production should then allow predictions of both breeding and movements of birds. Nix computed indices of plant growth based on climatic indices of light, temperature, moisture, and evapotranspiration; from these, and the limited data available on bird movements and breeding cycles at specific sites, he constructed maps of plant growth and predicted bird movements.

The results of Nix's analyses are summarized in Figure 9–3. In Figure 9–3A is indicated the timing of the maximum quarterly plant growth index values for Australia and New Guinea. In the north the maxima occur in the summer, coinciding more or less with the summer rainy season. In the south, maxima are more evident in the spring when the temperature warms up, but before the impact of the summer dry season; highest growth indices occur earlier in the more Mediterranean southwest. In general peaks of breeding activity by both residents and invading migrants coincide fairly well with peaks of plant productivity. Figure 9–3B indicates the predicted overwintering areas and seasonal movement patterns of birds. In general the most favorable wintering areas are those with a temperature regime and enough available moisture to permit plant productivity sufficient to sustain relatively high bird densities. Again, from the information available on bird movements, the predictions seem to match reality reasonably well.

Some more specific points also emerge from the analysis. First, temperature was a more important factor in southeastern Australia than it was elsewhere. Even though the moisture regime is favorable for most of the year, cold winter temperatures inhibit plant growth. This presumably accounts for the major migrations out of the region in the winter and for the notable altitudinal migrations that also take place within the region. Second, the southwest is a virtually self-contained area with early spring breeding and a localized wintering center on the coast. Third, the large central arid zone is subject to episodic rainfall with consequently more erratic bird movements. Nevertheless there is apparently a significant north-south component to these movements, with more southerly

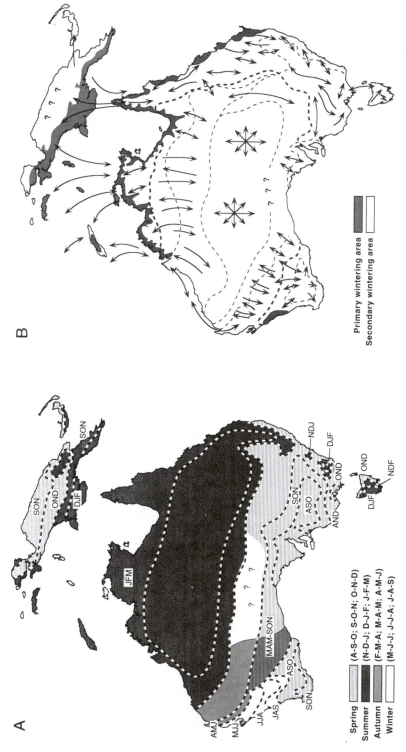

Figure 9–3. *A*. The timing of maximum quarterly plant growth indices for Australia and New Guinea. Letters refer to months. In the north maxima occur in the summer coincident with the rainy season. In the south maxima are more likely in the spring before the onset of the summer dry season. Growth is somewhat earlier in the Mediterranean southwest. *B*. Predicted overwintering areas and overall seasonal movement patterns of Australian migrant birds based on the seasonal plant growth indices. (After Nix 1976.)

breeding and a northerly retreat in the autumn and winter. This bias in the direction of movements reflects the action of tropical summer and temperate winter weather systems.

Finally, Nix's analysis perhaps explains some migratory movements that are otherwise difficult to interpret. For example, the dollar bird (a roller), *Eurystomus orientalis*, and the buff-breasted paradise kingfisher, *Tanysiptera sylvia*, both have populations that migrate from New Guinea to Australia to breed. These species have a "leapfrog" migration pattern with wintering north of the mountain spine of New Guinea and migration to Australia over resident populations in the somewhat drier areas south of the central mountains (Schodde et al. 1975). A glance at Figure 9-3A reveals that plant growth peaks in southern New Guinea in the summer, reflecting the fact that this region is more seasonal than other parts of the island. To invade this area would require competing with the already resident birds; the result is that selection evidently favors overflying this region and crossing the Torres Straits to Australia, where competitors are absent. The general issue of what sorts of selection might favor migration in particular species of bird in the first place, as well as how it might favor patterns like leapfrogging, will be discussed further below.

Birds are of course not the only organisms displaying seasonal migrations, and a list of participating biota would span organic diversity from fungal spores (e.g., in Rainey 1978) to the great whales. In terms of total biomass involved, insects are at least equal to birds, although except for a few butterflies and occasionally locusts, their movements are not nearly so noticeable. Two relatively well-studied New World insect migrants are the monarch butterfly and the large milkweed bug, *Oncopeltus fasciatus*. The annual migratory cycle of the monarch was summarized in Chapter 3; the cycle of the milkweed bug is very similar (see also Chapter 6). Both species are specialists on milkweed plants (*Asclepias* spp), and in eastern North America both have a two-step migratory movement in the spring (Dingle 1982; Brower and Malcolm 1991). The first step involves movement to the various species of milkweeds that flower in the early spring in the coastal plain around the Gulf of Mexico, the most common of which are *A. viridis* and *A. humistrata*. These southern milkweeds, however, set seed and die back in early summer so that the butterflies and bugs must move on. This they do by continuing northward where a vast crop of milkweeds, especially *A. syriaca*, begins flowering in late June and early July. After one to two generations on these northern milkweeds, both migrants undergo an adult reproductive diapause under the stimulus of short day (Dingle 1974; Barker and Herman 1976 and Chapter 6) and undertake their return movements south. For these species migration allows the exploitation of the prolific crop of summer milkweeds and the building up of large populations of insects. The size of these populations probably results in a net selection differential favoring migration in spite of mortality en route or suffered during the overwintering period. Both species are tropical in origin and cannot survive northern winters, even in diapause. It is the evolution of a diapause occurring concurrently with migration, however, that allows escape to the south in the autumn (Chapter 6 and below).

Round-trip seasonal migrations spanning several generations occur in several species of noctuid moth and in various leafhoppers and planthoppers. These have received a great deal of attention because most are major agricultural pests that invade crops as they are planted in spring or early summer. In eastern North America, for example, the fall armyworm moth, *Spodoptera frugiperda*, invades the northeastern and north central United States each spring via transport on southerly wind systems and builds up enormous populations in agricultural acreage (Sparks 1979; Johnson 1987). The same pattern is observed with the armyworm *Mythimna* (=*Leucania*) *separata* and the rice planthoppers *Nilaparvata lugens* and *Sogatella frucifera* in eastern Asia, where they invade rice fields. Northward-moving weather systems take these insects as far north as Korea on the mainland and across the Sea of Japan to the Japanese islands (Chen et al. 1989; and reviewed in Dingle 1989). In Australia four species of *Heliothis* (=*Helicoverpa*) moths invade agricultural areas in New South Wales in spring and summer from overwintering areas to the north and west (Farrow and McDonald 1987; Fitt 1989), and two species of *Heliothis* migrate into the southern United States from Mexico and Texas (Raulston et al. 1986). Late in the season many of these seasonally migrant moths and homopterans are trapped in situ by the onset of polar weather systems, because none of the species discussed can diapause either as adults or juveniles.

The resulting die-off of large numbers of insects led Rabb and Stinner (1978) to speculate that these migration systems might be the equivalent of a "Pied Piper" phenomenon in which each year the migrants were lost to the population. The migration system, they postulated, might be the result of the quite recent establishment of large areas of crops, providing temporary but rich resources for the migrants. Such a system would present a dilemma because such losses would result in severe selection against migration (Walker 1980). The dilemma may be more apparent than real, however, because more recent studies indicate that there can be considerable return movement on the winds of southward-moving weather systems to wintering areas where breeding can take place. This is true for both moths (McNeil 1987; Chen et al. 1989) and for planthoppers (Taylor and Reling 1986b; Riley et al. 1991).

Another consideration is that migration in these species may be adaptive for finding ephemeral habitats over large parts of the range. Johnson (1987), for example, suggests that *S. frugiperda* may have evolved migration to locate areas in Central America where there has been recent rain, producing suitable breeding habitats. This species is thus seen as being very similar in its migratory strategy to the African armyworm, *S. exigua*, discussed in Chapter 3 (Rose et al. 1987). This is also analogous to mechanisms proposed for insect migration in Australia (Drake and Farrow 1988, 1989). In these models the assumption is that even if losses do occur as the result of polar displacements, they are balanced by the reproductive success of the migration system in tropical or subtropical habitats. To the extent that return movements occur, they would add to selection favoring migration; in regions where there are large populations developing on crops, this additional selection potential could be, and probably is, considerable. The seasonal availability of resources is thus likely of major import for the migrations of these insects.

At the other extreme in body size from migrating insects are the mysticete or baleen whales; yet like the insect (and bird) species discussed above their migrations are at least in part resource driven (Dawbin 1966; Lockyer and Brown 1981). The most important factor influencing the migration of the large baleen whales such as the blue (*Balaenoptera musculus*), fin (*B. physalus*), sei (*B. borealis*), and humpback (*Megaptera novaeangliae*) whales is food availability. Polar seas are enormously productive, compared to those in temperate and especially tropical regions, because of the high solubility of CO_2 in cold water. This productivity, however, is largely limited to summer months because of the extreme cold and low light levels of polar winters. In the Antarctic the primary food of whales is the euphausiid crustacean *Euphausia superba*, commonly known as krill, which can reach such extraordinary densities that the sea may appear pink from the hemoglobin in their semitransparent bodies. In the Arctic there are also swarming euphausiids, but shoaling fish and large concentrations of *Calanus* copepods are additional major food sources.

In contrast there is very little food available in tropical waters, but the thermal environment is favorable for energy balance. The result is that baleen whales spend the winter months at low latitudes because, with little food available anywhere, the gain in heat balance becomes the primary factor. Selection for migration thus seems to center on the balance between food availability and heat loss with the deposition of fat in the form of blubber during the polar summers followed by a period of near starvation but energy conservation in warm waters.

This selection is further reinforced by the reproductive cycle. Newborn calves have extremely thin blubber providing little insulation; birth in warm waters allows the suckling calves to devote energy to growth and buildup of blubber rather than maintaining body temperature against an extreme thermal gradient. Lactation is also very costly, with an energy drain some 12–15 times that of pregnancy (Lockyer 1978) so that during periods when food is scarce, lactating females also benefit from tropical seas and reduced heat loss. Much of the age and sex variation in the migratory patterns of mysticete whales can be accounted for by the reproductive cycle; pregnant females, for example, spend the longest time on the feeding grounds, and suckling females and their calves are the last to leave tropical wintering areas (see Chapter 11). Whale populations or stocks are generally confined to one hemisphere, which raises the question of why migration doesn't occur from pole to pole. The critical factor here may well be the birth and suckling of the young, possibly both because of the favorable thermal environment of the Tropics and because the distances involved would overly stress the migratory capacities of newly weaned young. This stress would be intensified by the lack of food in tropical waters.

It should be noted that in contrast to the mysticetes, extensive migrations are uncommon in the odontocetes or toothed whales. Migration does seem to occur in the sperm whale (*Physeter macrocephalus*) and the white whale (*Delphinapterus leucas*) and perhaps some other species, but they do not seem to have the regularity of the migrations of the large mysticetes. These whales are primarily fish and squid feeders, and their food generally does not seem to show the marked seasonal fluctuations in abundance seen in krill or other food of the baleens. Selection for migration is thus weak, if present at all. The white whale is restricted to the polar regions and several other small odontocetes are also confined to temperate or polar latitudes. These species all give birth to their young in cold waters and so have evolved efficient heat-conservation mechanisms in the young, unlike the mysticetes.

Another interesting seasonal migration in the marine environment is that of the spiny lobster, *Panulirus argus*, off the Bahama Banks between the Bahama Islands and Florida (Herrnkind 1980, 1985, 1991; Kanciruk and Herrnkind 1978). During the summer months the Bahamas population of *P. argus* is scattered over the shallow bottom (2–10 meters deep) of the great and Little Bahama Banks in a variety of habitats from coral to sea grass flats. The water at this time of year is clear and warm ($\sim 30\,^{\circ}$C), and the primary activities of the lobsters are feeding and growth. In the autumn, changes in weather systems bring storms and wind that result in increased turbulence, increasing murkiness, and reduced water temperature, which can decline precipitously because the shallow sea loses heat rapidly when it becomes turbulent. At this time the lobsters begin a remarkable behavior known as queuing in which long lines of animals form and move across the ocean floor, each animal maintaining touch contact with the one in front, using antennae and feet. Laboratory experiments indicate that it is water movement, which greatly intensifies in the autumn, that triggers the queuing behavior. The queues of lobsters move westward off the Banks, and then turn southward along their western edge to move down the Florida Straits to warmer water where they spend the winter. Return movements in the spring are by single individuals that may spawn first at the edges of the Banks before spreading over the interior shallows.

Although the role of queuing in the migration system of *P. argus* is still not very well understood, the behavior does appear to have two major advantages. First, for all except the lead animal, drag is greatly reduced, allowing more rapid movement with less energy cost. Because movement can be up to a kilometer an hour, such costs could be considerable. Lead lobsters do not realize an energy saving, and perhaps because of this, the lead exchanges at frequent intervals. Second, the queues are an effective defense against predation. When under attack the animals in a queue rapidly spiral into a rosette formation with spiny antennae projecting outward. Observations suggest that such a defense can be very effective and that single individuals are much more vulnerable. The whole queuing system may be the key to exploiting rich resources on the Bahama Banks because it allows rapid escape from the Banks in the autumn when temperature drops.

Approximately every 10 years or so autumn water temperatures can drop rapidly to such a degree that lobsters die while molting or become moribund and more liable to predation. In these years the rapid escape allowed by queuing, triggered by water turbulence preceding maximum temperature drops, would be a distinct advantage. Furthermore, the wholesale movement of single lobsters would probably attract predators, and the defensive behavior facilitated by queuing would be an additional benefit. It thus may well be that it is the evolution of queuing behavior that has allowed the seasonal migration to the shallow Bahama Banks because it allows rapid escape in the autumn.

A final example of interesting seasonal migrations are those of many species of fish into the floodplain areas of tropical rivers of both Africa and South America (Lowe-McConnell 1975; Goulding 1980). The rainy season can bring enormous fluctuations in water levels several vertical meters in extent, with the result that much of the area through which a river passes will be inundated for periods as long as 6–8 months. The patterns for the two main tributaries of the Amazon are shown in Figure 9–4; the differences between the two rivers arise from differences in the timing of the rainy season as the Inter-Tropical Convergence Zone (ITCZ) passes from the south over the basin of the Rio Madeira to north over the basin of the Rio Negro (see Chapter 5). Each of these rivers themselves has several tributaries, and Michael Goulding (1980) spent two years studying the fishes of the Rio Machado, a clearwater tributary of the Rio Madeira, whose waters are generally turbid and more nutrient rich. Prominent among the migrant fishes were species of large characins, a varied group whose best-known examples are the various piranhas. Unlike piranhas, however, most characins are vegetarian.

The migratory movements of several species of large (up to several kilograms) characins can be somewhat complicated but can be generally summarized with the following pattern. As the waters rise in the Rio Machado between roughly the middle of November to the middle of February, the characins move downstream to the Rio Madeira where spawning takes place. After spawning, they migrate back

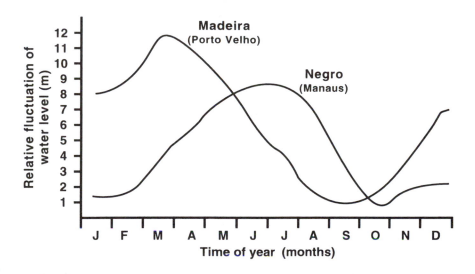

Figure 9–4. The seasonal rise and fall of water levels in two tributaries of the Amazon River in Brazil. (After Goulding 1980.)

up the Rio Machado at the height of the water level in the inundation zone in the floodplain. This floodplain is generally covered with forest, and the fishes now become foragers in this flooded forest, feeding extensively on the seeds and fruits of trees. These latter fall into the water where many float and are transported away from the parent tree. In a very real sense, the fishes exploit the migration systems of the trees, and they have well-developed teeth and jaws allowing them to crunch even hard and thick-walled seeds. They may also assist tree migration by passing some seeds through their guts. The juvenile fish probably move downstream in the Rio Madeira to feed in the floodplain of that river before attaining adult size and moving back to the Rio Machado, although the movement of these young fish is largely speculative. Suffice it to say, these characins, and also other fish groups, are remarkably adapted through both migration and feeding apparatus to take advantage of the extensive annually flooded areas of forest in the Amazon Basin.

The Evolutionary Ecology of Seasonal Migrations

All the above examples of seasonal migration illustrate a further point, that among the various taxa there can be conspicuous differences in the proportion of the taxon that is migratory. A question important to understanding the evolution of migration systems is, then, what causes are behind these differences? To take birds first, it is apparent that in some groups only one or a few species are migrants. In rollers (*Coraciidae*) and bee-eaters (*Meropidae*), there is only a single migrant species at each end of the taxon distribution. The European bee eater, *Merops apiaster*, and the European roller, *Coracias garulus*, migrate into Europe from overwintering areas in Africa where other resident species occur, and the dollarbird, *Eurystomus orientalis*, and rainbow bee-eater, *Merops ornatus*, migrate from Southeast Asia and New Guinea into Australia. All other bee-eaters and rollers are largely sedentary or local in their movements. In contrast to the limited number of migrants in these two bird families, the vast majority of shorebirds or waders are migrants to breeding areas in the Nearctic and Palearctic. Among insects large milkweed bugs and monarch butterflies are the single long-distance seasonally migrant species from among their respective groups, while among noctuid moths of the genus *Spodoptera*, most are impressive migratory travelers.

As with most complex behavioral ecologies and apparent in the examples already discussed, migration systems seem to have evolved as the result of a diverse set of selective factors that vary in importance in the different taxa, depending on ecologies and life histories. In shorebirds it is likely that the total area of available habitat, as it influences both food and nest site availability, may be the most important consideration in the evolution and maintenance of migration. Most shorebirds breed in habitats that are extensive and rich in food supplies compared to areas where they spend the winter. By far the majority breed in the high Arctic, which covers a vast area and produces an enormous biomass of potential insect food during the very brief summer nesting period. Even those species that breed on grasslands or other interior habitats in the Temperate Zone have available an area that is large relative to the wintering grounds. Only a few shorebirds breed in the Tropics or on the temperate margins of continents. The vastly increased nutrient and energy demands of breeding are probably difficult, if not impossible, to sustain on the basis of the resources available on the strands, estuarine mudflats, or grasslands of the continents south of the usual northern breeding areas. Certainly shorebirds can have a considerable impact on the density of infaunal prey, especially if they stay around for very long (Schneider and Harrington 1981; Wilson 1991). The availability of nest sites is no doubt also a major factor. Temperate and tropical feeding areas, not all that extensive to begin with, are also frequently subject to inundation during the tidal cycle, making nesting possible only at their margins. Predation likely enters into the equation as well, and mammalian predators increase in abundance and diversity at lower latitudes. A somewhat parallel situation to that of shorebirds occurs in *Spodoptera* moths; they too exploit greatly expanded seasonal breeding areas brought about either by the rainy season in the Tropics or the Temperate summer.

For the large milkweed bug, the monarch butterfly, and the spiny lobster, *P. argus*, the key to migration seems to center on the development of certain physiological or behavioral traits. The two insects belong to Neotropical genera, the other members of which are not migrants (Dingle 1978; Dingle et al. 1980; Ackery and Vane-Wright 1984). The key to the successful adoption of a migratory life history is probably the acquisition of the ability to diapause (Dingle 1978; Tauber et al. 1986), which the nonmigrant species are unable to do. In the case of the milkweed bug, the enabling mechanism has apparently been the ability to shut down the juvenile hormone production of the corpora allata in response to photoperiod (see Chapter 6). Milkweed bugs in general can respond to starvation by entering a diapause-like state which includes ceasing to mobilize resources for reproduction and greatly reduced metabolic rates (Dingle 1974); these resources are mediated by the corpora allata (Johannsen 1958). The evolution of a photoperiod link to the corpora allata meant that short days could produce the same effect. Because the advent of reproduction shuts down migration in milkweed bugs, the ability to turn off the reproductive system with short days means that the large milkweed bug now has a lengthy period to migrate to overwintering areas. The key to successful migration was thus not a means of invading northward in the spring but rather the means to escape southward in the autumn. Similarly the adult monarch also diapauses in short days (Herman 1981), permitting the long autumn migration so characteristic of this species.

For the spiny lobster the key response seems to be queuing behavior. Among the numerous species of *Panulirus* in the tropical seas, only *P. argus* displays this response. As with the two migratory insects, the constraint is apparently fall migration. In this case the requirement is an expeditious departure from the Bahama Banks when storms from the north presage the possibility of a rapid drop in water temperature that could prove fatal. Queuing facilitates such a departure. Again the limitation is not the ability, or rather lack of it, to invade but rather the evolution of the mechanism of escape once a seasonal migratory invasion has occurred.

Turning once more to birds, the species that have received the most attention regarding their migration are the large numbers of terrestrial birds, about 400 estimated by Karr (1980), that breed exclusively in the North Temperate Zone but winter in the Tropics. In addressing the question of why some species migrate and some do not, I shall focus on passerines, because it is for this group that most data are available. One approach, pioneered by MacArthur (1959), as discussed above, is to examine the habitats occupied. MacArthur, Willson (1976), and Herrera (1978) concentrated on the habitats of birds in their breeding areas. Bilcke (1984), on the other hand, proposed that the proportions of migrants in the various habitats in breeding areas were determined by the proportion and distribution of habitats in the wintering range. His conclusion was based on a comparison of migrants in Europe and eastern North America. In the former region most migrants occur in areas of scrub and open forest with vegetation of intermediate height; few occur in closed forest (Fig. 9–5). These migrants winter in Africa, which is dominated by the same sorts of habitats, and it is in these savanna and open-woodland habitats

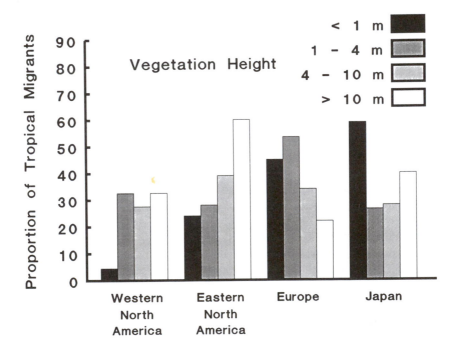

Figure 9–5. The proportion of tropical migrants occurring in different habitats, classified according to the height of the vegetation, from four regions of the Northern Hemisphere. Proportions add to more than 100% because migrants can occur in more than one habitat. (Based on data from Helle and Fuller 1988 and Mönkkönen and Helle 1989.)

that they spend their time while overwintering (Moreau 1972; Lack 1990). In fact in one study in West Africa, Palearctic migrants were never, in 20 years of studies, observed to enter primary forest, although they were common at forest edges, in open fields, and in man-made forest gaps (Brosset 1990). In the Neotropics, an area where primary forest covers a much larger proportion of the area than in Africa, many migrants do occupy tall primary and secondary forest (papers in Keast and Morton 1980), and the proportion of migrants in Temperate forests is also relatively high (Fig. 9–5).

Bilcke's analysis has been extended by Helle and Fuller (1988) and Mönkkönen and Helle (1989), who include measures of vegetation height and extend the database to include East Asia and western North America (Fig. 9–5). The vegetation height analysis confirmed Bilcke's thesis for eastern North American and European migrants. Two different patterns emerge in western North America and in eastern Asia, although the much fewer habitats sampled makes caution advisable in their interpretation. In western North America migrants are more or less evenly divided among the habitats with vegetation over one meter in height; this pattern may result from a more limited overwintering distribution than occurs in eastern North American migrants. These western migrants apparently winter almost entirely in a narrow strip along the Pacific Coast from the Mexican state of Sonora south to Guatemala (Hutto 1985). Vegetational diversity here is a good fit to the diversity in the breeding ranges. Similarly in eastern Asia, tropical regions are largely divided between forested (Wells 1990) and open areas with little in the way of shrub or savanna habitats. This would fit the distribution of migrants observed in Japan. Overall, then, there does seem to be a good correspondence between winter and breeding habitats occupied by migrants based on comparisons across continents.

This conclusion, however, still begs the question of which species, among those occupying a given overwintering habitat, are the ones that migrate and which are the ones that stay and breed in the Tropics. The question is directly addressed by Cox (1985), who then tests his model using the migration patterns of parulid warblers, a large proportion of which migrate between temperate North America and the Neotropics. Cox begins by noting that the ratio of juvenile survival to adult survival in species breeding in the Tropics is low, suggesting severe competition for food to feed the young and perhaps also for nest sites protected from a diverse set of predators. Adult survival can be as high as 80–90% per annum, including survival of migrants, compared to 40–50% for Temperate Zone residents. All these data indicate that many more birds can be maintained in the Tropics than can be raised. On the other hand, clutches of both residents and migrants at higher latitudes are larger than are those of tropical residents. The overall fecundity of migrants is less than that of Temperate Zone residents because the migrants have less time available to them for breeding. Because it is based on this presumed balance between competition and time available for breeding (so influencing fecundity) as the major cause for the evolution of migration, Cox calls his model the time allocation and competition theory.

The evolution of migration, and its form, is seen as taking place in four major steps. The first step in the process (stage I) occurs when tropical species expand their ranges into the subtropics, where climate and seasons are more variable. The selection driving this initial expansion could arise from a number of sources including climate, resource variability, competition, or predation, and it is likely that more than one are involved. Once range expansion occurs, these same factors may favor partial migration, involving further expansion of a portion of the population into areas that are seasonally favorable for breeding with retreat back to a main area during the remainder of the year (stage II). This stage would also be favored by unstable habitats leading to reduced site tenacity, as postulated also by Alerstam and Enckell (1979) (see Chapter 11 for further discussion of the evolution of partial migration). Such unstable conditions are generally not present in the Tropics, but they are and probably have been in the past in a region that extends from the southern edge of the Mexican Plateau northward into Texas and the interior deserts of the southwestern United States. Because of physiographic diversity and climatic change through time, geographical isolation promoting speciation was also probably frequent. Cox thus argues that the Mexican Plateau is the likely staging area for the evolution of migration in the New World, especially in groups that now have a high proportion of migrant species or populations.

Stages III and IV of Cox's scheme involve the development of long-distance disjunct migration by and further differentiation of partially migrant populations to still higher latitudes, coupled with the competitive displacement of lower-latitude residents and short-distance migrants. Although increased temporal or spatial seasonality of climate, as might occur with an overall trend toward a colder regime, would favor the evolution of long-distance disjunct migration, Cox favors competition as the most important driving factor. In an extension of earlier ideas (Cox 1968), he postulates that the continued invasion of other tropical species into the staging area resulted in the elimination of many partial migrants via inter-specific competitive interactions, leaving the migratory populations as the survivors. The ultimate consequence of competition, habitat change, and other factors might be a species such as the Kirtland's warbler (*Dendroica kirtlandii*), now endangered and limited to a tiny population (circa 200 or so pairs) with very restricted breeding and wintering distributions.

Because they are a speciose group with a high frequency of seasonal migrants, the parulid warblers seem a good taxon on which to test the model. First, long-distance migrants have fewer subspecies per species than residents or partial migrants, consistent with the elimination of resident populations during the course of the evolution of disjunct migration. Second, the northern limits of the ranges of partial migrants exceed those of residents, as predicted; the centers of distribution are still in the Mexican Plateau, however, also as predicted. The northern limits of residency for other tropical families that have evolved long-distance migration are similar to those of parulids. Finally, the theory predicts that breeding ranges of migrants should push north from the Mexican Plateau, and wintering ranges should push south, with intermediate populations eliminated. That being the case, regressions of the northern and southern limits of the ranges against the latitudinal interval between them should meet (at the point of zero interval) at a latitude that coincides approximately with the middle of the Mexican Plateau. If, on the other hand, there was a tropical origin of migrants, the point of crossing should be at a tropical latitude. A multiple origin should lead to crossing lines averaging out to an intermediate latitude, but with much higher variance nearer the origin; that is, at earlier stages of evolution.

In Figure 9-6 are the computed regression lines for the ranges of all the migratory subspecies of parulid warblers. As is apparent, the regression lines cross just before intercepting the ordinate at a latitude of about 29°N or roughly in the center of the Mexican Plateau. Furthermore, there was no tendency for variances to increase at smaller latitudinal intervals between breeding and wintering ranges. The evidence thus favors the Mexican Plateau as the staging area for the evolution of parulid migration systems. The heterogeneous climatic history and physiography of the region, promoting both speciation and migration, also goes a long way toward explaining why there is a large fraction of migrants in a diverse group such

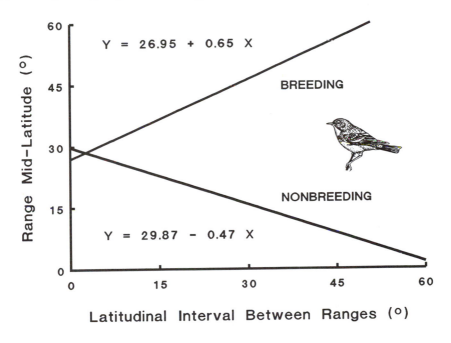

Figure 9-6. The computed regression lines for the latitudinal midpoints of the breeding and nonbreeding ranges of migratory parulid warblers in North America plotted against the latitudinal intervals between the midpoints. The lines cross at a latitude of approximately 29°N or roughly the mid-latitude of the Mexican Plateau. This suggests the Plateau was a staging area for the evolution of migration. (From Cox 1985; individual data points omitted.)

as the parulid warblers. As Cox points out, there now need to be tests of the various aspects of the theory and also tests using other groups of species. On the face of it, a similar theory based on the Sahara-Sahel region of Africa might well explain much of the evolution and diversity of the Palearctic-African migration system, but its explicit development and testing remain a challenge for future research.

Levey and Stiles (1992) also consider New World bird migration systems, using an approach different from cost-benefit analysis. They ask if there are specific ecological traits that make some species more likely to be long-distance migrants than others. To address this question, they examined the habitat and diet characteristics of Costa Rican land birds, an avifauna that shows much evidence of seasonal movements within the Tropics. The hypothesis tested was that Nearctic migrants shared many ecological characteristics with Neotropical groups showing high frequencies of migratory movements on a smaller scale. Seasonal movements within the Tropics were postulated to predispose those groups displaying them to migration out of the Tropics. Ecological attributes hypothesized to select for migration were high temporal and spatial variation in the resource base. Such variation was thought more likely among species feeding on fruits or flowers as opposed to insects or vertebrates, and among species living in the forest canopy or on the forest edge rather than in the vegetation of the interior. Birds in canopy or edge habitats and with fruit diets should be less buffered against seasonal changes or climatic extremes.

The data summarized in Tables 9-2 and 9-3 indicate that both diet and habitat are highly correlated with degree of altitudinal migration, the most common form of migratory behavior in these species. With respect to diet, for example, some 56% of nectarivorous species are migrants, the highest proportion observed (Table 9-2). Nectar is likely a resource particularly susceptible to transient fluctuations. Among frugivores, 34% are migrants. These proportions for fruit and nectar feeders contrast sharply with the 9.7% and 6.7% migrants among predators feeding on either small or large prey, respectively. With respect to habitat (Table 9-3), by far the highest proportion of migrants, about 26%, occur among species that inhabit the forest canopy or edge. Both sets of data support the hypothesis that specific ecological attributes are associated with increased tendencies toward seasonal

Table 9-2. Association Between Diet and Seasonal Altitudinal Migration in 346 Resident Forest Birds on the Caribbean Slope in Costa Rica

Diet Category	Total Species	Migrant Species	Percent Migrants
Small insects	145	14	9.7%
Fruit/seeds	94	32	34.0%
Vertebrates/large insects	75	5	6.7%
Nectar	32	18	56.3%

SOURCE: From Levey and Stiles 1992.

Table 9–3. Association Between Habitat and Altitudinal Migration and Habitat in Costa Rican Caribbean Slope Forest Birds

Habitat	Total Species	Migrant Species	Percent Migrants
Understory	88	6	7.3%
Canopy/edge	242	62	25.6%
Aerial	16	1	6.3%

SOURCE: From Levey and Stiles 1992.

migratory movements. Evidently these movements are adaptations to track fruit and flower availability, especially in more exposed and more seasonally variable habitats.

To fully support the Levey-Stiles hypothesis, Neotropical groups that are generally frugivorous and inhabit canopy or forest edge should be the groups in which Nearctic migrants occur. This does indeed seem to be the case. Families that are exclusively insectivorous and live in forest understory such as the Furnariidae (ovenbirds, spinetails, foliage-gleaners) and the Formicariidae (antthrushes and antpittas) contain no long-distance migrants. In contrast in families like the Tyrannidae (New World flycatchers) and Vireonidae (vireos) migrants are common. Some species in both taxa frequently consume fruit, especially the migrant species, and most tend to be forest edge (flycatchers) or canopy (vireos) species. In six of eight frugivorous families examined, migrants evolved, while none did so in any insectivorous forest interior families. It is also worth noting that many long-distance migrant species are insectivorous on the breeding grounds but revert to a concentration on fruit in the wintering areas. As a result Levey and Stiles suggest that one of the factors selecting for migration in the long-distance species is the rich protein source of Temperate Zone insects because fruits are notoriously low in protein content. Note also that competition, as in Cox's (1985) model, is not precluded here; but, if it occurs, it is likely to be a major factor primarily with a frugivorous feeding mode because tropical resident frugivores also increase insectivory during breeding. The competition would occur for protein (insects) to support reproduction.

These results suggest some interesting parallels between long-distance migrants in both the New and Old Worlds. As we have seen in Chapter 6, Bairlein (1990) and others have observed increasing frugivory in Palearctic migrants departing the Temperate Zone in the autumn and en route. It would be interesting to know if this switch reflects a high incidence of frugivory among related tropical residents. Several authors, as noted earlier in this chapter, also report that, like their New World counterparts, Palearctic migrants tend to be species of more open habitats and do not occupy dense forest while on the wintering grounds. There thus seem to be strong parallels between the habitat characteristics of Nearctic and Palearctic migratory species.

The major exception that does not fit the Levey-Stiles ecological traits model is the parulid warblers. Contrary to the predictions of the model, species in this group are both largely insectivorous and highly migratory. The best explanation for their high frequency of migration may be that a more northern origin in the Mexican Plateau meant that the primary selection resulted from highly seasonal habitats. Cox's competition-time allocation model is also more likely to apply because selection resulting from these factors could be more extreme in a more severe environment. In contrast to migrant parulids, tropical resident warblers are both insectivorous and denizens of wet forest understory; therefore they do fit the ecological traits model. Suffice it to say the Cox and Levey-Stiles models are far from mutually exclusive, and the relative degree to which each applies undoubtedly is a function of ecological circumstance.

Summing Up

Ultimately, long-distance seasonal migratory movements are driven by the resources available in the breeding habitat. The rich food supplies of the vast Arctic and North Temperate land masses or of the southern oceans, to cite prime examples, are present only for a limited period. To exploit these resources requires not only the ability to reach them, but also to escape with the onset of winter. For migratory birds, whales or other long-lived organisms, this factor has resulted in the repeated evolution of long round-trip journeys often accompanied by sophisticated navigation to precise breeding and wintering sites (Chapter 8). In insects such as the monarch butterfly and the large milkweed bug, the evolution of a link between photoperiod and the corpora allata permitted a short-day–induced diapause, effectively extending the life span to allow escape migration and overwintering in less severe climatic conditions. As a result the milkweed crop of temperate North America became available to them. The evolution of an escape mechanism may also be central to the ability of spiny lobsters to exploit resources in the shallow waters of the Bahama Banks.

Many long-distance seasonal migrants reflect a migratory trend in their respective taxa that has also led to the evolution of migratory behavior within the Tropics. Thus armyworm moths migrate between rainfall-induced seasonal habitats, and the northward movement of the fall armyworm in the southeastern United States is probably an extension of a pattern that already exists in the Caribbean and Mexican populations of the species. The advent of agriculture has provided extensive seasonally exploitable habitats north of the ancestral range. A similar mechanism may be at work in the oriental armyworm moth and in the rice planthoppers of East Asia. Monarch butterflies in Costa Rica migrate back and forth between the dry Pacific slope and the more humid Caribbean lowlands (Haber 1993). Neotropical bird families with a high proportion of intratropical migrants are also the ones represented among the species invading the Nearctic to breed.

The high frequency of migrants within an avian family seems to be a function of feeding mode and habitat. In the New World tropics a relatively high proportion of frugivores and nectarivores are migrants, while among insectivores migrants are generally poorly represented. Parulid warblers are the exception among the latter, but they apparently evolved farther north on the Mexican Plateau in a region of spatially and temporally more variable habitats. Here competition and time allocation may be more important selective factors in the evolution of migration. More variable and climatically harsher regions and habitats seem to select for an increased incidence of long-distance seasonal migration in the avian and insect faunas of both the Old and New Worlds. A further similarity of the two avian migrant faunas is that the migrants occupy similar habitats in both breeding and wintering areas, and in many cases also while en route. In a very real sense, in fact, seasonal migrations may be considered a special case, although a well-represented one, of migrations in response to ephemeral habitats. These latter will be treated more fully in Chapters 11 and 13.

10

Migration to Special Habitats

There are a number of migratory phenomena that involve movements to sites that provide specific requirements. Among the more familiar examples are the migrations to restricted breeding locations by marine vertebrates of terrestrial origin. Thus albatrosses and marine turtles wander widely over the world's oceans but return to breed on islands or beaches that are mere specks on the surface of the globe, driven by the needs of their young to hatch on land. The opposite need occurs in amphibians, whose young require water. This demand for water leads to migrations that take breeding adults to a range of aquatic habitats, some ordinary like temperate ponds, others exquisitely exotic like the bromeliad nursery pools of certain tropical tree frogs. Bromeliads may also serve as nurseries for crabs (Hartnoll 1963), of which many, like amphibians, spend their adult stages in terrestrial habitats but, with some notable exceptions, must return to water to release their eggs. The most studied of migrations to restricted breeding sites are those of diadromous salmonid fishes, as we have seen already at several places in this book. The origin and evolution of diadromy will also be taken up in this chapter.

Breeding requirements, however, are not all that drive migration to specific sites or habitats. Among the special needs of certain life cycle stages may be a requirement for shelter or the fulfillment of unusual physiological demands. We saw in Chapter 7 the requisites supplied to overwintering monarch butterflies by the high-altitude refuges of the transvolcanic mountains of Mexico. A diverse array of other insect species also seek refuges through migration to and from diapause sites. Hibernating amphibians, reptiles, and mammals also need refuge, and finding it often involves migration. Likewise many aquatic birds are flightless during the molt and must find places that provide both refuge and the resources for the high energetic demands of regrowing flight feathers.

All the above migrations have in common the fact that they take the migrants to places supplying specific needs that cannot be met elsewhere. Many are synchronized at least in part with seasonal changes, but they are sufficiently different from the usual run of seasonal movements that a separate treatment seems justified. For this reason, I consider them apart from the more usual seasonal migratory patterns described in the previous chapter. The examples discussed should illustrate the somewhat special nature of these migrations to restricted sites and habitats and the nature of the selection that may be acting on them.

Refuging Migrations

Among the more interesting of migrations to refuges are the molt migrations of waterfowl and some other aquatic birds (Jehl 1990). These movements occur after the breeding season when large numbers of birds from over a wide area concentrate at highly localized sites. At these locations the birds are flightless for periods of several weeks or more while regrowing the major flight feathers (primaries) following their molt. Such movements are independent of fall migration and, in fact, often are in the opposite direction from the latter (Salomensen 1968). Also, not all age and sex classes are involved, but usually only postbreeding adults. A good example occurs in Australian shelducks (*Tadorna tadornoides*), which concentrate on Lake George in eastern New South Wales from an area that apparently includes New South Wales, Victoria, South Australia, and Tasmania, at least judging from the recoveries of banded birds that have left Lake George following reacquisition of flight capacity (McKean and Braithwaite 1976; Fullagar et al. 1988). The area of recoveries is shown in Figure 10-1.

Where refuges are good, large concentrations of birds can occur. The two chief requirements for a good refuging site seem to be a protected or relatively isolated habitat and a rich food supply, because the long flightless period requires a predator-free environment and the replacement of the flight feathers is energetically demanding. When the right conditions occur, thousands of birds may gather for the duration of the molting period. A protected site occurs at Mono Lake in California, which is rich in invertebrate prey, and it is occupied by hundreds of thousands of eared grebes, *Podiceps nigricollis*, with some birds remaining as long as 7 months (Jehl 1988). At the Waddenzee, sheltered behind the West Frisian Islands off the Northwest Coast of the Netherlands, some 200,000 European shelducks (*Tadorna tadorna*) spend the molt period.

Another good example of migration to sheltered sites is the diapause migration of many insects. We have seen cases of these in milkweed bugs (*Lygaeus* and *Oncopeltus*) in Chapter 3 and in the monarch butterfly migrating to physiologically suitable diapause sites in Mexico or California (Chapter 7). But diapause need not involve overwintering only; a general pattern of migration to high-altitude summer estivation sites to escape hot dry conditions is characteristic of many noctuid moths (Oku 1983a). In Australia millions of bogong moths (*Agrotis infusa*) migrate every spring through Canberra and other nearby towns on their way to the Brindabellas and the mountains in Kosciusko National Park, where they spend the summer estivating in caves (Common 1954). Their high fat content made them a favorite food of the local aborigines. In northern Japan, species such as the cutworms *Euxoa sibirica* and *Ochropleura triangularis* emerge as adults in early summer and migrate to estivation sites at 1800 meters on the Fudotai Plateau of Mount Iwate.

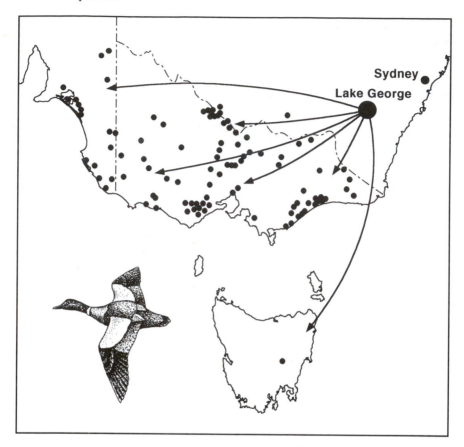

Figure 10–1. Recoveries of Australian shelducks banded at Lake George in New South Wales during their molting period. The distribution suggests a molt migration to Lake George from a wide area in southeastern Australia. (After Fullagar et al. 1989.)

Here they display specialized thigmotaxis, cued by long days, which facilitates movements to sheltered sites under stones. Moths taken from such sites usually did not fly in laboratory flight tests, or if they did, flew only in short bursts followed by obvious thigmotactic behavior (Oku 1980). In late summer and autumn, short days stimulate the reappearance of evening flight behavior and a return to lowland breeding areas; thigmotaxis is no longer present. There are thus marked behavioral differences between migrating and estivating moths. Depending on species and conditions, there is a continuum of behavior leading to the extreme migration-thigmotaxis-estivation syndrome. This syndrome seems to be selected for in species occupying xeric larval habitats such as those living in more open areas. Forest species live in relatively moist habitats and can estivate in situ, if they do so at all (Oku 1983a; Dingle 1989). Other species that migrate to high-altitude diapause or estivation sites are ladybird beetles such as *Hippodamia convergens* in western North America and the pentatomid bugs *Aelia* spp. and *Eurygaster integriceps* in the Middle East (Rankin and Rankin 1980a; Brown 1965).

As an example of migration to a sheltered site in a completely different taxon, the red-sided garter snake, *Thamnophis sirtalis parietalis*, hibernates for the winter in limestone sinkholes in Manitoba in aggregations of thousands of individuals (Aleksiuk 1976). Navigation and homing to these hibernacula are well developed, suggesting that the behavior is under relatively strong selection. The favored sites evidently provide considerable protection from freezing because they fill with snow, and the large aggregations of snakes also generate concentrations of metabolic heat. Like the overwintering sites of the monarch butterfly, these sheltered sinkholes probably serve the dual purposes of freeze protection and energy conservation.

Migration to Restricted Breeding Sites

Specialized breeding requirements are among the strongest factors selecting for migration. We can begin our discussion with a look at amphibians, where the invasion of terrestrial and semiterrestrial habitats, combined with the need to return to water to breed, provide the circumstances leading to the evolution of migration. Migrations are of two major sorts: a movement to water to propagate young and an exodus of newly metamorphosed juveniles invading the terrestrial habitat. In the leopard frog, *Rana pipiens*, overwintering adults in Michigan move to ponds to reproduce, followed later in the spring by young frogs, which also move to ponds—probably for reasons relating to water balance, because they do not breed (Dole 1965, 1967). After breeding, adults move back to terrestrial habitats where they establish home ranges from which they then migrate to hibernacula in the autumn. When tadpoles metamorphose into young frogs in midsummer, they leave the pond either by scattering in several directions or in some instances by defined routes, which can involve the frenetic exodus of thousands of froglets from ponds on rainy nights.

Similar sorts of migrations occur in red-spotted newts, *Notophthalmus viridescens* (Gill 1978a,b), and wood frogs, *Rana sylvatica* (Berven and Grudzien 1990), in Virginia. A characteristic shared by both species is that adults are highly site specific, returning repeatedly to the same breeding pond. Berven and Grudzien, for example, marked over 11,000 adult wood frogs during the course of their study, and every single one returned to the pond in which it was marked. Strong homing to breeding ponds has also been observed in several other amphibians (reviewed in Dingle 1980). Juveniles leaving ponds for the first time, however, sometimes moved to other ponds to reproduce; again, this was true of both species. In the wood frogs, 21% of marked newly metamorphosed males and 13% of juvenile females were recaptured in ponds other than those from which they emerged, with the longest distance moved being over 2500 meters. Some of Dole's leopard frogs moved as far as 5 kilometers. These juvenile amphibians are directly analogous to salmon "strays" that return to breed in streams different from where they were born (Quinn 1985 and Chapter 11).

Newts and wood frogs differ, however, in their population structure. Newts are organized into "metapopulations" with high reproductive rates in a few ponds; from these some juveniles migrate to ponds where their reproductive success is low. Movement from high to low areas of successful breeding maintains the newt metapopulations at a relatively constant size. In contrast, the wood frogs do not maintain constant population size but rather exhibit considerable annual fluctuation. They approximate a genetic neighborhood model (Wright 1932) in which populations are linked by gene flow yet can respond to local environmental conditions. In both cases selection probably maintains some straying of juveniles, because such strays could be at an advantage in any years when there was large-scale reproductive failure in the original home ponds.

In the autumn many Temperate Zone amphibians migrate to specific hibernation sites. In Europe two species of water frog segregate by site, with *Rana ridibunda* hibernating exclusively underwater and *R. lessonae* on dry land (Tunner and Nopp 1979). The hybrid species, *R. esculenta*, formed by combining the genomes of the two parental species (see Chapter 14), can hibernate in either condition. On the margins of the Neusiedlersee at the border between southeastern Austria and Hungary, Tunner (1991) marked on the eastern side of the lake over 2000 *R. lessonae* and over 700 *R. esculenta* that were migrating in autumn to overwintering hibernation sites. Of the 15 frogs recovered, 11 were hibernating nearby (~1 km), but four were recaptured at a dry-land site 15 km away at the southern end of the Neusiedlersee. One *R. lessonae* individual made this journey in 10 days or less, based on the interval between marking and recapture. It is also possible that 15 km is a minimum estimate because some frogs from elsewhere could have been captured and marked while moving through the capture area. There is not enough known about the biology of these frogs to ascertain whether homing to hibernation sites near the natal area is occurring, but whatever the reasons for it, this long-distance movement in so short a period represents quite a remarkable performance for a frog.

Paralleling the evolution of migration in amphibians are the movements of a variety of semiterrestrial crabs and hermit crabs that must also return to an aquatic medium to reproduce. The hermit crab *Coenobita clypeatus* on the Caribbean island of Curaçao, for example, spends most of its life several kilometers inland but must migrate over these distances to release eggs in the sea (de Wilde 1973). The blue land crab, *Cardisoma guanhumi*, also of the Caribbean region, can migrate similar distances (Gifford 1962), and the red land crab, *Gecarcinus lateralis*, covers up to several hundred meters in Bermuda (Wolcott and Wolcott 1982). These return migrations to the sea are generally synchronized to the lunar cycle so that eggs are released during high tides. Once land was successfully invaded, it is relatively easy to understand why a return to saline waters might be necessary to assure egg development, given the complexities involved in the evolution of new reproductive pathways; what is not so easy to see is why these crabs and hermit crabs took up a semiterrestrial existence in the first place. The initial steps may have been via invasion of estuaries which, because of runoff from the land, are more nutrient rich than the ocean (Norse 1977). Wolcott and Wolcott (1985) postulate several possible factors, including escape from predation and competition, that seem logical but are untested. They do note that movement inland seems relatively unconstrained by the ability to tolerate low salinities or by the energetics of locomotion. About all that is readily apparent is that we need much more information before we can fully understand these little-studied migration systems.

It might be particularly useful to explore the coevolutionary relation between egg size and terrestrial invasion in crustaceans (Dingle 1980). Terrestrial crabs returning to the sea produce thousands of small eggs, like their marine relatives. Some species, however, are wholly terrestrial or can reproduce in freshwater. These latter two groups produce few large eggs and often carry attached young (examples in Bliss 1968; Abele and Means 1977; Abele and Blum 1977). An interesting question, then, is the possible relation between competition, reproductive or developmental constraints, and the evolution of terrestrial lifestyles with or without migration.

The reverse situation, in which there is a marine lifestyle but a return to land to breed, occurs in pelagic birds such as albatrosses and shearwaters, in pinnipeds, and in sea turtles. The various species return to specific breeding sites on islands, including remote oceanic islands, or on continental margins, but otherwise spend their lives mostly or entirely at sea. Banding and tagging studies reveal that the same individuals usually return to very precise nesting or birthing locations year after year (Carr 1967; Fisher 1975; Reiter et al. 1981; Jouventin and Weimerskirch 1988; Wooller et al. 1990, Sydeman et al. 1991); a more difficult question to answer is whether they are returning to their own place of birth. Many seabirds and pinnipeds do return to breed at the natal site but for sea turtles, which may take 30 years to reach sexual maturity, the lack of a long-lasting tag persisting from hatch to breeding has left open the question of a return to the birthplace for egg laying.

Meylan and colleagues (1990) addressed the issue of natal homing in the green turtle, *Chelonia mydas*, with mitochondrial DNA analysis. This species nests on beaches throughout the Caribbean region and on Ascension Island in the South Atlantic. Meylan and colleages sampled the DNA from turtle populations nesting on different beaches and found that on the basis of genotype frequencies rookeries were largely distinct, except for two colonies in Florida and Costa Rica. Their results strongly suggest that gene exchange between colonies is rare and therefore that natal homing by the breeding females is very likely. New breeding beaches do get colonized, however, so there is some straying, and the founding of a new colony may account for the genotypic similarity between Florida and Costa Rica beaches. Although other hypotheses for homing to specific beaches, such as social facilitation, cannot be completely ruled out by this mtDNA study, they are very unlikely, based on the genetic data. In any case strong homing does occur, suggesting that selection acts to ensure return to a breeding site where oneself was born and survived to breed. As we have seen, this selection for homing to a natal site seems to be a general principal of migration to breeding locations that meet specific requirements.

Homing and straying also play important roles in the migratory life histories of pinnipeds. The general pattern for a number of species is breeding at a restricted site on an island or a protected shoreline followed by migration, often accompanied by dispersal, away from the rookery to feed in more open ocean waters (reviews in Orr 1970; Baker 1978; Dingle 1980). There may be separation of adults and juveniles and of the sexes at this time. Northern fur seals (*Callorhinus ursinus*), for example, breed in the summer on the Commanders, Pribilofs, and other islands in the Bering Straits. Following this, the population moves offshore, with the males wintering in the Gulf of Alaska, but females and juveniles migrating southward for up to 5000 kilometers (Orr 1970). Similar patterns with local variants are seen in other pinnipeds from both the Pacific and the Atlantic. In each case adults then home to the rookeries at the beginning of the next breeding season, often displaying very precise arrival times to begin the new reproductive cycle.

The role of straying has been demonstrated for northern elephant seals (*Mirounga angustirostris*) by LeBoeuf and colleagues (1974). These animals breed along the coasts of California and Baja California from December to February. The adults leave the rookeries in March, while the pups usually depart in May. After the departure of adults, immatures arrive on the beaches and may stay for 3 months. During the summer adult and subadult males return to molt, and these are again replaced by immatures until the breeders return in late November and early December. Prereproductives and adults overlap very little, and the former wander widely during the pre-adult period. It is this widespread movement of subadults that can lead to the founding of new colonies. LeBoeuf and colleagues observed the arrival of a subadult male on the Farallon Islands off San Francisco at a small beach not previously occupied. This individual was followed 3 weeks later by two cows, one of which gave birth to a pup. In subsequent years young males fought among themselves for breeding territories and fathered pups at ages when they would have been excluded from breeding in established colonies, and the colony continued to grow (Huber 1987; Sydeman et al. 1991). The founding of new colonies by straying subadults is a general rule among pinnipeds; other examples are reviewed by Baker (1978). In spite of the role of straying in colony origins, however, there seems to have been very strong selection for precision of return to a breeding location, a situation we also noted for amphibians.

The megascale movements of seabirds and pinnipeds are at one end of a rather lengthy continuum of migrations to ecologically specific breeding sites. At the opposite end of the distribution, in terms of both the size of the organisms and the distances traveled, are the microscale phoretic migrations of a group of parasitic mites of the genus *Poecilochirus* that are restricted to burying beetles (*Necrophorus* spp.). The beetles locate dead mammals or birds, bury the carcass, and lay eggs on the interred corpse; both sexes remain in the brood chamber and care for the young, who consume the carrion the parents have provided. The mites also feed on the carrion, and their life cycles are synchronized to those of the beetles; in order for the mites to survive and reproduce, they must move with their hosts from brood chamber to brood chamber.

The migratory behavior of *P. carabi*, restricted to the beetle *N. vespilloides*, was studied in Germany by Schwarz and Müller (1992). The mites the beetles bring with them to a newly established brood chamber are in the penultimate larval, or deuteronymph, stage. Soon after arrival they molt to adults and begin to reproduce. Six days after the females produce their eggs, the earliest deuteronymphs appear; these are the pre-adults that will climb aboard a departing beetle for phoretic transport to the next beetle nursery crypt. Schwarz and Müller captured beetles emerging from the brood chamber to seek another corpse and noted that the males departed first, about 8 days after the mites commenced egg laying. In so doing they bore 50% or more of the deuteronymphs that would mature in the chamber (Fig. 10-2). Female beetles followed about 3 days later bearing about 35% of the mites, leaving the remaining 15% or so to attach themselves to the beetle larvae when the latter left to construct puparia.

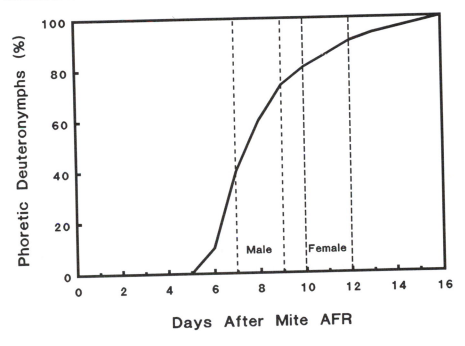

Figure 10-2. The relation between the production of phoretic deuteronymphs by the mite *Poecilochirus carabi* and the departure of their male and female beetle hosts. The first mites are ready to migrate about 6 days after birth (mite AFR on abscissa). The male beetle departs 7-9 days after the first mites are born and carries at least half the phoretic deuteronymphs. The female beetle departs a few days later and carries most of the rest. The few remaining mites must depart on beetle larvae leaving for pupation sites. (After Schwarz and Müller 1992.)

As both the male and female beetles are present in the chamber when the first half of the new deuteronymphs are ready to depart, how do the mites select the male, who is the first to leave? The answer turns out to be that the embarking mites cue in to the predeparture behavior of the beetles. Of the two sexes, the female spends most of her time on the top of the carcass, while the male spends most of his time on the periphery or in the soil. In experiments, the male was replaced with a second female. When this occurred, one of the females moved off the carcass after a few days and left the crypt, invariably carrying all available phoretic deuteronymphs when she did so. One case of sex role reversal was also observed, and in this case, too, the now peripheral female departed with the first batch of migratory deuteronymphs.

The departure of mites on first the male and then the female beetle both reduces intraspecific competition and assures the earliest opportunity to reproduce. If they waited for the larval beetles to mature, mites would perforce be delayed some 20 days more plus the time it takes to complete pupation. There is as a result strong selection for the ability to board a departing adult beetle. A further complication arises, however, because of the competition among mites as a consequence of the often heavy mite loads carried by individual beetles. Departing males, for instance, carry around 100 mites or about a quarter of their own weight. Young beetles recently emerged from puparia carry far fewer mites because the 15% or so deuteronymphs remaining after departure of both adults are divided among about 30 beetle larvae. Here the mites take further advantage of their hosts' behavior. Male beetles of all ages aggregate at carcasses and emit pheromones to signal females. At these gatherings the mites move among beetles redistributing themselves at much lower densities. The specialized phoretic migratory system of *P. carabi* has thus coevolved with the life cycle of the host to include not only the migration itself but also a series of behaviors to guarantee the proper timing of phoresy and reduced competition among siblings.

Diadromy in Fishes

Diadromy in fishes is a special case of migration to specific types of habitat. It is by far the best-studied example of this sort of migration, largely because of the commercial importance of many diadromous species, such as eels, trout, and salmon. For this reason, I treat diadromous migrations in their own section here. Diadromy is a general term that includes movements both from freshwater breeding sites to the sea for growth and development or *anadromy* and from marine spawning and nursery areas to freshwater where maturation takes place or *catadromy*. Anadromy is also sometimes used for species migrating from lakes to streams. A third subcategory is *amphidromy*, which applies to species migrating from either the sea to freshwater or vice versa, but that spend only part of the growth and maturation phase of the life cycle there before returning to the breeding habitat. All these forms of fish migratory life cycles are extensively defined and discussed by McDowall (1988).

There has been much speculation and analysis concerning the origins and subsequent evolution of the different forms of diadromy. From this some general trends emerge, although, as strongly emphasized by McDowall (1988), within these trends there are variations and exceptions driven by local ecologies and selective regimes. With respect to the illumination phylogenetic distributions can provide, taxa of fishes fall into three broad and informal groupings (Table 10–1). The first is represented by the lampreys, sturgeons and their allies, anguillid eels, and salmonids, with the latter including salmon and trout, smelts, and the somewhat bizarre salangids or icefishes of the northwest Pacific region. In all of these taxa the frequency of diadromy is high, and in the catadromous anguillids it is 100%, with the various species ranging over both hemispheres. It is also worth noting that the relatively low 24% anadromy of lampreys is biased by the fact that in several cases there are species pairs represented by one anadromous species and a very closely related exclusively freshwater form evidently derived from it (Hardisty and Potter 1971; Potter 1980). Thus the brook lamprey, *Lampetra planeri*, is thought to be a nonmigratory offshoot of the anadromous river lamprey, *L. fluviatilis*, and in Australia *Mordacia praecox* is evidently a derivative of *M. mordax*. If these species pairs are considered as single species units, an assumption that does not result in bias when considering other fish families, the percentage of anadromy in this group increases considerably.

The most obvious point about the taxa that fall into group one is that they all tend to be phylogenetically primitive, with radiations that probably date back to the earliest radiations of the agnathans and teleost fishes. Therefore, it is probable that diadromy also has an ancient origin (McDowall 1988). It is also likely that the

Table 10–1. Distribution of Diadromy Among Some Taxa of Fishes

Taxon	Diadromous Species	Total Species	Percent	Type*
Lampreys[†]	9	37	24.3	A
Acipenseridae	11	27	41.4	A
Anguillidae	15	15	100.0	C
Salmoniforms[‡]	44	94	46.8	A
Galaxiidae	8	36	22.2	Am
Clupeidae	32	180	17.8	A,C,Am
Mugilidae	14	70	20.0	A
Goliridae	45	800	5.6	A,C,Am
Cottidae	6	300	2.0	A,Am
Scorpaenidae	1	300	0.3	A
Soleidae	1?	117	0.8	A?

*Anadromous (A), Catadromous (C), Amphidromous (Am)
[†]Includes Petromyzontidae (33 species), Geotriidae (1 sp.), and Mordaciidae (2 spp.)
[‡]Includes Salmonidae (68 spp.), Osmeridae (12 spp.), and Salangidae (14 spp.)
SOURCE: From McDowall 1988.

current spawning habitats are also the ancient ones, with the eels arising in salt water and the lampreys, sturgeons, and salmonids reflecting a freshwater origin. The argument for a freshwater origin of the salmonids goes back to Tchernavin (1939) and has since been supported by Hoar (1976). Certainly, given the current distribution of the group and the frequency of landlocking within the group, a freshwater origin seems plausible, although not necessarily conclusive.

Gross (1987) argues that anadromy evolved from a fish that was originally a freshwater wanderer, eventually moved into estuaries or the sea perhaps to take advantage of the increased productivity of estuaries, became amphidromous, and ultimately became anadromous. A roughly reversed process taking a marine species toward freshwater would similarly account for catadromy. The argument is a logical one, but as McDowall (1988) points out, there is no explicit evidence that diadromous fish come from groups otherwise represented by migrants or wanderers. This is contra the situation seen in birds migrating into the Temperate Zone, where there is a good deal of comparative evidence that they were derived from migrants within the Tropics (Levey and Stiles 1992 and Chapter 9). McDowall also cites cases where Gross's model seems inapplicable, an example being the New Zealand inanya, *Galaxias maculatus*, in which catadromy appears to have evolved from an amphidromous species that spawned in freshwater.

The taxa in the second and third groups in Table 10–1 show progressively decreasing frequencies of diadromy among the species represented. Two of the group-two families, the salmoniform Galaxiidae and the Clupeidae (shads and herrings), are also of ancient origin, and in these there are cases where diadromy has evolved in species groups. The frequency of anadromy is high in the Mugilidae (mullets), but most of the species are so little studied that it is not known if all, most, or only a few populations within the different species are committed to freshwater breeding. In the gobies and all the families in the third group, diadromy is rare to very rare, even in families like the Soleids that contain several migratory species. Across these taxa it probably has a completely independent origin, as a function of local ecologies, in every case in which it occurs. Patterns of origin and the subsequent evolution of diadromy are apparently very diverse in all the taxa of the second and third groups. So much so, in fact, that the lack of uniformity precludes any general theory that would account for all of them.

Phylogenetic origins aside, the evolution of diadromy in any taxon is most likely a function of the resources available in the marine and freshwater habitats. This evolution is apparently driven by selection for more efficient adult feeding and growth in one environment (or increased winter survival in some cases) and more successful breeding in another. Diadromy from this perspective is part of a life history syndrome that allows a fish species to occupy more favorable conditions at each stage of its life cycle (Dingle 1980; McDowall 1988). An optimization approach would also stress that the benefits of migration must exceed its costs. It is worth noting in this regard that a strategy of not migrating is not ultimately limiting, as many species in predominantly migrating families or genera and indeed many populations within otherwise migratory species are to varying degrees sedentary. Often this means the evolution of a life-history syndrome different from the migrant, a subject I shall take up in Chapter 12.

The resource argument gains credence from an analysis of the geographic distributions of anadromy and catadromy among fishes. Gross and colleagues (1988) examine latitudinal frequency changes in the two types of migration with respect to the relative primary productivities of marine and freshwater environments (Fig. 10–3). The underlying assumption of the analysis is that food availability will be a function of standard measures of productivity expressed as grams of carbon fixed by photosynthesis per square meter per year. As the data displayed in Figure 10–3 indicate, there is a shift in relative productivities at about 40° latitude. At high latitudes oceanic waters exceed fresh in the amount of carbon fixed, while at lower latitudes the reverse is true. The increased productivity of freshwaters is particularly evident at latitudes below about 20° where freshwater habitats exceed marine by six- to eightfold. Gross and colleagues predicted that anadromy would be likely to evolve where food was more available in the sea as indeed it should be at higher latitudes. Pacific salmon smolts, for example, may experience a 10–50% increase in daily growth rates during the first week of

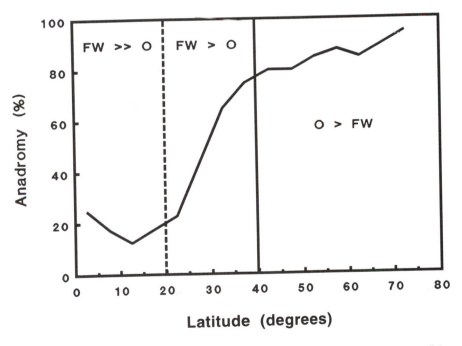

Figure 10–3. Relative proportions of anadromy and catadromy among diadromous fishes as a function of the productivities of ocean (O) and freshwater (FW) environments plotted as percent anadromy. Below 20° in the Tropics, freshwater habitats are more productive than the oceans and catadromy predominates; the reverse is true at latitudes above 40°. (Data from Gross et al. 1988.)

oceanic life. As predicted, above 40° latitude anadromy is by far the dominant mode of diadromous movement, occurring in 80% or more of the species. Conversely, between 20° and the equator in the region of peak excess freshwater productivity, around 80% of the species are catadromous.

If nutrition were the only factor involved in selecting between the two modes of migration, the transition in frequencies should manifest itself across the 40° productivity boundary. That it occurs instead between 20° and 40°, where freshwater habitats are only somewhat more productive, suggests that other factors are involved as well. One could be the sheer volume of the sea relative to freshwater, thus potentially providing more food even with somewhat lower primary productivity (indirectly a nutritional factor). Other factors such as fish movement across latitudes after crossing the ocean-freshwater boundary, competition, or predation may also exert an influence. In spite of potentially complicating selective agents, however, the general trend across latitudes is robust, and the food availability hypothesis is strongly supported.

Summing Up

What I have outlined in this chapter are migrations to meet distinct habitat requirements at specific periods of the life cycle. Among these requirements are shelter and special conditions for breeding. With respect to the former, the unique situation in waterbirds—the molt of the flight feathers means a long period of flightlessness with a high energy demand for feather regrowth—means the birds must seek often quite restricted sites. Where suitable sites occur, as at Mono Lake in California or Lake George in New South Wales, very large numbers of molting birds may congregate. The hundreds of thousands of eared grebes at Mono Lake are a case in point. Other examples of sites providing unique combinations of requirements are the high-altitude diapause sites of monarch butterflies in Mexico and of noctuid moths in Australia and Japan and the hibernacula of garter snakes in western Canada and of water frogs in eastern Austria. The selective benefits of all these shelters are obviously very high because the organisms using them make impressive migratory journeys to seek them out.

In the case of many breeding migrations, the movements apparently reflect at least in part the inability to escape completely from ancestral requirements for reproduction. Thus many amphibians and terrestrial crustaceans must return to an aquatic medium to release their eggs or larvae and sea turtles, pelagic marine birds, and pinnipeds must return to dry land to lay their eggs or rear their young. For the remainder of their life cycles they forage widely over vast areas of the world's oceans, often exhibiting remarkably precise navigational abilities to guide them in their returns to the rookeries. The evolution of diadromy in fishes, although not quite so clear, also seems to reflect by and large a return to ancestral spawning sites, following the evolution of a foraging mode in the opposite medium. Thus the preponderance of the evidence suggests that salmonids arose in freshwater and anguillid eels in the sea. The situation with several other taxa is more confusing, however, especially in those cases where the group is represented by only a very few diadromous forms and no consistent pattern of anadromy or catadromy. Finally, in the case of the phoretic mite, *Poecilochirus carabi,* its specialized breeding habitat in the brood chamber of its host burying beetle requires that it migrate as the beetle does or face extinction when its breeding site disappears.

Many of the migrations discussed in this chapter have a seasonal component, but it is not seasonality that defines them. Monarch butterflies, for instance, shelter in diapause sites in the winter while Japanese noctuids do so in the summer, yet they are similar in that both involve lengthy flights from breeding habitats to diapause sites and a return. Movements of waterfowl to molting aggregations may be in the opposite direction to the usual direction of autumn migration even though the two migrations may overlap. Among diadromous fish some spawn in the spring with egg hatch in the fall, while others reverse the two events; it is the need for the special spawning areas, coupled with the benefits of growth and maturation elsewhere, that provides the selection for the particular migratory patterns. Diadromy and the other patterns we have seen are sufficiently different from other migratory systems to require their own conceptual description and analysis.

11

Migration under
Ephemeral Conditions

Many habitats are not constant for very long when viewed over either time or space. The most obvious way in which they vary is with seasonal changes, and as we have seen in Chapter 9, migratory life cycles synchronized to the seasons have frequently been the product of selection. Changes also take place, however, as a consequence of ecological succession or of unpredictable variations in climate. Because of this many habitats are ephemeral, and their exploitation requires either that the organism escape by some form of suspended reproduction or dormancy for very long periods or by moving between habitats as the occasion demands. The banded stilt of Australia (Chapter 3) is an example of a creature that does both, migrating to the ultimate of ephemeral lakes after periods of reproductive inactivity that can be measured in years. Equally long periods of dormancy are found in the seeds of desert plants or the eggs of crustaceans inhabiting temporary water bodies. Organisms capable of long-distance movement have usually opted for migration in spite of the potential costs of investment in migratory capability that can occur at the possible expense to growth and reproduction. Perhaps even more importantly, they migrate in spite of the risk of failing to find a new habitat (Farrow 1990). As the examples in this chapter will indicate, the costs and risks have not prevented the repeated evolution of extraordinarily successful migratory strategies for exploiting ephemeral resources.

Migration in Ephemeral Habitats: Theory

The theoretical basis for the hypothesis that migration evolves to keep pace with changing habitat structure (Southwood 1962) has been considered extensively from a number of perspectives (Roff 1974a,b, 1975, 1986, 1994; Van Valen 1971; Järvinen 1976; Hamilton and May 1977; Levin et al. 1984). The general result of these studies can be illustrated with a simple probability model (Roff 1990). Imagine a species with populations in two habitats that persist in their suitability for growth and reproduction from one generation to the next with probability p. The probability that a habitat will persist to the next generation is p^2 and that it will persist for t generations is then p^t. As t increases, there is an ever greater chance that the habitat will be unsuitable for a generation, and the population in it will go extinct. Now if there is no migration, extinction of the species requires only that each habitat becomes unsuitable at some time during the interval t. With migration between the habitats, species extinction occurs only if the two habitats become unsuitable simultaneously, an event less likely to occur. The probabilities of the species persisting over the interval t are given by $1-(1-p^t)^2$ without migration as compared to $(1-[1-p]^2)^t$ where migration occurs. The significance of this can be seen in a numerical example in which p=0.5 and t=3 generations; in this case the migratory species is almost twice as likely to persist as the nonmigratory one. In a more realistic situation with p=0.95 and t=100 generations, the migratory species is 66 times more likely to persist and this likelihood increases to 6890 times more likely with a more unstable environment where p=0.9. The likelihood of extinction is further reduced by the addition of more habitat patches.

One can add still more realism to the models by incorporating costs or risks of migration, carrying capacities, and population growth rates; these can influence thresholds and proportions migrating (Roff 1975, 1994; Kaitala et al. 1989). Increasing the cost of migration or decreasing the mean value of the population growth parameter, λ, the ratio of population size at time t+1 to that at t, decreases the proportion of migrants in the population. In general increasing the mean carrying capacity, K, or decreasing the variances in either λ or K also decreases the proportion of migrants. In addition, as the number of habitat patches declines, there will be increasing selection for migration. In summary, in a heterogeneous environment migration will evolve as a function of the cost of migration, the number of habitat patches, and the means and variances of λ and K (Roff 1990, 1994).

For insects, empirical support for the relationship between ephemeral habitats and the increase in the frequency of migration was provided a long time ago by Jackson (1928) for ground-dwelling weevils (*Sitona hispidula*) and later by Brown (1951) for aquatic heteropterans (Corixidae). The major empirical summary, however, is that of Southwood (1962). In his review Southwood distinguished two categories of habitats, "permanent" and "temporary." The former habitats were ones that tended to last a relatively long time and had high carrying capacities; they included woodlands and other perennial plant communities, salt marshes, heathlands, and large lakes and rivers and their fringing wetlands. The temporary habitats were ephemeral in time with relatively low carrying capacities, and included plant communities in early successional stages and small or shallow water bodies. Comparisons among different arthropod taxa revealed that the proportion of migrants was negatively correlated with habitat persistence. An example is given in Table 11-1, which shows that for British Macrolepidoptera, there are more migratory species in habitats at the temporary end of the scale than there are for habitats at the permanent end. In woodlands, there were even species in which females had extremely short nonfunctional wings (brachyptery). Southwood presented no statistical tests for the trends indicated, but subsequent tests where there were sufficient data generally indicate a significant correlation in the predicted

Table 11–1. The Number of Species According to Habitat and Host Plant of British Macrolepidoptera with Number of Species Having Brachypterous Females and Number of Migrant Species Indicated

	Permanent		Temporary	
Habitats	Woods	Moors, woods marshes, heaths	Hedgerows gravel pits	Arable land
Host plants	Trees	Bushes, perennial climax plants	Non-climax	Annuals, weeds
Species	310	273	116	58
Species with female brachypters	13	—	—	—
Migrants	None	1	22	17

SOURCE: After Southwood 1962.

direction (Roff 1990). In a more recent summary and analysis Southwood (1977) expressed the relation between migration and habitat as H/τ where H = the length of time a habitat remains favorable and τ = generation time. When the ratio approaches unity, migration is favored; short habitat durations also select for rapid life cycle turnover and relatively high fecundities ("r-selection"). A more recent theoretical analysis by Roff (1994) produces similar conclusions.

Migration in Arid Environments

Among the most ephemeral of all habitats are patches of rainfall-induced new plant growth occurring within the arid regions of the world. Deserts and savannas themselves remain stable over long periods, but the growth, reproduction, and survival of organisms within them depends almost without exception on the distribution of rainfall. This is very likely to be erratic, although often with a long-term tendency to be seasonally correlated. The main short-term characteristic of rainfall in arid regions, aside from its scarcity, is its unpredictability. Monitoring of rainfall at several desert locations over the surface of the globe reveals that, in general, the less the rainfall at a site, the greater the positive skewness in a plot of rainfall amount against frequency (Barry and Chorley 1987). In other words the less the average precipitation, the more likely will there be many years of little or no rainfall for every year of above-average rainfall. When the latter do occur, a few will provide large amounts of rain leading to highly favorable conditions. Across years, there will also be a major spatial component, with different parts of a large desert region receiving varying amounts of moisture. As the above theoretical arguments would predict, many desert and dry-country organisms have adopted migratory lifestyles. In earlier chapters we saw examples such as the red-billed quelea and the banded stilt, but among the organisms that have adapted migratory movements to the patterns of rainfall in arid habitats, none has done so more successfully than the group of grasshoppers known as locusts.

Five species have been the focus of most studies of locust biology and migration because frequent outbreaks of high-density swarming populations often result in considerable crop damage over wide areas. These five are the desert locust, *Schistocerca gregaria* of the drier regions of Africa and Asia; the migratory locust, *Locusta migratoria*, roughly sympatric with *S. gregaria*, but more likely to occur in wetter areas like flood-plains and oases, and also present in Australia; the red locust, *Nomadacris septemfasciata*, and the brown locust, *Locustana pardalina* of Africa; and the Australian plague locust, *Chortoicetes terminifera*. Several other species of grasshopper and locust around the world such as the Senegalese grasshopper, *Oedalus senegalensis*, of West Africa, the lesser migratory grasshopper, *Melanoplus sanguinipes*, of North America, and the species of *Schistocerca* in the New World sometimes occur in very high densities, but not with the frequency or (usually) with the swarming characteristics of the typical locusts.

Locusts display in extreme form a process known as *gregarization* (Uvarov 1966, 1977), a response to increased population densities. It occurs to varying degrees among species and appears to be the result of repeated contacts among individuals Uvarov (1966). The most extreme cases of gregarization occur in *S. gregaria* and *L. migratoria*, but it is apparent also in other locust species. When locusts (*C. terminifera* is an exception) grow and develop under crowded conditions, their nymphs become much darker because of increased melanin deposition; they also become patterned in brightly contrasting yellow and black colors. When solitary or at low densities, on the other hand, the nymphs are pale green or fawn colored with an increasing green tendency at higher humidities (e.g., Dingle and Haskell 1967). Morphological characters are also affected. Crowded adults, the so-called *gregaria* phase, are larger and display different body proportions from their uncrowded or *solitaria* counterparts. The wings of *S. gregaria*, for example, are two times the length of the femur of the hind leg in the *solitaria* form but 2.3 times the length in *gregaria* individuals. Other differences include a wider head, a shorter pronotum with a somewhat depressed crest, reduced sexual dimorphism, and higher metabolic rates in the *gregaria* form. Life history differences in *S. gregaria* include a reduced fecundity in the gregarious morphs but more rapid adult maturation and earlier reproduction. The latter theoretically gives the *gregaria* populations higher potential population growth rates (Cheke 1978), but is likely to be counteracted by a longer hopper (nymphal) period. The *gregaria* hoppers, on the other hand, are larger and likely to survive longer (Farrow and Longstaff 1986).

These morphological and physiological differences between crowded and uncrowded morphs are great enough that in the desert locust they were at first considered separate species. The influence of crowding was worked out in the early part of the twentieth century by B.P. Uvarov (1921), who demonstrated that the two forms were a single species and developed a theory of "phase polymorphism," with the isolated and crowded forms designated the *solitaria* and *gregaria* phases, respectively. The term "polymorphism" is now better reserved for genetic differences, and the term "polyphenism" has been coined for phenotypic differences that are environmental in origin (e.g., Shapiro 1976). Hence the density effects in locusts are now referred to as *phase polyphenism*, with the full ramifications described in Uvarov's two-volume review of acridoid biology (Uvarov 1966, 1977).

The most conspicuous differences between *gregaria* and *solitaria* locusts, however, occur in behavior. These differences begin to show up as early as the second instar when the nymphs or "hoppers" begin to show mutual attraction and form up into bands that may contain many thousands up to (on some occasions in the desert locust) probably millions of individuals. These bands of hoppers are characterized by a behavior called "marching" in which the entire band moves off across country, with the cohesiveness of the band maintained by the mutual attraction among the hoppers composing it. Streams of hoppers that are diverted from the main band by objects or uneven terrain soon are attracted back toward it and again merge. Similarly small bands will move toward large ones and merge with them. The attraction is almost certainly visual, although olfactory responses cannot be completely ruled out. Hopper bands tend to move more or less in a straight line, often downwind, but can be diverted by terrain features. Once started in a given direction, a course may be maintained by moving at a fixed angle to the sun. Ellis (1951), for example, kept *L. migratoria* bands in the laboratory moving in a circle by having them march around a light in the center of an arena. Hunger levels evidently determine when the band pauses to feed because bands whose individuals have full guts continue marching, whereas when guts are half empty, the band will readily stop and begin feeding on a suitable food source. The total distance moved by a hopper band can be very impressive, with some bands moving tens of kilometers during the second through the fifth instars before finally eclosing to the adult. Larger bands usually move farther as indicated in Table 11–2.

When the nymphal *gregaria* locusts eclose to adulthood, the extreme mutual attraction among individuals continues, and the insects form into enormous swarms displaying a form of social behavior (Kennedy 1951; Farrow 1990; Rainey 1989). Swarm cohesiveness is maintained by the attraction mechanisms of the locusts within it. Individuals that leave the edges of the swarm soon turn and reenter it, and those left behind on the ground as the swarm passes over them takeoff and catch up to it; thus the integrity of the swarm is sustained by the responses of individual locusts. Although the headings of flying individuals within the swarm may be in a number of different directions, the overall direction of movement of high-flying swarms is usually downwind. A low-flying swarm will move primarily upwind so long as wind speeds aren't excessive; high winds cause flight to cease altogether. To an observer viewing the swarm from outside, the movement is thus a rolling one in a more or less consistent direction (Fig. 11-1). At the leading edge of the swarm, locusts settle and feed. As they do so the remainder of the swarm passes over and in turn reaches the leading edge and settles. As the trailing edge

Table 11–2. Summary of Movements of Hopper Bands of *Schistocerca gregaria*

Instar	Size of Band (m²)	Displacement (m) in 24 hr	Displacement (km) during Hopper Life
I	32	90	—
V	400	180	3.2
I	160	180–270	—
V	2,000	360	6.4
I	6,400	450–540	—
V	80,000	900	16.0
II-III	160,000	1,800	—
V	—*	1,800	32.0

*Band is same as that in line above but size not estimated in V instar.
SOURCE: After Uvarov 1977, from data in Ellis 1958.

is left behind, its constituents takeoff, catch up, overfly feeding compatriots, and settle once more when they become the leading edge. In good habitat, the locusts in the swarm spend most of their time feeding or resting and the swarm moves slowly. Feeding by swarms can strip the vegetation quite literally down to the bare ground, leaving a path of devastation so aptly described in the Book of Exodus: "And the east wind brought the locusts . . . they covered the face of the whole earth . . . and there remained not any green thing."

If green vegetation is unavailable in the country through which a swarm is passing, it will lose contact with the ground and continue downwind for a considerable period and distance (lower panel in Fig. 11–1). Swarms of this type were common during the 1954–55 desert locust outbreak (see below). Fresh flushes of vegetation caused by local rainfall will result in the swarms settling again. The odor of fresh or damaged vegetation will attract locusts in the laboratory, but it is not clear how much of a factor odor is in the field (Uvarov 1977). In addition to supplying energy and nutrients, fresh vegetation has high concentrations of the plant hormone gibberellin, necessary for egg maturation in locust females (Ellis et al. 1965). In addition to its influence on locust reproduction via plant growth, rainfall also influences egg development via soil moisture because the eggs, which are oviposited by the females just below the soil surface, require moisture for development.

The relation between meteorologic factors and locust swarming, as revealed by observations of *S. gregaria*, was the basis of a pioneering model of locust behavior produced by R. C. Rainey (1951), whose unique contribution was relating locust outbreaks and weather patterns. In Rainey's theory the gregarious phase of *S. gregaria* was an adaptation for long-range migration to new habitats. Swarms were carried downwind to areas of wind convergence along the Inter-Tropical Convergence Zone (ITCZ, see Chapter 5); in these regions of convergence in the fronts along the zone, rainfall was likely to occur, providing the ecological conditions necessary for reproduction and the subsequent growth of locust populations. In this scheme the swarming behavior was seen as an adaptation to

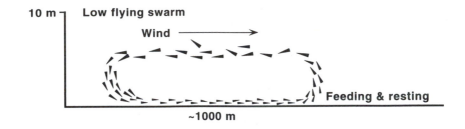

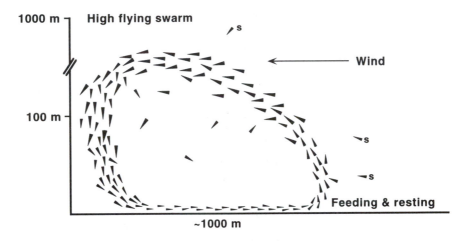

Figure 11–1. Characteristic patterns of movement and orientation in high- and low-flying rolling locust swarms. Low-flying swarms move mostly upwind so long as wind speeds are not too high (*upper panel*). When green vegetation is not present swarms may lose contact with the ground and fly at considerable heights moving downwind (*lower panel*). S = stragglers orienting back into the swarm. (After Uvarov 1977 and Farrow 1990.)

ensure that the locusts were carried downwind to zones where wind fields met and produced rain. The theory was tested in an all-out international effort in the locust plague years 1954–55 when locust populations were monitored on the ground and swarms were followed by aircraft over much of the northern two thirds of the African continent (summarized in Rainey 1989). The data on locust numbers and movements were then compared to meteorologic conditions over the same region, including local wind fields at the times of swarm movement. Much of the data seemed to confirm the Rainey hypothesis, for indeed many swarms were observed to travel downwind, to shift direction simultaneously with wind shifts, and often to arrive in areas of recent rainfall. The literature on locust and insect migration in the following decades reflected the apparent confirmation of the Rainey theory (e.g., Johnson 1969; Dingle 1980; Taylor and Taylor 1983).

Closer inspection of the data gathered from many years of research on several species of locust, however, reveal inconsistencies in the Rainey model and a much more complex set of adaptations to low-rainfall environments, especially when locusts other than *S. gregaria* are considered (summarized in Farrow 1990). First, *solitaria* individuals also migrate and in many cases fly for much greater distances than *gregaria* forms. For this reason, the *gregaria* phase can no longer be seen as a form specialized to carry out the migratory portion of the life cycle. Second, a major discovery, enhanced by subsequent radar studies, was that solitary locusts migrate at night (Davey 1953; Schaefer 1969). This discovery revolutionized studies of acridoid behavior because it drew attention away from the daytime swarms and pointed the way to studies demonstrating that the majority of long-distance movements were by individual locusts (and grasshoppers) flying after dark. Third, most migratory flight occurs during disturbed weather, and to the extent that it occurs downwind, it does so because migrants are carried downwind during unstable meteorologic conditions. There is no particular relation to the ITCZ, and in fact rains that occur along the ITCZ are likely to occur on the equatorial side of the zone up to hundreds of kilometers from where converging winds would concentrate migrant locusts. The association of locust migration with rains thus has more to do with synoptic weather disturbances than with the ITCZ. Fourth, there are some periods when migrant locusts appear to fly *upwind* to suitable breeding areas so that they exert control of their movements above and beyond simple wind transport. We have already seen examples of this for West African *L. migratoria* in Chapter 5. And finally, even when transported by winds at night, radar has revealed that the migrants exhibit considerable mutual alignment and collective orientation, strongly suggesting that they are influencing their own track or, in other words, navigating. The control of track by flying locusts was specifically excluded in Rainey's model. The data from other times and places also revealed that much of the swarm behavior of *S. gregaria* during 1954–55, on which Rainey tested his hypothesis, was quite exceptional and peculiar to that particular time period.

An example of the considerable complexity of *S. gregaria* movements is illustrated in Figure 11–2, taken from Farrow (1990) for populations in West Africa. In these populations most of the migration resulted from the movements of individuals, not swarms. From May to July emerging adults fly south with the prevailing northeast tradewinds toward light rainfall areas at the leading edge of the ITCZ, which is shifting northward at this time. The heavier rains of the southwest monsoon behind the ITCZ are unsuitable for locust breeding, because they create soils too high in moisture content and high humidities in which the locusts become increasingly susceptible to pathogens. Following breeding in July, there is a reverse migration to the north, this time largely with the northernmost prevailing winds of the monsoon. In October and November there is a hiatus in breeding as the ITCZ retreats southward, and the locusts again come under the influence of the northeast trades. There may be some population displacement during this time, but whether this constitutes migration is uncertain. From January to March cold fronts bring storms with rainfall in from the Atlantic to the northwest, again providing suitable

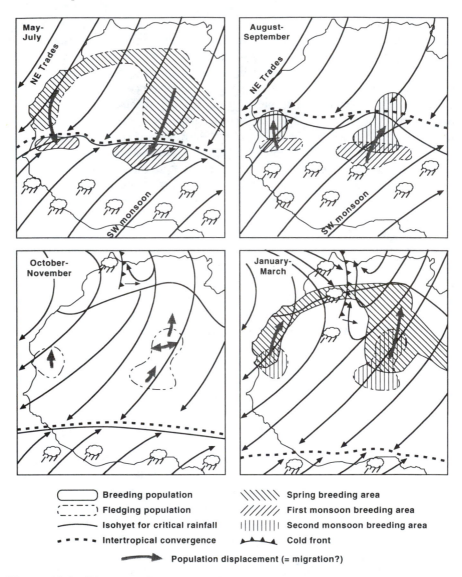

Figure 11-2. Diagrammatic representation of the seasonal movements of different generations of the desert locust, *Schistocerca gregaria*, in West Africa, showing the complexity of the migration patterns in relation to the Inter-Tropical Covergence Zone and the monsoonal and winter rains. (Modified from Farrow 1990 partly after Popov 1965.)

breeding conditions. The locusts now migrate to the north *against* the prevailing tradewinds to their spring breeding areas, and the cycle begins again. Only rarely do *gregaria* phase swarms form and only under particularly favorable conditions (not too little or too much rainfall and with just the right timing). Note also that the cycle depends on rainfall from two sources: the ITCZ and the monsoon in the south and winter storms from the Atlantic in the north.

When other species of locust are examined, it is also apparent that their migratory syndromes do not fit the Rainey model very well (Popov and Ratcliffe 1968; Fishpool and Popov 1984; Farrow 1990). The Australian *Nomadacris guttulosa* and the Sahelian tree locust, *Anacridium melanorhodon*, for example, diapause during the duration of the dry season. There is thus a more or less immediately post-fledging migration to diapause sites. Following diapause, there is a second migratory flight, triggered by high temperatures and humidities preceding the rains, to breeding sites in areas of rainfall. In both species the largest scale movements are by night flights rather than swarm flights by day. The behavior is thus similar to that of other insects that migrate to and from diapause sites as we saw in the last chapter. *Locusta migratoria* females, on the other hand, mature without delay upon emergence and migrate straightaway with obligatory postteneral flights. In very favorable habitats, high densities may lead to the production of swarms of *gregaria*, but unlike *S. gregaria*, these swarms rarely lose contact with the ground and so do not travel great distances. The only other locust which does cover long distances while in the *gregaria* swarming phase is the Australian *Chortoicetes terminifera*, but its patterns of behavior are otherwise quite different from *S. gregaria*.

Like *S. gregaria*, *C. terminifera* forms large swarms of mutually interacting individuals, but it differs from most other locusts and grasshoppers in that the *gregaria* phase is not morphologically distinct. A further major difference is that even in the swarming phase, large-scale migration is by single individuals migrating at night, not by swarms moving across the countryside by day. The proportion of a population taking off and migrating is, in fact, uninfluenced by density. After a night of migration by individuals, these locusts may re-aggregate, again forming characteristic swarms. Perhaps even more than other acridoids, *C. terminifera* is stimulated to migrate during disturbed weather; a case of such migration with a succession of outbreaks in southeastern Australia in 1978–79 is shown in Figure 11–3. Here there is clear downwind movement arising from the weather systems, but on other occasions there is clear displacement toward the southeast on what are evidently compass-directed upwind movements. The important point about *C. terminifera* is that populations move (and plagues thus spread) by the migrations of many locusts flying as individuals, not while displaying characteristic swarm behavior. In fact with the possible exception of *S. gregaria*, and then only some of the time, this is generally true of all locusts and grasshoppers that have been studied. The solitary phase is thus the migrating phase, not the gregarious phase.

This, then, leaves the question: Why swarm? According to Farrow (1990) the movement of locust swarms is best considered *extended foraging* behavior because feeding responses are not suppressed as they are in migration. Quite the contrary, for as we have noted, the individuals in a rolling swarm usually spend most of their time on the ground feeding and resting. The swarms are the consequence of a directional foraging strategy that allows individuals to move away from centers of

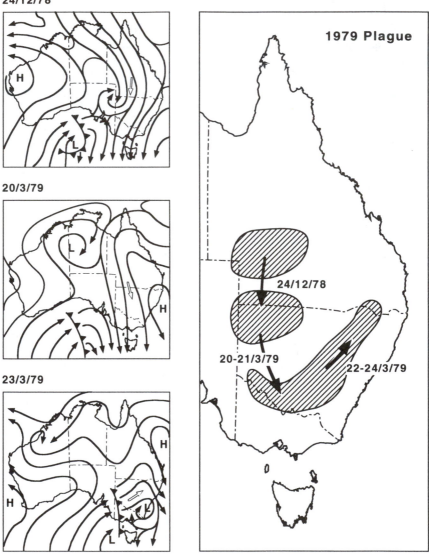

Figure 11–3. The movements of the Australian plague locust, *Chortoicetes terminifera*, during outbreaks in 1978–79 in relation to winds. The left-hand panels show the synoptic weather maps for 3 days during which locust migration occurred as shown in the right-hand panel; open arrows in left panels show primary wind directions. (Redrawn from Farrow 1990.)

high density where vegetation may already have been consumed and so avoid intraspecific competition for food. Cody (1972) provides a quantitative description for such a feeding strategy in desert finches. Moving in a more or less straight line, as finches and locusts do, also avoids any doubling back and recrossing already grazed areas as would sometimes occur if movements were random. Locusts also avoid contamination with fecal matter known to contain certain pathogens. Swarming may also lead to predator confusion or swamping as postulated also for North American periodical cicadas (Karban 1982), but given other selective factors likely to be operating, it is doubtful if this is a major factor in locusts. The bright colors of gregarious hoppers may well be an adaptation to enhance the visibility and hence mutual attraction, so promoting the feeding strategy of marching.

With their extreme swarming tendencies, *S. gregaria* and *C. terminifera* are the quintessential locusts. They seem able to exploit ephemeral habitats in arid to semi-arid regions much more effectively than any other acridoids because they are able to take advantage of favorable conditions occurring patchily over very wide areas but for a limited period. The swarming-foraging and individually migratory aspects of the life cycle, allowing them to concentrate in favorable areas, combined with their capacity for rapid reproduction and high rates of increase are what allow locusts to exploit patchy habitats so effectively. Swarming also gives them their unique capacity to reach outbreak or plague conditions (Farrow and Longstaff 1986). Put in another way, it is their very effective capacities for movement that allow them to reduce the variance in the stringent environments of the drier regions of the world (cf Leggett 1985).

Locusts are not the only insects to use a migratory lifestyle to adapt to dry habitats. A number of moths of the family Noctuidae have done so as well and can attain very high densities, making them, along with locusts, major agricultural pests in the drier regions of the world. Prominent among these highly successful moths are various species of *Heliothis* (some have been reclassified as *Helicoverpa*) and *Spodoptera* that are abundant in dryland cropping systems in North America, Africa, and Australia (Sparks 1979; Raulston et al. 1986; Fitt 1989). Many species of *Spodoptera* are known as armyworms because of the large crop-destroying aggregations of larvae that sometimes form. In the African armyworm (*S. exempta*) there is a seasonal progression of outbreaks associated with the rainfall occurring with the passage of the Inter-Tropical Convergence Zone. The progression extends from southern Tanzania into South Africa and from northeastern Tanzania into Kenya, Somalia, Ethiopia, and even as far as Yemen. For some 50 years it has been strongly suspected that the seasonal outbreak pattern was a function of migration, and recent evidence from a number of studies including large-scale mark-recapture, radar tracking, and detailed laboratory studies of migratory behavior have confirmed this (Parker and Gatehouse 1985a,b; Gatehouse 1986, 1987a, 1989; Rose et al. 1987; Pedgley et al. 1989; Wilson and Gatehouse 1993).

The African armyworm migratory pattern is dependent on sprouting grasses following rainfall. This species breeds throughout the year, a pattern made possible because even in the dry season sporadic rains occur, especially along the coast of

East Africa, and provide the necessary moisture to maintain some suitable grassy habitat. Other pockets of suitable habitat during periods of low rainfall exist in some highlands and in marshy areas associated with lakes and permanent or quasi-permanent water courses. The tendency to migrate is increased by high density, which also induces changes in caterpillar morphology and behavior analogous to the phase polyphenism of locusts, but even at low densities some genetically programmed migration occurs. This migration tends to occur downwind and is somewhat dispersive, although there is a net inland movement. Because convective storms do not occur at this time of year, there are no meteorological events capable of concentrating moths. The result is that moths remain at low density in the *solitaria* phase in which caterpillars are green and feed well-hidden at the bases of grass stems. During this dry season period, there is probably selection to reduce migratory potential because the risk of leaving a habitat patch is likely greater than that of remaining in one in which successful growth and maturation have taken place.

With the coming of the rains produced by the southward-moving ITCZ, the situation changes. The exact nature of the change is a function of the considerable spatial and temporal variation in the rainy season. Some of the downwind-migrating moths will encounter the first rainstorms, which are very likely to produce convergent winds that concentrate moths, in contrast to the dispersive winds of the dry season (Fig. 11–4). Furthermore, the rain tends to wash the moths out of the air. The net result is often a very high concentration of moths laying eggs in an area that is favorable because rains have just occurred there. These concentrations often result in outbreaks of "armyworm" larvae feeding voraciously on local crops. In East Africa these outbreaks usually occur not far inland from the coast because of the prevailing wind direction at the onset of the rains. The high-density larvae undergo a "phase transformation" to a dark *gregaria* caterpillar that is highly visible because in contrast to the green low-density *solitaria* larva, it feeds in the open at the tops of grass stems. These larvae thus display a phase polyphenism somewhat similar to that occurring in locusts. In the adults maturing from *gregaria* larvae, migratory behavior is more likely to be expressed, and each night large numbers of moths take off on downwind migrations that can cover considerable distances because the moths will fly for long periods. The greater migratory tendency allows the moths to escape overcrowded conditions, where they are subject to increased incidences of disease and predation, and to locate habitats where earlier rains have made conditions suitable for further reproduction and growth.

In years or in locations where the early rains are poor, the cycle of takeoff, downwind flight, and concentration by the occasional convective storms is likely to continue. The probability of being concentrated and of landing in a suitable habitat is a positive function of the time spent airborne, in turn a function of the prereproductive period (PRP). Because *S. exempta* displays an oogenesis-flight syndrome (Chapter 6), the longer the PRP, the longer the moths fly. These conditions of poor or erratic rainfall would thus be expected to select for high

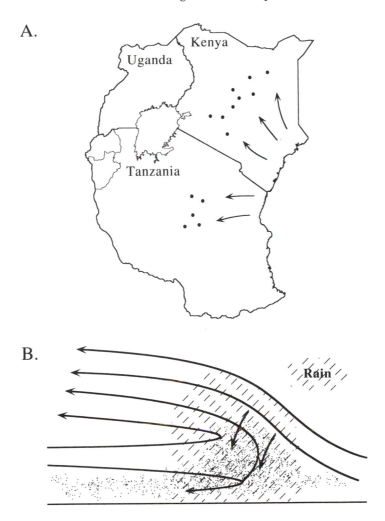

Figure 11–4. *A*. Inland downwind migrations of African armyworm moths in East Africa. These migrations are likely to encounter rainstorms at the beginning of the wet season that result in moth concentrations. *B*. Generalized diagram of moth concentration (stippling) resulting from convergent winds (*long arrows*) in a storm. The result is a fallout of moths (*short arrows*) in an area of recent rain.

migratory potential, including a longer PRP. In *gregaria* generations, the resulting synchronous production of large numbers of highly migratory moths provides the basis for a severe armyworm season later in the rains. Further secondary outbreaks can occur from the concentration of moths arising from previous outbreaks or from the overlay of moths of different origins arriving simultaneously or sequentially at a single location.

If, on the other hand, the early rains are abundant and widespread, migrating moths are quite likely to reach satisfactory habitats without concentration by storms. Under these conditions, the moths can spread out from dry-season habitats without the necessity for long-duration migratory flights, and even more so than in the dry season, moths will increase their chances of survival by staying put. Populations will remain at low to medium densities in the absence of concentrating meteorologic effects, and larvae will be mostly *solitaria*, with severe outbreaks at a minimum. Selection will now shift toward favoring lower migratory potential and shorter PRPs, because early reproduction and growth are more beneficial than migration when good habitats are extensive. Eventually, under either of the above rainfall scenarios, the dry season returns, and there is a dieback to favorable patches, the extent of the population contraction depending on the severity of dry conditions. The whole migratory system is genetically programmed and phenotypically modified to allow *S. exempta* to find suitable habitats, keyed to the varying degrees these landscapes are ephemeral, in the seasonal cycle peculiar to eastern Africa. The frequencies of the different levels of migratory capability are set by the directions and intensities of natural selection across both time and space.

The above model of the migratory life cycle of *S. exempta* makes two predictions. First, at sites with infrequent rainfall associated with the onset of the rainy season moths will show high migratory potential and long PRPs, whereas with abundant rainfall the opposite will be true. Second, early in the rainy season when rainfall and habitats are scattered, moths will have higher migratory potential and longer PRPs than later in the season when rainfall has become more widespread. Wilson and Gatehouse (1993) tested these predictions by examining PRP variation among populations in relation to both amount and frequency of rainfall in the month the rains began and hence so did moth migration. A negative correlation was expected between PRP and rainfall because with more rain there is less migration. Because rainfall at other times has much less influence on larval habitats or even none at all, no correlation was expected between rainfall in these months and migratory behavior. Both these expectations were realized. The negative correlations between mean PRPs of females from different populations collected in the field and reared under identical conditions in the laboratory are indicated in Figure 11–5. Whether rainfall is measured as the mean amount for the month of rainy season onset or as frequency of significant rainfall during that month, the negative correlation between PRP and prevalence holds. These results provide good support for the model and suggest that selection is indeed producing moths with higher migratory potential where habitats are more ephemeral and unpredictable.

Another denizen of arid environments that makes effective use of ephemeral habitats generated by patchy rainfall is the emu (*Dromaius novaehollandiae*) of Australia (Davies 1984). These large flightless birds inhabit virtually the entire Australian continent, especially its arid and semi-arid regions. Like its compatriot the banded stilt (Chapter 3), it is apparently a true nomad because birds banded in Western Australia have been recaptured in an essentially 360° circle around the

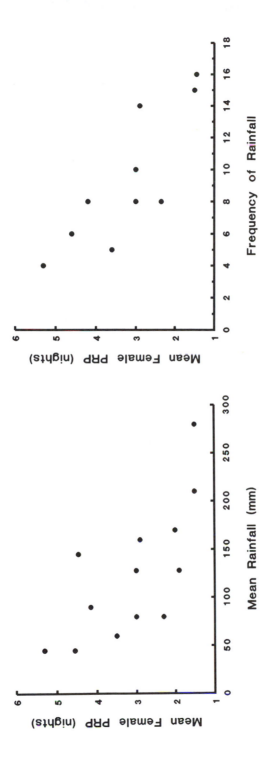

Figure 11–5. The relation between pre-reproductive period (PRP) and the amount of rainfall in the first month of the rainy season in females of the African armyworm moth, *Spodoptera exempta*. A negative relation between PRP and rainfall is evident whether PRP is plotted against mean rainfall (*left panel*) or frequency of significant rainfall (*right panel*) suggesting selection for longer PRPs where there is also stronger selection for migration. (After Wilson and Gatehouse 1993.)

banding points, with a scattered enough distribution of recovery sites to suggest no preferred direction of movement. Rather, emu movements seem to be a function of the erratic rainfall pattern common to the region, although there is a tendency to move southwest (i.e., poleward) out of the rangelands in winter and back again in the summer, the opposite of the usual north-south pattern keyed to season. These movements are driven by rainfall, which tends to come from northern tropical cyclonic systems in the summer and southern Antarctic storms in the winter. If, however, rainfall is frequent at a particular site during a year, the local emu population will tend to remain there.

In his studies of emu movement in Western Australia, Davies was able to take advantage of the fact that an extensive fence system, built to keep emus out of agricultural areas, has been constructed eastward 190 km from the coast north of Geraldton, extended nearly 600 km southeastward, and attached to a fence extending some 800 km northward (Fig. 11-6). These fences are effective in trapping the birds on the east and north sides if they are moving west or south and similarly on the opposite sides if moving in the reverse directions. A consistent observation from surveys using surface vehicles or aircraft is that large numbers of emus on one side of the fence are significantly statistically correlated with high rainfalls on the opposite side. In addition, movements of large numbers south in the winter are correlated with lack of summer rain in the north, as well as heavy falls of winter rain in the south. In 1970 the pattern of rainfall was reversed and heavy rains from tropical systems fell to the south of the fence, and in this year the pattern of emu movement also reversed, with the birds moving south toward the heavier rainfall areas in the summer. These observations confirmed the tendency of emu movements to be driven by the distribution of rainfall. Although nomadic in pattern because the timing and direction are unpredictable, the emu movements appear to be truly migratory, rather than simply extended foraging, because the journey toward recent rainfalls sometimes, at least, involves passing through good feeding grounds. In other words, stimuli that would normally arrest movement, the presence of suitable forage, appear to be temporarily suppressed.

The adaptive advantage of moving to regions of recent rainfall is obvious, for these are, as for locusts and armyworms, the regions that supply energy and nutrients from new plant growth. An interesting question, though, is how the emus find these areas. Do they, like locusts (and other insects) move more during disturbed weather? The data do not exist to address this question, partly, at least, because it doesn't seem to have been asked of this situation. A potential benefit of comparative studies and analysis across taxa is apparent here. There is no reason, a priori, to assume that emus move as groups to rainfall areas; they may be each individually moving (another potential parallel with locusts). However suitable areas are located, there are potential cues available, including large cloud banks visible as much as 300–400 km away, and it appears that emus may orient toward them because they seem to shift direction when the position of the cloud banks shifts. The sound of thunder also carries for a considerable distance. Evidence for orientation based on these cues is tenuous at best, however, and there is much work

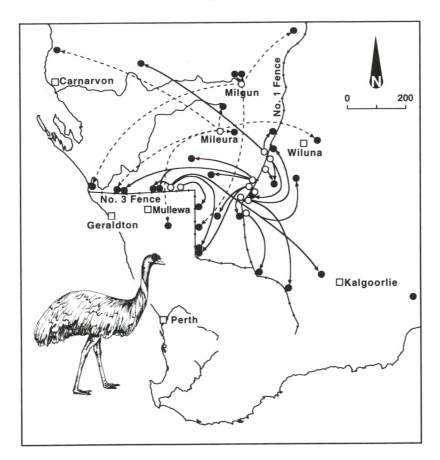

Figure 11–6. The position of the fence system in Western Australia designed to keep emus out of agricultural areas. Also shown are the banding (*hollow circles*) and recovery (*solid circles*) sites of emus. (Redrawn from Davies 1984.)

that needs to be done. Given the size and mobility of emus, the ingenuity of investigators is likely to be challenged.

Many other Australian arid land birds have been assumed to be nomadic, but there are reasons to be somewhat cautious in this assumption (Nix 1976; Davies 1984; Ford 1989). Two of the allegedly "classic" Australian nomads are the budgerigar (*Melopsittacus undulatus*) and the black honeyeater (*Certhionyx niger*). The budgerigar, however, shows a distinct north-south pattern with marked increases in numbers in the south in spring and summer and a retreat northward in the autumn and winter (Wyndham 1982). Likewise the black honeyeater shows regular north-south movements (Ford 1978). The budgerigar varies greatly in abundance from year to year in the exact sites where it is found, and both species—especially the black honeyeater, which can become very scarce—fluctuate considerably in overall abundance. Changes in abundance and apparency can easily give the impression of nomadic movements. Other species like the white-browed woodswallow (*Artamus superciliosus*) regularly irrupt to breed in numbers outside their normal range, and still others, like crimson chats (*Ephthianura tricolor*), appear to be more truly erratic and therefore nomads. Australian dry-land birds seem to have evolved a number of strategies, varying in their combinations of seasonal and opportunistic movements, to take advantage of the varying aridity and rainfall of the regions they occupy (Nix 1976).

Migration in Other Ephemeral Habitats

A number of northern birds have long been thought to be nomadic in their movements in a manner related to their food supply. Prominent among these are northern finches such as crossbills (*Loxia* spp.), the waxwing (*Bombycilla garrulous*); and predators like the snowy owl (*Nyctea scandiaca*), whose populations and, allegedly, movements are thought to be related to the population cycles of their small mammal prey. Some northern finches are clearly irruptive and nomadic, but there may be a predictable component of early winter movement to the southward and a return later to traditional breeding areas. In this they resemble in reverse the south-north pattern to the annual cycles of some of the Australian dry-country birds we have discussed. Some northern species, however, have left a record of clearly irruptive movement, like the common crossbill (*L. curvirostra*), which colonized newly established pine plantations in the British Isles as a result of winter irruptions (Newton 1972). In all the cases of so-called nomadism in northern finches and waxwings, it may be extended foraging that is going on, rather than migration. Indeed this is what the ornithological literature implies when the movements of these birds are discussed, although crossbills undergo an annual fat deposition cycle that strongly suggests migration (Berthold 1993). Suffice it to say the whole issue of nomadism in these boreal species could stand closer scrutiny with the distinction between migration and extended foraging in mind, as I have discussed them in this book.

In the case of one of the raptors thought to be nomadically irruptive, the pattern of movement seems rather to be a north-south seasonal one. This is the snowy owl. Kerlinger and colleagues (1985) used North American Christmas count data to plot the movements of this arctic breeding species during the winter in Canada and the northern tier of the United States. Snowy owls were one of the raptors thought to be cylically irruptive in synchrony with supposedly cyclic fluctuations in small mammals. Kerlinger and colleagues' analysis failed to support this notion. They divided the region of alleged irruptions into sectors and performed time-series analyses on the data within and across sectors. They found a regular southward migration of owls especially into the Canadian prairie provinces and the northern Great Plains of the United States. What appeared as irruptions were likely to be the result of high variability among counts where owls were scarce in any case. It is thus likely that the snowy owl, at least in North America, is a fairly conventional intracontinental north-south migrant. Similar analysis is needed for other supposed nomads to sort out the relative contributions of regular seasonal migrations, nomadic migratory or foraging movements, and population fluctuations to the changes in numbers observed.

Although caution is necessary in the interpretation of these avian nomadic movements, they clearly occur in some species as the cited examples indicate. It is thus appropriate to ask what factors select for nomadic migrations in avian species like the emu, the crossbill, and the banded stilt of Australia described in Chapter 3. Selective factors are clearest in the case of the banded stilt. Its breeding requirements (large shallow lakes with masses of brine shrimp) are such that they occur very irregularly and scattered over a wide area. Under these circumstances a life history that includes nomadic migration over relatively long distances, followed by opportunistic breeding, is the only viable one. Similarly, when low local food abundance reduces adult survival, nomadism will be favored (Cohen 1967; Taylor and Taylor 1977). Malte Andersson (1980) has considered the more subtle case where mortality was independent of nomadism or site tenacity, the two alternatives he analyzed. Also examined were the cyclicity of food production in a given area, with the perfectly cyclic condition described by having exactly each tth season "good," where t is a fixed interval, and the ratio of adult to juvenile mortality. Andersson's model predicts that nomadism is more likely to evolve with cyclic as opposed to random food fluctuations and increasingly so with longer intervals between good years; that is, with increasing disadvantage to "waiting things out" to reproduce with the next good year. Under these conditions of cyclic food production, nomadism is favored by large clutch size and higher juvenile relative to adult survival. This is perhaps consistent with cases where seasonal and nomadic movements are combined (as may be true in the budgerigar), but the restrictive assumption of setting mortality independent of tactic may be too unrealistic. The clutch size prediction needs to be reexamined as well, because the snowy owl, Andersson's example of a nomad with a large clutch, turns out not to be a nomad after all, as outlined above. That having been said, it should be noted that Andersson's model is an interesting initial attempt to explore avian nomadic movements from a theoretical perspective and deserves followup. If nothing else, the questions raised by consideration of the model indicate the need for both a more fully developed theory of vertebrate nomadic migrations and greater iteration between theory and data.

Like many northern or dry-country species a number of forest bark beetles are also dependent on a resource that can fluctuate widely in space and time. These beetles depend on trees that are at a particular stage of development or are in a weakened condition as a consequence of drought or other forms of stress. Many bark beetles are notorious pests to the timber industry because they attack both standing trees directly and newly felled trees or even lumber in sawmills. Their lifestyle is to construct extensive galleries beneath the bark, where larvae develop, and then to emerge as adults and migrate to new host trees. These latter they find either by the odor of the trees themselves or by detecting pheromones released by other beetles that have already located trees and begun to construct galleries. Once a suitable host tree or log is located, large aggregations of beetles can develop. Johnson (1982) describes a congregation of the large brentid weevil, *Brentus anchorago*, on a dead *Bursera simeruba* tree in the dry forest of Costa Rica that consisted of several hundred mating and egg-laying individuals. The cluster lasted for over 2 months. A curious sidelight to the biology of these beetles, as well as some stag beetles (Cerambycidae), is the enormous size variation among individuals; differences may reach two orders of magnitude, suggesting much variation in the quality of the host tree, the genotype of a beetle, or both.

By far the best studied of forest beetles, because of their economic impact, are the bark beetles of the conifer forests of Palearctic and Nearctic regions. Two good examples of migratory life cycles to exploit ephemeral resources are the large pine weevil, *Hylobius abietis*, of northern Europe (Solbreck 1985) and the Douglas fir beetle, *Dendroctonus pseudotsugae* of western North America (Atkins 1959 et seq.). The pine weevil adults emerge in May or June and fly to newly cut or otherwise recently killed conifers, where the females oviposit in the roots. The larvae may take up to 4 years to develop so that by the time they emerge as new adults the resource on which they matured has largely disappeared. Migration is thus an integral part of their life-history syndromes. The flight behavior of the new adults is characteristically migratory, with three phases. First, beetles fly into the wind and toward the sky to gain height, probably in response to brightness. Second, once they reach altitudes of 50 to 100 meters they turn downwind and may fly up to 100 km, although shorter distances are more usual. Finally, there is a phase when the insects respond to host odors to locate oviposition sites. If they fail to locate suitable conifers, they can repeat the performance over several days, each time passing through the sequence of migratory actions.

A similar seasonal cycle is shown by the Douglas fir beetle except that the hosts are conifers where the beetles construct galleries beneath the bark. The females arrive at the trees first, responding to host odor; the males later, in response to pheromones in the frass produced by feeding females. Atkins (1961, 1966b) studied the flight of these beetles on laboratory flight mills so that he could monitor the sequence of behaviors during migration. There was an initial period of warm-up and standing on tiptoes (Atkins 1959) just prior to takeoff. This was followed by a period of flight that could not be interrupted by host odors or pheromones. After this period, the latter stimuli would cause flight to cease. The behavior of the Douglas fir beetle during its migratory flight therefore resembles quite closely that of the pine weevil studied by Solbreck. Both beetles in their behavior also closely resemble the sequences during migratory flight found by Kennedy for free-flying black bean aphids in a laboratory flight chamber (Chapter 2). In all three species there is inhibition during migration of responses that usually arrest flight, and a clear sequence of characteristic behaviors from takeoff to landing. Selection in all three cases has generated a response that takes an insect away from a deteriorating or otherwise unsuitable location to a new one with a resource supply. The programming of a long duration to the migratory flight greatly increases the area over which the search for new resources can take place.

Summing Up

Theoretical modeling has yielded the general result that migration should be incorporated into life histories when habitats are transient in time or patchy in space. The more ephemeral the conditions, the more migration will be favored by selection. Migration will also be favored as the number of habitat patches or the overall carrying capacity of the environment declines. Selection for migration is also influenced by the means and variances of the population growth parameter, λ. All this adds up to an increasing tendency to evolve migratory life histories with an increasing degree of environmental uncertainty.

Dry-country habitats epitomize this uncertainty, and the drier the conditions the greater the variation in rainfall. The organisms that have perhaps most characteristically adopted a migratory life-style to successfully cope with desert conditions are certain species of locusts and grasshoppers. The elaborate phase polyphenism of the desert locust, *Schistocerca gregaria*, is the best-known example, but other acridoids have adopted similar lifestyles. The extremely gregarious swarming crowded-phase *gregaria* is superb at exploiting concentrated flushes of green vegetation because its behavior keeps it well concentrated where this resource exists. The relatively asocial phase *solitaria* develops under uncrowded conditions and is the migrant form that allows these locusts to exploit the more usual patchy or transient desert habitats. This model of solitary migrants and gregarious foragers is the opposite of the model of locust life cycles usually presented where the *gregaria* is considered the migrant. The gregarious phase is migratory in the Australian plague locust, *Chortorcetes terminifera*, but it migrates at night, the time during its diel period of nongregarious and nonswarming behavior.

In many ways similar to the locusts, the African armyworm moth *Spodoptera exempta*, exploits a highly patchy set of dry-season habitats by migrating. When rains are erratic, moths can be concentrated by scattered storm outflows causing outbreaks. Under these conditions, moths develop a *gregaria* phase that is more migratory, in contrast to locusts. In years of favorable rains, moths do not become concentrated by storm outflows, but rather are more evenly distributed over favorable habitats. Under these conditions a less migratory *solitaria* phase predominates. Dry-country birds that use migration to exploit patchy rainfall are the emu and the banded stilt of Australia. The stilt is typical of several dry-country birds in its rapid reproduction when breeding conditions are favorable.

Not all ephemeral habitats occur in dry country. Bark beetles and weevils of both tropical and temperate forests require trees in just the right condition. Because of this, the resources they need for reproduction and growth and development of larvae are likely to be quite patchy. As a result they, too, have evolved a distinct migratory phase to their life cycles. The suppression, during migration, of responses that usually attract beetles to trees, allows individuals time to explore a wide area in search of a suitable host, rather than stopping at a nearby host that may, like the one from which they emerged, be declining in suitability as a result of previous infestation. Northern finches like the crossbill that depend on coniferous cone production also show irruptive movements that may be migratory. Other supposed nomads, however, may in fact be more or less traditional north-south seasonal migrants.

In all the above cases migration is at least an advantage and usually a necessity if an organism is to survive in transient and ephemeral environments. The problems presented by such habitats have been solved in various ways, but all have in common the incorporation of a migratory interval as a fundamental part of a successful lifestyle. Many of the species concerned have a significant economic impact, making an understanding of these migratory life cycles fundamental to mitigating the consequences (Chapter 15).

12

Behavioral and Life-History Variability in Migration

Migration is rarely a unitary phenomenon even in those species usually regarded as classic migrants. Rather, there is much variation in performance not only among species, but also among populations and individuals. Broadly, this variation can be divided into three types, but with considerable overlap between categories. First, there are those species and populations that show behavioral or life-history variation associated with migration in the absence of any overt qualitative morphological or physiological differences. Second, there are species and sometimes populations that to varying degrees retain the morphological capability for migratory movement only during the migration period, often remobilizing energy and materials into reproduction at the termination of this period. Wing muscle histolysis in insects is a good example. The most extreme form of variation is the third type, occurring in species that are polymorphic (genetic variation) or polyphenic (environmental variation) in the morphology directly associated with migratory activity. This is exemplified by those insects in which populations may fluctuate in the proportions of winged and wingless individuals and in many plant species with heteromorphic seeds. In this chapter, I shall concentrate on variation of the first type, deferring cases of morphological contrasts between migrants and nonmigrants to the next chapter.

Among species that show no qualitative variation for migratory capability expressed in morphology or physiology, the variability that is observed can take several forms. These can be manifested in threshold, timing, pathway, or distance traveled and can be phenotypically or genetically based (or both). Variability is apparent across the full range of taxa and often give useful insight into the way that natural selection is acting to mold migratory life histories. Here I focus on the phenotypic expression of differences involving migration-related traits; a full discussion of gene influences is presented in Chapter 14.

Population and Species Differences

Differences among populations and closely related species occur both with respect to the presence or absence of migratory activity and in the amount or type of migratory activity that takes place. A clear illustration of such differences occurs in milkweed bugs of the genus *Oncopeltus* studied by Dingle and colleagues (1980). All species and populations are macropterous, with full powers of flight. The genus is worldwide, but the majority of species are confined to the New World, including all in the apparently rapidly evolving subgenus *Erythrischius*, which includes the species studied (Slater 1964). Among the latter, some populations of one species, *O. fasciatus*, are long-distance migrants that invade North America from subtropical refuges each spring in a manner very similar to that of the monarch butterfly (Chapter 3). Other populations of *O. fasciatus* occur throughout Mexico and Central America south to Panama and on Caribbean islands south to Martinique; these are largely sedentary, moving only between local milkweed patches where breeding can occur throughout the year. Other species of *Oncopeltus* are sympatric with *O. fasciatus* throughout the tropical and subtropical portions of the latter's range. Dingle and colleagues (1980) examined the flight performances of three migratory and three nonmigratory populations of *O. fasciatus* as well as the performances of three sympatric nonmigratory species—all, like *O. fasciatus*, members of the subgenus *Erythrischius* and therefore, presumably, quite closely related. Flight was analyzed in bugs induced to fly tethered in the laboratory by lifting them free of the substrate to trigger the flight reflex and then monitoring the duration of the subsequent flight. The flight performances are compared in Table 12–1.

The comparisons reveal major differences in flight performance among the populations. In the northern migrant populations of *O. fasciatus* from Iowa, Michigan, and Maryland about 25% of the individuals flew for more than 30 minutes, the duration taken as the operational definition of migration in these experiments (other experiments revealed that once a bug flew for 30 minutes it was

Table 12–1. Tethered Flight Performances of Samples from Migratory and Nonmigratory Populations of *Oncopeltus* Milkweed Bugs

Species of Population	No Flight	Number Flying for		Percent > 30 min
		0 < 30 min	> 30 min	
Migratory *O. fasciatus*				
Michigan	0	237	74	23.8%
Iowa	0	141	44	24.5
Maryland	0	40	13	24.5
Nonmigratory *O. fasciatus*				
Puerto Rico	6	53	3	4.8
Guadeloupe	13	29	2	4.5
Mexico	0	98	0	0
Tropical nonmigratory species				
O. cingulifer (Trinidad)	0	88	2	2.2
O. sandarachatus (Trinidad)	0	58	2	3.3
O. unifasciatellus (Colombia)	7	128	0	0

SOURCE: From Dingle et al. 1980.

likely to fly for several hours). This proportion may seem low at first glance, but two points should be noted. First, the bugs received no stimulus to fly other than loss of tarsal contact; even adding a light breeze on the head with a simple fan increases the proportion of long flyers considerably. Second, the bugs were maintained in a long-day environment, which leaves open only a small pre-reproductive "window" for migration (see Chapter 6); because the bugs were flight tested only once, this window could have been missed. Thus, the 25% migrants is a conservative estimate. The important point, however, is not the absolute proportion of migrants but how the proportions in these known migratory populations compare with those observed in the nonmigrants from tropical and subtropical regions. Here the data are unequivical. In none of the other populations of *O. fasciatus*, nor in any of the other species, did the proportion of long-flying bugs exceed 5%. A further difference was the fact that in some of the nonmigrants no tethered flight could be induced, whereas all migrant bugs flew for at least a brief period. In some cases, therefore, there may also be a difference in flight threshold between migrants and nonmigrants. Within each sample there is considerable variation in flight performance, but notwithstanding, the migrants on average fly considerably longer under the conditions of these experiments. Because all bugs were reared under the same environmental conditions, the differences among them are due to gene differences; and as we shall see in Chapter 14, genetic variation is sufficient to allow quite rapid responses to selection acting on flight and other aspects of a migratory syndrome.

Another example in insects in which laboratory flight performance reinforces field observations occurs in the North American lesser migratory grasshopper, *Melanoplus sanguinipes*, studied by Lynne McAnelly (1985, McAnelly and Rankin 1986). She sampled three populations: one from Colorado with a low tendency to migrate in the field, one from Arizona with an intermediate tendency, and one from New Mexico with a high tendency. The amount of migration paralleled increasing aridity of the sites, and therefore increasing ephemerality of suitable habitat patches. Under these conditions selection should favor migratory movement (Chapter 11). The New Mexico site was in fact an old mine tailing that had only recently been invaded by (migratory) grasshoppers. The results of the flight tests are shown in Table 12–2, and the tethered-flight behavior reflects the field observations. Only 5% of the Colorado animals were migrants, based on a criterion of 60 minutes of flight, while 58% of the New Mexico grasshoppers could be so classified. Although the exact proportions showing 60-minute flights varied, the differences among populations were maintained for up to four generations of laboratory rearing in a Colorado versus New Mexico comparison. Laboratory rearing eliminates the possibility of maternal effects derived from field-reared mothers and demonstrates that the among-population differences, as with *Oncopeltus fasciatus*, are genetic. In the case of *M. sanguinipes* natural selection has produced gene differences among populations, with migratory genotypes favored as sites become more arid and habitat patches more ephemeral.

A situation somewhat parallel to that occurring among the species and

Table 12–2. Differences in Tethered-Flight Performance Among Populations of the North American Lesser Migratory Grasshopper, *Melanoplus sanguinipes*

Flight Duration (Minutes)	Arizona 1982	Colorado 1982	New Mexico 1983
0–55	109	141	21
56+	43	7	29
% Migrant	28%	5%	58%*

*New Mexico different at p < .001 (using χ^2) from both Arizona and Colorado.
SOURCE: From McAnelly 1985.

populations of *Oncopeltus* milkweed bugs occurs with seed migration in eight weedy species of North American Asteraceae (composites). The seed in these species is a small hard achene to which is attached a parachute-like pappus for aiding in wind transport. Andersen (1992) studied the "settling velocity" of these seeds in the laboratory to get an estimate of the relative lengths of time these seeds would remain airborne. A higher velocity would indicate more rapid settling and hence less time spent aloft. Whole achene-pappus units were dropped down a Plexiglas tube, and their descent timed. The fall times were then converted to settling velocities by dividing by the length of the tube. The means and coefficients of variation are given for these velocities in Table 12–3 along with the origin and life history of each species.

Andersen found statistically significant variability at all levels of his analysis that included differences among species, among plants within species, and even among inflorescences and seeds within plants. The variation in settling velocities among species likely reflects differences in the importance of migration and the ability to disseminate seeds widely to the life histories of the plants studied. A glance at Table 12–3 reveals a tendency for introduced annual species to produce seeds with lower settling velocities and a proclivity to stay airborne longer as a result. These are classic weeds and colonizers of open ground. Their ability to produce seeds capable of a longer migration with, in all likelihood, wide dispersal over available habitats is probably in part what allows them to compete so successfully with native species. A positive role for migration and the elaboration of structures to assist wind transport in the success of weedy annuals is noted frequently in the plant literature (Venable and Levin 1983; Bazzaz 1986). The data in Andersen's study certainly support this notion. The variability noted within species and individuals may represent a form of risk spreading (den Boer 1981). In heterogeneous environments of the sort occupied by weeds, a selective advantage may accrue to those parent plants that can disseminate propagules over the widest possible area.

Variability also occurs among seeds that use animal transport. Several species of woodland herbs possess fatty and protein-rich elaiosomes on their seeds to facilitate removal by ants. In this case seeds are not so much dispersed by the ants as concentrated in favorable germination sites. Hughes and Westoby (1992)

Table 12–3. Settling Velocities, Coefficients of Variation (CV), and Ecological Characteristics for Eight Species of Weedy Composites

Species	Origin	Life History	Mean Settling Velocity (m · sec⁻¹)	CV
Heterotheca grandiflora (Telegraph weed)	Native	Biennial	1.02	0.420
Picris echioides (Oxtongue)	Introduced	Biennial	0.702	0.450
Chryopsis villosa (Golden aster)	Native	Perennial	0.514	0.140
Aster exilis (Slim aster)	Native	Annual	0.373	0.434
Taraxacum officinale (Dandelion)	Introduced	Perennial	0.307	0.207
Conyza bonariensis (Horseweed)	Introduced	Annual	0.291	0.250
Sonchus oleraceus (Sow thistle)	Introduced	Annual	0.251	0.546
Senecio vulgaris (Common groundsel)	Introduced	Annual	0.248	0.289

SOURCE: From Andersen 1992.

removed the elaiosomes from some seeds and compared their removal rates with those left intact. Not surprisingly the presence of elaiosomes increased removal rates. What was more interesting was that different ant species responded differently to whether seeds were clumped or single on a plant and to elaiosome size/seed size ratios. The experiments clearly demonstrated that in these plants that depend on ants for the migration of their seeds, it is the variability in the behavior of the transporting insects that matters. Ant behavior is therefore an important source of selection in the evolution of myrmecophory.

A bird whose variable migration among populations in nature has been confirmed with laboratory studies is the blackcap warbler, *Sylvia atricapilla*, of the western Palearctic, studied by Peter Berthold and his associates (Berthold 1985, 1988, 1991a; Berthold and Querner 1981; Berthold et al. 1990). The blackcap is a widespread species divided into many geographic races. These races migrate for varying distances and with different routes; populations breeding in Finland, for example, migrate south-southeast and then south to wintering quarters in eastern Africa, while populations from southwestern Germany migrate southwest and winter mostly in Spain, with possibly some spillover into adjacent North Africa. Berthold captured birds from four populations, southern Finland, southern Germany, southern France, and the Canary Islands, and bred them in the laboratory. The nestlings of all groups were hand raised under identical simulated natural light conditions so that there were no differences in the environments experienced during rearing. The migratory restlessness (Zugenruhe) of these hand-raised birds was then assessed in the laboratory in cages that allowed automatic recording of bird activity (described in Chapter 4) and in a light:dark regimen of

LD12.5:11.5. This cycle approximates the photoperiod the birds would experience under natural conditions in late summer–early autumn at the time wild populations would be migrating. Noctural migratory restlessness is a good index of migratory behavior.

The measurements of migratory restlessness plotted over some 200 days of recording indicate a strong positive correlation between restlessness and the distance flown during migration by the natural populations (Fig. 12–1). The Finnish birds are the most active, reflecting the fact that they migrate the farthest, from the Baltic to eastern Africa. The German birds are next in activity level, followed by the French. The latter population is only partially migratory (next section), with some birds remaining more or less sedentary during the winter while others migrate, and this division in migratory tendency is reflected in the pattern of restlessness. Lastly, the Canary Island birds are essentially residents throughout the year, and they display the least amount of nocturnal activity. The fact that all the birds were hand reared in a common environment suggests that the observed pattern among the populations is the consequence of gene differences and results from selection. As we shall see in more detail in Chapter 14, this is indeed the case. As with the lesser migratory grasshopper and the migratory milkweed bug, the differences in migratory performance measured in the laboratory are due in

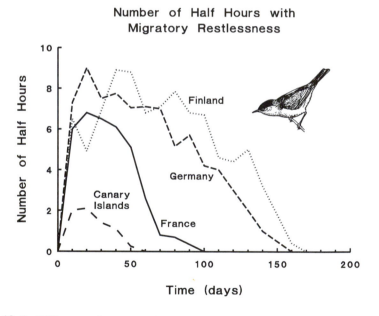

Number of Half Hours with Migratory Restlessness

Figure 12–1. Differences in nocturnal migratory restlessness between blackcap warblers from the Canary Islands and three locations in Europe. The duration of migratory restlessness in laboratory cages, expressed as number of half hours of occurrence, reflects the distance each population migrates in nature. (Redrawn from Berthold and Querner 1981.)

large part to genetic programs and not primarily to the environmental conditions either during development or at the time of migration.

When other species in the genus *Sylvia* are added to the comparison, there is also a good correlation between laboratory-measured nocturnal restlessness and migratory distance (Berthold 1984a, 1985; Fig. 12–2). There are quite large differences between the species. Those that occur in southern Europe are either sedentary or engage in very little migration; these include spectacled and Marmora's warblers (*S. conspicellata* and *S. saida*). Those that breed in northern Europe like the garden warbler (*S. borin*) migrate long distances to winter throughout central Africa. The pattern of migratory restlessness shown by hand-raised birds of each species is a good match to the migration distance of the parent population of that species in nature. Once more the fact that the differences are maintained in birds reared in a common environment indicates gene differences for the migratory trait among the species.

A particularly interesting case of population variation in movement behavior occurs in orcas or killer whales (*Orcinus orca*). In the inshore waters of British Columbia on the west coast of Canada, over 300 individual orcas have been identified, using photographic records that allow recognition of individuals based on natural markings on the dorsal fin and back (Bigg et al. 1986, 1987; Kirkevold and Lockard 1986; Heimlich-Boran 1988; Ford 1991). These orcas can be divided

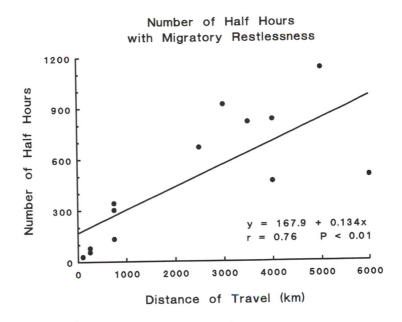

Figure 12–2. The relation between migratory restlessness in the laboratory and migratory distance among species of the Palearctic warbler genus *Sylvia*. Note that restlessness correlates strongly with distance traveled. (Redrawn from Berthold 1985.)

into two basic types, residents and transients, which overlap in range but are completely socially isolated from each other. They differ in morphology, social behavior, feeding habits, and most importantly for our purposes here, in movements. Residents live in stable social groups or pods of 5 to 50 individuals, each of which appears to share its own matrilineal descent; two or more pods form a clan that is characterized by the sharing of common call types and other patterns of acoustical communication. Residents also appear to travel less, are more predictable in occurrence, and feed primarily on fishes. In contrast, transients occur in smaller groups of only one to seven members, are generally less vocal, and appear to specialize in feeding on marine mammals. They also travel much farther and often go unobserved for long periods. There is not enough known about these transients to tell whether long-distance movements are truly migratory, with, for example, temporary suppression of feeding, or whether they are, rather, extended foraging movements in pursuit of wide-ranging prey, some of which, like seals and gray whales, are certainly migratory (reviewed in Baker 1978). The biological reasons behind the resident-transient distinctions are also unclear. These intelligent animals would certainly be capable of establishing the differences by learned traditions regarding movements, but it is not known whether or not they in fact do so, nor the extent to which natural selection might be acting on gene differences. The situation is a fascinating one and certainly deserves more study from the perspective of assessing just what kind of movement behavior is involved and how it functions in the life cycles of these elegant animals.

Finally, in this section on differences among populations and closely related species, it is worth noting that variations in movement do not occur independently, but rather correlations between migration and other traits are common. A good case in point occurs in populations of chinook salmon (*Oncorhynchus tshawytscha*) studied by E. B. Taylor (1990) in British Columbia. Taylor examined eight populations; four of these were "stream-type" migrants, with juveniles remaining in streams for a year or longer before migrating to the sea, and the remaining four were "ocean-type" migrants, which journey to the sea during the first year, often directly after hatching. Both groups were positively rheotactic—that is, they swam upcurrent—in laboratory tests for the first 2 to 3 months of life, but after that the ocean fish no longer responded positively to currents, consistent with their seaward movement at this time. The stream populations did not cease positive rheotaxis until they were a year old. Fish that remain in the streams also hold territories, and the stream fish were much more aggressive, although the ocean fish that migrated at 2 to 3 months did show more aggression than those departing downstream immediately. There was thus a positive correlation between levels of aggression and length of stay in streams. In the third trait compared, ocean fish grew faster than stream fish. All these phenotypic differences were expressed in fish reared in a common laboratory environment and so appear to have a genetic basis and hence to reflect adaptive divergence in the two types of fish. The maintenance of the two different life histories probably indicates a trade-off between the growth gains of early oceanic migration and reduced risks in the seaward journey for the older and more robust stream fish. Selection has then also favored suites of traits appropriate to each of the juvenile lifestyles.

The observation that ocean fish grew faster than stream fish in Taylor's study of the chinook is consistent with other studies showing that the more migratory the fish, the faster it grows and the larger the size it attains. This might be in part, at least, a consequence of the fact that the energetic cost of migration decreases with increasing size (Roff 1988). The size and growth differences between migrants and nonmigrants, however, seem to hold even in fishes that are small. Snyder (1991) compared sizes and growth rates of migratory and nonmigratory sticklebacks (*Gasterosteus aculeatus*) from the Navarro River system of California. Landlocked inland fish from the interior region of the Navarro drainage were both smaller and slower growing than conspecifics migrating from coastal estuaries upstream for breeding. The differences in growth are shown as a function of the length of fish in Figure 12–3. Growth rates in both populations declined with size, but the estuary population always grows faster, and so achieves a larger size as a result. The largest migrants are about 60 millimeters in length, while the largest of the landlocked fish reach only about 50 mm. The rearing of both groups under identical laboratory conditions demonstrates that selection has produced gene differences between them.

Size differences would also be expected to influence other life-history characters in fishes. That this is indeed the case has been shown nicely for

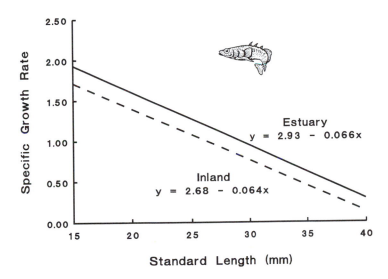

Figure 12–3. Linear regression lines of specific growth rate against size for esturary (migrant) and inland (nonmigrant) sticklebacks from the Navarro River in California. Specific growth rate = log L2 – log L1/t × 100 where L1 and L2 are the initial and final lengths and t = time in days. Slopes do not differ, but the estuary fish grow faster and reach a larger size. (After Snyder 1991.)

migratory and nonmigratory brown trout (*Salmo trutta*) in the English Lake District by J. M. Elliott (1987, 1988, 1989). These trout came from two small streams, or becks in the local parlance. The nonmigratory population was isolated in Wilfin Beck by a waterfall that prevented invasion from Lake Windermere of any conspecifics. The migratory population spawned in Black Brows Beck, from which they could proceed downstream either to estuaries or the sea. Life-history differences between the two populations are summarized in Table 12–4.

Several differences between the two sorts of brown trout are apparent. The first is the fact that migratory fish are larger, as we have already seen for other fishes. In this case the differences are manifested not only in body length but also in weight. Migratory fish are more robust, with a 25-cm migrant some 50 grams heavier than a nonmigrant of comparable length. The larger migrant female produces about twice as many eggs as the nonmigrant and these eggs are larger because although twice the number, they are triple the weight of nonmigrant eggs. The larger size at birth and in the fry stage is translated into larger adults in the migrants; these in turn produce more and larger eggs with a greater reproductive investment, as determined by the proportion of energy devoted to egg production. Genotype is a major contributor to the population differences because these persisted even when trout were reared under similar conditions in the laboratory. Access to the sea in the migrants has resulted in selection for larger fish and greater reproductive effort, no doubt sustained by the greater food resources available in the sea. A further consequence seems to be density-dependent selection on the fry born to migrants. The greater coefficients of variation for the sizes of first-year fish of this population arise from the much greater variation in numbers of eggs produced in a given spawning season (Elliott 1989). The landlocked trout,

Table 12–4. Differences in Life History Characters Between Migratory and nonmigratory brown trout (Salmo trutta) from Two Streams (Becks) in the English Lake District

Character	Wilfin Beck (Nonmigratory)	Black Brows Beck (Migratory)
Fry fork length*	19.3 ± 0.2 mm	26.2 ± 0.3 mm
Fry wet weight[†]	83.1 mg	139.3 mg
Wet weight of 25-cm female	142.0 g	194.4 g
Eggs laid by 25-cm female	239	512
Wet weight of eggs	14.1 g	50.8 g
Length of spawning females	18–28 cm	26–45 cm
CV wet weight[‡]	51–57%	20–51%
% energy to egg production	16–17%	30–37%

* Mean for trout hatched from two egg samples in the laboratory + SE
[†]Weights of fry born to females 25 cm in fork length.
[‡]Coefficients of variation for sizes of first-year trout.
SOURCE: From Elliot 1987, 1988, 1989.

in contrast, seem to be subject to little variation in first-year size and to selection that is primarily a function of age.

Comparisons of anadromous and sedentary populations as well as comparisons across species were used by Hutchings and Morris (1985) in their analysis of salmonid life histories. Using the multivariate statistical technique of principal components analysis to assess covariation among life-history traits, they found a primary syndrome ranking salmonids from large, early-maturing semelparous forms, like species of *Oncorhynchus*, to the opposite of small, late-maturing iteroperous individuals, represented by some char (*Salvelinus*). There was a secondary trend from anadromous to exclusively freshwater forms. The clustering of species on the basis of life history matched quite closely the taxonomy of the group, suggesting a phylogenetic association of traits as well as adaptation to local conditions. The association of certain life history traits with anadromy implies that migratory behavior and the evolution of these life histories is closely linked. One of the consequences is that the anadromous lifestyle allows increased growth rate at sea, larger size, and higher egg production. But it also incurs costs, including an exhausting migration up swift-flowing streams to spawning sites. Very long, exhaustive migrations probably select for semelparity, because the probability of spawning again is low due to high costs, and for large size for increased energy efficiency (Roff 1988); this is exactly the situation seen in most Pacific salmon (*Oncorhynchus*). The degree of iteroparity is a likely compromise between alternatives that are a function of the ratio of adult to juvenile survival (Schaffer 1979).

This ratio is an important feature of the analysis of the migration of the American shad, *Alosa sapidissima* (Carscadden and Leggett 1975; Shoubridge 1978). If adult survival is high relative to juvenile survival, iteroparity should be increasingly favored. Further, environmental variance that influences adult survival should select for an earlier age at first reproduction and increased fecundity; the converse should be true if environmental variance has a greater impact on juvenile survival (Schaffer 1979). These predictions generally held up across populations of the shad. In the St. John's River in Florida where environmental variance was low and temperatures for juvenile survival were favorable over a long period, the population is semelparous with a high fecundity of around 410,000 eggs per spawning. This result also reflects the high energy expenditure during the upstream-spawning migration due to the relatively high water temperature. Shad are particularly sensitive to energy costs during migration because they do not feed once in freshwater. The St. John's is thus harsh for adults, but benign for juveniles. At the opposite end of the shad's range, in the Miramichi River of New Brunswick in eastern Canada, environmental variance was high, but lower temperatures reduced the cost of migration while increasing the mortality risk to juveniles. Here iteroparity is the rule, and fecundity is reduced to 266,000 eggs per lifetime. Similar patterns were observed in Pacific rivers. On both coasts, variation was also noted as a function of the length of the migratory journey.

The situation in Pacific rivers is especially interesting because shad were introduced into West Coast streams from the Hudson, Susquehanna, and Potomac Rivers about a century ago (Shoubridge 1978). In these Atlantic rivers 20% to 40% of adults are repeat spawners, age at maturity is 4.6 years for females and 4 years for males, and lifetime fecundity averages 300,000 to 350,000 eggs. In the recently evolved populations in Pacific rivers, 32% to 77% are repeat spawners, age at maturity varies from 4.0 to 4.5 years in females and from 3.3 to 3.8 years for males, and mean lifetime fecundities range from 321,000 to 500,000 eggs. This rapid evolution of life-history syndromes adjusted to location suggests genetic variation for the traits and strong selection. As with salmon, migratory and life-history characters are also coevolved.

Intrapopulation Variation

In addition to instances where migration and life histories differ among whole populations, there are numerous cases where migratory behavior occurs only in some fraction of a population, with the remainder sedentary or moving merely locally. This phenomenon of intrapopulation variation in movement is known as partial migration. A further variant, differential migration, involves the movement of varying distances by different individuals and will be discussed further in the next section below. The two sorts of migratory behavior patterns are not mutually exclusive, and indeed the migratory portion of a partially migrating population or species is quite likely to move for different distances depending on circumstances. There are documented instances of partial migration in a wide array of taxa from insects to fish to birds (Swingland 1983; Lundberg 1988), and the phenomenon may be even more widespread than generally realized. To take one case in point, a recent intensive trapping, mark, and recapture study of the supposedly sedentary short-toed treecreeper (*Certhia brachydactyla*) in Germany revealed that a significant fraction of the individuals caught at the trapping site near Lake Constance were in fact migratory (Bauer and Kaiser 1991). As studies expand, additional cases in birds and other taxa will very likely be discovered. The phenomenon seems to be an important life cycle variant, and its causes therefore important to discern.

In at least some species there seems to be a strong genetic component to partial migration. Two examples in birds are the blackcap warbler, *Sylvia atricapilla* (Berthold and Querner 1982; Berthold 1988; Berthold et al. 1990), and the European robin, *Erithacus rubecula* (Biebach 1983). Blackcaps were captured in partially migratory populations from southern Europe and the western Mediterranean region and brought to the laboratory. When offspring from the field-caught birds were reared under constant conditions, they still divided into migratory and nonmigratory fractions when tested for migratory restlessness. The higher proportion of migrants in the parent population, the greater the fraction of migrants appearing in their hand-raised offspring, and if high- and low-restlessness populations were crossed, the offspring were intermediate. Finally, and most convincingly, the frequency of migrants could be increased by artificial selection. All these genetic experiments will be covered in some detail in Chapter 14; suffice it to say here that they provide convincing evidence that there is a strong genetic influence on partial migration in the blackcap. Similarly, laboratory breeding experiments in the robin demonstrated a high proportion of migrant offspring when parents were migratory and a low frequency of migrants when parents were mostly resident, again indicating a significant genetic component.

In the case of the robin, at least, field studies suggest there are also strong environmental influences on whether or not an individual migrates. In a study of robins in Belgium, Adriaensen and Dhondt (1990) found that sex of the birds, especially, was a major factor, with almost all females migrating but varying proportions of males remaining over the winter as residents. Social dominance by resident males probably accounts for much of this difference, and in fact, in many species of partially migrant birds, dominance seems to be a major factor in determining that by and large it is juveniles and females that migrate (reviewed in Gauthreaux 1982). In the robin the frequency of the resident males is also influenced by the habitat; in parks and gardens most males are resident, but in woodlands some 70% of the breeding males were migratory. In addition, across the geographic range of the robin the proportion of migrants decreases from north to south (Cramp and Simmons 1988).

In terms of both survival and breeding success, there is a tremendous advantage accruing to residents over migrants. Survival for resident males was 50% as compared to 17% for migrants. Even in particularly severe winters, resident survival was no lower than that of migrants. Breeding success was positively correlated with date of settlement on territory, with resident males often established as early as the end of January. Early settling residents experienced 74% mating success in contrast to about 44% for migrants and only 19% for even later settling birds of unknown status. Overall the probability of breeding was between two and four times higher in residents than in migrants.

With all the apparent disadvantages of migration, it is appropriate to ask: Why migrate? Adriaensen and Dhondt argue logically that the partial migration system of the robin represents a conditional strategy with unequal payoffs. It is conditional in the sense that there are two alternatives, migration and residency, with the option chosen depending on circumstances. In this case the options have unequal payoffs because the resident strategy is by far the more successful. The migrants then, are "making the best of a bad job," and the payoff is not in migration per se but derives presumably from a particular migrant being better off migrating than staying. This also implies that an individual should change from the migrant to the resident option if it can. None were observed to do this, but because of the low survival of migrants, there was probably little chance of a bird surviving long enough to have the opportunity to change. Residents would also be difficult to displace because of the head start they acquired each spring. Adriaensen and Dhondt also noted that this sort of environmentally driven conditional strategy is not necessarily incompatible with gene differences between migrants and nonmigrants as found by Biebach (1983). It does suggest, however, that in the field, the expression of a particular behavior is environment dependent (a genotype by environment interaction). The differences found in the laboratory under constant conditions are therefore likely to be differences in threshold of response. The same may be the case with the blackcap warbler, but in the absence of field studies of survival and breeding success in migrants and residents, the relative contributions of genes and environment remain to be determined.

A bird in which migrants are known to change behaviors to become residents is the European blackbird, *Turdus merula* (Schwabl 1983; Lundberg 1985, 1988; Schwabl and Silverin 1990). In partially migrating populations of blackbirds, females are more likely to migrate than males and juveniles are more likely to migrate than adults. Lundberg's (1985) study of blackbirds in southern Germany showed that with respect to agonistic interactions, males were almost always dominant over females, and adults similarly usually dominated juveniles. During the winter males actually could gain weight from fat deposits, but resident females showed much less ability to do so. It is likely that this difference was a consequence of the control of resources by males. Migrant individuals again seem to be forced into a making the best of a bad job by the dominance of males and adults. There may be, in addition, an action of selection favoring a genetically based tendency to migrate as juveniles, because these young birds put on migratory fat deposits that the adult males, especially, do not. Schwabl (1983; Schwabl and Silverin 1990) noted that migrants showed a much greater tendency to become residents than the reverse; this was true for both adults and juveniles. For example, about 85% of resident adult males remained residents in succeeding years, whereas over half of the migrant males became residents rather than remaining migrants. Juveniles were not quite so successful at holding resident status or in becoming residents following a year when they migrated, but they clearly displayed the same trends.

In completely migratory populations a dominance-driven conditional strategy can appear in the wintering area. Rappole and colleagues (1989) studied wood thrushes, *Hylocichla mustelina*, at a wintering site in the state of Veracruz, Mexico, using both banding and radiotracking. The same birds returned in successive years and fell into two categories, wanderers and sedentary individuals. The former moved more than 150 meters from the point of capture, were not aggressive toward other wood thrushes, and foraged mostly in second-growth areas, often in company with other wanderers. Sedentary birds moved less than 150 meters from point of capture and held territories, mostly in primary forest, from which they excluded other thrushes; they suffered lower mortality than wanderers. Birds can and do change categories, especially from being a wanderer to holding a territory. This latter fact plus the clear advantage to territory holding in terms of mortality imply that the wanderers, like partially migrating robins and blackbirds, are making the best of a bad job and are pursuing a conditional strategy. Given the opportunity, they cease wandering and take up the sedentary, territorial option.

A case where partial migration may not be a conditional strategy but rather a mixed evolutionarily stable strategy, usually abbreviated to the acronym ESS, is that of the giant tortoises (*Geochelone gigantea*) of Aldabra atoll in the Indian Ocean (Swingland and Lessels 1979; Swingland, 1983). The concept of an ESS was developed from game theory models largely by John Maynard Smith (see Maynard Smith 1976, 1982 for summaries) and is defined as a strategy—the ESS—that when common in a population, cannot be displaced by any alternative "mutant" (i.e., genetically based) strategy. A mixed ESS differs from a conditional strategy, such as seems to be present in European robins and blackbirds, in two ways. First, payoffs for each alternative, here migrating or staying, should be the same; second, if one alternative increases in frequency, its payoffs should be reduced or, to put it in another way, its lifetime reproductive success should be negatively frequency dependent. A mixed ESS is thus a situation in which the ratio of alternative behaviors is evolutionarily stable because if the ratio changes, so do the payoffs and the balance is restored. The classic case of a mixed ESS is the 1:1 sex ratio, because if the ratio deviates from unity, the relative fitness of the more numerous sex declines. Although strictly speaking they are limited to a haploid genetic system, ESS models have been very successful in providing insights into behavior.

In Aldabran giant tortoises a portion of the population migrates to the coast at the beginning of the rains. The migrants differ from nonmigrants in being male biased, in being longer and narrower, and, in females, in having more pre-ovulatory ovarian follicles. Carapace shape is a heritable character in giant tortoises (although not directly tested in Aldabra), suggesting that migrants and nonmigrants may, at least in part, be genetically influenced. Whatever the basis for the difference among individuals, be it genetic, environmental, or very likely both, a given tortoise seems to be consistent across years in migrating or not. The payoff to migrating is a richer food source leading to enhanced reproduction following the rains; the disadvantage is the scarcity of shade at the coast, which the tortoises must have. In the heat of the day overheating is fatal. Because of differences in shade between coast and inland, the lifetime reproductive success of migrants and nonmigrants may not differ much. More of the smaller females than larger males die at the coast, suggesting there is competition for the available shade. The nature of the trade-off is frequency dependent because the greater the number of migrants, the less the available food and shade. This frequency dependence plus the consistent behavior of individuals from year to year imply that the differences among tortoises arise from a mixed ESS at the population level in the form of a genetic polymorphism, although two other alternatives, namely the aforementioned conditional strategy or a learned response, cannot be completely ruled out. The tortoises are a good example of how difficult it can be to sort out from among the alternative hypotheses.

One way to try to sort from competing or alternative hypotheses is by modeling and simulation. This was the approach used by Kaitala and colleagues (1993) in developing an ESS model for partial migration. In their modeling efforts they considered a population of birds (although the model is not necessarily confined to birds) that reproduce during summer. A part of the population overwinters at the breeding site, while the remainder migrate elsewhere to overwinter and return the following year to the breeding locale. These migrants face a higher risk of dying during migration but a lower risk of starving during the winter. Because of this, they are assumed to suffer density-independent mortality. The sedentary birds do not risk death during migration, but do face a higher risk of starvation, with competition for food increasing at higher densities. Therefore, they are assumed to suffer density-dependent mortality, and this mortality would also be frequency dependent because it would be influenced by the proportion of migrants. There is some ambiguity in the ESS concept with respect to fluctuating populations inhabiting a finite number of patches in a randomly varying environment, because a chance run of a particular condition over several time intervals can result in favoring a strategy that is inferior under more common conditions (Levin et al. 1984). This problem has been dealt with by Ellner (1984, 1985) based on the computation of expected growth rates of "invading strategies," and Ellner's approach with some modifications was used by Kaitala and colleagues in the development of their model for subsequent simulations. For the latter, various parameter values were used that were consistent with what is known about locally interacting and stable bird populations; for example, the egg-laying capacity of a given female was assumed to be five eggs, and the local population size at which density independent growth is reduced by half—a measure of carrying capacity—is taken as 100 birds.

The second point arising from the analysis is that the reproductive successes of the two alternative behavioral types do not need to differ in order to produce partial migration. This is evident in Figure 12-4B where even when the reproductive success of migrants is equal to that of nonmigrants, a small fraction of the

Three major points emerge from the analysis and simulations. First, density-dependent overwinter survival is the single most important requirement for the maintenance of partial migration in a population. This is clearly shown in Figure 12-4A. In this figure the proportion of migrants in the simulated populations is plotted against the overwintering population density under two sets of conditions. In one the expected mean survival under density independence for the nonmigrants, $\mu_n=0.5$, is much higher than the expected survival under density independence for the migrants, $\mu_m=0.2$. In the second case, the two values are $\mu_n=0.7$ and $\mu_m=0.6$, or in other words, the expected mortalities don't differ very much, and there is thus not much disadvantage to migration. Under the first set of conditions density exerts a very strong effect, with the frequency of migration increasing markedly as density increases. The effect is even more dramatic under the second set. Here the proportion of migrants starts at just over half, as would be expected if there was little to choose between alternatives, and rises sharply as density begins to shift the balance of mortality. It also follows that if density-independent mortality of migrants is less than that of nonmigrants, then only migration will be a stable ESS, and conversely, if the reverse is true, there will always be some sedentary individuals. As Kaitala and colleagues point out, this prediction is quite amenable to direct testing, at least in principle.

The second point arising from the analysis is that the reproductive successes of the two alternative behavioral types do not need to differ in order to produce partial migration. This is evident in Figure 12-4B where even when the reproductive success of migrants is equal to that of nonmigrants, a small fraction of the

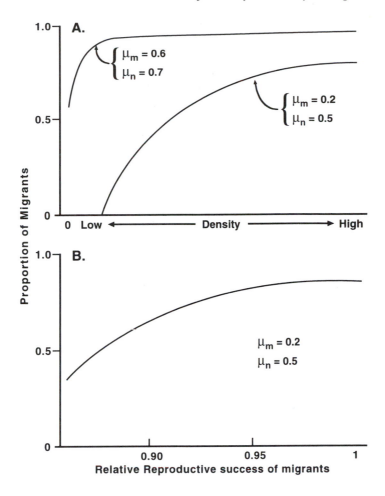

Figure 12–4. Some results of ESS models concerning partial migration in birds. *A*. The influence of density on the proportion of migrants in a population under two conditions that differ in the expected mean overwinter survival of migrants (m) and nonmigrants (n). *B*. The influence of the relative reproductive success of migrants on the proportion of the population migrating. (Modified from Kaitala et al. 1993.)

population is still sedentary. In terms of the model, differential mortality is the critical factor that drives the system and permits the coexistence of the two strategies. Also indicated in Figure 12–4B, however, is that the magnitude of the ESS fraction of migrants can be affected by relative reproductive success; not surprisingly, as the relative success of migrants in reproduction declines, so does the proportion of individuals displaying that option. Recall that the migrants in partially migrating European robins and blackbirds do indeed suffer reduced reproductive success; below, in fact, that which seems necessary to balance the fitness of remaining sedentary. The model predicts that because migration is a best of a bad job alternative, some additional controlling factor should drive it, and indeed this seems to be the case, with dominance serving this function. The model also predicts that individuals will change to migratory behavior as an inevitable consequence of improved reproductive success, and once more, such changes have been observed, at least in blackbirds.

The third point to come out of the study was that environmental stochasticity, in turn affecting density-independent mortality, had very little influence on the ESS solution and certainly not enough influence for the evolution and maintenance of partial migration. Any potential influence was effectively overridden by the density-dependent overwinter survival. Furthermore, random variability in breeding site had no influence. This outcome was somewhat surprising because it so contradicted many previous models and assumptions (Cohen 1967; Lack 1968; Lundberg 1988). The notion that stochastic effects on the resident population might be responsible for maintaining genetic dimorphisms (Biebach 1983; Berthold 1984b) was also not supported. Once again the paramountcy of overwinter survival was evident. The fact that this survival is so important in this ESS model and that it depends so heavily on density dependence suggests that the influence (or not) of density dependence should be carefully looked for in real-life partial-migration systems. What, for example, might drive partial migration in the absence of density dependence, as might occur, say, in some insects? Might those insects that can migrate *and breed* elsewhere be avoiding the competitive consequences of density dependence, as suggested by the Taylors (Taylor and Taylor 1983)? In those cases where migratory behavior is so clearly influenced by genotype, what is the contribution of density and its related social consequences to the expression of the behavior, i.e. to genotype by environment interactions? Kaitala and colleagues call for good field data on the causes of mortality patterns and differential reproductive success, and this call is certainly justified. Such information will be vital for an understanding of the evolution of both partial and complete migration patterns.

Variation in Routes and Patterns

Populations and individuals vary not only in whether or not they migrate but also in the distances they travel, the routes they follow, the timing of arrival or departure, and the pattern of behavior during the migratory journey. Variation in distance traveled is generally termed differential migration and has received considerable attention, especially in birds. One of the most extensive studies is that of Ketterson and Nolan (1976, 1979, 1983) on the dark-eyed junco (*Junco h. hyemalis*)in eastern North America. This subspecies, the "slate-colored junco," breeds in Canada, along the northern edge of the United States, and at higher altitudes in the Appalachian Mountains south to Georgia. Striking in the junco pattern of overwintering is a latitudinal cline in sex ratio with males predominating in the northern part of the winter range and females in the southern, as determined by extensive banding and capture data. Female juncos, in other words, migrate farther than males. Across the northern tier of the United States from southern Maine to southern Minnesota, wintering populations are 70–80% male (about 65% in south-central Ontario); across the southern tier from South Carolina to Texas, populations are 70–80% female. Young birds of both sexes predominate to the north of the adults, indicating that adults migrate farther, although this contrast is not as clear for males as it is for females.

Ketterson and Nolan (1983) propose a multifactorial model to account for the differential migration observed in juncos with each age and sex class settling for the winter where there is an optimal balance of several selective pressures (Fig. 12-5). Males, which must establish territories for breeding, have more to gain from early arrival on the breeding grounds than females and so winter farther north, even though winter mortality is higher there (Ketterson and Nolan 1982). Because older males are usually dominant over younger, they can afford a slightly later return; they can thus move farther south to take advantage of milder winters and, presumably, reduced overwinter mortality. Younger birds may also be minimizing the risk of death during migration itself. Females move still farther south because for them there is far less benefit to arriving in the breeding areas early and, presumably, also because they both gain milder winters and are more likely to reduce the severity of density-dependent mortality. As with juvenile males, young females probably migrate less far to reduce the risks of longer flights. The net result of this population partitioning by age and sex is a north-south gradient of younger to older males followed by younger to older females.

Sex differences in distances traveled are not restricted to birds. A particularly interesting case is seen in American eels (*Anguilla rostrata*) (Helfman et al. 1987). Female eels mature at a length greater than 45 cm, while males seldom exceed this size. Length and age at maturity are positively correlated with latitude and distance from the Sargasso Sea, where the eels are presumed to spawn. This means that it is the females that predominate upstream in rivers and their headwater lakes, the more so the farther north one samples. Males, on the other hand, are the more abundant sex in the southeastern United States, where they are generally restricted to estuarine habitats. The precise mechanism for the differential migration is unclear; the earlier settlement of male larvae, differential mortality, and even a sex change to female in larger or older larvae have all been proposed and are not mutually exclusive. Helfman and colleagues further propose that selection may be acting differently on the two sexes. Males may be selected for early maturity in highly productive habitats that promote rapid growth; they are time, energy, and risk minimizers, forgoing migration upstream to minimize risk and energy costs because they have little to gain from the upstream migration. Females in contrast, benefit (or at least suffer fewer costs) from achieving large size, even at the expense of slower growth in less productive freshwater habitats, because large size means higher fecundity.

The differential migration of the sexes in the American eel would also

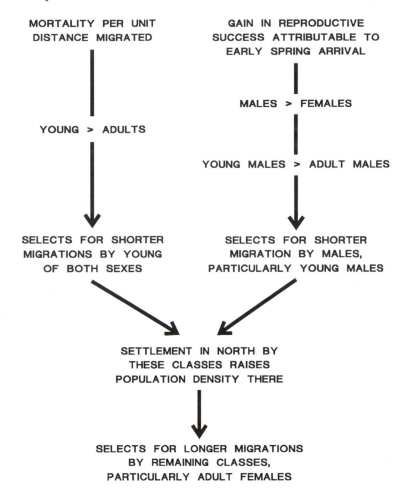

Figure 12–5. The model proposed by Ketterson and Nolan (1983) for the differential migration by sex of the dark-eyed junco in North America.

contribute to a panmictic (random) mating system because assortative (nonrandom) mating would be very unlikely between individuals maturing in the same place or even in the same year. Panmixia is an advantage as a bet-hedging life-history syndrome, that is, one where parents are selected to produce offspring of mixed genotypes to "cover all bets" in unpredictable habitats. This is the situation faced by drifting eel larvae (Chapter 5), whose place of settlement and maturation is largely determined by currents. Other aspects of eel life histories contributing to panmixia are the diversity of juvenile habitats across latitudes and river characteristics, and the spawning at a single site (presumably) by eels from all ecological histories. Panmixia and the other attributes of eel biology all are likely to interact to produce a successful generalist species. Helfman and colleagues thus introduce ideas, that although speculative, may help to explain many heretofore puzzling features of *Anguilla* migratory life histories.

Among mysticete (or baleen) whales there may also be segregation by age and sex, in this case with respect to both location and time spent on low-latitude wintering or high-latitude feeding grounds (Lockyer and Brown 1981). The minke whale (*Balaenoptera acutorostrata*), especially, shows very uneven representation of the sexes in both hemispheres in certain areas and months, although details and adaptive value, if any, remain to be determined. Better understood are the segregating patterns of humpback whales (*Megaptera novaeangliae*). In the Southern Hemisphere the southward migration in the spring is led by newly pregnant females and immature males, followed in turn by interreproductive females and mature males, and finally by females early in lactation accompanied by their calves. On the feeding grounds there appears to be general mixing before the return migration is begun by the lactating females and calves, which spend only 4 months in the Antarctic feeding areas. The young are apparently weaned in the warm-water wintering grounds, an aspect of the life cycle in which humpbacks differ from other mysticetes (except minke whales), which wean calves while still in the feeding areas. Following the mothers and young, the other humpbacks return northward in roughly the reverse order of their arrival in the Antarctic. This timing seems to be a consequence of the fact that conception occurs at low latitudes so that it behooves them to return to the wintering areas for mating. Pregnant females and immature males have the most to gain from the longest time spent procuring energy and nutrients.

Individual humpbacks also vary their migration routes and the locations where they spend the summer or winter. North Pacific humpbacks were identified by the pattern of the underside of the flukes (see Chapter 4), and 61 individuals were identified in different years (Darling and McSweeney 1985). The number of whales traveling between various areas is illustrated in Figure 12–6. By far the majority, 51 whales, wintered in the Hawaiian breeding grounds and summered in the bays and sounds of southeast Alaska. A further eight, however, summered in the area of Prince William Sound and another near Vancouver Island, British Columbia. Whales did not necessarily return to the same summer area each year as one was found in different years in southeast Alaska and Prince Rupert Harbor, B.C. Likewise for the wintering areas, as one male was found in successive years in the area around the Revillagigedo Islands off Mexico and in Hawaii in successive winters. Earlier studies by Japanese investigators (Nishiwaki 1966) have revealed that some summering Alaskan whales also winter off the coast of Japan. In the common wintering area off Hawaii, whales spending the summer in different places do mate, so the North Pacific humpback population is undoubtedly a single entity or "stock." Given the intelligence and swimming capacity of these large cetaceans, it is not surprising that they display such flexibility in their migrations, but it would be interesting indeed to know what factors determine the various routings.

Both laboratory and field studies of insects reveal considerable variability in

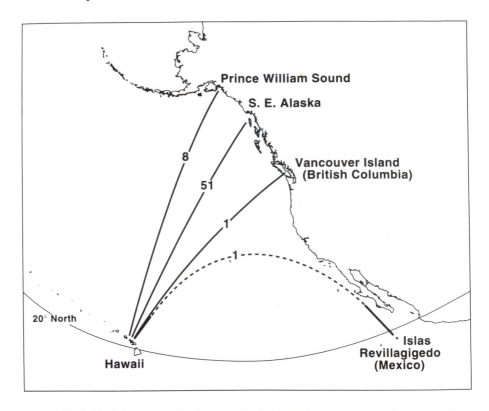

Figure 12–6. Variation among Pacific humpback whales in summer and winter areas. The lines connect destinations but not actual routes. Solid lines connect summer areas with the wintering area off Hawaii. The dotted line indicates an individual wintering at different places in different years. (Modified from Darling and McSweeney 1985.)

duration or timing of migratory flight. A characteristic of tethered-flight tests is that most individuals fly for only brief periods, with a few flying for long duration (see, e.g., Table 12–1). This produces a highly skewed distribution of flights (Johnson 1976; Dingle 1980). Some of the variation is no doubt due to the artificial nature of the flight tests, but as the laboratory flight performances do reflect behavior in the field (Gatehouse and Hackett 1980; Dingle 1985), some adaptive basis for the differences is likely as well. Davis (1980) examined several adaptive and nonadaptive hypotheses and concluded that short flights are probably adaptive most of the time for most individuals because many suitable habitats for colonization are close at hand. The conclusion seems reasonable but has yet to be rigorously tested. In one particular well-studied case, the African armyworm moth, *Spodoptera exempta* (Riley et al. 1983; Gatehouse 1986, 1989; Gatehouse and Hackett 1980), both laboratory and field studies (using radar and other methods) reveal an early evening peak of flight activity by newly emerged moths and a second peak much later that trails off toward dawn. There may also be a brief dawn peak, but flight ceases quickly with the advent of daylight. The net result of these different flight peaks will be a differential distribution of migrants downwind and greater coverage of available habitats. Presumably both spatial and temporal habitat heterogeneities could contribute toward selection differentials maintaining the variability in the timing of departure from the emergence sites.

An especially interesting environmentally variable component influencing seed migration distances occurs in the Jeffrey pine (*Pinus jeffreyi*) of western North America (Vander Wall 1992). In this conifer the seeds are transported in two stages. In the first, they are carried and scattered by the wind, usually only for a distance of 5 meters or so from the parent tree. Subsequently they are carried up to 60 meters farther by the yellow pine chipmunk (*Tamias amoenus*) and some other animals and cached as food stores. The seeds must be buried for germination to take place, so the transport and concentration in cache sites by the chipmunk is an important stage in the life history of the Jeffrey pine. Up to 95% of the seeds are transported by rodents after carriage by the wind, so variability in the behavior of the rodents or other seed-caching animals can have profound impact. Vander Wall goes on to argue that a one-stage concept of seed migration is too simplistic, because an important step with regard to distance traveled, the degree of concentration or dispersal of seed, and influence on germination and subsequent seedling growth is the stage of animal transport.

Other cases where habitat heterogeneities may maintain variation in migratory behavior concern the straying, or movement outside the usual pathways, observed in several migrant fishes. As examples, sticklebacks, *Gasterosteus aculeatus*, are frequently captured far out in the ocean at a time when they should be breeding in estuaries or freshwater (Quinn and Brodeur 1991); and spiny dogfish sharks, *Squalus acanthus*, which usually migrate for about 500–800 km along the continental shelf of western North America, may occasionally move 7000 km across the Pacific Ocean (McFarlane and Beamish 1986). In salmon the rate of successful homing to natal streams is remarkably high. Nevertheless a small proportion, which varies among species and populations, fails to home, but rather enters adjacent streams. Leggett (1984) and Quinn (1984; 1985, Quinn and Brodeur 1991) argue that such straying may be an important component of the evolutionary ecology of these fish, and Quinn, in turn, suggests that straying is not an aberration, but rather that homing and straying are in dynamic balance.

This balance is maintained by the characteristics of both the spawning rivers and the life histories of the species in question. The two most important characteristics of streams in this context are their stability and complexity. In a river that is relatively unchanging over long periods, salmon will evolve physiological and behaviorial adaptations appropriate for that river and will by and large express high and consistent reproductive success; disasters are by definition rare. In contrast, an unstable river would at intervals subject salmon to reproductive disaster. In the former case selection should favor homing and in the latter more straying. Evidence seems to confirm this assumption. Sockeye salmon (*Oncorhynchus nerka*) in the Babine Lake system of British Columbia, for example, displayed nearly 100% homing when captured and placed at the mouths of large stable streams, but only 87% homing when the natal streams were small and unstable. The figures were 93% and 53% respectively when the fish were displaced elsewhere in the lake (McCart in Quinn 1984). Similarly, sockeye salmon breeding in long rivers with much habitat diversity have a higher homing tendency than pink salmon (*O. gorbuscha*) breeding in short simple streams feeding directly into the ocean. Presumably stream complexity favors the evolution of complex adaptations to the diverse stream habitat and confers a greater advantage to homing. Finally, a species such as the chinook salmon (*O. tshawytscha*) that varies greatly in age at maturity is more likely to home, because it is effectively already straying in time; a natural disaster occurring in a home stream would affect only a single year class, sparing others to breed in succeeding years. The entire model is summarized in Table 12–5.

Kaitala (1990) has used an ESS approach to do simulation studies of homing and straying in salmon and reaches three conclusions. First, straying should occur, at least in some fraction of fish, whenever there is any chance of reproduction in the non-natal river. Second, the smaller the relative size of a natal river, the smaller the proportion of homing fish. Third, an increase in the uncertainty of reproduction, as in an unstable stream, also decreases the fraction homing. These conclusions are quite similar to Quinn's, although it should be noted that Quinn's model specifies a variable genetic basis for the traits, whereas this ESS model does not incorporate the possible role of genetic heterogeneity in the salmon populations.

It is worth noting as well that not only do salmon vary in whether or not they home to the natal stream correctly, they may also vary in the routes they follow through the ocean when returning to fresh water. Sockeye salmon from the Fraser River system in British Columbia can follow two different routes around Vancouver Island (Hamilton 1985). The salmon return from the northwest, and in most years they make their way to the Fraser River along the southwest coast of Vancouver Island and up to the mouth by following a path through the Juan de Fuca Strait around the southeast tip of the island (Fig. 12–7). Only a few fish take the "northern diversion" through Johnstone Strait along the northeast coast of the island

Table 12–5. Predicted Levels of Variation in Pacific Salmon (Oncorynchus) Homing as a Function of Stream Characteristics and Variation in Age at Breeding. Variation decreases from left to right in the table. Within ranking of salmon species, there is also much population variation

Seletive Factor	Predicted Level of Homing
Stream stability	Chinook > Sockeye > Coho > Chum ≅ Pink
Stream complexity	Sockeye > Chinook ≅ Coho > Chum ≅ Pink
Age class variation	Chinook > Sockeye ≅ Chum > Coho > Pink

SOURCE: From Quinn 1985.

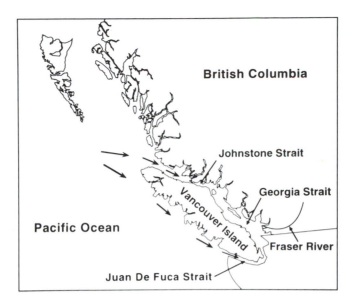

Figure 12-7. Two routes taken by sockeye salmon returning to the Fraser River of British Columbia at the south end of the Strait of Georgia. The northern route through Johnstone Strait occurs in a larger portion of the population during long-duration spawning runs occurring when water temperatures are higher than normal.

and on through the Georgia Strait to the Fraser. When the spawning run is of long duration, however, a much larger proportion of fish take the northern diversion. These lengthy runs tend to occur when Pacific water temperatures are higher than normal, and this warming in turn is especially likely when there is a large tropical El Niño event. Especially large diversions occurred in four major El Niño years and probably reflect a tendency of the salmon to prefer cooler waters, which in those years occur on the inshore side of Vancouver Island. It is likely that staying in cooler water results in considerable energy savings for the salmon (see Chapter 6).

Straying is a phenomenon that is not unique to homing salmon, for it is also common among birds, where it is usually termed vagrancy. The popularity of "rare bird alerts" and "hot lines" to guide birders to species outside their normal range attests to the frequency of vagrants. These are especially frequent where peculiarities of landform or weather systems are likely to concentrate migrants; the various peninsulas, islands, and bays along the coast of California are examples of geographic features that result in concentrations. A long-term program operated by the Point Reyes Bird Observatory on the Farallon Islands about 50 km west of San Francisco monitors the passage of migrant birds, which often land on these islands, and has provided vivid documentation of the occurrence of vagrants (De Sante and Ainley 1980; DeSante 1983). Over a 10-year period from 1968–1977 some 2,177 individual vagrants of 74 species were recorded. Most of these were species that breed in the northern part of North America east of the Rocky Mountains and regularly migrate southeast in the autumn and northwest in the spring, but birds from both southeastern and southwestern North America and even a few from the Palearctic were also represented. In orientation cage studies of autumn vagrant blackpoll warblers (*Dendroica striata*) captured in the Farallones, DeSante (1973) concluded that an important factor contributing to vagrancy in this species was mirror-image misorientation; that is, rather than migrating to the southeast, these birds were migrating orthogonally to the southwest. The majority of all migrant birds landing on the Farallones are first-year juveniles in the autumn and second-year birds in the spring, and it is reasonable to suppose that "mistakes" in orientation are more likely in these young birds. These are no doubt strongly selected against, but they provide the potential variation, if genetically based, for the evolution of new migratory pathways if habitat changes permit these to happen. Such an instance may have occurred in the wet spring of 1993 when several pairs of parula warblers (*Parula americana*), whose normal breeding range is the eastern and southeastern Unites States, bred in coastal California (personal observations).

The evolution of a new pathway is exactly what seems to have occurred in the blackcap warbler, *Sylvia atricapilla*, of central Europe. East of a "migration divide" running roughly through Prague, in the Czech Republic, and central Austria, blackcaps migrate southeast to the eastern Mediterranean before turning south to their wintering areas in Africa; to the west of this line birds migrate southwest to winter in Spain and the western parts of North Africa (Helbig 1991a). In the past 20 to 30 years, however, there has been a large increase in the number of blackcaps wintering in southern England and Ireland, presumably as a result of increases in bird feeders, and possibly also in various garden shrubs that retain fruit during the winter (Leach 1981; Bland 1986; Berthold and Terrill 1988). There is very suggestive evidence to indicate that the Irish-British overwintering population represents central European birds that have developed a new migratory pathway by migrating to the west or west-northwest rather than following the usual routes to the southwest or southeast.

The necessary variation in orientation to provide the basis for the evolution of an east-west migration is seen even in blackcaps from populations migrating in the usual directions, because some of these birds, when tested in orientation cages, show preferences for westerly orientation in the autumn (Berthold 1992). The offspring of birds captured in Britain do show a westerly orientation response in the autumn when tested in orientation cages, in contrast to southwest German birds tested at the same time, which orient to the southwest in these cages as expected (Berthold et al. 1992). Birds captured while wintering in Britain thus show an orientation response that would get them from central Europe to their wintering areas. Further orientation evidence suggests that the breeding area of at least some of these birds may be in central Austria. Birds from this region, taken as early nestlings near the city of Linz and hand raised, showed a westerly orientation in autumn consistent with migration to the British Isles (Helbig 1991a). Similarly, hand-raised birds from southwestern Germany and from eastern Austria oriented to the southwest and southeast, respectively, as would be expected from recoveries of birds banded in those areas. Helbig postulates that Linz is in a hybrid contact zone between blackcap populations to the west and east, and that in this zone there is strong selection against any hybrids because the latter migrate due south (as demonstrated by laboratory crosses (Helbig 1991b)) and would have to cross both the Alps and the Mediterranean. Selection against hybrids may then have reduced intraspecific competition and allowed a more rapid establishment of the British wintering birds in the contact zone than elsewhere.

The British winterers may also benefit from earlier arrival at the breeding grounds. When German birds are kept under photoperiods simulating winter conditions in Britain, they begin vernal migratory activity earlier than birds kept under simulated day lengths matching the traditional wintering area of Spain (Terrill and Berthold 1990). Because the distance to central Europe is also less from Britain than from Spain, calculated arrival times on the breeding grounds for the birds wintering in the former location are much earlier on average and could substantially enhance breeding success. The British birds should also be better acclimatized to the possible inclement weather occurring in early spring. To sum up, even though much of the evidence is by necessity circumstantial, there is now a strong case for the recent evolution of a new migratory circuit for blackcaps, with breeding in central Europe and wintering in the British Isles. This is proof of the point, stressed by Leggett (1984, 1985), Quinn (1985), and others, that variation in migratory systems is worth studying, and that strays or vagrants are important in their own right and are not simply wastage in systems that are imperfectly tuned by natural selection.

In addition to variation in orientation, there can be intraspecific variation among migrant birds in their behavior along the migratory route. Bluethroats, *Luscinia svecica*, for example, migrating from the Scandinavian Arctic to Asia show age differences in stopover time at sites in central Sweden (Ellegren 1991). Migration of adults was more concentrated in time, and stopover duration was shorter than for juveniles. Also, in contrast to juveniles, adults did not lose mass at the stopover and started accumulating fat earlier, suggesting that they were more efficient in preparing for migration.

In another example of intraspecifiic variation along the migratory pathway, five species of North American wood warblers (Parulinae), wintering in or passing through the Yucatan Peninsula of Mexico, segregate by sex according to habitat, and three additional species likely do so as well (Lopez Ornat and Greenberg 1990). Mist-netting studies revealed that males occupied more mature habitats while females occurred more in second growth or other early successional areas. There was no evidence that females were "floaters," with habitat differences the result of male dominance, but rather the sexes seemed to occupy preferred habitats. Further support for the preferred habitat hypothesis comes from studies of hooded warblers, *Wilsonia citrina*. In this species, when resident males were removed, females did not move into the areas they had previously occupied (Morton et al.

1987), and laboratory hand-reared birds showed distinct preferences for plant height and habitat structure consistent with their habitat segregation in the field (Morton 1990). The repeated pattern across species of wood warblers suggests that habitat selection is not arbitrary because males are consistently associated with forest and females with more open habitats.

Summing Up

Migratory behavior and life history patterns occur with a variety that matches that of the taxa across which migration occurs. It is not surprising that migration varies among species, reflecting other differences in adaptations and lifestyles, but it is also the case that there is much variety at lower taxonomic levels. Studies of organisms as contrasting as milkweed bugs, grasshoppers, weedy composites, fish, and birds reveal patterns among populations that can exquisitely reflect overall adaptations to particular environmental challenges. Experiments performed in common environments strongly imply that indeed there are adaptive differences in genotype frequencies established by the action of natural selection.

Population differences in migratory behavior are correlated with more or less predictable syndromes of life-history traits, perhaps most clearly exemplified among the fishes. Anadromous populations contain large individuals that generally produce large numbers of eggs. If they must expend a great deal of energy and run high risks to move up long and swiftly flowing rivers, they also tend to be semelparous. This is the pattern seen in Pacific salmon (*Oncorynchus*). Landlocked populations, on the other hand, are smaller and produce fewer eggs. They certainly expend less energy reaching spawning grounds and may also run fewer risks; both conditions select for iteroparity. The larger size of the migrants may derive in part from an energetic constraint; energy use is more efficient in larger individuals. Their larger size can also be more readily acquired if they are exploiting the richer resources of the ocean. The evolution of new migration–life history syndromes among populations of shad introduced into the rivers of western North America strongly supports the notion that these patterns are adaptive and genetically based.

Even within a population some individuals may migrate whereas others may not, the condition known as partial migration. The phenomenon is especially apparent in birds; dominant individuals, usually adult males, may maintain year-round residency, while subordinate birds, mostly females and juveniles, may migrate. Here adaptation may not be the key. Rather migration may be a "best of a bad job" or conditional strategy that is the best available alternative if staying would result in exclusion from favorable habitats by more dominant individuals. This idea gains added appeal from those cases where migrants switch to residency when provided an opportunity. There may also be cases where partial migration is a mixed strategy with neither movement nor residency favored over the other. Balancing selective conditions have possibly selected for such a strategy in the giant tortoises of Aldabra Atoll. A modeling study based on the theory of evolutionarily stable strategies (ESSs) suggests that density-dependent survival of overwintering residents (in the case of birds) is the single most important element for the maintenance of a mixed partial-migration strategy. This mortality is, of course, very difficult to assess in nature. Other important points to emerge were that differential reproductive success between residents and migrants is not necessary to maintain the system and that environmental stochasticity influencing density-independent mortality had little or no effect on the ESS solution. ESS modeling has its limitations—it ignores most of genetics, for example—but in this case it does point out some important things to look for in analyzing complex systems like partial migration.

Individuals differ not only in whether or not they migrate; when they do migrate, they vary in the distances traveled or the routes taken. Often the differences occur between the sexes and presumably reflect sex differences in the benefits and costs of migration. Among birds it will likely pay for males to arrive at the breeding grounds early to establish territories, and this means greater benefits to shorter migration distances as seen in juncos. Male American eels also move less far than females, but in this case they stand to benefit more than females by early maturity. Estuarine habitats with rich resources provide the wherewithal for rapid growth. Slower growth and delayed maturity mean, for females, higher fecundity; they migrate farther into river systems where the richness of resources matters less. In insects sex differences are less obvious, and the advantages of spread over all habitats of heterogeneous environments may be what selects for differential movements.

Finally, there is much variation in the routes traveled. The occurrence of vagrants is abundantly reported in the ornithological literature. In one case, that of the blackcap warbler in Europe, the presence of genetic variation in migratory direction has resulted in the evolution of a new route. Within the past half-century a population has become established that migrates not from central Europe to Spain for the winter, but rather to southern Britain and Ireland. This new pathway has apparently been selected for by the increased winter resources of feeders and fruiting garden shrubs and perhaps also by the earlier arrival on the breeding grounds permitted by the more northern photoperiodic cycle and the shorter route. Straying in salmon is apparently an important element in population dynamics because it permits colonization of new spawning streams. ESS models suggest it will be favored in more heterogeneous conditions because as the possibility of breeding failure increases, so does the advantage of producing young that return to streams other than the natal one. This example is one of the best among many to demonstrate that rather than being the consequence of "error variance," the diversity of population and individual patterns among migrants can be an important outcome of natural selection.

13

Polymorphisms and Polyphenisms

In the previous chapter I discussed behavioral and life-history variation in migration unaccompanied by distinct and overt changes in morphology or physiology. In this chapter I shall discuss those instances in which overt changes do occur, concentrating on phenotypic expression of differences. The analysis of genotypic variation will be covered fully in the next chapter. The first of the overt changes covered concerns those species in which the morphological capability for migration is retained only for the migratory period, often with remobilization of energy and materials to enhance reproduction when migration ceases. This is commonly observed in insects; species in several families and orders histolyze the wing muscles during sedentary periods and divert the breakdown products into egg production (Nair and Prabhu 1985). In most cases this breakdown is irreversible, but in some insects the muscles can be regenerated (Bhakthan et al. 1971). The process of terminating migratory capability is carried to its logical end point in ants, termites, some crickets, and some aphids, all of which shed their wings after completing migratory flights, usually in company with wing muscle histolysis. A vertebrate parallel to insect flight muscle breakdown is the degeneration of the swimming muscles of Pacific salmon (*Oncorhynchus* spp.) as energy and protein are diverted to mate and breeding site acquisition and gamete production at the end of the upstream migratory journey.

The greatest contrasts in form and function between migrants and nonmigrants occur in those species that are polymorphic or polyphenic with respect to the morphology associated with migratory activity. In the older literature the term polymorphism was used to indicate discontinuities in form without distinguishing between genetic and environmental contributions to the differences. Where genetic contributions could be identified, the term "genetic polymorphism" was sometimes used. More recently, polymorphism has been reserved for cases where differences between morphs are gene based with the term polyphenism applied where distinct forms are produced in different environments (e.g., Shapiro 1976). Both types of variation are common and widespread across all kingdoms of life. Among migrants both polymorphisms and polyphenisms are especially apparent in insects, which display variation in wing structure, and in plant seeds, which take a variety of forms. There is a broad taxonomic distribution in insect pterygomorphisms with variation in the length of the wings (macroptery, or long wings, vs. brachyptery, or short wings), in the possession of wings (macroptery vs. aptery, or the absence of wings), or in several cases in both conditions. Other examples of migratory polymorphisms and polyphenisms are scattered over the plant, animal, and fungal kingdoms but are especially evident in heteromorphic seeds. Because insects and plant seeds are the best-studied examples of migratory heteromorphisms, I shall focus on these two groups of organisms in this chapter. A discussion of avian wing-form variation, in the context of constraints on migration, was presented in Chapter 7.

Pterygomorphism in Insects

Among insects variation in the flight apparatus, or pterygomorphism, is a widely expressed form of variation in migratory capability. It includes histolysis of the wing muscles, which alone has been documented in more than 50 species in at least 20 families and eight orders (Johnson 1976), dealation of the wings, and a broad array of variation in alary form including complete aptery. Both genes and environment contribute to the variation in migratory performance as outlined at the beginning of the chapter. It is useful to consider both sources together in the case of insects, because as we shall see, a full spectrum of variation encompassing both can be seen even within the same species. Insects thus display a quite remarkable and very fascinating degree of flexibility in the evolutionary development of their pterygomorphic syndromes. Some idea of the diversity present in both wing and wing muscle morphology can be seen in Table 13–1, which indicates environmental influences on pterygomorphism in a selection of species.

The flexibility of wing muscle histolysis is evident in the ecology of the hemipteran genus *Dysdercus*, commonly known as cotton stainer bugs. These bugs are widely distributed in both the Old and New World Tropics, and their common name of "cotton stainer" comes from the fact that many species will feed on the seeds of commercial cotton and, in the process of feeding, introduce a fungus that stains the fibers in the bolls. They are thus often of considerable economic importance. In the wild, species of *Dysdercus* are seed predators on many host plants of the Malvales from trees with large dehiscing fruits such as the baobab (*Adansonia*) and kapoks (*Thespesia, Sterculia, Ceiba*) to a variety of herbs and bushes (*Gossypium, Hibiscus, Sida*). On most of these hosts, and especially on the large trees, the bugs arrive and colonize after the start of dry season fruiting. Once they start feeding on the seeds, the immigrant females histolyze the wing muscles, begin to develop large numbers of eggs, and often reproduce explosively (Dingle and Arora 1973; Fuseini and Kumar 1975; Derr 1980a). The protein from the histolyzing wing muscles is remobilized and transferred to the oocytes (Nair and Prabhu 1985). Males sometimes histolyze, but more often retain the flight muscles, presumably to move about in search of females, a tactic made possible by their relatively low reproductive investments.

Table 13–1. Environmental Influences on Flight Muscle and Wing Form in Selected Insects

Species	Environment	Influence	Reference
Hemiptera: Gerridae			
Gerris spp.	Increasing daylength	Increased brachyptery	Vepsäläinen 1978
	Increasing temperature	Increased macroptery	Vepsäläinen 1978
Limnoporus canaliculatus	long to short day	Increased macroptery	Zera & Tiebel 1991
Hemiptera: Other			
Dysdercus spp.	Feeding	Wing muscle histolysis	Dingle & Arora 1973
			Derr 1980
Cavelerius saccharivorus	Crowding	Increased macroptery	Murai 1977
Homoptera: Planthoppers			
Nilaparvata lugens	Crowding, starvation	Increased macroptery	Kisimoto 1956
Prokelisia marginata	Crowding	Increased macroptery	Denno & Grissell 1979
Homoptera: Aphids			
Aphis fabae	Crowding mother or progeny	Increased alates	Shaw 1970
Acyrthosiphon pisum	Crowding, old mothers	Increased alates	Sutherland 1969
			MacKay & Lamb 1979
Megoura viciae	Crowding, senescent host leaves	Increased alates	Lees 1966, 1967
Coleoptera			
Leptinotarsa decemlineata	Long day	Wing muscle development	de Kort 1969
Dendroctonus monticolae	Low moisture	Wing muscle regeneration	Reid 1962
Lepidoptera			
Orgyia thyellina	Short day	Wingless female produced	Kimura & Masaki 1977
Orthoptera			
Gryllodes sigillatus	Long day, crowding, high temperature	Increased macroptery	McFarlane 1966
			Mathad & McFarland 1968
Pteronemobius fascipes	Intermediate photoperiod	Increased macroptery	Masaki 1973

Two examples of seasonal cycles in *Dysdercus*, *D. voelkeri* in West Africa and *D. bimaculatus* in Central America, are illustrated in Figure 13–1. Even though the two species occur on separate continents, the similarity of tropical dry and wet season patterns resulting from the movements of the ITCZ (Chapter 5) renders the seasonal population patterns of the bugs similar as well. Populations on a given tree reach peak abundances as fruits are dehiscing because females are histolyzing flight muscles, remaining in place, and reproducing. There is a rapid decline in the populations toward the end of the dry season when lack of available moisture reduces the amount of feeding; bugs therefore retain the wing muscles and depart. Departure rates increase as the dry season intensifies, and the bugs are subjected to increasing food and moisture stress. Some of these departing bugs may succeed in colonizing other late-fruiting trees, but most scatter and maintain low numbers on forbs such as *Hibiscus* or *Sida* (Fuseini and Kumar 1975 for *D. voelkeri*) or on the forest floor as revealed by their attraction to cotton seed baits (Derr 1980 for *D. bimaculatus*). Starvation results both in retention of the wing muscles and reproductive diapause until migration to trees at the start of the next dry season

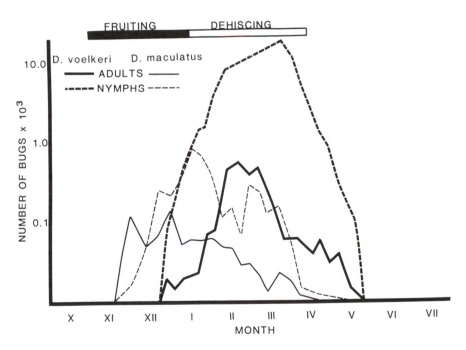

Figure 13–1. Seasonal cycles of *Dysdercus* bugs on two continents, *D. voelkeri* on *Ceiba pentandra* in Ghana and *D. bimaculatus* on *Sterculia apatela* in Costa Rica. Fruiting occurs at the onset of the dry season, which reaches its height as fruits are dehiscing. In the absence of moisture or food, reproduction ceases, wing muscles are retained, and the bugs migrate, resulting in a population crash. (From Dingle 1982; data from Derr 1980a and Fuseini and Kumar 1975.)

fruiting cycle. This starvation can be caused by a lack of either food or water, because the bugs partially digest seeds externally and so require water to feed.

Depending on the nature of the host plant, species of *Dysdercus* may differ in the tactical details of their migrations, as shown for three African species by Dingle and Arora (1973). *D. fasciatus* feeds on the fruits of the baobab, a large tree that produces an enormous seasonal fruit and seed crop and hence a "feast or famine" situation for the bugs. When fed, females of *Dysdercus fasciatus* never fly, but immediately begin muscle histolysis and produce their first clutches of over 100 eggs about 5 days after becoming adults (at 28°C). When they are starved and retain wing muscles, individuals of this species show the longest duration migratory flights. *D. fasciatus* is the most opportunistic species of the three analyzed, with the earliest reproduction when conditions are favorable and the strongest migratory flights when they are not. The other two species, *D. nigrofasciatus* and *D. superstitiosus*, are more catholic, feeding on a variety of annual and perennial malvaceous host plants. Even when fed, some females of these species can still fly for a short period, so allowing some movement to new hosts at all times. Reproduction is correspondingly delayed, with females beginning oviposition at about 9 days after adult emergence (28°C) or some 4 days later than *D. fasciatus*. Again, however, complete starvation results in retention of fully functional wing muscles and a switch to the migratory mode. These general themes of migration when starved and wing muscle histolysis with reproduction in the presence of food are seen also in other species of *Dysdercus* in both Old and New Worlds and in some other seed-feeding bugs such as *Oxycarenus* spp. (Adu-Mensah and Kumar 1977) and *Neacoryphus bicrucis* (Solbreck and Pehrson 1979).

Across species and continents the association of *Dysdercus* with malvaceous host plants involves a relation between migration-colonization life history patterns and the distribution and abundance of resources (Derr et al. 1981). In both Africa and the Americas, it is the largest species in the genus that epitomizes the migrant colonizer syndrome. These species are specialists on trees that produce large but short-lived concentrations of oil-rich seeds widely separated in both space and time (e.g., *D. fasciatus* in Africa). Large size is associated with increased rates of egg output in *Dysdercus*; it may also permit greater survival during stressful periods, as would occur during long migratory flights or over a long season without seeds (Dingle et al. 1980), and increase energetic efficiency (Roff 1991 and Chapter 6). Intermediate-sized *Dysdercus* are generalist feeders on shrubs and annuals that are both more evenly distributed and less productive of seeds. Migration and egg output are correspondingly less dramatic. The smallest species considered had the lowest reproductive rates and an essentially noncolonizing life history on weedy annuals. The migration-colonization syndrome of large size, rapid wing muscle histolysis, and explosive reproduction, involved in specializing on large but ephemeral seed crops of big trees, goes a long way toward explaining why it is the largest species of *Dysdercus* that are the notorious cotton pests. Like the wild hosts, a field of cotton is a concentrated source of oil-rich seeds without, however, many of the stresses and constraints imposed by time and distance. The life histories of these large *Dysdercus* species are essentially pre-adapted to exploit the cropping situation (see also Chapter 15).

Other noteworthy instances of a strategy of muscle histolysis are forest bark beetles, of which the North American genera *Ips* and *Dendroctonus* are good examples (see also Chapter 11). In *Ips confusus* (Borden and Slater 1969; Bhakthan et al. 1970, 1971) muscle degeneration begins in both sexes after the beetles attack a tree and proceeds rapidly until 90% or more of the volume is lost, leaving a bag formed by the sarcolemma containing lysosomes, tracheoles, and a few granules; there is thus little structural organization. Muscles of males degenerate to a somewhat lesser degree than those of females. Unlike *Dysdercus*, *I. confusus* can regenerate the wing muscles, if it proves necessary to move to a new host. In a related bark beetle, *Dendroctonus monticolae*, regeneration is triggered by reduced moisture in the galleries (Reid 1962 and Table 13–1). Regeneration proceeds both by using the remains of the old flight muscle and by the formation and then differentiation of new myoblasts with the whole process taking about 13 days to complete. The ability to regenerate flight muscle gives *I. confusus* greater flexibility than *Dysdercus*. This flexibility is probably an adaptation to a somewhat more uncertain food supply engendered by the requirement of trees in just the right condition to attract beetles.

The next evolutionary step from wing muscle histolysis is modification of the wings themselves, including their complete elimination. At one end of the scale, with only small differences between wing morphs, is a soapberry bug, *Jadera aeola*, studied on Barro Colorado Island (BCI), Panama, by Tanaka and Wolda (1987). Species of *Jadera* feed on host plants in the Sapindaceae or soapberry family. They are quite closely related to *Dysdercus* and like them feed on the seeds of plants varying in growth form from trees to shrubs and forbs. On BCI the bugs feed in enormous numbers on the dropped seeds of sapindaceous trees during the latter part of the dry season. During the wet season, *J. aeola* apparently enters a reproductive diapause until the following dry season, but its ecology during the period is very poorly understood. Suffice it to say, however, no nymphs are in evidence during the rainy period.

At the beginning of the dry season, the bugs surviving from the previous breeding season reproduce, and the maturing nymphs feed on the ripening seeds present at this time. These nymphs eclose to the adult stage in both long-winged and short-winged forms; the differences in wing length are illustrated in Figure 13–2. Short-winged bugs also tend to be smaller overall. Shortly after they emerge as adults, the short-winged bugs can fly very short distances (circa a few tens of centimeters), but they soon histolyze the wing muscles and reproduce. Long-winged individuals, on the other hand, possess strong powers of flight and leave the site where they emerge, usually without reproducing. They are evidently capable of diapause whereas the short-winged bugs are not. Other differences between the two forms include earlier maturation and reproduction in the short-winged forms (relative to those long-winged bugs that did lay eggs) and the production of more and smaller eggs by the short-winged females. Oviposition behavior also differs;

A. Long Winged Short Winged

B.

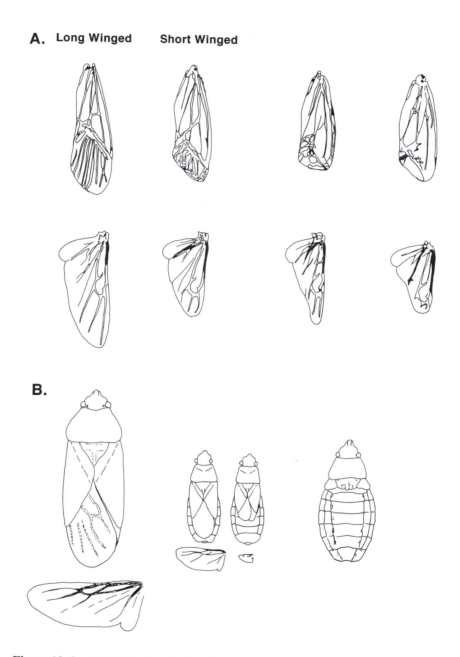

Figure 13–2. *A.* Variation in wing length among individuals of the tropical soapberry bug, *Jadera aeola*, from Barrow Colorado Island, Panama. (Redrawn from Tanaka and Wolda 1987.) *B.* Variation in wing length among species of Mediterranean lygaeid bugs. From left to right: long-winged *Lygaeus equestris*, polymorphic *Horvathiolus gibbicollis*, and apterous *Apterola kunkeli*. (Redrawn from Solbreck 1986 and Solbreck et al. 1990.)

the long-winged females lay their eggs in the canopy, with the eggs then dropping to the ground, while their short-winged counterparts oviposit on the ground itself.

The earlier maturation and reproduction and higher fecundity of the short-winged forms suggests an adaptation to a limited breeding period, and indeed Tanaka and Wolda indicate this is the case. The long-winged offspring of the females that diapaused over the previous wet season evidently depart the feeding sites and themselves diapause, resulting in only one generation per year. The short-winged offspring, in contrast, reproduce to form a second generation, which will survive to depart and diapause in those years or in those locations where there is a lingering but short-term supply of sapindaceous seeds that have not sprouted, rotted, or been destroyed by seed predators. The production of both kinds of offspring by the post-diapause females thus constitutes a mixed or bet-hedging strategy with the possibility of leaving additional descendants in good conditions, but a reserve of immediately diapausing adults for poor years or sites. The control of wing form is not known, but all offspring of short-winged pairs were long winged. This leads me to suspect that long-winged mothers can produce either long- or short-winged offspring, depending on as yet unknown environmental influences, but that short-winged mothers can produce only long-winged offspring. The production of short-winged offspring by short-winged parents would be strongly selected against as the latter cannot leave or diapause and seeds will no longer be available. Such maternal effects, with one form of female producing mixed offspring and the other not, also occur in other insects and are an effective way for natural selection to design mixed strategies (Mousseau and Dingle 1991 and Chapter 14).

At a stage beyond *J. aeola* in the evolution of pterygomorphism is the salt marsh planthopper (*Prokelisia marginata*) a homopteran of the coasts of North America (Denno 1985; Denno et al. 1985). Whereas some flight was possible in the short-winged form of the soapberry bug, the brachypter of *P. marginata* is completely flightless. This insect occurs on only one host plant, the perennial grass *Spartina alterniflora*, but the grass varies according to where it occurs in the marsh. Plants in the high marsh are generally shorter and of lower nutritional quality, but they remain structurally intact over the winter and provide thatch in which planthoppers can survive. Taller, more nutritious plants occur along tidal streams, but these are destroyed by winter storms and the action of ice, precluding overwinter survival. The seasonal cycle and the role of the two wing forms in *P. marginata* is illustrated in Figure 13–3. The brachypter reproduces earlier than the long-winged macropter, giving it a net reproductive advantage, and survives all year in the short grass of the high marsh. It suffers a disadvantage at midsummer, however, because of the poor nutritional value of the high marsh *Spartina* and because planthoppers become extremely crowded, producing smaller and less fecund individuals. When crowding occurs, macropters are produced, and these migrate to the tall grass of the streamside habitats. Here they take advantage of nutritious host plants and build up dense populations, in turn producing still more macropters to migrate and actually select nutritious grass. At the end of the season

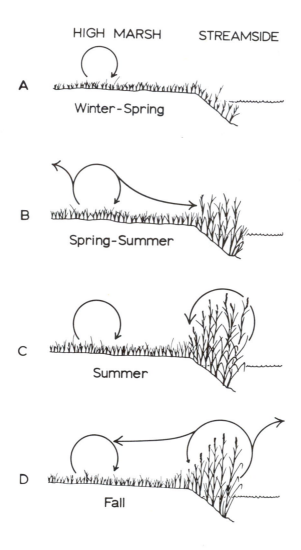

Figure 13–3. The seasonal migratory cycle of the salt marsh planthopper *Prokelisia marginata* on the east coast of North America. *A.* Overwintering individuals and offspring of first-generation adults remain in the high marsh. *B.* As streamside *Spartina* grass grows tall, it is exploited by migrants, but some adults of both brachypters and macropters remain in the high marsh. *C.* Large populations develop in streamside growth, while smaller populations develop in the high marsh. *D.* In the fall streamside populations produce mostly migratory macropters, some of which return to the high marsh to produce the overwintering generation. (From Denno et al. 1985.)

in autumn, the macropters migrate back to the high marsh and join the brachypters in the overwintering generation. In temperate salt marshes about 80% of adults are macropterous, but this frequency is reduced in the much less seasonal marshes of Florida, where there is only 10% macroptery.

Modification of wing morph frequency by population density is common among pterygomorphic insects (Table 13–1). In general, crowding will indicate deterioration of the habitat, and there is then an increase in the proportion of macropters that can leave and colonize new habitats. The phenomenon has been well studied in two major plant pests of Asia, the oriental chinch bug (*Cavelerius saccharivorous*) on sugarcane (Murai 1977; Fujisaki 1985) and the brown planthopper (*Nilaparvata lugens*) on rice (Kisimoto 1956). As with *P. marginata* the brachypters have a reproductive advantage, and the macropter serves as a means to escape and colonize elsewhere. The same outcome is produced via a maternal effect in aphids, where the crowding of parthenogenetic wingless females causes them to increase the frequency of winged young (Table 13–1 and reviewed in Mousseau and Dingle 1991). Winged mothers will not produce further alates, the reverse of the situation with *J. aeola*, above, but here selection would favor winged forms producing the more fecund wingless apterae when they found a new host plant. Older wingless mothers of the pea aphid (*Acyrthosiphon pisum*) are also more likely to produce alate young, because as they age so does the host plant, making it less nutritious (MacKay and Lamb 1979).

Comparative studies of species and of populations within species have been particularly revealing of strategies of pterygomorphism. Denno and colleagues (1991) examined the relation between wing form and habitat persistence in 35 species, representing 41 populations, of delphacid planthoppers (Homoptera). They were able to gather information from both their own studies and from the literature on how long, in terms of number of generations, different populations were able to persist in the habitats they occupied. With this information they were then able to plot percent macroptery as a function of habitat duration (Fig. 13–4). The results show that in habitats of short duration, the frequency of macropterism is much higher. Furthermore, a direct comparison of two species, one from a temporary habitat (*Prokelisia marginata*) and one from a more persistent one (*P. dolus*), demonstrated that the species from the temporary habitat developed macropterous individuals much more readily when crowded (see also Fig. 13–3). Both species also increased wing morph production under interspecific crowding, with *P. marginata* again more sensitive to population density (Denno and Roderick 1992). Denno and colleagues (1991) postulate that short-duration habitats select for macroptery in two ways: by favoring habitat escape and by dictating the availability of mates. The latter will be in short supply because they, too, will be escaping.

A second good example of the insights generated from comparative analysis is Christer Solbreck's study of 12 species of Mediterranean seed bugs in the family Lygaeidae (Solbreck 1986; Solbreck et al. 1990). Wing development varies extensively among these species, seven of which are macropterous, three pterygomorphic, and two completely apterous (Fig. 13–2B). The bugs also differ

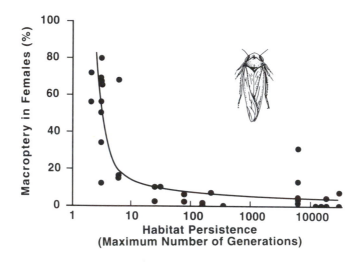

Figure 13–4. The relation between macroptery (here in females) and habitat persistence measured in generations for 35 species of planthoppers representing 41 populations. The occurrence of the long-winged forms declines dramatically in persistent habitats. (Redrawn from Denno et al. 1991.)

in life-history traits that are correlated with wing form. First, both the fully winged species and the winged individuals of the dimorphic species are larger than brachypterous or apterous individuals or species. As a corollary to this size difference, among the monomorphically long-winged species development time is independent of body size, a relation also characteristic of *Oncopeltus*, another lygaeid, and *Dysdercus* (Dingle et al. 1980; Derr et al. 1981). In contrast, dimorphic and apterous species show a positive relationship between development time and size, with larger forms taking longer to develop (Fig. 13–5A). The differences also reflect much higher growth rates in the monomorphic macropters, with growth rates positively correlated with size. The macropters in addition display high rates of production of relatively small eggs in a relatively short life span. At the opposite extreme the small monomorphic apters slowly produce a few large eggs over a long life, with at least one species, *Apterola kunckeli*, capable of living for months. The larvae hatching from these large eggs are quite resistant to starvation, while the relatively small larvae of macropters are not (Fig. 13–5B).

The wing forms and correlated life histories of these lygaeids reflect their respective ecologies. The long-winged species feed and move about on the host plants where wings would be an advantage for frequent foraging flights (Waloff 1983). These species are also migratory, and the early, rapid production of eggs is an advantage when colonizing new habitats. Their larger size probably also buffers them against the stresses of migration and possible delays in finding new hosts to colonize (Dingle et al. 1980), and once in good habitats their rapid growth

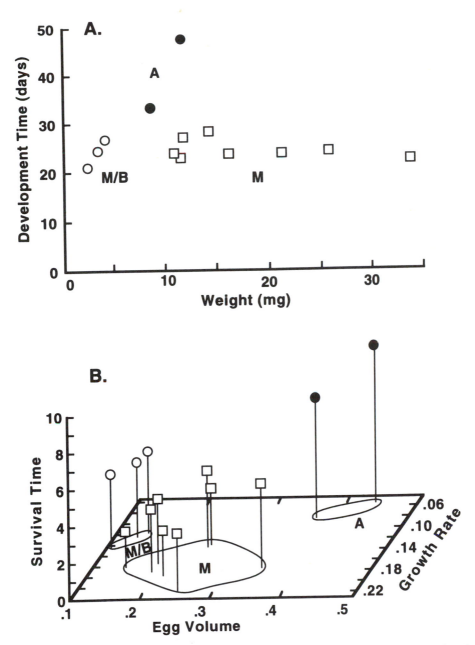

Figure 13–5. *A.* Relation between weight of the last-instar larva and development time in Mediterranean lygaeid bug species displaying different wing morphs. Solid circles and letter A = apterous; hollow circles and letters M/B = wing dimorphic; squares and letter M = long winged. *B.* The relations among survival time (days), egg volume (mm³), and growth rate (per day) of starved first-instar larvae of the bugs shown in *A.* Symbols as in *A.* (Redrawn and modified from Solbreck et al. 1990.)

rates would be an advantage. In dimorphic species, the brachypters reach adulthood earlier and have a shorter preoviposition period than macropters, and so are apparently adapted to exploit currently favorable conditions. The macropters would allow migration out of the area if conditions deteriorated. Finally, the two apterous species are ground living and feed on seeds that have fallen from the host plants and occur in dry, stable habitats of low productivity. They have long lives, slow growth, large eggs, and nymphs resistant to starvation, all characteristics that would allow them to overcome environmental vicissitudes or, in other words, to be able to survive "gaps in time" (Solbreck et al. 1990), rather than to escape in space, when the environment is temporarily unfavorable. Wing reductions and the sort of life history found among these Mediterranean seed-feeding lygaeids seem to be characteristic of areas stable over evolutionary time, for they also occur with high frequency in the evolutionarily old faunas of southwest Australia and South Africa (Slater 1975, 1977).

The relation between habitat, lifestyle, and pterygomorphism is further evident in the aquatic water striders or pond skaters (Hemiptera: Gerridae). The evolutionary ecology of wing form has been particularly well studied in this group of bugs as a result, especially, of the work of Kari Vepsäläinen (1973 et seq.) and Arja Kaitala (1987 et seq.) in Finland and Daphne Fairbairn (1986 et seq.) in North America. Depending on season length, gerrids have either one (univoltine) or two (bivoltine) generations per year. In most cases they overwinter in diapause either at protected sites away from the water bodies that are their breeding habitats or by sheltering beneath objects or vegetation on shores or banks. In the spring they move back to the water surfaces and breed. Within this generalized pattern populations and species display an array of wing development syndromes ranging from fully winged at all times to completely wingless. These syndromes and patterns of voltinism are a function of habitat characteristics, the length and climatic regime of the breeding season, and interactions of these with niche breadth and population density (Vepsäläinen 1978).

The variations in wing morph and life history are outlined in Table 13–2. Thus populations of *G. rufoscutellus* and *G. thoracicus* in highly temporary habitats (top of the table) are uniformly long winged while populations of *G. lacustris* and G. najas in isolated lakes or permanent pools are uniformly wingless. The pattern reflects Southwood's (1962) conclusion that migrants will be characteristic of temporary habitats with nonmigrants, in this case wingless and therefore flightless adults, able to persist only in habitats that are long lasting (Table 11–1). The catholic *G. lacustris* has populations characterized by polyphenism, polymorphism, and monomorphism, reflecting its geographic distribution and wide range of habitats. In general, predictable environments (in the case of gerrids, predictable drying of the water body) will select for polyphenism, and unpredictable environments will select for genetic polymorphism as a form of bet-hedging (Ziegler 1976). The contrast is seen in populations of *G. odontogaster* and *G. lacustris* in the relatively long and warm summers of central Europe. The former live in temporary habitats likely to dry up in warm years, and the frequency of the long-winged morph increases over the course of the summer under the influence of photoperiod (Vepsäläinen 1971); in the autumn, the offspring of the summer generation are all long winged and migrate to diapause sites. *G. lacustris*, in permanent habitats, is genetically polymorphic, with some individuals wingless, and so confined to their natal area, and others winged, allowing colonization of possibly more favorable habitats or escape in the case of catastrophe. The life histories of gerrids seem to be maximally flexible as both long- and short-winged morphs are known from at least some populations of all species studied (Vepsäläinen 1978). Habitats can exhibit morphism cycles as populations go extinct, are reestablished by long-winged immigrants, and finally evolve to the appropriate morph equilibrium under the influence of the local selective regime (Järvinen and Vepsäläinen 1976).

The patterns of gerrid morphology and life histories summarized in Table 13–2 provide a broad overview of the respective behavioral ecologies. Of necessity the table glosses over details, but the subtleties of specific syndromes can be interesting and are worth considering with some examples. Arja Kaitala (Kaitala 1987, 1988, 1991; Kaitala and Huldén 1990) has looked at the relation between food deprivation and wing muscle histolysis, reproduction, and longevity in *G. thoracicus*, *G. lacustris*, and *G. odontogaster*. As with all gerrids these three species are predators feeding on aquatic insects and crustaceans that they can catch at the water surface or, more usually, on insects falling into the water and becoming wet or trapped in the surface film. Food deprivation in the wild is likely to occur with different frequencies among different populations of gerrids, depending on habitat and climate. Current food shortage is unlikely to predict future availability because gerrids cannot readily "overfish," depending as they do on outside inputs of prey. All three species diapause over the winter, as winged adults in the populations studied by Kaitala, and migrate to water bodies in the spring. Once in the water bodies some individuals histolyze the wing muscles, while others do not and so retain the ability to search (via extended foraging, ranging, or migration) for new breeding habitats if conditions deteriorate. Males are less likely to histolyze than females, and the following discussion is confined to the females.

Kaitala initially compared populations of *G. thoracicus* and *G. lacustris* occurring near each other in southern Finland. When food was plentiful, *G. thoracicus* females displayed a high reproductive capacity and a short lifetime, and when food was scarce reproduction fell but longevity increased significantly (Fig. 13–6). In contrast, *G. lacustris* responds to food scarcity with decreases in both fecundity and life span. In Finland *G. thoracicus* is an inhabitant of temporary rock pools and brackish water bays of the Baltic Sea, two habitats likely to be asynchronous in their favorability. In warm summers, the rock pools are likely to dry up, but the bays are favorable; in cool summers the water in bays is likely to be too cool for breeding, but the rock pools don't dry. Food availability in either case is, however, likely to be much more uncertain. As an apparent adaptation to habitat instability, some, but not all, females of *G. thoracicus* histolyze the flight muscles when they arrive in the breeding areas in the spring. These females lay

Table 13–2. Relation among Habitats, Wing Form, and Life Cycles in European Water Striders (Hemiptera: *Gerridae*)*

Habitat	Wing Form	Example
Temporary ponds	LW[†], univoltine	*Gerris rufoscutellus* in Finland
Ponds subject to drying in long warm summers	LW, multivoline	*G. thoracicus* in central Europe
Small temporary unproductive rock pools, small brackish bays	LW, long lived, interreproductive flights	*G. thoracicus* in southern Finland[‡]
Temporary ponds, streams; cool summers	Seasonal polyphenism; LW winter (diapause), SW summer generation	*G. thoracius* and *G. odontogaster* in Finland
Temporary ponds, streams; long warm summers	LW diapause (overwintering) generation of strong flyers, polyphenic summer	*G. odontogaster* in central Europe
Ponds, medium-sized lakes and streams; cool summers	Seasonal polyphenism; LW winter (diapause), SW summer generations	*G. argentatus* and *G. paludum* in Finland
Ponds, medium-sized lakes and streams; long warm summers	LW winter (diapause), polyphenic summer generations	*G. argentatus* and *G. paludum* in Europe
Rivers, lakes; cool summers	Polymorphic winter, SW summer generations	*G. lacustris* in Finland
Ponds, medium sized lakes, rivers	Seasonal polymorphism and polyphenism; LW winter, SW summer	*G. lacustris* in parts of Finland and Europe
Rivers, lakes; long warm summers	Polymorphic winter and summer generations	*G. lacustris* in Central Europe
Large lakes, rivers; cool summers	Polymorphic, univoltine	*G. lateralis* in Finland
Large lakes, rivers; long warm summers	Polymorphic, bivoltine	*G. asper* in central Europe
Streams; warm summers	Polymorphic	*G. najas* in Poland
Isolated permanent ponds; cool summers	SW; univoltine	*G. lacustris* in southern Finland
Isolated pools of permanent streams; cool summers	SW; univoltine	*G. najas* in Finland
Isolated bogs	SW	*G. sphagnetorum* in Finland

*Modified from Dingle 1980 with data from Vepsäläinen 1973, 1974, 1978. Arrangement is in approximate progression from completely long-winged to fully short-winged populations.
[†]LW = long-winged (macropterous); SW = short-winged (brachypterous) or wingless (aptery)
[‡]See text for detailed account of *G. thoracicus* in southern Finland.

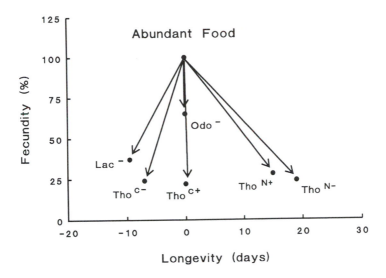

Figure 13–6. The change in relative fecundity and longevity as a function of food stress in three species of *Gerris*. Under abundant food, fecundity is 100% and longevity is normalized to 0. The effect of low food in terms of percent of maximum (abundant food) fecundity and increased or decreased longevity is expressed as the lower point for each species or population. Lac = *G. lacustris*; Odo = *G. odontogaster*; Tho = *G. thoracius*. The pluses and minuses indicate flyers (no histolysis) and nonflyers (histolysis), respectively. C and N indicate central and northern populations. (Redrawn from A. Kaitala 1991.)

60–100% more eggs than those that do not histolyze, but the latter retain the ability to change habitats even though the trade-off is reduced fecundity. An additional element of flexibility occurs when food levels are low because females can trade off reduced fecundity for longer life and a chance to reproduce later either when food levels are restored in the current breeding place or for those retaining wing muscles, at another place where food may be more available. In brief, *G. thoracicus* females seem capable of switching between classical "r and K strategies." The relatively permanent habitats of *G. lacustris* have not selected for similar flexibilities. In this species some females also histolyze, but with no gain in fecundity (although histolyzing females do reproduce earlier). With reduced food levels, there is no trade-off between fecundity and longevity as both are reduced under the stress of starvation.

Adding *G. odontogaster* and a central European (Hungary) population of *G. thoracicus* to the equation makes matters still more interesting. The *G. odontogaster* were, in this case, from a Finnish population that occurs generally in less permanent habitats than *G. lacustris*, but there is some overlap with the latter species. Trade-offs were compared among the three species and between females that histolyzed the wing muscles and those that did not. The results are summarized in Figure 13-6. In the *G. thoracicus* populations the nature of trade-offs among flight, life span, and fecundity differ between northern and central Europe, reflecting the greater environmental uncertainty in the north. In the Finnish population, as we have seen, reduced food levels reduce fecundity, as expected, but also result in a greater life span; the effect is stronger in females that histolyze the wing muscles but is evident in those that retain flight capacity as well. In the Hungarian populations of *G. thoracius*, which in addition to occurring in more stable habitats are also multivoltine, there is reduced fecundity and either reduced life span (histolyzers) or no change in life span (no histolysis). This population thus resembles *G. lacustris*, which also occurs in more permanent habitats, in showing reduced fecundity but not increased life span when faced with food stress. Finally, *G. odontogaster*, a species more or less intermediate in choice of habitat between northern *G. thoracicus* and *G. lacustris*, reduces egg production with starvation, but not as much as more permanent habitat species; it does not alter life span. It differed conspicuously from *G. lacustris* in that it histolyzed wing muscles only when food was scarce, whereas *G. lacustris* histolyzed when food was abundant, thus resembling *Dysdercus* and many other histolyzing insect species. In more permanent habitats, *G. odontogaster* seems to use its wing muscles as a reserve to tide it over unfavorable periods. It is apparent from these analyses that phenotypic plasticity is overall an important life-history trait in these water striders. Looked at as a function of the cost of flight, the price is paid in terms of longevity in a temporally unpredictable environment (retaining wing muscles = shorter life) and in terms of decreased fecundity in temporally stable habitats (retaining wing muscles = reduced fecundity). The evolutionary ecology of migration in these gerrids is thus much more complex than simply its presence in temporary habitats and its absence in permanent ones.

A further complication occurs because there is a positive correlation between flight threshold and the frequency of the winged morph as revealed by comparisons both within and among species of gerrids (Fairbairn 1986, 1988; Fairbairn and Desranleau 1987; Fairbairn and Butler 1990). In eastern Canada near Montreal *Aquarius* (*Gerris*) *remigis* is a species of streams and impoundments along the streams; the frequency of macropters in this population is very low, usually on the order of only 1% or so. Observations of marked individuals indicated that displacement from the site of marking was only a few tens of meters in both morphs with no difference between them. Furthermore, most of the winged overwintering females of this species histolyzed the flight muscles; and even of those that did not, the threshold for flight, as measured using tethered flight on individuals brought to the laboratory, was very high. Why, then, are macropters retained at all? When life histories were assessed, it was revealed that pre-diapause macropterous females reproduced earlier than apterus females; by contrast, post-diapause macropters begin to oviposit after the apters and have lower overall fecundity. Fairbairn (1988) suggests that macropters thus may be at a selective advantage in warm, but rare, habitats that favor pre-diapause reproduction because of the possibility of a second annual generation. There may also be the occasional need to migrate. Apters, with their higher overall reproductive capacity, will be favored in the normal cool habitats. In northern California, *A. remigis* occurs on temporary streams that cease flowing, except for a few remnant pools, during the summer dry season. These populations breed in winter and spring, during the rains, and either take refuge on larger streams during the summer (macropters) or on the remnant pools (apters). The apters that survive have an advantage because they can reproduce before the macropters, which must fly back to the streams, but they risk extinction in dry years. Here the frequency of macropters is high, flight thresholds are low, and flight can be prolonged (Kaitala and Dingle 1992).

Comparisons among species also reveal the trend toward more migration among winged individuals when they occur with increased frequency. This trend is summarized for four species of gerrids from eastern Canada in Table 13–3. The gerrids were tested for flight in the laboratory using tethering, and flight thresholds were indexed on a scale from 1 to 10 where 1 indicated the test insect flow immediately upon being lifted from the substrate, intermediate steps involved repeated contact with the substrate, wind on the head, and so forth, and the final 10 indicated the insect would not even open the wings without the aid of the experimenter. As the proportion of macropters in the population declined, the threshold for flight rose, as did the proportion of macropters that histolyzed the wing muscles. Determination of the distances moved by marked individuals in the field showed that *G. buenoi* migrated on average more often than *A. remigis* and that *L. dissortis* and *G. comatus* migrated more than *G. buenoi*, but were not different from each other. Among those that did migrate, there were no apparent differences in distance traveled. These species and other Temperate Zone gerrids

Table 13–3. The Relation Between Pterygomorphism and Flight Characteristics in Four Species of Gerrid from Eastern Canada Field studies indicated decreasing migration from *L. dissortis* to *A. remigis*

	Limnoporus dissortis	*Gerris comatus*	*Gerris buenoi*	*Aquarius (Gerris) remigis*
Percent macropterous	100	70+*	50+*	<3
Phenology	Univoltine[†]	Bivoltine	Bivoltine	Univoltine[†]
Flight threshold ($\bar{x} \pm$ s.e)[‡]				
Spring	3.0 ± 0.5	6.2 ± 0.5	9.2 ± 0.7	9.8 ± 0.2
Summer	3.8 ± 0.9	4.6 ± 0.6	6.0 ± 0.2	5.9 ± 0.5
Autumn	—	3.8 ± 0.4	6.8 ± 0.9	—
Percent histolysis in overwintered macropters	2	0–18[§]	46.7	73.3

*Wing reduction primarily in non-diapause summer generation.
[†]*L. disortis* and *A. remigis* may be partially bivoltine in suitable habitats and warmer summers.
[‡]Flight threshold is an index of the difficulty of inducing flight in laboratory flight tests (higher value = more difficult)
[§]0% in a spring sample, 18% in a summer sample.
SOURCE: From Fairbairn and Desranleau 1987.

migrate primarily in the early spring when they emerge from overwintering diapause sites and search for breeding habitats (although see California *A. remigis* discussed above). It is at this time that differences between species are most strongly expressed. Another interesting point that comes out of these studies is that the correlations among frequency of macropters, wing muscle histolysis, and flight threshold imply a constraint, at least in the short term (Zeng 1988), on the evolution of pterygomorphism and that this constraint will favor either fixation of the brachypterous or apterous phenotype or the maintenance of relatively high frequencies of macropters (Fairbairn and Butler 1990). This issue of correlational constraints will be raised again when I discuss the genetics of migration in Chapter 14. Finally, one caveat needs to be mentioned. All laboratory flight tests on the Canadian gerrids involve measures of flight threshold and not estimates of duration. Because migrants may have high thresholds (Rankin and Burchsted 1992), flight thresholds must be interpreted cautiously with respect to what they reveal about migratory tendency.

The Evolution of Flightlessness in Insects

The advantage to having wings to move about and, if necessary, to migrate is obvious, and indeed entomologists often argue that one of the reasons for the enormous diversity of insects in the world is their ability to fly. There is nothing to say they must fly, however, and the existence of species and populations with reduced wings or the absence of wings as illustrated in the previous section suggests that some sort of trade-off exists between flight and other components of fitness (Roff and Fairbairn 1991; Roff 1994). It is appropriate to ask at this point what these might be. We have received a hint in the studies of Kaitala (1991), cited above, showing increased survival in some wingless gerrids in the presence of food shortage. Also, in aphids there is a trade-off between lipid storage and gonad size (Dixon et al. 1993); alates store more fat, but apters direct more resources to gonads. We now need, however, to examine the issue of trade-offs in a broader perspective. By examining cases where flight and long-distance movement have been selected *against*, we may perhaps use them to illuminate the mirror image of why flight and migration have been selected *for*.

Recent reviews have examined the relative fitnesses of the two morphs by comparing data on life histories across species (Dingle 1985; Roff 1986; Roff and Fairbairn 1991). The general trends are illustrated in Figure 13–7 which shows differences between long-winged and short-winged (including both apterous and brachypterous) morphs with respect to four life-history traits in a selection of species. In two of these traits, development time and adult longevity, there are no consistent trends, and in about half the instances there are in fact no differences. For age at first reproduction and fecundity, on the other hand, statistically significant differences among morphs are apparent, with a strong tendency for the short-winged females to begin ovipositing earlier and to produce more eggs than their long-winged counterparts. There are some exceptions to this for fecundity. For example, the long-winged morph produces more eggs in some beetles such as *Ptinella* spp. (Ptiliidae) (Taylor 1978) and the carabids *Calathus cinctus* and *C. melanocephalus* (Aukema 1991), and in the moth *Orgyia thyellina* (Sato 1977). In the latter species, however, the short-winged female lays larger diapause eggs and displays higher reproductive effort (= no. of eggs × egg size). Data on egg size are lacking for the beetles. But in spite of these exceptions, the fecundity and age at first reproduction statistics clearly indicate a gain in Malthusian fitness for the short-winged relative to the long-winged morphs. The cost to wings and wing muscles can thus be assessed in terms of this fitness. The data from which Figure 13-7 was compiled are for females, however, and it is less clear what the cost might be for males, because their reproductive contribution, and hence Malthusian fitness, is difficult to measure. There is also a presumptive advantage for males to remain macropterous and flight worthy to facilitate a search for females. Short-winged males do gain in testis and fat body size, however, and in the oriental chinch bug (*Cavelerius saccharivorous*) gain a mating advantage (Fujisaki 1992); these facts plus the high frequency of the apterous morph in males of at least some species certainly suggest a cost to the retention of wings even in males. Overall, then, the very real advantages to flight, and the very important advantage of migration by flight for escape and colonization, must be assessed in fitness terms against the equally real costs of retaining the flight apparatus.

Kaitala and colleagues (1989) have attempted such an assessment for the water strider, *G. thoracicus*, using an approach based on evolutionarily stable strategies (ESSs). The concept of an ESS was introduced in Chapter 12 in our discussion of partially migrating Aldabran tortoises and partially migrating birds (Kaitala et al. 1993). Recall that there is a strategy, the ESS, that when it is common, no alternative genetically based strategy can increase in the population. Once again, there is some ambiguity with respect to fluctuating populations inhabiting a finite number of patches in a randomly varying environment, as noted above with models for partial migration (Kaitala et al. 1993), so that Ellner's (1985) approach based on computation of expected growth rates of "invading strategies" was also used for *G. thoracicus*.

Kaitala and colleagues first examined a two-habitat model of rock pool versus brackish sea bay. Either or both were assumed to be stochastic and unpredictably favorable with respect to the distribution of females over the habitats. Females could either histolyze wing muscles and lay all their eggs in one place or retain them and "spread the risk" by laying in several places. Under deterministic density-dependent conditions, an ESS strategy meant that females always used both habitats; environmental uncertainty changed only the fraction of females occupying the habitats. Further, strategies maximizing arithmetic or geometric mean fitnesses (i.e., average population levels) were not ESSs, and the fraction of females moving to different habitats maximized the average per capita reproductive rate in both habitats but not the total population size. In a more complex multiple rock pool model, the facultative response of muscle histolysis to population density was

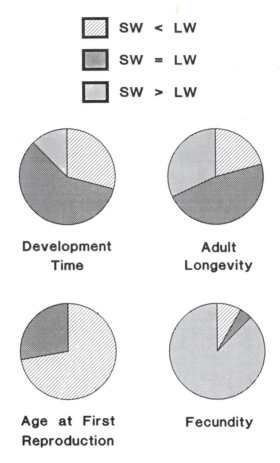

Figure 13–7. Life history differences between long-winged (LW) and short-winged (SW) morphs in a selection of 15 insect species for development time and longevity, 22 species for age at first reproduction, and 25 species for fecundity. With respect to development time and longevity, differences are not significantly different from a 1:1 ratio; for age at first reproduction and fecundity, differences are significantly different from 1:1. (Redrawn from Roff and Fairbairn 1991.)

always an ESS with respect to 100% histolyzing or never histolyzing. This conclusion derives from the observation that when competition between siblings becomes a factor in small rock pools, females do better by retaining flight and distributing offspring over many pools. If, on the other hand, the overall population level is high so that competition cannot be avoided by flight, females do better by histolyzing because all rock pools are occupied anyway. This situation held in both constant and stochastic environments. Although only a limited set of uncertain environmental conditions were considered, the analyses of these models do point out some interesting potential relations among density dependence, environmental variation, and the relative costs in flight-cum-migration pterygomorphisms. Given the extensive facultative responses to crowding in insects expressing pterygomorphism (Table 13–1), density dependence is a factor that would seem to deserve more theoretical analysis with respect to specific migration patterns. Of particular interest, in view of the ESS models, is the relation between pterygomorphism, density in local patches, and overall population (or metapopulation) levels.

On a broader scale, Derek Roff (1990; 1994) has considered the sorts of environments that might favor the evolution of flightlessness. Speculation on the subject goes back at least to Darwin, who noted that beetles from the Atlantic island of Madeira displayed a high frequency of flightless forms and postulated that a flightless morph would be more fit on an oceanic island than a winged form because the latter would frequently be blown out to sea to perish (Darwin 1859, p. 104). The difficulty with Darwin's hypothesis, however, is that it doesn't take scale into account; for a small, winged insect an island is a large place, and it is far more likely to land somewhere else on the island than be blown out to sea. Roff (1990) actually looked for a relation between the occurrence of insects on islands and winglessness and found there was none. The absence of a relation is shown clearly in Figure 13–8, which plots frequency of flightlessness for insect species for mainland areas against the value for comparable groups on islands of

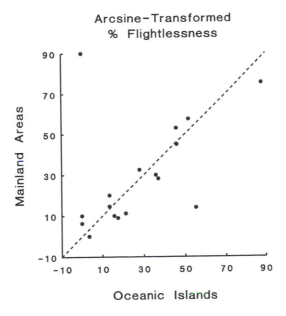

Figure 13–8. The relationship between proportion of flightless insect species on oceanic islands versus mainland areas for comparable groups in areas of approximately equal latitude. The dashed line is the line of equality; no increase in flightlessness on islands in evident. (Redrawn from Roff 1990.)

approximately equal latitude; there is no tendency toward greater flightlessness on islands.

There was, however, a strong tendency toward higher incidences of flightless morphs at both higher altitudes and latitudes. This is shown in Figure 13-9, which indicates the percentages of flightless carabid beetles against latitude for both high islands and low islands. There are more wingless species in highland areas, and there is an increasing frequency of flightlessness at higher latitude in both high and low island groups. Roff interprets these results to mean a greater tendency to evolve flightless morphs in more temporally stable habitats, because successional habitats of the same type persist longer at higher elevations and latitudes; this conclusion is supported by simulations (Roff 1994). Studies of succession along latitudinal and altitudinal gradients definitely support this tendency in habitat persistence. There was also an increase in flightlessness across habitats on a gradient of persistence; woodlands, for example, contained more short-winged forms than old fields, supporting the earlier conclusions of Southwood (1962). One interesting exception was the tendency among arboreal species to retain flight, presumably because of the advantage of flight over walking in large three-dimensional spaces such as tree canopies (Waloff 1983).

Although, overall habitat heterogeneity explains most of the evolution of diversity of wing form in insects, other factors can be involved as well (Wagner and Liebherr 1992; Roff 1990, 1994). For example, insects with incomplete metamorphosis (nymph to adult) are more likely to evolve flightlessness than those with complete metamorphosis (larva to pupa to adult), presumably because the latter frequently undergo major niche shifts between larva and adult, making the retention of the capability for wide-ranging movement a distinct advantage. Wing reduction is also common where energetic costs of flight are high such as on subantarctic islands with cool temperatures and high winds. The high cost of desiccation during flight may account for the many cases of wing reduction in desert insects. In general, however, there is a paucity of data with which to address the role of factors other than habitat heterogeneity in the evolution of flightlessness. There is also a need to address the partitioning of flight itself between migration and various foraging activities, and how this might influence the evolution of wing form.

Seed Heteromorphisms

In many groups of plants there is discontinuous morphological variation in seeds, known as heteromorphism, that in many ways is analogous to the pterygomorphism of insects. In most cases heteromorphisms are associated with differential capabilities of seeds to migrate ("dispersal" in the botanical literature). These differential migratory tendencies are also correlated with germination times; seeds that have a high migratory capability generally germinate more quickly than nonmigrant seeds (Venable and Lawlor 1980; Fenner 1985). Another general

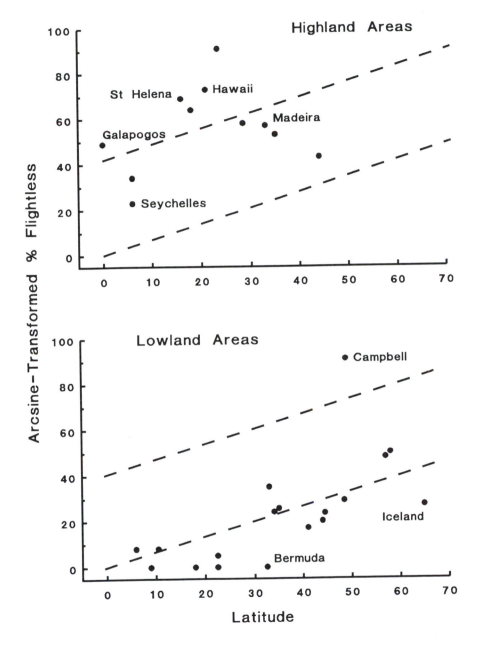

Figure 13–9. Percentage of flightless carabid beetles on islands as a function of altitude and latitude. The dashed lines are the regression lines for high (> 600 m) and low (< 600 m) islands. Flightlessness is significantly greater on high islands and at higher latitudes. (Redrawn from Roff 1990.)

characteristic of seed heteromorphisms is that they often occur in taxa with flowers that are also heteromorphic, so that where different evolutionary possibilities already exist with respect to flower form, they are also increased by variation in seed morphology (Silvertown 1984; Venable 1985). The two most common ways in which this further increase in evolutionary possibilities is realized occur in the heteromorphic ray and disk seeds among the composites (Asteraceae) and in the contrasting seeds produced by the generally cryptic, often selfed, cleistogamous flowers and the generally showy, outcrossed chasmogamous flowers of species like jewelweed (*Impatiens*; Balsaminaceae).

The flower heads of many familiar composites consist of a central disk of small, cryptic florets surrounded by a ring of much more conspicuous flowers bearing showy petals. The black-eyed susans of horse racing's Preakness stakes are a well-known example. The "black eye" of each inflorescence consists of numerous dark brown miniature florets with the bright orange-yellow of the "eyelashes" made up of the broad petals of the ray flowers. Most composites are far less showy, but nevertheless produce two types of seeds. Each floret produces a hard-coated achene or one-seeded fruit. Those seeds deriving from the disk often bear a feathery structure or pappus that serves as a parachute (Fenner 1985). In contrast, it is frequently the case that the ray flowers produce an achene with no pappus and a much heavier seed coat. This coat may be smooth or covered with spines and hooks that result in transport attached to the fur of mammals (Sorensen 1978 and Fig. 13–10).

Venable and Levin (1985) examined the ecology of achene heteromorphism in the composite *Heterotheca latifolia*, an annual growing on disturbed soils over much of eastern and central North America from northern Mexico to New York. This species exhibits an achene dichotomy with a disk fruit possessing a thin wall and a pappus and a ray achene with a thick wall and no pappus (Fig. 13–10). Both are about 3 millimeters long and weigh about 730 micrograms, but the ray achenes are apportioned more into the thick wall, while the disk fruits have a 60% heavier embryo. The pappus-bearing disk achenes also germinate quickly, with almost 100% doing so as soon as they contact soil moisture. The rate of germination for the ray achenes is much lower, although it does increase as the fruits age. They also germinate if the heavy wall is cut or removed so it is this structure that constrains germination. The migratory achenes are thus consistent with the general trend in seed heteromorphisms for the migrants to germinate sooner. Flower heads become smaller as the season progresses and concomitantly produce a higher proportion of ray achenes. Up to a threshold size larger plants generate flowers producing more disk achenes. This seems to be an adaptation promoting seed migration because achenes from taller plants on average are carried farther from the parents. Overall, however, migration distances are not great, with a mean distance settled from parent plant varying from 0.52 meters for a ray achene from a short plant to 1.65 meters for disk achenes from tall plants. The longest recorded migration distances for disk achenes were on the order of 20 meters, but there is probably the occasional wind-aided transport of even greater distances on

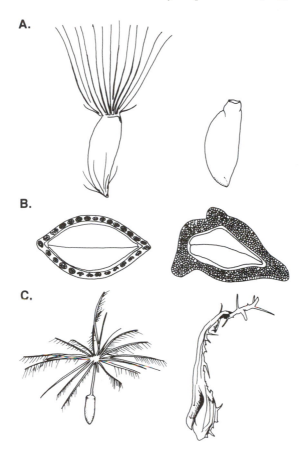

Figure 13–10. Seed heteromorphisms in two weedy composites, *Heterotheca latifolia* (*A&B*) and *Picris echioides* (*C*). Disk seeds in each case bear a feathery pappus; ray seeds are thick-walled in *H. latifolia* and bear assorted spines and hairs in *P. echioides*. (*H. latifolia* redrawn from Venable and Levin 1985; *P. echiodes* drawn from a photograph in Sorensen 1978.)

particularly gusty days. This latter, of course, would be virtually impossible to follow.

Venable and Levin conclude from their studies that the seed heteromorphism of *H. latifolia* represents a division between a relatively high risk migration strategy (disk achenes) and a relatively low risk strategy of spreading germination over time (ray achenes). The migrants must reach a suitable site, but if they do, they germinate at the first favorable rain. Under most circumstances this is probably the best life history to follow, but if the early rain is followed by drought conditions, the strategy will be disastrous. In that case the fallback strategy is represented by the ray achenes, which can remain dormant for extended periods protected from xeric conditions by the heavy seed coat. This dual bet-hedging strategy thus represents a compromise between growth and perseverance, and it resembles similar bet-hedging strategies in pterygomorphic insects such as the water striders of temporary streams in California (Kaitala and Dingle 1992).

Results somewhat similar to those of Venable and Levin were obtained by McEvoy and Cox (1987) in a study of ragwort (*Senecio jacobaea*) in western Oregon. Ragwort is also a composite producing heteromorphic ray and disk achenes. The disk seeds bear not only a pappus but also rows of small spines or trichomes that assist transport in the fur of mammals. McEvoy and Cox marked and recovered seeds from ragwort plants growing under varying conditions at both coastal and inland sites. In this case the disk achenes are released from the plant shortly after they mature and are transported away from the plant. The ray fruits, on the other hand, remain on the parent for some months. As in *H. latifolia* the achenes don't move very far, with 89% of them recovered within 5 meters of the plant of origin. Long-distance movement is thus of rare occurrence. When it does occur, it is probably by the disk achenes either in the occasional suitable winds, by secondary transport along the ground, or in mammalian fur. Plant height increased distances traveled but was much modified by site conditions such as vegetation density and humidity, both of which were negatively correlated with density. The authors stress, therefore, that achene geometry is only one of several factors that influence distance traveled, and secondly, that migration is not by a single mechanism. Recall that a similar point was made for pine seeds that are transported by both wind and seed-caching mammals (Vander Wall 1992 and Chapter 12). Further, like *H. latifolia* (Venable and Levin 1985), ragwort seems to be following a bet-hedging strategy with disk achenes providing opportunities both to retain the home site and colonize new sites in the short term, while their ray counterparts provide the conservative alternative of establishing locally at some time in the future. In this way the plant can average out unpredictable environmental variation.

A slight variation on the theme of wind/animal transport vs. delayed germination occurs in another weedy composite, bristly oxtongue (*Picris echioides* L.) (Sorensen 1978). The small flower heads of this plant produce about two dozen disk achenes and half a dozen outer ray seeds. The disk achenes bear a pappus, but are smooth walled. It is the outer achenes from the ray flowers that bear hairs and spines making them liable to animal transport. The two achene types are illustrated in Figure 13–10. The role of the hairs and spines was confirmed in a laboratory experiment in which 300 of each type of seed were scattered on the floor of an enclosure over which a pair of mice were allowed to move. The spiny ray achenes stuck to all four feet and to the base of the tail of each mouse, but none of the smooth disk seeds were transported in this manner. Because of the burrowing and caching habits of the assorted field mice and voles that can transport seeds, the habitats reached by this form of transport may be quite different from those accessible by wind alone. No differences in germination times were found between the two types of seeds so the dimorphism in this case is apparently confined to the transport system. Bet-hedging in this case evidently relates only to short-term environmental patchiness, not long-term temporal variation.

Seed heteromorphisms may be associated not only with wind or animal transport but also with movement by water. In a comparative study of heteromorphic seeds in 48 species of the genus *Spergularia* (Caryophyllaceae), Telenius and Torstensson (1991) made several interesting observations. The heteromorphism in these plants is associated with the presence or absence of "wings" on the seeds, and they noted, first, that in the 11 species in which dimorphic seeds occurred, these seeds were on average larger than those of species that were monomorphic. This counterintuitive observation led them to postulate that it was the development of seed wings that allowed the evolution of larger seeds, facilitating their movement away from the parent plant. The heteromorphism may in this instance reflect seed size variation rather than different criteria for the selection of seed migration. A second interesting observation of this study was that the 11 species with a proportion of winged seeds were perennials, which may be a reflection of a trend to larger seeds in perennial than annual plants. Finally, these 11 species also occupied primarily wet or marshy habitats. The presence of wings may thus have evolved not for transport by wind, but rather to facilitate floating and secondary transport by water.

Water transport may also have been a selective factor in the evolution of flower heteromorphisms in the jewelweed, *Impatiens capensis* (Schmitt et al. 1985). In this plant, seeds—which are not dimorphic—are expelled from a capsule ballistically, with the largely outcrossed seeds of chasmogamous (CH) flowers thrown farther than the obligately self-fertilized seeds of the cleistogamous (CL) flowers. Schmitt and colleagues measured seed migration distances from a standard height of 50 centimeters and found that both the mean and variance were greater for CH seeds, although very few seeds of either type traveled more than a meter from the force of expulsion alone. The position of CH flowers higher on the plant and on the ends of nodes rather than close to the stem, as in CL flowers, also increases the mean distance thrown. Jewelweeds thrive in marshy areas or along the banks of streams, and a further consequence of greater expulsion distance for the CH seeds is that they are more likely to be secondarily transported by water, because they are buoyant. If they are, they will, of course, reach germination sites farther from the parent and experience greater risk of not germinating at all. The net result in this system is that the heteromorphism of the flowers results in the differential migration of seeds that are not themselves dimorphic. A further aspect is that those offspring arising from selfed CL flowers, and therefore genetically like their parents, are likely to germinate in the parental microenvironment, whereas outcrossed CH offspring will end up in sites that are more likely to be different. The bet-hedging strategy in this case has apparently evolved to include the genetic system.

In our discussion of seed heteromorphisms it is apparent that migratory capability interacts with at least two other important characteristics of seeds, size and dormancy. What this means is that migration cannot be considered alone, but rather must be evaluated as part of a syndrome of traits that includes these two characters (Venable and Brown 1988). This syndrome serves three population dynamic functions in spatially and temporally varying environments: bet hedging, escape from crowding, and escape from competition with siblings or parents. The exact form the syndrome takes will depend on the contribution of each of these factors to the environments experienced by the plants. For example, increasing heterogeneity by increasing the number of habitat patches and the variance in habitat permanence creates the possibility of reducing risk by continuously colonizing new areas. This would select for the classic strategy of high migratory capability, small seeds, and low dormancy seen in annual weeds. Departures from this extreme form of environmental uncertainty would select for different relative investments in the three traits making up this syndrome. Further, an evolutionary breakthrough in one trait may lead to a different evolutionary trajectory in another with which it is correlated. Larger seeds in *Spergularia* when wings are present may be a case in point (Telenius and Torstensson 1991). A consequence of syndromes, however, is that traits do not evolve in isolation, but if they share genes in common, as the notion of adaptive syndromes implies, there will be trade-offs among coevolved characters (see also Chapter 7). In the evolution of the various syndromes involving seed heteromorphisms (and insect pterygomorphisms) both the fact and the character of these trade-offs become especially obvious. The nature of the genetic "blueprints" underlying migration and migratory syndromes is the subject I shall address in the next chapter.

Summing Up

Migration confers considerable benefits to organisms because it in effect allows them to even out environmental heterogeneities. It permits escape when conditions are unfavorable and colonization of new habitats when they become available. In spite of these benefits, however, it also imposes costs. Energy and materials devoted to migratory apparatus and effort are lost to growth and reproduction, and there is a risk entailed by departing a habitat to locate a new one. These risks include both mortality en route and the possibility of failure to find a suitable place to settle. Migratory organisms must thus strike a balance between departing and staying, between choosing life "here" or "elsewhere" (Southwood 1977; Solbreck 1978). The variability in migratory behavior that we have seen in the preceding chapters is a reflection of the action of natural selection in maintaining that balance. The most obvious manifestation of that variability is the occurrence of migratory polymorphisms and polyphenisms.

Insects are the group in which such heteromorphisms are perhaps the most apparent because of the frequently occurring dichotomy between the presence or absence of wings. The dichotomy need not be overtly expressed, however. In many insects wings are present but wing muscles are not, so that the wings cannot serve for flight. The loss of wing muscles may be facultative, as we have seen for both African and tropical American cotton stainer bugs and for migratory aphids. In these taxa and others the wing muscles break down and their proteins are diverted to reproduction after migratory flight has located a new habitat. In other insects, such as some gerrids, wing muscles may never be present even though wings are. The degree of wing reduction can also be very variable. The tropical soapberry bug, *Jadera aola*, is pterygomorphic, but the wings of the brachypterous form are only a little reduced compared to the wings of the long-winged morph. In many other insects the distinction is much more dramatic: either there are fully functional wings or there are none at all. The most completely studied of the pterygomorphic insects are the water striders (Gerridae) of Europe and North America. In this taxon the great diversity of aquatic habitats occupied has produced an equally diverse set of tactics and strategies that range from fully winged 100% flying species and populations to ones that never have wings and so never fly (Table 13–2). Many gerrids and other insects have evolved facultative pterygomorphism or polyphenism. Environmental conditions that predict habitat deterioration, especially crowding, induce the winged morph.

Studies of water striders and other insects such as planthoppers (Homoptera) are also revealing the nature of the trade-offs among the different expressions of wing form. The presence of wings allows escape when habitats deteriorate and the exploitation of habitats that are ephemeral. The energy and material requirements for flight can be assessed relative to reproduction. Wingless individuals cannot fly, but they often mature more rapidly, reproduce earlier, live longer especially under stressful conditions, and make a greater investment in reproduction (Fig. 13–7), all of which confer increased Malthusian fitness. It should also be noted, however, that in those species that are migratory and bear wings throughout life, selection has frequently acted to reduce the physiological costs of migration. In groups like migratory grasshoppers these costs can be surprisingly small (Rankin and Burchsted 1992). There are thus various evolutionary pathways for maintaining the balance between the costs and benefits of migration.

The evolution of flightlessness provides a conceptual mirror image to the evolution of a lifelong functional flight apparatus. Extensive comparative studies reveal that flightlessness evolves where habitats are stable for a very long time, the opposite to the unstable environmental state that leads to the incorporation of migration into the life history. Winglessness is thus increasingly common in locations where habitats are long lasting; examples include high latitudes and high mountains, where cooler climates prevent rapid environmental change, and large lakes whose size prevents drying even during drought.

Among plants migratory heteromorphisms are characteristic of the seeds of several different families. Mostly these are polyphenisms with the type of seed determined by the form of the flower that produces it; heteromorphic flowers characteristically give rise to heteromorphic seeds. In composites it is often the case that disk florets produce a thin-walled achene that also bears a parachute-like pappus that facilitates wind transport. Ray florets, on the other hand, produce an achene that is thick-walled and bears no pappus, although in some cases the seed coat may possess spines and hooks that promote attachment to the fur of mammals. If the ray achenes are not equipped with transport devices, they fall near the parent, and the thick wall must be broken or scarred before germination can occur.

The migratory disk achenes germinate quickly when suitable soil and moisture conditions are encountered. The seed heteromorphism thus provides a bet-hedging strategy that averages out environmental variation in time and space.

Other seed heteromorphisms include the evolution of winged and unwinged seeds and differential ballistic transport associated with chasmogamous (CH) and cleistogamous (CL) flowers. The development of wings seems to be associated with the evolution of larger and heavier seeds in *Spergularia*, a caryophyllaceous genus inhabiting fields and wet areas. The close association between winged seeds and species growing in wet areas also suggests that wings may have evolved in conjunction with water transport. In jewelweed (*Impatiens*) ballistic expulsion of seeds from outcrossed CH flowers carries them for greater distances than does expulsion from selfed CL flowers, so that it is offspring of qreater genetic variance that reach new habitats. Here, too, water transport may be involved in seeds from CH flowers.

The coevolution of migratory characteristics with traits like seed size and dormancy indicates that seed heteromorphisms represent migratory syndromes. This means that selection acting on one trait will also influence the others, on the assumption that the traits share genes in common. Proximate environments also influence seed transport with density of surrounding vegetation, plant height, and relative humidity among the variables having marked effects. The occurrence of syndromes and the influence of the environment point out the need to examine interactions between genes and environments in producing complex adaptations in response to selection. It is to genetic influences that we now turn.

14

Evolutionary Genetics of Migration

As we have seen in the previous chapters, migration is a dynamic and diverse process with a great deal of intriguing variation expressed across all taxonomic levels and in all aspects of migratory behavior. Running through the diversity is a common thread of adaptation to changing habitats, but there seems to be no limit to the forms adaptation can take. As seasons change, habitats deteriorate, or new environments become available, many organisms migrate, but the distances moved may be long or short, orientation may be simple or complex, and all, most, or only a few of the individuals in a population may take part. From the perspective of the organism, migration renders the environment more equable (Leggett 1985) and so is an important homeostatic behavior. For this reason one would expect relatively intense selection for optimal movement strategies. The observed variety of distinct patterns present in these indicates that natural selection has shaped considerable diversity, and new configurations are continually evolving.

To understand the migratory patterns we observe, we must analyze two sources of variation and their interactions. The first is the environmental "templet" (Southwood 1977) that molds migratory behavior via natural selection; the second is the genetic structure upon which selection acts. Without underlying genetic variation traits cannot respond to selection, and the influence of genes is thus fundamental to the evolution of migratory syndromes. On the other hand environments determine the expression of genes, and so it is equally important to consider the environment in which a syndrome is evolving. Because different phenotypes may be produced when a particular genotype experiences different environments, the interactions between genes and environments, known as reaction norms when the expression of given genes or genotypes is measured across several environments (Stearns 1989; van Noordwijk 1987, 1990), are a third source of variation that is basic to adaptation and evolution. These three important contributors to the evolution of migration, and indeed any trait, will be considered in turn in this chapter.

For convenience in discussing the analyses of genetic variation underlying migration, I have somewhat arbitrarily partitioned them into three different approaches. The simplest of these is the determination of whether a trait displays a genetic component. Typically samples of two or more populations (or species) that differ with respect to a trait are reared in a common environment to see if the differences persist and are therefore due to gene differences. Because these methods are especially frequently used in studying variation in plants, they are usually referred to as "common garden" experiments. We have already seen some examples of these in previous chapters. Often the experiments are extended by crossing individuals of the two populations and observing the trait in the hybrid progeny. The second general approach is to search for the effects of single genes or, at most, a very few genes. Often these investigations take the form of analyzing the influences of the presence or absence of a particular isozyme known to be a part of a metabolic pathway related to the trait; for example, one involved in releasing energy for flight in the case of migration. Most complex traits like migratory behavior, however, are a consequence of many genes. The analysis of polygenic influences requires quantitative genetic methods that allow partitioning of phenotypic variation into sources arising from the genes, the environment, and genotype by environment interactions. I shall present examples derived from each of these approaches and try to elucidate how they might contribute to our understanding of migration and migration syndromes. The latter, as we shall see, result from the fact that no gene influences one trait alone, but rather all genes are pleiotropic with the result that traits, and importantly, combinations of traits are the result of several genes acting in concert to produce the phenotype we see. Genetic variance, genetic correlations arising because genes influence more than one trait, and genotype-environment interactions will all influence the rate and direction of the evolution of migration syndromes (Dingle 1991a, 1994).

Population Differences

Four sets of results from common garden experiments are illustrated in Figure 14–1. In these experiments different populations of migratory species were reared in a common environment, tested for migratory performance, and then crossed to assess the performance of hybrids. The first case is that of the cowpea weevil,

---➤

Figure 14–1. Gene differences between populations in migratory behavior as revealed by common garden experiments. *Callosobruchus maculatus* (cowpea weevil): Percent of migrant morphs in migratory (M) and largely nonmigratory (I) populations from West Africa and their hybrids. Females are listed first in the crosses, and there is a maternal bias in the hybrids (Messina 1987). *Melanoplus sanguinipes* (lesser migratory grasshopper): Percent long-flying individuals from a relatively nonmigratory Colorado population (CO) and a migratory New Mexico population (NM). Females listed first (Rankin et al. 1986). *Microtus agrestis* (European vole): Number of passages between compartments of a laboratory chamber in migratory northern (N) and nonmigratory southern (S) Swedish populations and their hybrids (H) (Rasmuson et al. 1977). *Sylvia atricapilla* (blackcap warbler): Migratory restlessness in migratory German and nonmigratory Canary Island populations and their hybrids (Berthold and Querner 1981). (Modified from Dingle 1991a.)

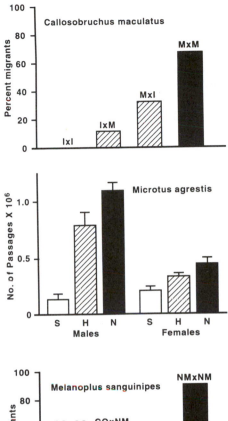

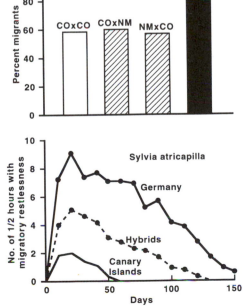

Callosobruchus maculatus, a pest of stored legumes (Messina 1987). Differences in population density produce two winged morphs of this beetle; one of these is migratory and results from crowding, and the other is sedentary. Populations of *C. maculatus* vary in their propensities to produce the migrant morph, and these propensities are retained when the beetles are reared under similar conditions, demonstrating gene differences. The results for two populations and their hybrid are shown in Figure 14–1. A strain from Ibadan, Nigeria (I) produced no migrants, while one from Maroua, Cameroun (M) produced about 65% migrant morphs. Crosses between the populations produced intermediate frequencies, suggesting that the genetic contribution is additive. Additive genes are those that specifically contribute to the resemblance among relatives (as contrasted to genes that are no more likely to cause resemblance between relatives than resemblance between any members of a population) and will be discussed more fully when we consider quantitative genetics below. Usually additive traits are the result of polygenic effects, and such is probably the case here. Because they reveal resemblance among relatives, additive traits also imply a high potential for evolution, an interesting point in the case of these weevils because the polymorphism may have evolved in the few thousand years since humans have been storing legume seeds. The nonmigrant morph displays higher reproductive potential than the migrant and may be selected to exploit rich supplies of seeds, while the latter can locate new resources when crowding signals overexploitation in situ. Note also that there is a bias in the hybrids toward the female parent. Such biases are common and indicate a maternal effect or nongenetic inheritance of a trait from the mother. They, too, will be discussed further below.

Variation in flight performance also occurs in the lesser migratory grasshopper, *Melanoplus sanguinipes*, of North America (McAnelly and Rankin 1986; Rankin et al. 1986). Field observations and tethered-flight tests revealed that insects from a Colorado population are less migratory than those from a site in New Mexico, which also show a strong tendency for long-duration flights in the laboratory. Differences between these populations persisted for four generations and were not influenced by temperature, photoperiod, food level, or crowding; rather the only consistent source of differences was population origin, suggesting a major component of genetic control. These differences are shown in Figure 14–1, which also indicates the performance of hybrid offspring. In this case the offspring are not intermediate between the parents in performance but instead strongly resemble those from Colorado. Variation, in other words, does not seem to arise from additive effects primarily but rather reflects a strong bias or dominance deviation toward Colorado.

Population differences in migratory behavior are not limited to insects. Two cases from warm-blooded vertebrates are illustrated in Figure 14–1. In the first case, Rasmuson and colleagues (1977) studied two Swedish populations of the European vole, *Microtus agrestis*. The northern (N) population undergoes classic microtine cycles of population abundance and is highly migratory, especially in the case of young males at high densities. The southern population (S) is neither cyclic nor migratory. Migration was assessed by monitoring the passage of voles among the compartments of a laboratory test chamber, with northern animals moving more than southern and males moving more than females. Hybrids were intermediate, again suggesting an additive genetic component, but with some dominance deviation toward the migrant parents, especially in the case of males. In the second case, Berthold and Querner (1981) studied nocturnal migratory restlessness in populations of the blackcap, *Sylvia atricapilla*. Here populations from Germany (migratory) and the Canary Islands (nonmigratory) displayed big differences in restlessness. Their hybrid offspring were intermediate, once again suggesting additive effects in the genetic contributions to migratory behavior.

Blackcaps are long-distance migrants with accomplished orientation mechanisms that allow them to reach population-specific wintering and breeding areas. As with many other birds, the young of the year are able to find the overwintering area without any assistance from conspecifics, suggesting a genetic basis for specific orientation directions. This hypothesis was tested by Helbig (1991b) using two blackcap populations from central Europe, one migrating southwest to Spain and the other southeast to the Middle East before turning south to Africa (see Chapters 8 and 9). Helbig crossed the two populations and found that the direction of orientation of the F_1 offspring was almost exactly intermediate between the directions of the parents and significantly different from both, suggesting a major additive component to the trait, as was also the case with migratory restlessness. There is also some evidence suggesting that orientation and restlessness are genetically correlated (see below) and share some genes in common. Berthold and colleagues (1990) crossed migratory blackcaps from Europe with a nonmigratory population from the Cape Verde Islands. The F_1 offspring not only showed some migratory restlessness but also a preferred migratory direction consistent with the European parents. Either both populations share a common directional preference, expressed only with restlessness, or direction and restlessness are transmitted together. Although these two alternatives could not be distinguished by the experiments, it seems evident that at minimum the expression of a directional tendency is correlated with the expression of migratory restlessness.

Traits directly relating to movement and orientation are only some of the elements that make up migration. In addition to these a number of other correlated traits are likely to be involved, contributing to an overall migratory syndrome. Migrating salmon, for example, move between fresh and salt water and must make compensating physiological adjustments. Foote and colleagues (1992) examined anadromous sockeye salmon and nonanadromous freshwater breeding kokanee (both *Oncorhynchus nerka*) with respect to their adaptability to seawater. Although both populations and their hybrids showed some ability to adapt, there was a clear order in the extent of their capabilities for doing so. The migrating sockeye were the first to show increased seawater adaptability in the spring of the second year of age and were the most tolerant of high salt concentrations. They were followed in order by the hybrids and kokanee. In addition to physiology, life history traits in salmon are also correlated with migration. Chinook salmon (*O. tshawytsha*) display two migratory phenotypes: "stream type" juveniles migrate to the ocean after a year or more in freshwater while "ocean type" juveniles depart for the sea during their first year. Taylor (1990) found three correlated responses to these two alternatives: first, stream types retained positive rheotaxis for a full year while ocean types did not; secondly, ocean types are less aggressive than the territorial stream types; and finally, ocean types grew faster. All these differences were expressed in a common laboratory environment, suggesting gene differences and probably adaptive divergence between the two migratory types. These results also hint at the complexity of migratory syndromes and their genetic components.

Another approach to sorting between genetic and environmental influences is cross-fostering, or the exchange of offspring between parents. This also controls for parental effects that result from being reared by a particular set of parents. Harris (1970) used cross-fostering in one of the first attempts to distinguish genetic from environmental influences in a migrant. His subjects were two species of gulls breeding in the British Isles. The herring gull (*Larus argentatus*) is largely a resident species in Britain while the closely related lesser blackbacked gull (*L. fuscus*) is a migrant in the winter to the Iberian peninsula and northwest Africa. Harris interchanged eggs from *L. argentatus* with those of *L. fuscus* and marked and banded all chicks shortly after hatching. Seven cross-fostered *L. fuscus* were recovered, of which five occurred well south of Britain, indicating normal migration. In contrast, of the 19 cross-fostered herring gulls recovered, 12 were scattered between Cape Finisterre and southern Portugal, well south of the normal range of British birds and of recoveries of several control chicks that had not been cross-fostered. These birds had evidently migrated like their foster parents, indicating a strong environmental component to migration as well as the genetic component suggested by the southern recoveries of cross-fostered *L. fuscus* chicks. The results are important because they so clearly reveal that both genes and environment can make contributions to migratory behavior.

One other case of population hybrids and migration is worth mentioning, one that is particularly interesting because the hybrid population occurs naturally. This occurs in European water frogs already mentioned for their long-distance (for a frog) movements at the Neusiedlersee in Austria in Chapter 10. As indicated there, the hybrid, the edible frog *Rana esculenta*, is a cross between *R. ridibunda* and *R. lessonae*. Hybrid *R. esculenta* females must mate with a male of one of the parent species, duplicating the *R. ridibunda* genome and casting out the original *R. lessonae* genome when they do (Schmidt 1993). They thus transmit only their *ridibunda*-derived genome to their offspring with the paternal contribution coming from the "parasitized" male of either *R. ridibunda* or *R. lessonae*. *Rana ridibunda* is usually absent from mixed populations of *R. lessonae* and *R. esculenta* so that the hybrid *esculenta* genome is maintained. The hybrid also migrates to hibernation sites either on land or in water whereas each parent species is confined to water (*R. ridibunda*) or land (*R. lessonae*) (Chapter 10). Tunner and Nopp (1979) examined tolerance to reduced oxygen in all three forms and found the hybrid much superior in its ability to withstand low-oxygen extremes, an ability that often leads to much greater survival over the winter. Various competition experiments also indicate that the hybrids are superior under extreme conditions of density and pond drying (Semlitsch and Reyer 1992; Semlitsch 1993). Heterosis for a number of traits, including migration and hibernation, thus apparently contributes to the continued maintenance of the hybridogenetic species.

Single-Locus Effects

Major differences among individuals with respect to migratory performance attributable to the effects of single Mendelian loci are relatively rare, a fact that should not be surprising considering the multiplicity of factors that enter into migration. Attempts to identify monogenic effects have focused on those instances where clear demarcations can be made between migrants and nonmigrants. Such is the case with wing polymorphism in insects (see Chapter 13), but even in this situation of clear morphological differences, control seems to be polygenic. Roff (1986), for example, surveyed 23 examples of wing dimorphisms and found only eight to be controlled by a single-locus, two allele system (five involved beetles). With one possible exception, the dominant allele yielded brachyptery. An additional case involves the discovery of a brachypterous morph, unknown in the field, in a laboratory population of the milkweed bug, *Oncopeltus fasciatus* (Klausner et al. 1981). In this case brachyptery was recessive.

Two other cases from lygaeid bugs have been analyzed in greater detail. Paralleling the observations in *O. fasciatus*, Solbreck and Anderson (1989) found recessive brachyptery in the Mediterranean lygaeid *Spilostethus pandurus*. They also looked for pleiotropic traits influenced by the alleles for wing length and found that the short-winged bugs had a longer and narrower thorax (wing muscles were still present) and a longer proboscis than the long-winged. There were no life-history differences observed between the morphs. As with the milkweed bugs, brachypters have not been found in nature. The second and very interesting case of monogenic control of wing length occurs in a second Mediterranean lygaeid, *Horvathiolus gibbicollis*, a species that is wing dimorphic in the wild (Solbreck 1986 and Chapter 13). As with other naturally occurring monogenic wing polymorphisms, brachyptery is dominant, as revealed by controlled laboratory crosses. The gene differences in wing length also resulted in correlated responses in life-history traits. First, the brachypters mature to adulthood more rapidly, and second, they begin to oviposit earlier and finish ovipositing later than the macropters, whereas the latter display a peak of oviposition associated with wing muscle histolysis following flight. The short-winged females also produce larger early eggs. The combination of shorter development time and larger early eggs likely gives the brachypters an advantage where the two forms must compete, but the macropters obviously have the advantage when it becomes necessary to colonize new habitats. The types of life-history differences observed between morphs of *H. gibbicolis* seem to be typical of wing polymorphic insects (Dingle 1985; Roff, 1986 and Chapter 13).

A second approach to monogenic influences on migration has been the attempt to detect particular enzymatic variants associated with migrants. Thus Myers and Krebs (1971) found differences in the transferrin (TF) and leucine aminopeptidase (LAP) loci in *Microtus* populations in Indiana between those leaving an area and those remaining. Other studies of voles, however, have failed to show differences in these loci between those staying or departing (Gaines and McClenaghan 1980). Unfortunately the lumping of all vole departures under the term "dispersal" means that many studies may be failing to distinguish migratory behavior from extended foraging or ranging (see also Chapters 2 and 6).

In insects there have been efforts to detect differences between winged and wingless morphs of a species in enzymes controlling flight metabolism. The principal target has been the glycerophosphate dehydrogenase (Gpdh) locus because of its particular importance in the metabolic pathway that maintains glycolytic flux and produces the ATP needed for flight (Sacktor 1974). Zera (1981) examined this locus in water striders that are both migratory and wing polymorphic (Chapter 13). He surveyed 11 species of gerrids displaying a full range of wing polymorphism, but even though the average number of allozymes in each species was high (5.36 ± 0.96), well above values for most other insects, there was no statistically significant association between Gpdh genotype and wing form for any of the populations or species examined. The Gpdh enzyme was found to be monomorphic among populations of the light brown apple moth (*Epiphyas postvittana*) of Australia, but there was population variation and variation associated with flight capacity in selected laboratory lines in the phosphoglucomutase (Pgm) locus (Gu 1991). The evolutionary significance is, however, not clear. Again, it is probably worth stressing that although an association between a particular enzyme variant and flight or migratory performance has a certain reductionistic appeal, there are so many factors involved in a complex behavior that genetic control of migration of a simple form is likely to be uncommon. It is far more likely that several genes will be involved, influencing various aspects of performance, and so it is to the analysis of polygenic systems that we must turn next.

Migration as a Polygenic Trait

Complex traits like migration are likely to be influenced by genes at many loci for two primary reasons. First, many factors contribute to the organization of migratory behavior and its accompanying physiological and life-history correlates, and second, polygenic systems allow the generation of many different response patterns to the action of natural selection (Roff 1986). The analysis of polygenic traits and the genetic variances and genetic correlations associated with them require the methods of quantitative genetics. These methods are covered extensively by Falconer (1989), and a rapidly growing literature is refining them for application to naturally occurring behaviors and adaptations (see, e.g., Boake 1994). Here I shall summarize these methods only very briefly, but with, I hope, sufficient clarity that readers can follow the logic behind their use and their contribution to understanding the evolutionary genetics of migration.

The basic quantitative genetic model assumes that differences in traits among individuals are the consequence of both genetic and environmental effects so that the mean phenotype in a population (P) equals the mean of genetic (G) and environmental (E) influences. Each individual in the population will have its own phenotype (P_i) in turn subject to genetic and environmental effects (G_i and E_i). The total phenotypic variation present is the sum of the differences between the mean values and all the individual values

$$\Sigma\,(\bar{P}-P_i) \;=\; \Sigma\,(\bar{G}-G_i) + \Sigma\,(\bar{E}-E_i)$$

Squaring both sides to get rid of negative values and dividing the sums of squares by the sample size gives the respective variances and covariances (as in performing analysis of variance)

$$V_P = V_G + V_E + 2\mathrm{cov}(G,E)$$

This is the basic equation of quantitative genetics and says simply that the phenotypic variance present (V_P) results from genetic and environmental variance plus a covariance term that indicates that genes may be expressed differentially in different environments (genotype environment interactions and correlations).

The genetic variance in turn can be broken down into components where

$$V_G = V_A + V_D + V_I$$

The first term, V_A , is the so-called "additive genetic variance" and is the genetic variance that is specific to the resemblance among relatives such that if a trait were completely additive, the mean expression of a trait for offspring of a particular cross would be exactly intermediate between the expression of the trait in the parentals (see, for example, the blackcap warbler in Fig. 14–1). Usually these traits are polygenic and arise from the contributions of many loci, each of small effect. Because it contributes specifically to resemblances among relatives, the additive genetic variance is particularly important for the evolution of a trait, for it best describes the potential for response to selection, at least over the short term (selection can change gene frequencies, which complicates responses over the long term; see, e.g., Barton and Turelli 1989). In this sense, V_A can thus be considered "Darwinian variance." The other two terms in the equation for genetic variance refer to dominance variance, V_D, or the tendency to resemble one parent more than the other, and the interaction variance, V_I, the variance expressing interaction among genes in which the expression of a particular allele depends either on what alleles are present at other loci or on chromosomal configuration or both. Because these two sources of variance do not contribute specifically to resemblance among related individuals, they are less important for knowing how a trait will respond to selection.

Because of the importance of V_A to evolution, it is important to estimate its proportional contribution to the total phenotypic variance, V_P, or in other words, one would like to estimate the value of the ratio of V_A to V_P. This ratio is the heritability, symbolized h^2. (This is the "narrow sense" heritability; the ratio of V_G to V_P is "broad sense" heritability.) In practice this can be estimated in the laboratory by holding the environment constant with a growth chamber or similar device so that both V_E and hence the $G \times E$ covariance term are nil, and V_P arises from genetic influences alone. Taking advantage of the fact that V_A contributes specifically to resemblance among relatives, one can estimate the heritability from measures of a trait made on relatives, the standard combinations being full-sibs, half-sibs, and parents and their offspring. Measures of full-sibs are complicated by dominance variance and family-rearing effects and so only broad-sense heritability can be estimated; these complicating factors can be eliminated by rearing half-sibs, which share only one, rather than both parents, although the experimental designs can be complicated (see Hegmann and Dingle 1982 for an example). Because the experimental design is more straightforward and usually less labor intensive, offspring-on-parent regression is more frequently used. In this case the slope of the regression of the measures of the trait in the offspring on the midpoints of the measures on the parents estimates the heritability. If the offspring are exactly intermediate—that is, if the trait is totally additive—the slope will equal unity, which thus sets the upper bound for heritability.

One can also take advantage of offspring-parent resemblances to estimate heritability from the response to selection. In this case it is called the "realized heritability." Its estimation depends on the intensity of selection, S, one applies to a population, measured as the difference between the mean of the original population and the mean of selected parents; the response to selection, R, the difference between the mean of the offspring of the selected parents and the original population mean; and the heritability h^2, because the additive genetic variance provides the degree of resemblance between offspring and parents. In a nutshell, then, $R = h^2 S$, and this simple relation is used to compute h^2, knowing R and S. Selection is being used with increasing frequency to estimate the proportions of additive genetic variance in natural populations including migrants.

Although the sample is still very limited, heritabilities have been estimated for some migration-related traits in an array of organisms (Table 14–1). Included here are two laboratory measures that have been used to index migratory performance—tethered flight in insects and migratory restlessness in birds (see Chapter 3). Wing form (macroptery vs brachyptery) and wing length in insects are also included because in the species indicated, they have been shown to be associated with migration. In all the cases listed in the table, heritabilities are fairly high and significantly different from zero, indicating that a large fraction of the genetic variance is the consequence of additive effects. This suggests that there is sufficient additive genetic variance to predict that were natural selection directional on any of the measured traits, response would be relatively rapid. If these species are at all representative, there exists considerable potential for the evolution of new migration syndromes, given appropriate selection, although to an extent this will depend also on genetic correlations discussed below. The observation that a new migration route to and from wintering sites in Britain and Ireland (Terrill and Berthold 1990, and Chapter 12) has evolved in the blackcap indicates that there was indeed additive genetic variance for migratory direction and other associated behaviors in this species (see also Helbig 1991b and above), as the data for migratory restlessness in Table 14–1 would suggest. The heritability estimates in the table should be treated with a certain amount of caution because they are for traits associated with migration and not necessarily on migratory behavior itself. Nevertheless they certainly hint at the evolutionary potential of migration in the various species listed. It will take data from many more species to know how generalizable these heritabilities are, although surveys of heritabilities from many traits and species indicate no shortage of genetic variation (Mousseau and Roff 1987; Roff and Mousseau 1987).

Migratory Syndromes

Estimates of the heritabilities of migration-related traits only get at part of the important genetic variation affecting migration. This is because migration and other important behavioral traits do not occur in isolation, but often show correlations with traits such as life history, morphological, or physiological characters that are expressed in the form of syndromes or "strategies" (Dingle 1986). Frazzeta (1975) calls such coordinated traits complex adaptations and likens them to the parts of an engine, all of which must function together to make the adaptation "work." As we have seen, a particular migratory syndrome may involve wing length, fat deposition, and timing of reproduction as well as movement. The causes of correlations among characters can be either environmental or genetic, and if we wish to determine how natural selection acts on a syndrome like migration, it is necessary once again to distinguish among sources of variation. The various methods discussed above for estimating heritability, for example, offspring-on-parent regression or selection, can also be used to estimate genetic correlations which represent the degree to which traits share genes in common. Methods are summarized in detail in Falconer (1989) and in papers in Boake (1994).

The logic behind estimating (additive) genetic correlations is similar to that behind estimating heritabilities, although here, instead of variances, covariances among traits are involved. For example, instead of examining the regression of say, size in offspring on size of the parents, one could look at the correlation or sharing of genes between flight in the offspring and size in the parents. Similarly, if there are shared genes, then selection for size should also yield a response in flight. The parent-offspring regression method is subject to large sampling errors, reducing precision (Falconer 1989). It thus requires large samples for accurate estimates, but it has the advantage of allowing the construction of a full correlation matrix for all traits measured. Selection gives much greater precision, because if it results in statistically significant differences between lines in traits other than the one

Table 14-1. Heritability Estimates for Migratory and Related Traits

Species	Trait	Heritability	Method	Source
Lygaeus kalmii (milkweed bug)	Flight duration	0.20–0.41	Regression	Caldwell and Hegmann 1969
Spodoptera exempta (armyworm moth)	Flight duration	0.50–0.88	Regression	Gatehouse 1986
Epiphyas postvittana (light brown apple moth)	Flight duration	0.43–0.57	Regression and selection	Gu and Danthanarayana 1992
Sylvia atricapilla (blackcap warbler)	Migratory restlessness	0.58–0.87	Selection	Berthold 1988
Erithacus rubecula (European robin)	Migratory restlessness	0.52	Regression	Biebach 1983
Laodelphax triatellus (small brown planthopper)	Wing form	0.27–0.36	Selection	Mori and Nakasuji 1990
Oncopeltus fasciatus (milkweed bug)	Wing length	0.49–0.87	Regression and selection	Dingle et al. 1988
Dysdercus bimaculatus (cotton stainer bug)	Wing length	0.51	Regression	Derr 1980

selected, then a genetic correlation with the selected trait is demonstrated. Because one is usually selecting on only one trait, however, it is possible to construct only the vector of genetic correlations with the selected trait. Selection does have the major advantage of usually yielding unambiguous results with smaller samples.

In my laboratory we have examined migratory syndromes in a comparative study of populations of the milkweed bug, *Oncopeltus fasciatus* (Palmer and Dingle 1986; Dingle and Evans 1987, Dingle et al. 1988; Palmer and Dingle 1989). Two populations were analyzed in detail. The first was a highly migratory one from Iowa that migrates into the upper Midwest in late spring and early summer and returns south in the autumn to overwintering areas in Texas and around the coast of the Gulf of Mexico. Its migratory pathway parallels that of the eastern North American population of the monarch butterfly, although the migrant milkweed bugs don't, so far as we know, penetrate as far south as Mexico (see Chapter 3). The second population was from the Caribbean island of Puerto Rico and is nonmigratory. The milkweed host plants of the bug on Puerto Rico bloom and set seeds throughout the year, and the insects need only to move among local patches of plants. Tethered flight tests in the laboratory confirm the field observations with considerable long-distance flight manifested by the Iowa bugs and very little by the Puerto Rico sample (Dingle 1978b and Table 12–1).

We first assessed genetic variance in these populations using offspring-on-parent regression in samples reared in a common environment of an LD14:10 photoperiod and a temperature of 27°C. These conditions would be experienced in nature at some time during the year by both populations. Heritabilities estimated from the regressions are given in Table 14–2 and indicate considerable additive genetic variance for some traits but not others. Also included are heritability estimates for some traits in bugs reared in another environment, showing that the expression of additive genetic variance can be environment dependent. Both

Table 14–2. Heritabilities Estimated from Offspring-Parent Regression Prior to Selection in Females from Iowa and Puerto Rico Populations of *Oncopeltus fasciatus* reared at LD 14:10 and 27°C

Trait	Iowa Heritability ±SE[*]	Puerto Rico Heritability ±SE
Wing length	0.87 ± 0.15[†] 0.55 ± 0.22[‡]	0.58 ± 0.15[†]
Head capsule width	0.71 ± 0.17[†]	0.61 ± 0.15[†]
Age at first reproduction	-0.20 ± 0.13 ns 0.25 ± 0.12[‡]	0.35 ± 0.10[†]
Clutch size	0.14 ± 0.08	0.13 ± 0.10 ns
Fecundity first 5 days	0.50 ± 0.17[†]	0.06 ± 0.11 ns

[*]Iowa data from Palmer and Dingle 1986.
ns = not significant
[†]p < 0.001.
[‡]Estimated at LD 16:8 and 23°C from half-sib analysis (Hegmann and Dingle 1982).

similarities and differences between the populations were observed in the various traits considered. There were high levels of additive genetic variation in the two morphological characters, wing length and head capsule width, in both populations; there were also statistically significant genetic correlations between these traits of 0.32 in Iowa bugs and 0.68 in the Puerto Rico sample. A difference between the populations occurred in fecundity, measured as the number of eggs produced in the first 5 days of the reproductive life of a female; that is, from the date the first egg was laid. This measure was chosen because it is early egg production that would be important for a migrant colonizing a new habitat. In the Iowa population there is a high and significant heritability for fecundity, and there was in addition a significant genetic correlation between fecundity and body size (head capsule width). Neither the heritability of body size nor its genetic correlation with size were significantly different from zero in the Puerto Rico bugs, suggesting that life-history syndromes might differ between the migratory and nonmigratory populations. The two populations also differ in the heritability of age at first reproduction in the common environment.

The possibility of important differences in the genetics of life histories between migrant and nonmigrant bugs was investigated further using selection. The trait initially chosen for selection was wing length for three reasons. First, it is easily measured and so makes the operation of selection experiments less labor intensive. Second, the heritability in both populations is high, indicating that there is likely to be a strong response to selection. Third, studies of several species have suggested a positive relation between wing length and migration. Selection was carried out for several generations in both populations using a within-family selection design to slow inbreeding (Falconer 1989). In this design the longest or shortest winged individuals in each family, in the respective selected lines, were chosen as the parents of the next generation rather than simply selecting some proportion of the longest or shortest irrespective of family. In this way each family is represented in the succeeding generation. Even though this design reduces selection intensity, the response to selection on wing length was rapid, as shown in Figure 14–2, which gives the responses to selection for both long and short wings and in control lines in which matings were random over the first nine generations of the experiment; selected Iowa bugs are illustrated in Figure 14–3. Both populations respond strongly with little difference between bugs from the two sources when selected in either direction. In all cases the change in wing length was on the order of 10–15% of the base population, and it was possible to select the smaller, shorter-winged Puerto Rico bugs up to the size of Iowa bugs, and the larger Iowa bugs down to the size of Puerto Rico bugs. (Subsequent experiments indicate the smaller Puerto Rico bugs are at an advantage when food is scarce as it sometimes is in their habitat [Dingle 1992]).

What was more interesting was that selection revealed important differences between the two populations in traits correlated with wing length (Fig. 14–4). In neither population was there a correlated response in age at first reproduction (AFR) to selection on wing length. Thus even though the heritabilities of AFR in

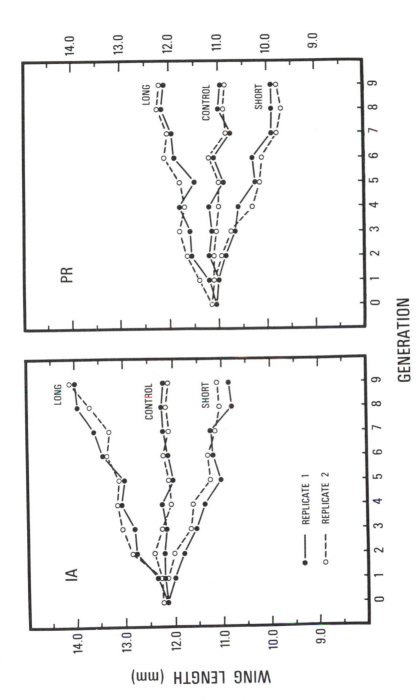

Figure 14–2. Response to selection on wing length in females of a migratory Iowa (IA) and a nonmigratory Puerto Rico (PR) population of the large milkweed bug *Oncopeltus fasciatus*. Each point is a mean wing length for each line in each generation. (From Dingle 1988.)

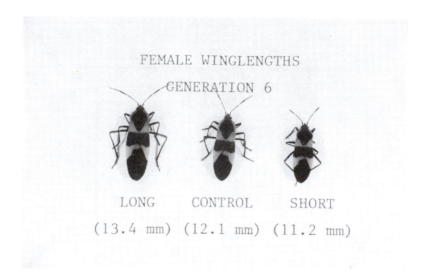

Figure 14–3. Milkweed bugs from each of the Iowa lines after six generations of selection.

the two populations seem to differ (Table 14–2), this is not reflected in correlation between AFR and wing length. The two populations did differ with respect to two other pairs of traits. The first of these pairs is wing length and fecundity. In the migratory Iowa population there is a strong positive response in fecundity to selection on wing length, but in the nonmigratory Puerto Rico bugs, there is no evidence for such a genetic correlation (Fig. 14–4B). (The apparent significantly reduced fecundity in the short-winged bugs of replicate II disappeared when later intervals of oviposition were considered [Dingle et al. 1988].) Second, there is also increased flight in the long-winged Iowa lines, whereas there is no change in the amount of long-duration flight in any of the lines from Puerto Rico (Fig. 14–4C). The association between wing length, flight, and fecundity is what might be expected for a migrant colonizer where a syndrome involving strong flight capability and the ability to produce eggs early in reproductive life would be an advantage when invading newly opened habitats as indeed the Iowa bugs do each spring and early summer. The nonmigratory Puerto Rico bugs do not encounter this periodic opportunity to colonize because habitats are continuously available and reproduction ongoing.

These selection experiments were then extended in the Iowa population to selection on flight itself, with the results indicated in Figure 14–5. Selection on flight for only two generations produced statistically different performances between lines selected for long-duration and short-duration flight, suggesting the availability of additive genetic variance for flight performance and confirming the earlier results of Dingle (1968). In addition this selection experiment demonstrated

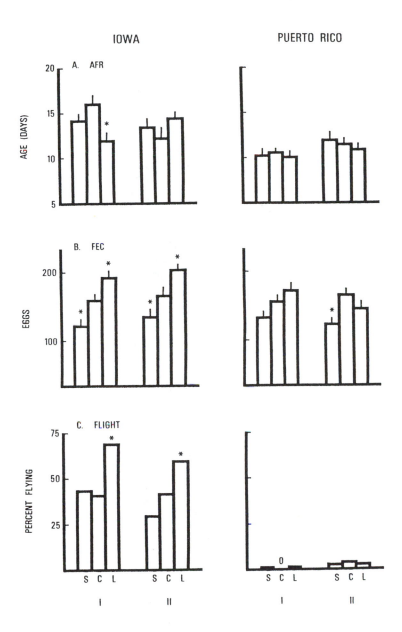

Figure 14–4. Correlated responses to selection on wing length in Iowa and Puerto Rico milkweed bugs. *A.* Age at first reproduction (AFR). *B.* Fecundity (FEC) or the number of eggs laid during the first five days of oviposition. *C.* Percent of bugs flying for more than 30 minutes in a tethered flight test (FLIGHT). Asterisks indicate a statistically significant difference between selected lines and controls. S, C, L indicate the short-winged, control, and long-winged lines for the first (I) and second (II) replicates. Standard error bars are present where appropriate. (From Dingle 1988.)

genetically correlated responses between flight and both wing length and fecundity; once again there was no indication of a correlated response involving AFR. The migratory syndrome involving positive genetic correlation among flight, wing length, and fecundity was again demonstrated in the migratory population. A consequence of the absence of genetic correlations involving age at first reproduction is flexibility of reproductive timing independent of migration. This genetic uncoupling of AFR from other traits means there are no constraints resulting from genetic correlations with respect to the timing of breeding. This would allow response to environmental uncertainty and would be advantageous in either population. What these selection experiments demonstrate is that not only do the two populations differ in terms of gene frequencies, but also in the ways their genomes are organized. There is a genetically coordinated system of flight and life history traits in the migratory Iowa bugs, but not in the nonmigrants from Puerto Rico. Natural selection has evidently acted both on particular traits and on the genetic correlations among those traits (Bradshaw 1986; Dingle 1994).

A syndrome involving genetic correlations among traits determining migratory tendency has also been demonstrated in the sand cricket, *Gryllus firmus*, by Fairbairn and Roff (1990). This species is wing dimorphic with macropterous individuals having long hind wings (the wings used for flight in crickets) and the ability to fly, while micropters have both the hind wings and the dorsal longitudinal flight muscles (DLM) reduced and are incapable of flight. In addition, macropterous females vary in their tendency to histolyze the DLM within the first few days following eclosion to the adult. Fairbairn and Roff selected lines for increased and decreased proportions of macropters, as well as maintaining unselected control lines, tested macropters from all lines for their tethered-flight propensities, and determined the extent of DLM histolysis. Flight propensity here was measured as flight threshold, using a graded series of stimuli to induce flight, starting with simple tarsal release (Chapter 13). An increase in the proportion of macropters was positively correlated, both genetically and phenotypically, with increased flight and the retention of functional DLM. Thus, in lines with a higher proportion of macropters, the macropters also had a lower flight threshold.

Fairbairn and Roff propose that these correlations form the basis of a migratory syndrome coordinated by juvenile hormone, which is known to influence all the

---

Figure 14–5. Correlated responses to selection on flight in migratory Iowa milkweed bugs. Bugs were selected for both longer (FL) or shorter (NF) flights for two generations. *A.* More males (M) and females (F) of the long-flying line made flights of over 30 minutes than bugs of the short flying line. *B.* Bugs of the long-flying line had longer wings. *C.* There was no difference in age at first reproduction between lines. *D.* Fecundity was higher in the long-flying line. Asterisks indicate levels of statistical significance for separate comparisons of males and females (***p < .001, *p < .01) and standard error bars are indicated. (From Dingle 1994 after Palmer and Dingle 1989.)

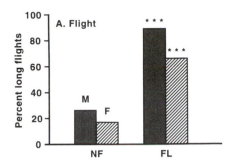

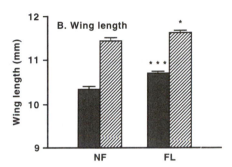

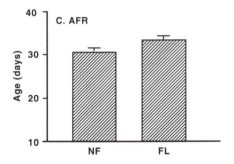

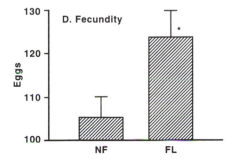

traits involved (Dingle 1985; Rankin 1991 and Chapter 6). Levels and timing of juvenile hormone production influence migration in different taxa of migratory insects (Rankin et al. 1986 and Chapter 6), and it is likely that genes influence migration at least in part through their control of hormone production. Migration in such cases could be a threshold trait expressed when hormone levels were sufficient, and these levels would then mediate the action of genes influencing the behavior. It is worth noting here, however, that direct tests of juvenile hormone influence on migration show that the hormone influences duration of flights rather than flight threshold, at least in the insects examined (milkweed bugs, ladybird beetles, and monarch butterflies—Rankin et al. 1986; Rankin and Burchsted 1992).

Both sibling analysis and artificial selection were used to analyze the relation between flight and life histories in the light brown apple moth (*Epiphyas postvittana*) of Australia by Gu and Danthanarayana (1992). This species appears to be generally sedentary, but both field and laboratory studies indicate that some moths are capable of making flights that would carry individuals long distances (Danthanarayana 1976a,b, 1983). These long flyers are especially likely to occur under harsh environmental conditions that produce smaller, lighter weight moths more likely to make long flights. The moth is a species primarily of the temperate areas of Australia so that most generations probably produce relatively few migrating individuals. Migration, rather, seems to serve as a mechanism of escape and colonization during occasional stressful periods. It is not part of a regular seasonal cycle in the same way that it is in milkweed bugs or monarch butterflies.

Both the sibling analysis and the selection experiments revealed genetic correlations between flight and various life-history traits in *E. postvittana*, and these differed in interesting ways from those found in the milkweed bugs discussed above. In these moths flight capacity was negatively correlated with body weight, the number of eggs laid during the first 5 days of adult life, and with total fecundity. In other words the migratory individuals tended to be smaller and less fecund. These smaller migrating moths also took longer to develop to adulthood, took longer to lay their initial eggs, but lived longer. Differences in the resulting computed life-history parameters between lines selected, respectively, for long- and short-duration flight are shown in Table 14–3. Population growth would clearly be greater where short-flying individuals predominated because r_m in this line was about 30% greater than in the long-flying line. In these moths it appears there is a trade-off between large, fecund, short-flying individuals readily able to exploit

Table 14–3. Population Parameters for Long- and Short-Flying Selected Lines of Light Brown Apple Moth Reared at LD 14:0 and 20°

	Long-Flying	Short-Flying
Net reproductive rate (R_0)	54.4	131.2
Generation time (T, days)	63.9	58.5
Intrinsic rate of increase (r_m)	0.63	0.83

SOURCE: From Gu and Danthanarayana 1992.

favorable habitats and small, less fecund, later developing and reproducing individuals with energy and metabolic resources devoted to flight and escape from unfavorable conditions to colonize elsewhere. In moths, as in the milkweed bugs, selection seems to have produced a migratory syndrome, but one with a rather different set of adaptations.

A particularly interesting case of a genetic correlation involving migration is reported by Han and Gatehouse (1991) for the Chinese population of the oriental armyworm moth, *Mythimna separata*. In this and other noctuid moths the females "call" by releasing a pheromone that results in males flying upwind and locating females for mating. These moths are highly migratory, but migration ceases when females begin to call. The pre-calling period (PCP) is thus an important element in the life history because it puts a limit on the duration of migration.

Han and Gatehouse first established that there was no significant influence of mothers on the PCPs of their daughters. They then established isofemale lines—that is, lines of moths each of which was descended from a single female—that displayed either early calling or late calling behavior with median ages at first calling of four and nine nights, respectively. Crosses between these lines suggested that calling age was influenced by the father far more than the mother. In Lepidoptera the females are the heterogametic sex with female gender the result of an XO genotype. Because calling age is influenced strongly by the male parent, it is likely carried on the X chromosome, which in these heterogametic females is inherited from the father. This would also explain the lack of any significant mother-daughter regression with respect to calling age. Some autosomal genes are apparently also involved, but their influence is much weaker than the very marked X chromosomal effect.

The migration of *M. separata* involves poleward movement in the spring and summer for several generations, largely on northward-moving wind systems, followed by a southward movement in the autumn, to a large extent on the northerly winds then prevailing (Chapters 5 and 11). The generational series of northward movements creates a genetic cline across latitudes in the duration of the PCP with the frequency of later-calling females increasing toward the north. This is because calling puts an end to migratory flights so that the longer the PCP, the longer the period of migration and the farther the migrants will move. The presence of the cline in PCP was confirmed by comparing moths captured in Nanjing, with a median PCP of five nights, with a sample taken to the north in Gongzhuling, with a significantly different median PCP of seven nights. A consequence of the cline is that in the autumn northern moths will have a longer PCP, in turn allowing a longer period for a return movement to the south, necessary if these moths are to survive. Because transporting wind systems are likely to be somewhat more uncertain in the autumn and because longer migratory flights are necessary than in spring to ensure reaching a favorable breeding location, the delayed PCP and correlated longer migration is likely to be advantageous. The system of X chromosomal inheritance will reduce the genetic load of early-calling moths during northward movement because perforce early-calling (and XO) females will not migrate far, if at all. Han and Gatehouse suggest that X chromosomal inheritance of PCP might be expected in other poleward migrating lepidopterans where a return migration at the end of a Temperate Zone breeding season might be important. It is worth noting that in many Lepidoptera, genes influencing development traits such as diapause are located on the X chromosome, possibly as a result of selection for differing developmental periods at different latitudes or in different seasons (Tauber et al. 1986; Danks 1987).

In two other migratory moths studied by Gatehouse and his students, genes influencing the prereproductive period (PRP) are also located on the X chromosome. These are the African armyworm, whose migrations have already been discussed in detail in Chapters 3 and 11, and the silver Y moth, *Autographa gamma* (Wilson and Gatehouse 1992; Hill and Gatehouse 1992). In both species longer PRPs mean longer migrations. Individuals of *A. gamma* migrate as far north as Britain from overwintering areas in North Africa and the Middle East. In the armyworm, selection and sib-analysis indicated a strong genetic component to PRP regulation in female moths, but not in males, consistent with an X chromosomal influence. In the silver Y, similar analyses indicated that the genes with the greatest influence on PRP were on the X chromosome, but there was also indication of gene influences on male PRP. This effect on males implies that some of the PRP genes are also located on autosomes. Results with both species are consistent with the prediction that genes influencing migration where a return migration is important are likely to be located on the X chromosome in lepidopterans.

A particularly interesting case of the influence of genes influencing migratory life cycles comes from the cottonwood aphid, *Pemphigus betae*, studied by Nancy Moran and her colleagues in Utah (Moran et al 1993). These aphids exhibit two sorts of life cycle. In the first they alternate host moving from the primary host, the cottonwood *Populus angustifolia*, to a summer root-feeding parthenogenetic phase on herbaceous secondary hosts, species of *Rumex* and *Chaenopodium*. The second type of life cycle occurs entirely on the secondary hosts without migration to cottonwoods by a winged morph that produces a sexual generation laying overwintering eggs. Different clones of *P. betae* vary in their propensity to produce the migrant forms. Moran and colleagues found that when reared in a common garden, clones that were produced by lineages that had migrated to cottonwood the previous year produced more migrants than clones from lineages remaining on *Rumex* or *Chaenopodium* roots the year before. There were thus genetic correlations among the traits distinguishing among clones.

Migratory syndromes involving genetic correlations are not confined to insects. Differences between migratory and nonmigratory populations in genetic correlation structures have also been found in sticklebacks, *Gasterosterus aculeatus* (Snyder and Dingle 1989). Populations of these fish from the Navarro River on the northern coast of California display two life-history patterns. Estuary populations spend the winter in saltwater and migrate into freshwater to breed, in a pattern similar to that shown by the species in other parts of the world (e.g., Tinbergen 1953). Upstream populations, on the other hand, are confined to restricted areas of the stream and so are largely sedentary. When reared in a common environment, the migratory fish reproduce at a later age (AFR), are larger at reproductive maturity, and have higher fecundities than the upstram populations.

These gene differences suggest adaptation for migration in the estuary overwintering fish. To estimate genetic correlations, full-sib families of both populations were divided and reared in both fresh and salt water (a "split-brood" experimental design) with the results shown in Table 14–4. In the estuary fish the only statistically significant genetic correlation is that for AFR with size at first reproduction in the saltwater treatment. All other genetic correlations were positive, but none were significantly different from zero. In contrast, in the upstream nonmigratory fish three statistically significant positive genetic correlations were estimated: AFR with size at first reproduction and size with clutch size in fresh water, and size with clutch size also in the saltwater treatment. The results thus reveal some differences in genome organization for life histories in migratory versus nonmigratory sticklebacks and also suggest some effects of different rearing environments on that structure. It should also be noted for these experiments, that because of limitations imposed by labor and space considerations, the correlation estimates were obtained only from an analysis of full-sibs. Such an analysis contains unknown biases from dominance and maternal effects due to a common rearing environment (Falconer 1989). Still, the results do imply differences in the genetic organization of life histories in the two stickleback populations.

There is also some preliminary evidence for genetic correlations involving migratory behavior in birds. Berthold and colleagues (1990) crossed a migratory population of the blackcap from southern Germany with a nonmigratory population from the Cape Verde Islands. In contrast to the migrants, the Cape Verde birds showed virtually no migratory restlessness in cages nor any orientation behavior. Of the 35 F_1 hybrids produced, 13 individuals showed migratory restlessness, indicating that such behavior could be transmitted into a nonmigratory population. Five of these birds (of seven tested) were active enough in orientation cages for orientation direction to be assessed. These birds oriented in a direction (NE-SW)

Table 14–4. Genetic Correlations ($\pm$ SE) for Life History Traits in Migratory (Estuary) and Nonmigratory (Upstream) Populations of the Stickleback, *Gasterosteus aculeatus*, When Reared in Fresh and Salt Water

Traits	Freshwater		Saltwater	
	Upstream	Estuary	Upstream	Estuary
AFR[*] × size	0.48[†] ± 0.11	0.31 ± 0.17	0.12 ± 0.14	0.56[†] ± 0.11
AFR × clutch size	0.26 ± 0.15	0.27 ± 0.22	0.25 ± 0.13	0.19 ± 0.17
Size × clutch size	0.82[‡] ± 0.05	0.31 ± 0.23	0.72[‡] ± 0.07	0.43 ± 0.17

[*]Age at first reproduction.
[†]$P < 0.05$ for test of $r_G = 0$.
[‡]$P < 0.001$.
SOURCE: From Snyder 1988.

appropriate to the German parentals. These experiments do not exclude the possibility that restlessness and orientation behaviors are transmitted independently, but some degree of genetic correlation seems more likely. The experiments of Helbig (1991b), discussed above, also suggest genetic correlations between migratory activity and orientation. The results, in any event, suggest that it would be worth looking in birds for migratory syndromes involving genetic correlations.

Genes and Environments

The phenotype produced by a genotype depends on the environment experienced by that genotype, and as indicated in the discussion of quantitative genetic models above, the terms 2covG,E and V_E incorporate this fact into the equation for the variance of the population phenotype, V_P. The differential responses of a genotype can be assessed by rearing that genotype in several different environments to observe the phenotypes produced; the patterns of response are known as the reaction norms. A single genotype can be studied in this way only in clonal organisms, but a reasonable alternative is to consider siblings as genotypes and divide broods across the environments to produce the reaction norms. Populations of a particular provenance or samples from selected lines (Dingle 1992) can also be treated as "genotypes" of individuals sharing at least some genes in common.

Reaction norms yield a great deal of information about potential for response to selection when selective regimes encompass more than one environment, as indeed real selective regimes do (Via and Lande 1985; Groeters and Dingle 1987, 1988; Stearns 1989; van Noordwijk 1990). For example, genetic correlations can be computed for the performance of a trait in two environments and may be positive or negative. If negative, it means that good performance in one environment may be offset by poor performance in another (a negative correlation) so that it may not be possible to evolve optimal performance in both. If genotypes differ with respect to the instances in which they do well or poorly, there is genotype by environment interaction, further complicating evolution toward optimum performance. Nevertheless, the genetic variation may lead to plasticity, resulting in flexibility in traits when environments vary over time or space (Groeters and Dingle 1987; van Noordwijk and Gebhardt 1987). The latter situation is apt to apply in migration.

Two examples of reaction norms involving insect migratory genomes are illustrated in Figure 14-6. The African armyworm moth displays density-dependent phase polyphenism. Crowded populations have dark, active larvae, develop rapidly (2–3 days faster than uncrowded larvae), and display much more long distance migratory flight as adults (Chapter 11). Extensive studies have revealed that long-duration tethered flights are more frequent in offspring of long-flying parental pairs (Gatehouse 1986; Woodrow et al. 1987). Offspring of these long flying pairs were divided between crowded and uncrowded rearing conditions; the crowded or *gregaria* sibs produced 70% long flyers as compared to only 6% for the uncrowded or *solitaria* sibs. The offspring of short flyers showed little tendency to undertake long-duration flights under either rearing environment. In other words there is a genotype by environment interaction expressed with neither "genotype" expressing long-duration flight when uncrowded, but only the "long-flying genotype" expressing long flights under conditions of crowding. Parker and Gatehouse (1985a,b) and Woodrow and colleagues (1987) suggest that the phase polyphenism is part of a migration syndrome that includes genes for long-distance migration that are expressed when populations reach high densities following the onset of the rainy season. Migration allows the moths to escape from crowding and to colonize new habitats made available by the flush of vegetation following the rains. Low population densities in effect signal conditions that are either not particularly suitable for migration or do not demand escape as a result of habitat deterioration. The genotype-by-environment interaction is thus an important element in an overall migration strategy in these moths.

The second example in Figure 14–6 shows the tethered-flight responses of the migratory Iowa milkweed bug population and the nonmigratory Puerto Rico population described above. In this case samples from each of the two populations were treated as different "genotypes" and tested at two temperatures as shown in the figure. As expected, the nonmigratory bugs showed little long-distance flight under either set of conditions. The migratory Iowa population, on the other hand, shows differential expression of migratory genes between the two temperatures. At

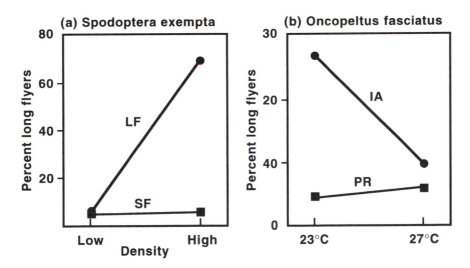

Figure 14–6. Genotype by environment interactions in the African armyworm moth (*Spodoptera exempta*) and the milkweed bug (*Oncopeltus fasciatus*). The long flyer (LF) genotype in the armyworm moth is expressed only at high densities; the short flyer (SF) genotype displays little flight at either density. In the milkweed bug migratory Iowa (IA) bugs fly at 23° but not 27° while the nonmigratory Puerto Rico (PR) bugs do not fly much at either temperature. (From Dingle 1994.)

23°C some 27% of the population expresses long-duration flight, operationally defined here as flights of over 30 minutes in duration; at 27°C only 10% of the population makes these long flights. Under cool conditions in the spring and fall migratory flight would thus take place. In midsummer, however, after the bugs arrive at northern latitudes and milkweeds are flowering and fruiting, migrants would tend to remain relatively sedentary. Again, the differential expression of migratory genotypes is likely to be an important part of the role migration plays in the seasonal cycles of these bugs. So far as I am aware, there are no studies of the potentially important across-environment genetic correlations involving migratory traits and the effects they may have in facilitating or constraining the evolution of migratory syndromes. This would seem to be a ripe area for future research.

Genotype-by-environment interactions are also likely to be important components of partial migration in birds. There is a good deal of evidence for the presence of genetic effects in the control of bird migration (Berthold 1988, 1991a,c and summarized above), yet it is apparent that in some species in which migratory genes are present in most individuals, only a small proportion of the "genetic" migrants actually migrate (Lundberg 1988; Adriaensen and Dhondt 1990). What this probably means, as also suggested by Adriaensen and Dhondt, is that migration is triggered by a threshold mechanism responding to some particular combination of environmental inputs. These latter may be responses to current habitat conditions or to behavioral inputs from other members of the population, as in the dominance models of Fretwell (1972) or Gauthreaux (1978, 1982). In such a conceptual scheme birds with low thresholds would always migrate while those with high never would; for intermediates the environmental conditions would control migration, which is to say, that it is the genotype-by-environment interactions that would matter. The thresholds would be set by natural selection acting on the migration syndrome and its genetic correlations both within and across environments.

Another type of environmental influence on traits results from the nongenetic transmission of environmental variation from parents to offspring. Usually these are transmitted by the mother in the form of maternal effects, but they are not limited to maternal transmission and paternal effects are known, especially in insects (reviewed by Mousseau and Dingle 1991). These nongenetic influences can be physiological—for example, large mothers may give rise to larger young independent of offspring genotype by laying larger eggs or providing more nutrients—or they may be behavioral, involving parental care in some way. A presumed example of parental care is the migratory behavior of young normally sedentary herring gulls when reared by migratory lesser blackbacked parents (Harris 1970 and above). Grandparental effects may also be present if grandparents produce parents that transmit nongenetic traits to the young. An example occurs in the guppy (*Poecilia reticulata*); large males produce genetically large female offspring that produce nongenetically large offspring in turn (Reznick 1981).

A maternal effect directly influencing migration occurs in many species of aphid in which the degree of maternal crowding determines the wing form displayed by the offspring. This effect is prominent in the parthenogenetic phase of aphid life cycles in which females give birth to live young. Typically, crowded mothers produce winged migratory offspring (see also Chapter 2), and uncrowded females give birth to wingless sedentary adults in the next generation (reviewed in Mousseau and Dingle 1991). Aphids are interesting because they undergo very rapid generational turnover with extensive telescoping of the developmental aspects of the life cycle. This telescoping involves the presence of the germaria of future offspring already in the older embryos of the current generation; that is, embryo daughters contain the germaria of granddaughters. This creates a situation highly favorable not only for the evolution of maternal effects, but also for effects lasting more than one generation (Blackman 1975). There are several cases of grandmaternal effects reported for aphid clones. An example concerning migration occurs in the pea aphid, *Acyrthosiphon pisum* (MacKay and Wellington 1977). In this case the age of the grandmother at the time that mothers are born influences whether those mothers will themselves produce alate or apterous daughters. Mothers born when the grandmothers were young produce predominantly wingless offspring, while those born of older grandmothers produce mostly winged migratory granddaughters. It is also generally true in aphids that winged mothers will produce only wingless daughters, so that migration does not occur in two successive generations. Finally, it should be noted that autumnal short days may influence aphid mothers to produce migratory sexual offspring that produce zygotic overwintering eggs, often on an alternative host plant (Blackman 1975; Mousseau and Dingle 1991).

There may also be genetic variance for maternal effects among clones or populations of aphids. Groeters (1989) examined differences in production of winged offspring under crowding in populations of the milkweed-oleander aphid, *Aphis nerii*, sampled from Iowa, California, and Puerto Rico. This aphid has a worldwide distribution in tropical and subtropical areas, including seasonal invasion of Temperate Zone regions such as the North American East and Midwest. Populations might, therefore, be expected to differ in their tendencies to produce winged offspring, depending on habitat and seasonal life cycle. Iowa aphids were expected to have a higher proportion of winged offspring, because populations in the upper Midwest cannot overwinter and must therefore be at least established by migratory individuals each spring. Puerto Rico aphids were expected to show the highest frequency of aptery, because large, continuously flowering and fruiting host milkweed plants (*Callotropis procera*) are available throughout the year, and the aphids should not need to move from these hosts. Colonization of new leaves and shoots should be possible simply by walking. California aphids should be intermediate to these extremes, because winter survival is possible but some movement between host plants, many of which are seasonal milkweeds, is probably also necessary for survival. These predictions were only partially realized. Iowa mothers, when crowded, did produce the most winged offspring, 37.7%, but Puerto Rico and California mothers produced 31.6% and 25.7% winged offspring, respectively, which was the reverse of the order predicted for those two populations. The results do, however, demonstrate the presence of genetic variation between the populations for the maternal effects influencing wing production. Variation was also found among clones within the populations, further indicating gene influences. There seemed to be no syndrome involving migration, morphology, or life history for any population of *A. nerii*. Where host plants can be exploited through migration and colonization, this is apparently accomplished through the production of greater numbers of colonists, not colonists with a distinct syndrome of life-history traits. These aphids thus differ from most other insects described above in the way they incorporate migration into the life cycle.

Summing Up

Studies of migratory behavior have revealed much in the way of genetic variation among the species that have been analyzed. Because these species are so few, it is hard to say at this point whether the information we have gained from them is generalizable to the genetics of migration in other species. Clearly, there is much work to be done. With the caveat that the data come from only a very few species, it is nevertheless evident that a good deal of genetic variation for migratory behavior is present among organisms and that this variation is more than sufficient for the continuing evolution of migration patterns. Genetic influences are revealed by common garden experiments elucidating population differences, by breeding experiments that uncover Mendelian loci, and by quantitative genetic analyses. All these methods indicate that migration is not isolated from other aspects of the life history, but rather consists of coevolved traits organized into complex adaptations or adaptive syndromes constructed for particular migration ecologies.

The most frequent and simplest experiments addressing gene differences are those where one or more populations are reared in the same environment or "common garden." Because the environment is the same, observed differences must result from genes. Populations of North American grasshoppers, Swedish voles, European blackcap warblers, and Pacific salmon are but four examples where intraspecific genetic variation in migration has been demonstrated. In the case of salmon not only behavior but also concurrent physiological responses to fresh or salt water differ, revealing the presence of migratory and nonmigratory syndromes. In the blackcap, changes in the winter environment in the British Isles have apparently resulted in the action of selection on the genetic variation to produce a new migration route within the last few decades. The blackcap migratory syndrome also involves both the timing of migratory behavior and its direction of orientation. Crosses between individuals of several species with migratory and nonmigratory genotypes suggest that much of the observed variation is contributed by additive genes, those that specifically contribute to resemblances among relatives. In a few cases, differences between migrants and nonmigrants may be the result of identifiable Mendelian loci. Such is the case with the lygaeid bug *Horvathiolus gibbicollis*, in which wing morph behaves as a single-locus effect.

Here too, however, there is a migratory syndrome, including life history as well as behavioral traits, that results from the pleiotropic action of the alleles.

Quantitative genetic methods, especially, have revealed the variety of migration syndromes. In milkweed bugs, moths, crickets, and stickleback fish migratory and nonmigratory populations differ in the genetic organization underlying life histories. Migratory milkweed bugs, for example, display positive genetic correlations of flight with wing length and fecundity; nonmigratory milkweed bugs do not. Migratory moths delay reproduction; in these female heterogametic insects, the genes influencing the migratory syndrome are associated with the X chromosome. In insects in general, gene action may be mediated by juvenile hormone. In an exception to the rule of migratory life cycles based on genetic correlations among traits, clonal milkweed-oleander aphids vary the proportion of migrants via maternal effects rather than incorporating a syndrome of correlated traits. The presence of syndromes in so diverse a set of migrants, however, implies that selection acts not only on individual migratory characters but also on the presence or absence of genetic correlations among traits and on the strengths and directions of those correlations (Bradshaw 1986; Dingle 1994). Because of this the course of evolution in migrants will be affected in complex ways. And needless to say, adaptive syndromes are not unique to migration-related traits.

Finally, genes and interactions among genes are expressed as a function of environments; the answers to important evolutionary questions concerning migratory adaptations will involve genes, environments, and the shape of reaction norms. The expressions of migratory behavior in some environments but not others in armyworm moths, aphids, and milkweed bugs are but three examples. To truly understand the evolution and function of any trait, we must understand all three components. We are beginning to learn much from the application of genetic methods to the development and evolution of migratory behavior and migratory syndromes, and these methods are also magnifying contributions from physiology and ecology. In spite of the often formidable barriers to studying the genetics of natural populations, the insights gained should be reason enough to persevere in sorting out the contributions of genes to migration.

PART FOUR

Applications and Implications

Migration is not only a fascinating subject in its own right, it also has some important implications for the quality of life of human populations. Foremost among these implications is the significance of migratory species to pest management and conservation biology. In Chapter 15 I briefly examine the importance of migrant crop pests, concentrating on insects. After first considering the economic impact of some of the world's major migrant pests, I consider those aspects of their biology that result in their evolving to pest status. I then move to a consideration of management solutions, stressing the necessity for combining an understanding of migration ecology with wide-scale efforts to keep track of the status of pest populations in the field. I conclude with an overview of three attempts at large-scale monitoring and management of migratory pests. These have met with varying success in their applications to pest management problems.

In Chapter 16 I discuss the relation between migration and conservation, focusing especially on three cases of particular interest in North American conservation biology. These concern the problems attending conservation efforts with respect to the monarch butterfly, Neotropical migrant land birds, and the chinook salmon of the Sacramento River in California. Each represents a different problem in conservation biology. The monarch butterfly is a species that is not itself in danger, but its migrant populations represent a unique lifestyle that is endangered, an "endangered phenomenon." A major difficulty with respect to Neotropical migrant land birds is assessing whether there really is a significant decline in populations. Accurate assessments of population trends are very difficult to make, and both evidence and interpretations of data conflict. The task of assessing needs and developing sound conservation programs is thus a formidable one. Finally, I discuss what is perhaps the most discouraging example of an endangered species. The chinook salmon is a valuable commercial resource, yet conflicting interests and poor water management have driven many populations to extinction or severe declines. In all three cases, a need to understand the basics of migration biology and their interactions with social needs and political realities is necessary if conservation efforts are to have any hope of success.

15

Migration and Pest Management

[I]t is becoming all too clear that the migratory activity of most crop pests has been grossly underestimated, and that insufficient attention has been directed at the control problems which this activity generates.

JOYCE 1981

The Problem

An extraordinary number of the world's major agricultural pests are long-distance migrants, and to successfully solve the problems they generate through economic losses, it is necessary to understand the role migration plays in their biology. One of the threads that is woven through much of the discussion in this book is the relation between migration and temporary habitats. This relationship goes a long way toward explaining the often overlooked migratory characteristics of most agricultural pests. Southwood (1962, 1971) has noted that not only are insect migrants likely to be denizens of temporary habitats, but also that most of our major crops are descended from ruderal ancestors evolved to colonize bare ground or other early stages of botanical succession. These epitomize ephemeral conditions, and so these ancestral crops were already likely to have associated with them a group of species well adapted to an itinerant lifestyle with high migratory activity. The provision of large, frequently monocultural stands of host plants in the development of agriculture has resulted in an explosion of the populations of those itinerant species that we now consider pests. These include insects, mites, molds and rusts, nematodes, and even vertebrates such as the house mouse (*Mus musculus*) and the quelea (Chapter 3). Although all of these cause severe crop losses, the greatest economic damage as a consequence of migratory activity is caused by insects, so I shall focus my discussion on them as the prime example of the problems of migration and pest management.

An indication of the impact a major crop pest can have is seen in species like the desert locust, *Schistocerca gregaria*, or the African armyworm moth, *Spodoptera exempta*, two insects that cause enormous and widespread damage. The desert locust is distributed throughout much of Africa, the Middle East, and the

Indian subcontinent; the African armyworm is mostly a pest in the drier areas of sub-Saharan Africa (Chapter 11), although its range extends eastward across much of Asia and even across the Pacific to Hawaii (Betts 1976). The scale of invasion of these two species is frequently described from biblical verse to the dry statements of pest management reports. Two such descriptions convey the magnitude of pest populations (Betts 1976). Radar observations near Delhi, India, in July, 1962 recorded densities of about one locust per 10 cubic meters up to an altitude of some 1500 meters over an area of approximately 1400 square kilometers. When one considers that a single locust eats up to its own weight of crop in a single day, it is no wonder that a swarm of locusts leaves "not any green thing," as noted so succinctly in the Book of Exodus. One swarm in the Horn of Africa was estimated to have been able to consume 80,000 tons of fresh vegetation a day or enough to feed 400,000 people for a year (Roffey 1985). Similarly, African armyworm caterpillars are frequently reported at densities on the order of 100 larvae per m^2, and there are occasional records of concentrations as high as 3,000 per m^2 covering areas up to several tens of square kilometers. Left unchecked, the damage caused by infestations of armyworms or swarms of desert locusts can reach staggering economic proportions in a very short time, with tragic human consequences.

Given the number of insects that can be flying and the potential transport and concentrating effects of weather systems (Chapter 5), invasions into crops can take place remarkably rapidly. This was demonstrated for the black cutworm moth (*Agrotis ipsilon*) in a mark-release and recapture study undertaken by Showers and colleagues (1989) in the central United States. They marked and released several thousand male moths in southern Texas and Louisiana. To the north of the release point were eight traplines 474 or 618 km long with 174 and 246 sticky traps, respectively, baited with the sex pheromone produced by cutworm females. Releases of moths in the south were performed when warm southerly winds were forecast to produce northbound airflows. Six males were recaptured, one in Kansas, one in Missouri, and four in Iowa, having traveled from 921 to 1266 km from their respective release points in 4 days or less. Examination of wind data from the National Weather Service indicated flights took place below 1300 m altitude. An important conclusion deriving from this and numerous other studies of migrant pests is that densities capable of severe economic impact can develop quickly, even though the pests must move long distances to reach the susceptible crop (Joyce 1976, 1981). The fact of pest migration forces pest managers to be aware of insect distribution and abundance over a very wide area, if control programs are to be successful.

Not all migratory insects are pests, not even all those most directly associated with ephemeral or temporary habitats. Partly this is fortuitous, owing to the fact that many migrants specialize on hosts that are of no agricultural or horticultural importance. The monarch butterfly and the large milkweed bug are confined to milkweed and so are appreciated rather than reviled for their migratory capabilities. The various cotton stainers (*Dysdercus*) feed on assorted malvaceous plants; it was only when one of these, cotton, became an important commercial crop that they became pests. Because of their specialization on the Malvaceae (and occasionally other Malvales), however, they are not the important and widespread pests that the locusts, armyworms, and other generalists are.

This last point gives us some insight into why certain insects become so widespread and successful as agricultural pests. Fitt (1989) lists four reasons why noctuid moths of the genus *Heliothis* are so successful at invading, colonizing, and exploiting agricultural ecosystems, in some species on a worldwide basis. Although Fitt confined his discussion to *Heliothis*, his reasons seem applicable to other migrant pests as well. First among them is polyphagy. On a global scale the various *Heliothis* have been recorded feeding on hundreds of plant species in dozens of families. The wide-ranging *H. armigera* is listed from a minimum of 60 cultivated and about 67 wild hosts, and it and *H. punctigera* together are recorded from 161 plant species in 49 families in Australia (Zalucki et al. 1986). Other pests may not be quite so catholic in their feeding habits, but polyphagy is a common lifestyle.

With respect to pest success, polyphagy confers three big advantages. First, populations may develop on several hosts within a region and sum to large numbers as a result. Second, populations may develop continuously over a long period by successively exploiting several hosts, both cultivated and wild, of differing phenologies. Third, even in times of stress populations can persist at low densities, because migrating females have a high probability of locating at least some suitable host plants that can sustain larval development. Even with variation in host plant suitability, polyphagy affords numerous opportunities for the survival, growth, and spread of pest populations.

The second major reason for pest success is migration itself, often combined with much mobility on a more local scale as well. Migration allows spatial redistribution, the location of distant habitats, and escape from stressful conditions such as cold, drought, or the senescence of host plants. As we have also seen for numerous migrants through this book, *Heliothis* spp. show specialized behavior that distinguishes migration from other forms of movement. Most especially this involves a characteristic vertical takeoff flight that carries the moths above the boundary layer and into prevailing wind streams. These latter can take the moths for long distances in a short time, as we saw above for moths of the black cutworm. For *Heliothis* engaged in more local extended foraging movements, flight lasts only 1–2 hours with frequent interruptions to imbibe nectar and oviposit. In desert locusts migration and extended foraging can take individuals equally far. Recall that in these insects it is the solitary-phase individuals that are the usual migrants, moving in the wind systems of the upper air at night, while the rolling gregarious swarms are the behavioral and ecological equivalents of the locally moving moths. If one of these swarms is picked up by a wind field, it can travel almost, if not as far, as the night-flying migrants. Both migration and extended foraging are components of mobility that contribute to the success of migrant pests.

The third contributor to this success is the occurrence of diapause or some other form of reduced activity (often called quiescence in the entomological literature). This is especially prevalent in principally Temperate Zone insects such as *Heliothis* spp., although some tropical pests also display a diapause or quiescence to some degree, as in some locusts (Uvarov 1966). For *Heliothis* the diapause is facultative and photoperiodically induced, as it is in most diapausing insects (Beck 1980). Perhaps more interesting, and more important from a pest perspective, is the fact that several species, such as the armyworms, *S. exempta* and *Mythimna separata* and the silver Y moth, *Autographa gamma*, show delayed reproduction if they are migratory (Han and Gatehouse 1991; Hill and Gatehouse 1992; Wilson and Gatehouse, 1992). This delay (an oogenesis-flight syndrome, Chapter 6) has the consequence of extending the migratory period and hence also of migration distances. Two of these species, *M. separata* and *A. gamma*, can exploit regions well to the north of their wintering areas by invading in the spring and escaping in the autumn. It is likely that a similar mechanism is at work in the fall armyworm, *S. frugiperda*, a summer invader of crops in the southeastern United States. Some tropical species of *Heliothis*, as well as temperate species subject to very hot summers, can also diapause at high temperatures (Fitt 1989) or during periods of drought.

Fourth, and finally, migrant pests display high fecundities. As many as 3000 eggs have been oviposited by single females of *H. armigera* and *H. zea*. Even the more usual figure of 1500 eggs is impressive, however, especially when it is realized that these eggs are laid during a reproductive lifetime that begins only a few days after adult eclosion and lasts a brief 8–10 days. This sort of life history has the potential to generate especially rapid population growth. High fecundities are not unique to migratory pests, but combined with other features of the biology of these insects, they contribute to the propensity to reach high densities capable of severe economic damage. The various species of *Heliothis* are capable of producing tens of thousands of moths per hectare (Stinner et al. 1977; Fitt and Daly 1990), and we saw at the beginning of this chapter what sorts of densities are reached with locusts and armyworms. Dealing with life history syndromes that involve high mobility and are capable of producing these sorts of numbers present formidable challenges to pest managers trying to stay ahead of the game.

Attempts at Management Solutions

First and foremost in developing successful pest management strategies is the need to understand thoroughly the biology of the insects concerned. This point seems obvious, but the checkered history of chemical pesticides tells us all too often of the problems that arise when a few basic facts of biology are ignored, with the most obvious problem perhaps being that of the frequent evolution of pesticide resistance. Kennedy and Way (1979), in their summary of a symposium on migrant pests in the southeastern United States, pointed out that greater effort has been expended on monitoring tools and the coordination of monitoring efforts than has gone into studying the migratory behavior of the pests being monitored. This point has been reiterated by McNeil and Roitberg (1985) and had been stressed in earlier papers by Wellington (e.g., 1977) and others. Wellington provocatively titled his 1977 paper "Returning the insect to insect ecology: some consequences for pest management" and showed how knowing details of the biology of the western tent caterpillar, *Malacosoma californicum*, a pest of forest trees, could profoundly influence control methods. The eggs laid by young females of the moth develop into offspring that are more vigorous and far more migratory than offspring of old females. The silken tents of these more active larvae are recognizable by their shape and can be the focus of control in which only these specific larvae are targeted. Such a focus is far more effective and at far less cost than the blanket attempts at control usually applied.

Kennedy and Way (1979) illustrated their point about studying behavior by showing how radar, rather than simply functioning as a monitoring tool, had provided a series of rather profound insights into the behavior of long-distance migrants. First, it has demonstrated that takeoff on migratory flights is a routine nightly activity in many insect pests from moths to locusts. Second, warm air at the surface often encourages emigration flights leading to insect movement in the "right" direction for successful colonization; north, for example, in southeastern North America or southeast in Australia. Third, the height reached by night migrants is controlled by the insects themselves. Fourth, by choosing a specific height, migrants are also effectively choosing the winds they will ride and the direction in which they will travel. Fifth, a remarkable discovery from radar tracking is that migrants are not simply scattering in all directions, but rather they show a common orientation even at night. In other words they are *not* "dispersing"; they are instead concentrating by heading in a specified direction. This has enormous consequences for pest management and forecasting because it indicates that large numbers of insects are likely to arrive in a crop system simultaneously, not scattered over intervals of time or space. Sixth, the migrants can control the duration of their flights and hence to some extent the distance of their flights. As part of their behavioral syndrome, migrants control their descent from the atmosphere as well as their ascent into it. These six behavioral observations give a very different picture of insect migrants from the one generated by assumptions of random takeoff and "passive" transport by the wind.

In addition to the behavior in flight, it is also important to understand the ecology of the migrants on the ground. This is nicely illustrated by studies of the Australian plague locust, *Chortoicetes terminifera* (Hunter 1986, 1989). Plagues of these locusts originate in the arid interior and move outward in the spring and summer to the agricultural areas of southeastern Australia. Research on the ecology of these insects indicates they prefer open plains of heavy clay soils along creek lines, usually bearing a cover of stones. Knowing this allows concentration of the search for developing locust populations to about 20% of inland Australia in a region lying mostly in the Channel Country of southwestern Queensland and in the northern parts of South Australia. Important to the survival and growth of the locusts in this region are the presence of green grasses, especially the perennial curly and barley mitchell grasses and the annual button grass. These grasses are in turn tied to rainfall. A single fall of 20 mm or more of rain along with a monthly mean maximum temperature above 23 °C allows sustained growth of barley mitchell grass; curly mitchell grass requires about a 40 millimeter rainfall. Button grass also grows under these conditions, but goes dry in about 6 weeks if there is no further rain. The mitchell grasses, however, last about 2 months, enough time for locusts to hatch, grow, and reach adulthood. Furthermore, adult locusts can accumulate the fat needed for migration on dry, but still green, mitchell grasses. In mitchell grass areas, therefore, locusts can complete development after a single substantial rainfall. Knowing these facts of locust ecology allows prediction of where plagues are likely to develop. Armed with this information the Australian Plague Locust Commission has successfully controlled locust outbreaks by spraying on average only about 1000 km^2 per year, a minuscule portion of the total inland range of the locusts.

Basic to applying ecological understanding to pest management is thus knowing where and when pest outbreaks of significant economic impact are likely to occur. The pursuit of this objective has led to the development of the so-called Synoptic Survey, involving extensive sampling, reporting, and mapping of the species in question (Joyce 1981). The information gathered and collated is then compared to that obtained over previous periods to determine if an outbreak of economic impact is likely to take place, and if so, to forecast where and when it is likely to develop. Key to the success of such forecasting based on a Synoptic Survey is a dynamic map of the changing distribution and abundance of the pest. Mapping as an ecological tool was first formally developed by Winifred Benchley in 1914 (Taylor 1986a); its use for following the population dynamics of a number of pest species was extensively advanced and refined by L. R. Taylor and the Rothamsted Insect Survey, which sampled insects throughout the British Isles and was later broadened to include much of Western Europe (Taylor 1979, 1986a). An example of the changing distribution of the aphid *Aphis fabae* in Europe in 1978, based on data collected by EURAPHID, an outgrowth of the Rothamsted Survey, is illustrated in Figure 15–1. The role of migration in the multiple invasions of the British Isles by this aphid can be clearly seen through the use of this dynamic mapping procedure.

In order to make correct forecasts of outbreaks of migrant pests, it is necessary to combine ecology with an understanding of regional meteorology. There are numerous examples where wind patterns or other meteorological conditions define the movements of major pest species. The passages of the Inter-Tropical Convergence Zone (ITCZ) in Africa determine the breeding success, movements, and subsequent outbreaks of the African armyworm and the desert locust, and the southerly winds prevalent in the spring and summer in the southeastern United States serve as the transporting vehicles for the fall armyworm and black cutworm moths, among others. The influence of local storms can also be important. In the Gezira region of the Sudan, just south and east of Khartoum, cold outflows of storms that typically occur just after sunset can double the density of airborne insects in as little as 2–3 minutes (Joyce 1981). Other examples of the relation between meteorology and migration are given in Chapter 5.

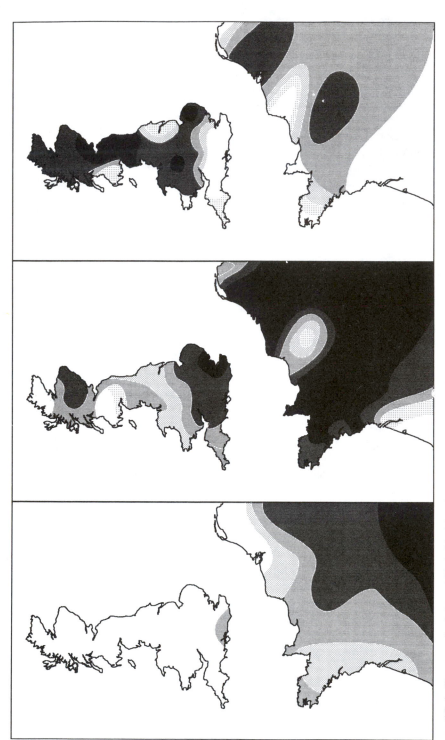

Figure 15–1. The spread of the aphid *Aphis fabae* in western Europe and Britain in spring and early summer as determined from data from a network of EURAPHID sampling stations. Aphid densities are indicated by degree of shading. High densities first appear in central France (*left panel*) and move northward as the season progresses. (Adapted from Taylor 1986a.)

There have been numerous attempts to improve understanding of pest movements and enhance forecasting through the use of models. Many of these are conceptual, and whereas they are often useful bookkeeping devices to order complex ecological events, their predictive abilities are apt to be limited. One problem is that so far they have largely failed to incorporate the differences in behavior between insects that are migrating and those engaged in ranging or extended foraging. Some recent conferences, however, suggest that the situation is changing (e.g., Isard 1993) and more useful conceptual models should be developed in the future.

In addition to conceptual models, there have also been attempts to generate more formal mathematical statements regarding insect movements. One approach is to assume a set of "rules" for movement and see if insect examples actually conform to those rules. Diffusion, random walk, and Markov chain models are of this sort and have been used effectively to describe within-patch foraging movements of insects, most notably by Peter Kareiva (1982, 1983a,b; Kareiva and Shigesada 1983). Gabriel (1977) attempted to apply these sorts of models to milkweed bug migrations, but the incorporation of realistic wind flows made the models extremely complex. A somewhat different approach has been used by Ron Stinner (Stinner and Saks 1985) in modeling the movements of *Heliothis zea* and the Mexican bean beetle, *Epilachna varivestis*, in North Carolina. Here the models were developed and the rules determined from observations of the insects themselves. The basic idea was that fields of crops exerted a certain attraction to moths or beetles and that the strength of the attraction was a function of the distance from the insect to the field. The *H. zea* moths moved much more freely than the beetles, and so the rules governing the population dynamics of the two species were quite different. The moths were influenced by the total area of fields planted to a crop, but beetle densities were primarily a function of field perimeters. The practical impact of this difference is that changes in field size (but not total area) would not influence *H. zea* populations, but would likely have an effect on the dynamics of the bean beetle. Even though differences between migrating and foraging behavior are not explicitly taken into account (although physiological "willingness" of the beetles to fly was incorporated), this modeling effort clearly shows that useful predictions can come from models in which the biology of the insects dictates the formulation. We can expect that incorporation of what we know to be true of the behavior of migrants, so far all too often ignored, will lead to further significant improvements in models pertaining to the population dynamics of migrant pests.

Coordination of Efforts

The longest history of attempts to monitor and forecast the activity of a major migratory pest is associated with the desert locust (Betts 1976; Roffey 1985). Systematic and centralized collecting and mapping of data on the dynamics of locust distributions were begun in 1929 by Boris Uvarov and Z. Waloff. By 1943, partly with impetus resulting from wartime operations in the Middle East, forecasts were issued monthly, and the organization that did so became the Anti-Locust Research Centre after the war, headquartered in London. Later, as the Center for Overseas Pest Research, it expanded its purview to include other pests. From 1958–1973 locust monitoring was supported through the various countries supporting the United Nations Food and Agriculture Organization. The FAO continues to coordinate the forecasting program through a group of regional organizations in Africa and the Middle East, and other locusts and grasshoppers are now included.

Locust forecasts are generated by assessing the distribution of the insects and estimating the likelihood of breeding and movement of various locations. The assessments are based on reports of swarms, bands of hoppers, and scattered locusts. These reports are submitted from a number of sources including staff of local and regional locust control offices and plant protection departments, local administrators, ships, and a number of other unofficial contributors. When locust populations are high, these can sum to thousands of sightings a month. The information is then mapped, and the maps continuously updated. Weather developments are determined by synoptic charts on which meteorological observations are plotted. From all this information the scale, distribution, and stage of development of locusts can be established, and the course of population growth and trajectories of probable movements forecast. Needless to say, all of this represents an enormous undertaking, but the procedures have produced quite remarkable success, especially when considering the patchiness and uncertainties of reporting. An example of the success over the years 1960–1965 is given in Table 15–1. The forecasts predicted whether there was a high, intermediate, or low likelihood of locusts appearing in an area. The proportion of times locusts in fact appeared matched the forecasts reasonably well, as can be seen by examining the totals. Out of 514 forecasts that locusts were very likely to appear, they in fact did so 85% of the time; when the forecasts suggested little likelihood of appearance, they did so on only 40% of the occasions. These forecasts and verifications are largely qualitative, but they are a major factor in mitigating the impact of the locust swarms because they have focused and made more effective efforts at control.

In Europe migrant pest forecasting has been developed via the Rothamsted Insect Survey begun in 1960 by L. R. Taylor (1979, 1986a). This was started with a few light traps at Rothamsted and by 1976 had grown to 126 sites scattered over the island of Britain and included both suction and light traps. By mapping the distribution and abundance of moths and aphids over time, as indicated by the collections in the traps, the fluctuations of populations in both time and space over the year can be plotted. An aphid warning system was started in 1963 in Great Britain and in 1977 a system known as Actaphid was initiated in France (Robert and Choppin de Janvry 1977). The British and French systems were expanded into EURAPHID in 1980, a system designed to sample some 350 or so aphid species with a network of about 40 suction traps in Britain, France, the Low Countries and Scandinavia. These are supplemented by radar in southeast England. We have already seen an example of the sort of map this system generates in Figure 15–1.

Table 15–1. Success Rate of Desert Locust Forecasts in Africa During the Years 1960–1965. Each prediction was a forecast that locusts would be present or breeding in a country

Overall Level of Locust Activity	Predictions of Probability		
	High	Intermediate	Low
Swarms widespread (August 1960-March 1961)	85% (110)[*]	45% (73)	12% (26)
Heavy infestations in east only (April 1961–July 1963)	88% (300)	62% (167)	34% (116)
Recession (August 1963–October 1965)	74% (104)	48% (142)	42% (242)
Total (August 1960–October 1965)	85% (514)	54% (382)	40% (384)

[*] Percent of times locust swarms actually appeared at a given predicted probability level out of the number of forecasts (parentheses); for example, out of a total of 514 forecasts that locusts would appear with high probability, they did occur 85% of the time.
SOURCE: From Betts 1976.

In North America the Pests and Weather Project of the University of Illinois and the Illinois Department of Energy and Natural Resources was developed to forecast the location, timing, and densities of pests migrating into Illinois crops (Achtemeier et al. 1987). The aim was to determine the source of the migrants, the spatial and temporal distribution of the insects in the atmosphere, the factors causing termination of flight, and finally, the population statistics of arrival. The primary target species was the corn leaf aphid, *Rhopalosiphum maidis*. The Project introduced a number of novel methods into the monitoring system, including traps placed on helicopters to sample aphids in the air and various forms of genetic "fingerprinting" to try to identify the sources of the aphids. Radar and ground trapping have also been used. So far there has been some success at tracing Illinois arrivals to probable sources in Texas, Oklahoma, and Arkansas. An outgrowth of the Pest and Weather Project has been an Alliance for Aerobiology Research (AFAR), an attempt to broaden the scope of research and forecasting to include much of eastern North America and to extend it to a greater variety of both insect and noninsect pests (Isard 1993). In spite of these efforts, however, and some earlier conferences on the subject of pest migration (Rabb and Kennedy 1979; MacKenzie et al. 1985), there has been little success in generating wide-area monitoring and forecasting in North America. At this writing in the fall of 1994, the success of AFAR remains to be determined.

Summing Up

The behavioral ecology of a high proportion of the world's major agricultural pests makes it imperative for those who would manage those pests to understand the role migration plays in their life histories. Movement is an integral part of the population dynamics of these species, making the spatial component of population change coequal in importance with the temporal element. The need to consider both spatial and temporal dimensions creates enormous problems for pest managers. Such considerations mean that a wide-area approach is needed, because in order to predict what is likely to happen "here," and so to institute effective control measures, one must know what is happening with respect to pest populations over a very wide region of "elsewhere" (Chapter 2). Although there are undoubted successes in wide-area forecasting, as we have seen with the desert locust (Table 15-1), it is obvious that much needs to be done. The acquisition of biological understanding may be the easy part, because the necessary regional and even global cooperation to achieve forecasting goals will require both international cooperation and money.

Not only must broad geographic areas be considered to gain biological understanding of pest dynamics, but so too must other aspects of ecologies. Most major pest species are polyphagous. This means that to locate and monitor populations a variety of crops and natural vegetation must be considered. This presents both problems and opportunities. Whereas it is true that a large number of host plants must be surveyed, knowing the synchrony between insect and plant phenologies can narrow the target considerably. If pest populations can be located on the hosts on which they survive times of stress, they can be the object of much more focused control measures. This is not easy when the area to be surveyed is large, but the relatively successful management of the Australian plague locust in the interior of Australia with concentrated control efforts (Hunter 1986, 1989) indicates what can be accomplished.

The unintended environmental consequences of pesticide use and the repeated evolution of resistance in target species has engendered a reassessment of pest management strategies. One of the consequences of that reassessment has been the focus on methods that selectively discriminate against particular classes of individuals in the population rather than simply causing indiscriminate mortality (Wellington 1977). One of the aims is to avoid unnecessary pesticide use by using chemical agents only where and when necessary in the face otherwise of losses that would exceed an economic threshold. This approach is one element of what has become known as integrated pest management or IPM. To be successful an IPM strategy must take migration into account. Success in control is likely to be far greater and at far less cost if migration can be accurately forecast and control measures initiated before takeoff so that populations cannot migrate and concentrate to reach economically damaging or even plague levels. As wise entomologists have repeatedly stressed, successful strategies must be based on a thorough understanding of the behavior and ecology of pest species, and for most this includes migration. There is no better reason why basic and applied research on migration systems must go hand in hand.

16

Migration and Conservation

The exponential rise in human population and the impact on the environment that growth has generated have resulted in profound consequences for the myriad species with which we share our planet. Many have gone extinct, many others have become scarce, while some, such as the migrant pests described in the last chapter, have increased in the newly created or expanded habitats. Alarmed by the loss of habitat and the extinction and reduced numbers of many species, we have responded with at first crude and now increasingly sophisticated attempts to prevent the extinctions, reverse apparent declines, and preserve a diversity of habitats. From its beginnings under the inspiration of Gifford Pinchot and others at the beginning of the twentieth century, the conservation movement has evolved from a series of relatively simple set-aside programs to the development of a science that attempts to guide the preservation of habitats and species through an understanding of their ecology. After a long gestation was born the science of conservation biology, which is now usually defined as the study of biological scarcity and diversity (Soulé 1986; Wilson 1988).

Migration is the key to the biology of several species that are becoming increasingly scarce or face the risk of extinction. In migratory organisms all the political, economic, and biological problems inherent in trying to generate sound conservation strategies are brought into sharp focus. Migrants may breed in one country, transit several others, and spend the nonbreeding season (or breed again) in yet another. Stopover sites may be as important as start and end points, and the resources necessary to support migrants anywhere along the line may be in conflict with economic or social aims. Thus, for example, migratory geese of several at-risk species feed on grasses and can cause considerable crop losses either en route or on the wintering grounds if the grasses they feed on happen to be sprouting cereal grains (Greenwood 1993).

Obviously migrants will be affected if breeding or crucial nonbreeding habitats are altered for human use. Rainforests and wetlands are currently receiving much attention as overwintering sites for migrant land birds and breeding sites for migratory waterfowl, respectively. The declining numbers of many North American dabbling ducks can almost certainly be traced to the draining of marshes for agriculture in the northern Midwest and Great Plains states and provinces

(Terborgh 1989). To cite one example, the state of Iowa has lost 99% of its wetlands, and several other states are not much better off (Reffalt 1985). Tropical dry forests have disappeared to an even greater extent than rainforests. Their loss can have an even greater impact than wet forest loss because they occur over limited areas yet support high concentrations of overwintering avian migrants (Terborgh 1989). Finally, the occurrence of many migrants over a wide area and in many habitats can make attempts to assess their status difficult at best. Not only is it difficult to know whether or not migrants are somewhere else if they are absent from a census site, but it is often in the political or economic interest of governments or industries to obscure population declines, if in fact they exist.

One of the major questions in conservation biology concerns the best design for reserves. The issues are complex and outside our scope here, but one directly relates to migratory behavior. This concerns whether the best conservation design is to establish a few large reserves or many small ones, perhaps connected by corridors of similar habitat. Most of the discussion revolves around whether species (mostly birds) prefer edge or interior habitats. For many species, however, especially insects and mammals, the most important issue may be movement behavior. If migratory behavior or ranging behavior in search of new breeding sites truly involves dispersal with scattering in several directions, then small reserves may be suitable or even preferable. If, on the other hand, movements involve concentration of those taking part, then large reserves may be better and/or it may be necessary to provide corridors. The willingness of migrants to cross areas of unsuitable habitat and their degree of site fidelity in breeding areas (especially in the case of migratory birds) are other factors that will need to be considered. Several authors are now advocating a metapopulation approach to population dynamics (e.g., Ricklefs 1992; Villard 1992). This brief and adumbrated account can merely highlight some of the biological uncertainties.

In this chapter I shall consider three case histories that illustrate problems concerning the conservation of migratory species. The monarch butterfly is a good example of a situation in which the species is not itself in danger of extinction, but that portion engaged in extensive migratory movements is. Lincoln Brower and his associates argue that conservationists should consider not only the decline of species, but also what they term "endangered phenomena" or spectacular and unique aspects of life histories such as the migration and concentrated overwintering aggregations of the monarchs (Brower and Malcolm 1991). The second case involves the many migratory North American songbirds that winter in the Neotropics. The apparent decline in numbers of several species has prompted John Terborgh (1989) to ask "where have all the birds gone?" As we shall see the answer to this question is far from a simple one, and it may even be the case that for at least some species, the declines are more apparent than real. Finally, I consider the case of the Pacific salmon on the west coast of the United States, where there can be little doubt of the impact of human manipulations of the spawning rivers on populations of these valuable commercial fish. Indeed the salmon may represent the most critical case of all, for if we cannot save species of such obvious economic benefit, what hope is there for those whose benefits to us may be far more subtle or whose right to exist may be only for their own sake?

The Monarch Butterfly

The extensive range of the monarch in the New World and its establishment in Australia and Hawaii mean that the species itself is at no risk of extinction. This is not the case for the two migratory populations of North America (Brower and Malcolm 1991; Malcolm 1993). Because of their specialized requirements with respect to suitable overwintering sites, these entire populations are concentrated during the winter into a few specific locations that in total are no more than tens of hectares in extent. This remarkable clustering makes these populations extraordinarily vulnerable to disturbances that change the characteristics of the aggregation sites. Because the probability of extinction of a species or population is inversely proportional to the number of refuges (Quinn and Hastings 1987) the risk to the migrant monarchs is high and the threat of anthropogenic activity is obvious and acute.

With the exception of a few small sites along the Gulf Coast of Florida, all overwintering of the monarch population of eastern North America takes place in the Transvolcanic Range in the states of Michoacán and Mexico in central Mexico (Brower and Malcolm 1991 and Chapter 3). Only 10 Mexican sites are known, and the 5 largest and so far least disturbed of these occur within an area that is only 800 square kilometers in extent in the high-altitude (3000 meter) summer fog belt dominated by forests of oyamel fir (*Abies religiosa*). Microclimate is the key to the survival of the butterflies (Chapter 6), so that any disturbance to the forests can pose a significant threat because even small microclimatic changes can tip the balance against the monarchs. The fact that the local economy is highly dependent on these forests puts the butterflies at considerable risk and poses a severe problem to any preservation efforts.

Both the Mexican government and private conservation groups have taken steps to protect the monarch overwintering sites. In 1977 the private conservation association Monarca, A. C., was formed with the express aim of working for practical solutions to saving the butterflies based on the realities of Mexican politics and the local economy and culture (Ogarrio 1993 and other papers in Malcolm and Zalucki 1993). In 1984 a trust was established by Monarca, A. C., in cooperation with the federal government that resulted in the purchase of 700 hectares in the monarch overwintering area. Joint action of Monarca, A. C., and the government also led to a presidential decree in October 1986 (Diario Oficial, 1986) clearly defining the extent and level of protection of five of the overwintering locations.

Unfortunately conservation efforts are hindered by a local community property system known as "ejido" and a complex mosaic of controls over the forest exerted by federal, state, and local government entities. The laws concerning land ownership make it extremely difficult for private conservation organizations to purchase blocks of forest critical to the survival of the monarchs. The situation is unlikely to change unless the Mexican government undertakes major revision of land ownership laws. Thus although Monarca, A. C., has acquired 700 hectares of critical habitat, the remainder of the overwintering area remains under the ejido system, which allows the local people or ejidatarios to work the land but not to sell or rent it. Most of the land is therefore not legally available for purchase and conversion to reserves. Increasing population pressure, low agricultural productivity, and lack of education make conditions for the ejidatarios difficult at best. This creates a situation of dependence on the forests for fuel and for extra income through the sale of wood for construction and papermaking (Snook 1993). In addition the forests are being cleared to expand agricultural land even within the boundaries of the protected reserve.

Monarca, A. C., has worked hard to include the ejidatarios in their butterfly conservation efforts. Many of the local people serve as interpretive guides to the two sites that are open to the public, and the local economy in general has been given a considerable boost by the influx of tourists, of whom some 40–50,000 visited the area during 1985–86 (Snook 1993). Tourists are themselves a source of impact on the forests because roads must be built, fuel burned to provide them with food, and visitor facilities constructed, but so far at least they are having far less effect than the continued use of the forest to support other aspects of the local economy. In spite of this, control of tourists has received far more attention than other factors that affect the forest in much more acute and negative ways. The net result, however, of the joint activities of the Mexican government and Monarca, A. C., has been considerable progress toward the preservation of the monarch colonies, but it is equally obvious that steady attrition of the forests makes the future survival of these migrants problematic. The Mexican monarchs epitomize the biosocial problems faced by conservationists. The efforts made on their behalf by the enlistment of good biology and good politics in support of good intentions creates an outstanding model program. It is a revealing irony that in spite of this, demographic and cultural realities still make the future of this remarkable migration phenomenon very tenuous. One can only hope that present and future conservation efforts succeed enough to keep the great orange wave of butterflies from being reduced to a trickle.

The situation may not be much better for the western population overwintering on the California coast. Here the butterflies overwinter in some 200 or so locations scattered along the Pacific from Humboldt in the north to San Diego in the south (Lane 1993). The largest aggregations number in the few tens of thousands, although there were apparently larger clusters in the past, and the butterflies use several native and introduced tree species, most notably the Australian exotic *Eucalyptus globulus*. A 1989 survey on the central California coast revealed several potential risks to the overwintering populations (Brower and Malcolm 1991). Many sites are on private land, and these are continuously threatened by real estate development, driven by the very high land values of oceanside and near oceanside properties, and several have succumbed within the last decade (Malcolm 1993).

Even the many sites included in the parks protected under various governmental jurisdictions are not immune from deterioration. Partly this is due to benign neglect, so that ecological succession makes sites unsuitable, and partly it is the result of active programs, including expanding recreational use and the well-intentioned removal of exotic *Eucalyptus* trees. There is an obvious need for public acquisition of more monarch sites and an equally obvious requirement for better management of sites already located on public lands (Brower and Malcolm 1991).

Two recent developments give some reason to hope that western populations may be at least maintained at some approximation of present levels (Allen and Snow 1993). In 1984 a master plan for habitat preservation and protection of monarch sites was initiated as the Monarch Project under the sponsorship of the Xerces Society, a conservation organization specifically dedicated to the preservation of invertebrate diversity. One of the activities of the Monarch Project is to work with landowners to secure protective easements on properties where monarchs cluster. Those who donate conservation easements retain ownership of their land but provide protection of the butterfly sites legally and permanently. The second development was the passage in California in 1988 of Proposition 70, a Parks and Wildlife bond act that provided $2 million to procure habitats with monarch overwintering sites. Combined with the Monarch Project, Proposition 70 has provided the backing to organize the preservation of monarch sites along the length of California. The Monarch Project is assisting in identifying critical sites and contributing to land use plans in towns with major aggregations, like Santa Cruz and Pacific Grove, and has begun a habitat revegetation project to enhance monarch occupation of locales in native Monterey pine forest. It also carries out education projects involving both the schools and the general public. Although there are still conflicts ahead with developers and other powerful interest groups, there is at least reasonable hope that the conservation efforts on behalf of western monarchs will be an environmental success story.

Neotropical Migrant Birds

More than 250 species of migratory bird breed in North America and spend the winter in tropical regions to the south (Terborgh 1989). In the last few decades American and Canadian ornithologists and birdwatchers have complained with increasing frequency that many migrants are no longer as abundant in familiar locations as they once were and that many species have disappeared from those sites altogether. John Terborgh (1989) recounts the bird diversity he knew as a youth around his family home in the Virginia countryside, and the absence of many species now, apparently in the face of suburban sprawl. Anecdotal observations such as this are backed up by at least some of the long-term census data available. Censuses from floodplain forest at Cabin John, Maryland, and upland deciduous forest in Rock Creek Park, Washington, D.C., both across the Potomac from Terborgh's Virginia home, show major declines and local extinctions of tropical migrants (Table 16–1). Of the six most common species at each site, all have declined to half or less of their former abundances between censuses taken in the 1940s and those of the 1980s, and three have gone locally extinct. Many more of the less common species also went extinct during the 40-year interval. Reports of declines on a larger scale are also evident. When asked if they had noted declines in the numbers of birds traversing the Gulf of Mexico, a large majority of an audience of Gulf States ornithologists reported that indeed they had (Sidney Gauthreaux, personal communication).

The reported declines in numbers of Neotropical migrants parallel the destruction of old-growth forests at both ends of and all along migratory routes. Even though in the northeastern United States forest is actually increasing, it is still mostly in relatively early stages of second growth (Terborgh 1992). Suburban sprawl and agriculture are still taking out many habitats, and some particularly avian rich habitats have virtually disappeared. As an example, California riparian areas support about 400 pairs of migrants per square kilometer—only eastern forests support more migrants—yet just 5% of this habitat remains (Barnes 1993). There were estimated to be around 17,000 pairs of yellow-billed cuckoos (*Coccyzus americanus*) in California riparian growth in the early part of this century (as gathered from the ornithologist Joseph Grinnel's records), but today only some 75–100 pairs remain. At the opposite end of the migratory route the destruction of rain forest to create open areas for subsistence agriculture, sugarcane fields, and especially cattle pasture (we shall see the impact of cows on salmon below) have eliminated large tracts of habitat known to harbor migrants during the nonbreeding season. The decline of the Bachman's warbler (*Vermivora bachmanii*) to probable extinction is likely due to the conversion of Cuban dry forest to sugarcane, although loss of canebreaks in the forests of the southeastern United States may also have contributed (Terborgh 1989; Sherry and Holmes 1993).

Table 16–1. Changes over 40 Years in Mean Number of Breeding Pairs per Census of the Six Common (in 1940) Species of Tropical Migrants at Each of Two Sites Near Washington, D.C.

Species	Cabin John, Maryland		Rock Creek Park	
	1940s	1980s	1940s	1980s
Red-eyed vireo	15.3	4.8	41.5	5.8
Northern parula	14.0	2.3	—	—
American redstart	13.3	0.9	—	—
Acadian flycatcher	7.3	3.4	21.5	0.1
Kentucky warbler	4.5	Extinct	—	—
Wood thrush	4.0	Extinct	16.3	3.9
Ovenbird	—	—	38.8	3.3
Yellow-throated vireo	—	—	6.0	Extinct
Scarlet tanager	—	—	7.3	3.5

SOURCE: From Terborgh 1989.

The destruction of tropical habitat also seems to be a particularly serious problem because the migrants from North America winter mostly in southern Mexico, Central America, and the islands of the Caribbean, an area an order of magnitude smaller than occupied by the breeding ranges. This crowding into a relatively small winter range would on the face of it make the birds particularly susceptible to habitat changes in the tropical wintering grounds. The scissor-tailed flycatcher (*Tyrannus forficatus*) epitomizes the situation. It breeds over most of Texas, Oklahoma, and Kansas with some spillover into neighboring states; all but a very few of these birds winter in Guanacaste Province, Costa Rica (Terborgh 1989 and personal observations), an area the size of an average Texas county. In Guanacaste much of the savanna habitat preferred by the flycatchers is being converted to irrigated pasture. Many other migrants winter in Costa Rica, a country with a generally excellent conservation record, where nevertheless about 80% of the original forest cover had been removed by 1983 (Terborgh 1989).

There are problems, however, with the "common wisdom" that Neotropical migrants are declining. First, we have no baseline against which to compare modern-day populations against those of 50 or 100 years ago, let alone any earlier period (Terborgh 1989). Thus, even if we are seeing declines, are they part of a temporary trend in long-term population fluctuations resulting from climate or other "natural" factors, or do they in fact represent a serious impact of human land use practices? Further confounding the issue is that the two possible causes are not mutually exclusive. Second, it is extremely difficult to census bird populations (or any other population) accurately, especially over a wide area and for a long time. We are thus dealing with very messy and incomplete data to the extent that we have data at all. Even data we have are subject to varying interpretations. Radar records seem to show declines in migrant numbers over the Gulf of Mexico over 30 years, but in the same period wind patterns may have changed, causing migrant streams to move farther west (Gauthreaux 1992; personal communication). Third, we don't know whether local fluctuations reflect long-term changes. This factor is in part determined by habitat. For example, declines are much more apparent in fragmented forest both north and south than in larger forest stands (Askins et al. 1990). Reasons vary but seem to include increased cowbird parasitism and increased nest predation from mammals that adjust well to humans, such as raccoons and skunks (Wilcove 1985), as well as habitat patch size itself. Finally, despite the fact that birds are better known than any other group of animals, we are still largely bereft of fundamental knowledge of their basic biology and life histories, including breeding ranges and migratory behavior (Dhondt and Matthysen 1993).

An example of the difficulties inherent in interpreting census data, even if one has it, is shown in Figure 16–1. The data are from mist-net captures of Wilson's warblers at the Palomarin Field Station of the Point Reyes Bird Observatory located just north of San Francisco, California (Peaslee et al. 1993), and show the number of warblers captured in mist nets annually over a period of 13 years. The lines on the graph indicate two alternative biological interpretations of the data. The solid

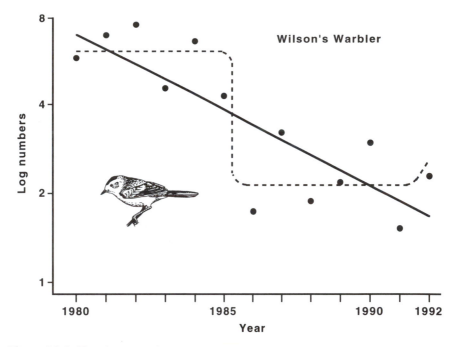

Figure 16–1. Two interpretations of yearly mist-net captures of Wilson's warblers at the Palomarin Field Station of the Point Reyes Bird Observatory. The solid regression line assumes an overall decline in population levels; the dotted line assumes steady levels interrupted by a sudden decline caused by major flooding in 1986. (Adapted from Peaslee et al. 1993.)

line (a regression line) assumes a more or less consistent downward trend in the population over the interval; overall habitat deterioration at either or both ends of the migration route is commonly cited as a probable cause for such a decline. The dotted line assumes that both very wet and very dry periods (Faaborg and Arendt 1992) have a negative impact on this warbler, which breeds in moist habitats subject to both flooding and drying out. In this case the population is viewed as steady between 1980 and 1985; as suffering a major decline in 1986 when there was a very wet late winter–early spring with considerable flooding; as remaining steady during 1987–1991 but failing to recover because these were drought years; and as beginning a recovery in 1992 when the rains returned. Unfortunately, with present data, there is really no way to determine which, if either, interpretation is correct. With luck, a few years of "normal" precipitation might suggest which is more likely (under scenario two, a recovery of the population would be predicted).

In an attempt to provide both a baseline and information on population trends, the Breeding Bird Survey (BBS) was organized in 1965 through the U.S. Fish and Wildlife Service under the leadership of Chandler S. Robbins (Robbins et al. 1986). This is a nationwide effort depending largely on the participation of volunteers. There are some 1800 survey routes, usually along rural roads, and a volunteer drives each route once a year during the breeding season in May or June. A route includes 50 stops at half-mile intervals at which the observer records all birds seen or heard during 3 minutes. In spite of obvious limitations due to variation in observer competence, a bias toward singing birds that occur in habitats near roads, and successional and other habitat changes, the BBS is the best and most comprehensive effort to date to estimate populations of North American birds at the time migrants are present. With proper analysis and attention to sources of bias, it should detect declines in numbers if they occur (Butcher et al. 1993). In an influential paper, Robbins and colleagues (1989) purport to show, based on BBS data, that indeed there has been a serious decline in Neotropical migrants, especially those breeding in forests, during the 1980s. This appears the more serious because there was actually an increase in the previous decade. A large interagency program, Partners in Flight (Peaslee et al. 1993), to promote research and monitoring in the United States on Neotropical migrants was largely justified on the basis of this paper (James et al. 1995).

Once again, however, there are difficulties in the interpretation of the data, some of which are the result of different methods of analysis. The problems are considered by James and colleagues (1995; see also Sauer and Droege 1992). The analysis of the BBS data by Robbins and colleagues (1986, 1989) was based on linear methods, whereas James and colleagues used nonlinear methods and also used more stringent criteria in deciding which data to use. From their analysis James and colleagues conclude that while there appeared to be an overall decline of about 6% over the BBS period in total numbers of Neotropical migrant paruline warblers, the group they analyzed, the median trend for the 26 species with the best data was an increase over the same period. Further, declines were not specifically evident in forest birds. Differences in methods of analysis thus led to quite different interpretations of the same data.

Still, not all the news was good. Significant declines were observed by James and colleagues particularly in two species, the prairie warbler, *Dendroica discolor*, and the cerulean warbler, *Dendroica cerulea*. This is in the face of surveys that seem to show the cerulean warbler may actually be expanding its range (Robbins et al. 1992). Another interesting observation to come out of the analysis of James and colleagues was that population declines were most severe in certain geographic areas, especially those associated with higher altitudes. Thus, in the southern Appalachians, the Ouachitas, the Adirondacks, the Blue Ridge, and the Cumberland Plateau, one or the other of the analytical methods used indicated that the average species had a higher than 0.7 probability of showing a decline. Because the foci of declines occur in mountainous regions, atmospheric contamination, which correlates with altitude, becomes suspect as a cause. The general conclusion of James and colleagues, however, is that even though some species are almost certainly declining and several are declining in some (high-altitude) regions, the supposed overall decline that is drawing much of the attention of conservation biology is based on weak scientific evidence at best.

In view of conflicting evidence and opinion, it is appropriate to ask whether there really is a conservation problem that concerns Neotropical migrant birds. First, it is obvious that some species are in trouble. Bachmann's warbler is probably extinct and Kirtland's warbler (*Dendroica kirtlandii*) is down to less than 500 pairs breeding in jack pine habitat in north-central Michigan. We seem to be too late to save Bachmann's warbler, but efforts to protect the Kirtland's from cowbird parasitism seem to be succeeding (Terborgh 1989). It is likely, however, that neither of these two species was very common in historical times because of specialized habitat requirements. Species like the cerulean warbler, prairie warbler, and certain other non-paruline migrants are more problematic. They seem to be declining, but the reasons are unclear. The cerulean warbler nests in mature forests usually near water, a habitat that is undoubtedly declining, but the prairie warbler nests in the sorts of second growth and scrub habitats that would seem to be increasing. A second point is that numbers of migrants are clearly decreasing in fragmented forest (Askins et al. 1990) as Table 16–1 shows. We lack a very good idea, however, of just how much of the breeding ranges of various forest species are occupied by fragmented forest as opposed to large tracts. Nor do we really know what "large" is or the degree to which successional stage might matter. Finally, as James and colleagues (1995) demonstrate, declines in populations appear to be occurring in certain physiographic regions characterized by higher altitudes.

One further point needs mention. There is no question that humans are having an enormous impact on the environment within the entire traveling range of Neotropical migrants. Very large tracts of habitat known to be used by migrants are disappearing at an alarming rate, especially in the Tropics. Even though we may not be detecting population declines, it seems unlikely that these major habitat changes are having no impact at all. Locally they certainly are. The Central Valley of California is an impoverished place because its riparian areas, and with them dense and diverse bird populations, are largely gone. It is also clear that species like Bachmann's warbler have been seriously harmed by anthropogenic changes. Even though these impacts seem few at present, it would be sad if we failed to heed the warning they provide to at least be alert to potential problems. Further, whether or not migrant birds are yet affected, it is obvious that large natural areas, valuable in their own right, are being destroyed. And as many authors stress (see, e.g., papers in Keast and Morton 1980, and Hagan and Johnston 1992) migrant species are integral parts of ecosystems and their biology can only be understood in the context of those ecosystems. To understand the one, we must also have the other.

Askins and colleagues (1990) make an excellent case for striking a balance between what they call the "Chicken Little Syndrome" of labeling every downturn in populations a major disaster and the "Tobacco Lobby Syndrome" of demanding impossibly rigorous statistical requirements for certainty. There are enough instances of population declines of migrant birds of greater or lesser severity to be worrisome. It is also the case that we are losing much habitat. Without claiming that the sky is falling, it would seem wise to err on the side of caution and to proceed on the assumption that at least some migrants and some important migrant habitats are in trouble and to design our conservation strategies accordingly.

To design those strategies, we need not only more information, but information specifically on the population processes inherent to the biology of migrants and of the role of migrants in ecosystems in all sections of their routes (Dhondt and Matthysen 1993; Nur 1993). Population changes are the result of adult survival, reproductive success, and recruitment of young into the breeding population. Monitoring programs such as the BBS do not deal with these factors, and so cannot provide insight into why populations may be increasing or decreasing. Three methods that can are constant-effort mist-netting, nest monitoring, and the study of individually marked birds (Nur 1993). As yet, however, long-term studies of this sort are few, in large part because they demand a large commitment of resources. Nevertheless they are necessary if we are to have any hope of doing more than guessing in outlining conservation policies. It will also be necessary to increase efforts to study migrants in ecosystems. The papers in Keast and Morton (1980) and Hagan and Johnston (1992) represent a good start but are only a start. Finally, in both population and community studies the large- and small-scale movement behavior of migrants must be factored into the population equations. Movements are capable of making migrants more flexible, but they can also make them more vulnerable. Only by a commitment to the sorts of studies outlined will we be able to sort out critical aspects of population biology and community ecology and formulate effective conservation strategies to preserve migrants and their habitats.

Pacific Salmon

Perhaps no other species so epitomizes the negative ecological consequences of human activity than the Pacific salmon of the western United States. As much as migratory bison characterized the pre-settlement Great Plains, so did the salmon define the rivers of the West Coast. The great salmon migrations that once extended from bank to bank in western rivers and streams are now reduced to a mere remnant. The cause is obvious; almost every western river is now dammed, either for hydroelectric power or to supply water to agriculture (Reisner 1993; anyone who wants to truly understand the relation between water, politics, and the growth of the American West should read Marc Reisner's magnificent book). Not only do the dams impose obstacle after obstacle on the way to the breeding sites, but the spawning streams have become silted and polluted from the backed-up lakes and the agricultural runoff.

A few examples suffice to illustrate. The Columbia River prior to the 1930s was the greatest salmon river in the world. Fifteen million fish migrated upriver every year to spawn. Grand Coulee Dam was completed in 1941. Today fewer than two million salmon make the run up the Columbia, fish ladders notwithstanding, and about half of the watershed's migrations are in danger of going extinct. The Snake River, a major tributary of the Columbia, in the late 1800s produced 1.5

million or more spring and summer chinook salmon; by 1960 this had been reduced to 125,000 (Fulton, 1968) and counts in 1989 and 1990 yielded only 21,244 and 26,524 fish (Fox 1992a). California salmon have fared even worse (Fig. 16–2). By the end of the 1970s there were some 1250 major reservoirs in California, and *every* major river but one had been dammed at least once (Reisner 1993). The Stanislaus River alone, on its short run from the Sierras to the sea, had been dammed no less than 14 times! Friant Dam on the San Joaquin River single-handedly wiped out a spawning run of 150,000 fish. Up and down the California coast, in rivers large and small, migratory runs of salmon are being nominated for endangered or threatened status under the provisions of the U.S. Endangered Species Act. Overall in the four states of Washington, Idaho, Oregon, and California, there are 214 stocks of Pacific salmon whose persistence is in jeopardy and 106 more that are now extinct (Nehlsen et al. 1991).

Perhaps no other single western water manipulation has had such far-reaching impact on salmon and the rest of the environment as the California Central Valley Project (CVP). Water is taken from the Sacramento River and its tributaries in northern California and transported hundreds of miles to irrigate the pastures, cotton, and alfalfa of the San Joaquin Valley in the south. Along the way a part of the water is siphoned off to supply farmers in the Sacramento Valley. Shasta Dam on the upper Sacramento River was completed in 1945 and Keswick Dam in 1950,

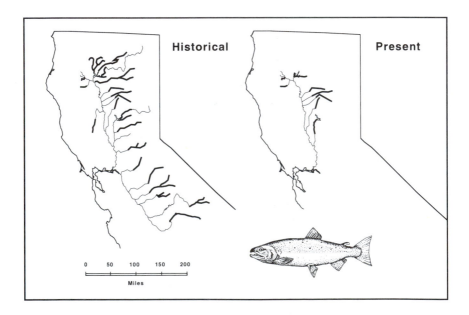

Figure 16–2. Spawning areas (*heavy lines*) of the chinook salmon in the Sacramento–San Joaquin River drainage of California in historical times and at present. (Redrawn from a map provided by Peter Moyle.)

effectively blocking salmon runs to the upper reaches of the Little Sacramento, McCloud, and lower Pit Rivers (Fox 1992b; Moyle et al. 1989) whose watersheds encompass some 6600 square miles. Trinity Dam was finished in 1964, closing off the Trinity River, the major tributary into the Sacramento from the northwestern part of California. What was perhaps the final blow to the salmon spawning runs was the construction of the Red Bluff Diversion Dam in 1967, impounding water below Shasta Dam for diversion into the Tehama-Colusa and Corning Canals, which supply irrigation water to the west side of the Sacramento Valley.

Getting upstream past dams is only half the problem for migratory salmon. Once spawning occurs, the young must be able to return downstream. An enormous impediment to the downstream migration in the Sacramento Valley is the battery of 300,000 horsepower pumps at the terminus of the Sacramento Delta just before the river enters San Francisco Bay (Reisner 1993). These feed the aqueducts transporting CVP water to the San Joaquin Valley to the tune of millions of acre-feet of water that would otherwise discharge into the Pacific (an acre-foot is an acre of water a foot deep). The juveniles migrating to the ocean are helpless to avoid the currents generated by the great sucking force of the pumps, and untold millions of hapless young are either converted to salmonburger or spewed into the aqueduct and turned into fish stew in the sun-heated irrigation canals. In wet years only about 20% of the outflow through the Delta is diverted through the pumps, and most of the salmon juveniles can avoid being sucked through them. In dry years, it's a different story because 50% of the flow (more in very dry years) is pumped southward, and the pumps wreak a terrible toll. Between 1986 and the winter of 1992–93, California experienced a period of devastating drought. In spite of this and in spite of the ever decreasing water storage behind the dams and flow down the river, in the "critically dry" years of 1987 and 1988 and the "dry" year of 1989 the agricultural clients of the CVP (and the parallel State Water Project) received every last drop of their water entitlements plus some extra, because water exports from the Delta actually *increased* from 1987 to 1989! The salmon, on the other hand, had their water flow reduced by 90%. In 1991 only a "March miracle" of rainfall late in the winter season prevented Shasta Lake from going completely dry. Also in 1991 only 191 spawners of the winter run of chinook made it to the Red Bluff Diversion Dam.

The Sacramento River is the only salmon river in the world with no less than four genetically distinct annual runs of chinook salmon (Moyle 1976 and personal communication). These runs represent the flexibility in salmon life histories (Chapter 12) matched by natural selection to the great habitat diversity of the pre-dam Sacramento drainage (Fig. 16–2). At the turn of each new year, the winter-run chinook migrate up the Sacramento. These are fish that are still short of being fully mature; they are silver-sided and their meat is firm, making them the most desirable fish from a commercial perspective. Once they enter the river, they mature rapidly and develop their gonads as they move through the cool water flows of winter up to the smaller tributary streams. They spawn in the early spring runoff in April and May, and the eggs incubate in the gravel during the summer in spring-fed waters that keep them cool. The young then return to the sea the following year. The spring run is the second annual run to enter the river, doing so from March through May during the primary period of snow melt. It may at one time have been the most abundant run in California (Campbell and Moyle 1991). As with the winter run the gonads are immature, and these, too, are excellent eating fish. The spring-run fish migrate into the tributaries and spend the summer in deep pools. They spawn from September through October, and the young spend a year in the streams before proceeding out to sea. Because they must spawn in the upper reaches of the Sacramento drainage, these winter- and spring-run fish are the ones most susceptible to the building of dams. The irony is that because they are still not mature when they enter freshwater, they are also commercially the most valuable.

In contrast to the winter and spring runs, fall-run fish are fully mature when they enter the river—"colored up" in the jargon of the salmon fisherman. Because they are almost ready to spawn when they leave the sea, they reproduce quickly in the lower reaches of the drainage as soon as the water becomes cool. These are big river spawners much favored by fish hatcheries, where the fact that they are already mature when caught means that holding time in the hatcheries is short. The conversion of energy and materials to gonads, however, means that the meat is already losing firmness when the salmon are caught by fishermen. The final and fourth population of the year is the late fall run. These enter the Sacramento beginning in October and sometimes continue until February. Like the fall-run fish, they spawn in the main stream of the river beginning in January and extending through March. The fry hatch usually in June and spend a year in the river before migrating to the sea. The fall and late fall runs are less likely to be blocked by dams, but they suffer greatly from the warm sluggish flows of drought years and from the diversion of water by the Delta Pumps.

In the years following the Second World War, the Sacramento sustained a harvest of several hundred thousand fish, increasing to over a million when conditions were right (Reisner 1993). Prior to 1967 when the Red Bluff Diversion Dam was built, roughly 200,000 adult salmon made the winter run up the Sacramento (Rectenwald 1989). This figure is anecdotal and a little uncertain, but there is little doubt as to what happened to the winter run after 1967 (Fig. 16–3). After falling to less than 60,000 fish following completion of the dam, there was a brief recovery to around 115,000 upstream migrants in 1969. The decline has been steady since, until the pitiful 191 fish made it to the dam in 1991. The late fall and spring runs, while never as large as the winter run, have fared no better. Only the fall run survives in anything like commercial quantities, because these are the fish that undergo most of their maturation at sea and spawn in the lower Sacramento. The fall-run fish returning in 1988 from the wet-year spawns 2 years earlier numbered close to 140,000, but numbers fell rapidly to only 40,000 by 1992. Similar low numbers were seen in the drought years of the 1970s, so the decline may be temporary. There can be little doubt, however, that even should the fall run survive, unless major shifts in water allotments are made soon, this would be the only run of the original four with commercial numbers of fish, albeit lower

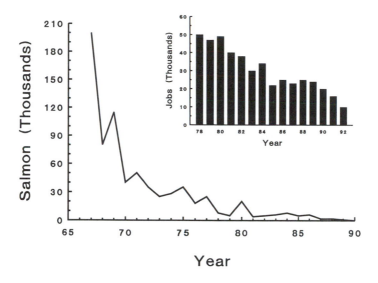

Figure 16-3. The decline in winter-run chinook salmon in the Sacramento River of California since 1967 and the loss of jobs associated with fishing in the state since 1978. (From data provided by Peter Moyle.)

quality fish, and two of the runs, winter and spring, may go totally extinct. It is little wonder that the fish sold as "poverty steak" for 10 cents a pound during the Depression (Reisner 1993) was selling in 1993 at my local supermarket not 15 miles from the Sacramento River for close to eight dollars a pound, on sale.

A social and economic cost has accompanied the environmental impact of the dams. One aspect of this cost, the loss of fishing jobs, is also shown in Figure 16-3. As late as 1978 there were some 50,000 offshore and onshore jobs associated with the salmon fishing industry in California. By 1992 there were only 10,000 jobs left. In the fall of that year more than 300 of the 350 or so salmon boats making up the fishing fleet sailing out of Fort Bragg, California, were for sale (Reisner 1993). Thousands of farm jobs were also lost when, as inevitably had to happen, water flows to the San Joaquin Valley were finally cut off after 5 years of drought. It is arguable that by maintaining water flows at normal levels or above through 1989, the CVP water managers succeeded in costing the California economy some thousands of fishing jobs in addition to the farm jobs later lost; it is also probable that the impact on farm jobs would have been much less had a sensible policy of reduced water flows to the San Joaquin Valley been followed. The farm jobs will return with the rains. The fishing jobs may be gone forever.

What can be done? Reflecting the culture from which they come, American fisheries managers have tried for a technological solution. Fish ladders, bypasses, and fish screens covering pump and diversion intakes have all been constructed. The declines in salmon from California to the Canadian border demonstrate that these are not working, and in fact the attempted technologies have probably hurt more than they have helped (Black 1993). Hatcheries produce large numbers of fish, but they are genetically inferior, usually come only from fall runs, and do poorly in the wild. The only long-term solution is a fundamental change in the way water is used in the American West. Current agricultural uses of water in California are heavily subsidized and make little economic sense. It takes 7 to 8 feet of water annually in the climate of much of California to produce grass; that translates into almost 50 thousand pounds of water to raise 1 pound of cow. These quantities of water go to a pasture crop worth about $100 million in California in 1985; that same year the economy of southern California's cities, which received less water than irrigated pasture, was close to $300 *billion* (Reisner 1993). As Marc Reisner puts it laconically: "California has a shortage of water because it has a surfeit of cows—it's really almost as simple as that." And because we have cows and cotton, we may, perhaps soon, have no salmon.

In spite of the formidable obstacles from an agribusiness industry accustomed to having its own way and a tradition in the American West of profligate and unregulated water use, there is some hope the situation may be turning around. Several factors are at work. Cities all over the West are increasingly recognizing the imbalance between their water use and their economic contribution and the water use and contribution of big agriculture. Many states have legalized water transfers whereby water can be sold by an allotted user to others lacking allotments, thus putting market forces to work that favor water conservation. There is increasing public and political awareness that water left in rivers may be more valuable than water taken out of rivers. The Endangered Species Act is being used with increasing effectiveness. Farmers and fishermen are now talking to each other about their common problems, including loss of jobs when water is wastefully used, and there is even serious discussion of dismantling some dams. Perhaps as a society we are learning what the migrations of the salmon can teach us, that the natural world is not simple, and we cannot easily bend it to our wishes with technology without unhappy consequences. In their transit from spawning to spawning the salmon occupy several habitats. We have disrupted so many that even with their extraordinary resiliency and life history flexibility, they have vanished or are in danger of doing so from many of their home rivers and streams. The big question remains: Can we all—ranchers, farmers, loggers, landowners, fishermen, city dwellers—cooperate for the mutual benefit of the salmon, and ourselves?

Summing Up

The three cases discussed in this chapter span a range of problems faced by conservation biologists trying to deal with migratory organisms. The monarch butterfly is not endangered as a species, but its populations that undertake long-distance movements are at risk. The migratory journeys are thus endangered phenomena that are likely to disappear unless measures are successfully undertaken to preserve them. Conservation programs that are in many ways models of cooperation between governmental and private sectors have been initiated to preserve overwintering sites for both eastern (Mexico) and western (California) populations of the monarch. Yet because of land laws, population growth, and powerful counter interests, even these model programs run considerable risk of failure. The monarch migration system in many ways exemplifies the many conflicting forces that threaten so many species and populations.

Neotropical migrant land birds present a rather different problem, in scale and in the difficulty of determining how great the problem really is. Although there is a good deal of evidence suggesting many of these migrants are declining, there is also much suggesting that many, if not most of them, are not. At least some that are declining or have even gone extinct are birds of limited distribution. Any organism is difficult to census on a large scale or over a long time, and this means it is extraordinarily difficult to acquire the data necessary to determine population trends. The Breeding Bird Survey has limitations but is a good start on monitoring. It cannot, however, provide information as to the causes of population declines or increases. For that we need both population studies of the birds and data on patterns of land use as it changes over both space and time. Migrant birds are also integral parts of the ecosystems in which they occur, and we need much more knowledge of those ecosystems. It is unlikely that a species by species approach to Neotropical migrant birds will be successful either from the perspective of straightforward data gathering or from that of designing long-term conservation programs and policies. To generate the necessary understanding, we shall need to integrate migration and other movements into both politics and ecology. The situation is summarized nicely by Morton (1992).

The salmon of the American West Coast are perhaps one of the most discouraging cases of the impact of humans on nature. These are valuable commercial fish supplying protein-rich food and thousands of jobs, yet in spite of their value, many populations have been driven extinct or nearly so. There is irony in the fact that much of the water needed for salmon survival has been diverted to support cows, a far more ecologically costly source of protein and one with some negative consequences for health. Indeed salmon are but one example of heretofore unknown health benefits of a resource now in short supply, one of many reasons advanced to support effective conservation measures. Because they are so valuable yet have been so decimated, salmon epitomize the clash between conservation, competing commercial interests, a misguided water ethic, and just plain greed. Against the collusion of government, water law, and private interests, the salmon and those who fish for them have had little chance. One can only hope that recent signs of a rethinking of the economics of water in the West will be more than just one more doomed effort to save a heritage.

Our record as stewards of our planet's biota has not been a very good one. This has been as true for migrants as it has been for other organisms, but there are nevertheless some signs of hope. Those of us who delight in the now appreciable numbers of migrating gray and blue whales that annually pass the coast of

California are quite aware of the success of some conservation efforts (Baskin 1993). The annual return of elephant seals to Ano Nuevo Point is another success story, for these animals were reduced to less than a hundred at the end of the last century. It is hunting pressure that affects whales and seals, however, and this is easily rectified. Habitat loss in the face of human population pressure is another matter altogether. Far more political will and international cooperation are needed to rectify population growth and needless habitat destruction, the more so because scientific data and theory to back up policy are still in short supply. Still, there is much we do know. The question is, do we have the will power to apply the knowledge we have?

17

Summing Up and Future Directions

The ability to move is one of the most important characteristics of living things. Migration in all its often astonishing variety is the most elaborate expression of that ability, and its prevalence in so great a diversity of organisms demonstrates that it has been repeatedly favored by natural selection over alternative lifestyles. Movements away from a habitat as occur during migration must be subject to considerable risks, yet time after time in organism after organism migration appears as an element in the life cycle. The prevailing theme of this book has been the attempt to understand the role of migration in the greatly diverse life cycles in which it occurs. Congruent with that theme I have outlined the forms that migratory behavior and migratory lifestyles can take and tried to shed light on why, so often, migratory episodes assume such importance in the evolution of life histories.

Faced with all the variety and complexity of migration and the diversity of taxa in which it occurs, any attempt to draw general conclusions presents a daunting task. In spite of the variety present, however, three major threads seem to stand out in the tapestry of migratory life histories. They can be expressed as follows: (1) Migration involves specialized behavior that is both qualitatively and quantitatively different from other types of movements; (2) migration is a syndrome integrating behavioral, physiological, morphological, and life-history traits; and (3) migration is an adaptation to shifting or patchy environments. I shall first summarize what I see as the evidence supporting these conclusions. Following these summaries I shall then conclude with a few thoughts on what I see as some important questions for future research and on what students of migration will need to do in order to go about providing answers.

Migration as Specialized Behavior

John Kennedy's important contribution to the study of migration was to define a set of characteristics that clearly distinguished migration from other types of movement behavior (Kennedy 1985 and Chapter 2). Establishment of these diagnostic features allows specific tests of whether a particular *individual* is a migrant, independently of any knowledge of its previous history. These features, according to Kennedy, include specialized behavior and physiology that involve persistent enhanced movement and the suppression or inhibition of responses relating to growth and reproduction. The manifestation of migratory behavior means that environmental inputs that would ordinarily stop movement, say stimuli signaling food or a germination site, fail to do so in a migrant. As a result migration is persistent; it is also straightened out, because straightening a movement increases the chances of reaching a new habitat. Other specializations include specific behaviors of departure and arrival, with the latter behavior primed by the migratory activity preceding it. Ongoing migratory behavior thus increasingly lowers the threshold for inputs that eventually stop movement and cause the migrant to settle. Because migration often takes place over a long distance relative to the journeys the organism usually makes, it is also likely to be supported by the reallocation of the energy needed to sustain it. All of these characteristics provide potential tests for determining if an observed individual is indeed a migrant.

The underlying assumption in adopting Kennedy's definition of migration for this book was that the characteristics that defined migration for winged aphids also did so for all other migrants, with some modifications to allow for obvious differences in biologies. Plants, for example, lack nervous systems so it is inappropriate to include nervous inhibition as an element in their migratory behavior; other migratory characteristics, however, should still be identifiable if the above assumption is valid. I think it is useful at this point to reflect back over the examples of migration we have analyzed and to examine our assumption of a generalizable definition as critically as we can. Do other migrants in fact exhibit the same symptoms of migratory behavior as do Kennedy's aphids, or is it necessary to define migration in a different way?

In Table 17–1 I list several examples drawn from the foregoing chapters that seem to indicate specialized behavior of migrants consistent with the definition adopted. At least some of these, such as salmon smolts and the arctic tern, have long been considered migrants by criteria other than the ones outlined by Kennedy. These clearly show suppression of the vegetative activities of maintaining bottom contact and station keeping in salmon smolts and feeding in terns, and so provide a quasi-independent test of the validity of our criteria for migratory behavior. Some other examples of migration in the table have heretofore been questionable. Thus crossbills and queleas have often been considered "nomads" and in a category distinct from migration because, among other things, they did not display round-trip movements with characteristic destinations. As indicated in the table, however, both deposit fat in the manner of migrants before departing on their journeys. This fact certainly suggests that selection treats them in much the same way it treats classic bird migrants, producing the physiological specialization we observe. Another example is the emu, again hardly a typical to-and-fro migrant, yet in bypassing suitable habitats on the way to sites of recent rainfall, it is clearly not responding to sensory inputs that will later result in the cessation of movement. The remaining organisms listed in Table 17–1 all in one way or another display

Table 17–1. Examples of Specialized Migratory Behavior or Physiology

Organism	Migratory Behavior
Large milkweed bug	Flight primes reproduction (Chapter 6)
Forest bark beetles	Flight primes settling (Chapter 11)
Desert locust	Migrates at night when nonmigrants inactive (Chapter 11)
African armyworm	Spiraling takeoff; reproduction delayed (Chapters 3 and 14)
Mantis shrimp larvae	Phototaxis enhanced; thigmotaxis suppressed (Chapter 3)
Spiderlings	"Tiptoe" takeoff behavior; ballooning promotes landing (Chapter 3)
Blackcap warbler	Migratory restlessness at night; sleep suppressed (Chapter 12)
Artic tern	Undistracted by food (Chapter 3)
Emu	Undistracted by favorable habitat (Chapter 11)
Crossbills; quelea	Deposit excess fat (Chapter 3)
Atlantic salmon smolts	Open-water swimming; suppress bottom contact (Chapter 6)
Sea turtle hatchlings	Undistractability ("frenzy") during movement out to sea (Chapter 8)
Tumbleweeds	Special separation cells in stem; seeds protected during travel (Chapter 3)
Jewelweed (*Impatiens*)	Specialized ballistic mechanism promotes departure (Chapter 13)

behaviors or physiologies that strongly suggest they are migrants in the sense implied by the definition of migration I have adopted. It is also worth noting that the table is not meant to be exhaustive; other similar examples of specialized migratory symptomologies are scattered throughout the book.

In spite of the fact that our definition does seem to work well in practice in distinguishing migration from other behaviors, some caveats are in order. Prime among these is that many of the presumptive examples of specialized migratory behavior both listed in Table 17–1 and discussed elsewhere are anecdotal. This is the case, for example, with the spiderlings, the arctic tern, and the emu. It is also partly true of the observations of the spiraling upward flight of the African armyworm, although such flight has now been observed in so many departing migrant insects as to be an accepted part of migratory activity.

A further caveat is that only rarely have the predictions explicit in our definition of migration been subject to direct experimental test, and where tests have been applied, they have been only partial. A relation between migratory and settling activity has been demonstrated in forest bark beetles because experiments show that flight lowers the responses to sensory cues that promote settling. We do not know, however, whether the relation between flight and settling is reciprocal as it is in aphids, because tests to ascertain if repeated partial settling responses prime flight have not been conducted. In the blackcap warbler and several other birds interaction between settling and migration seems likely because migrants in the laboratory retain nocturnal restlessness for longer than birds actively migrating in the wild. A reasonable interpretation is that the active flight reduces restlessness and renders the birds more likely to cease migrating and settle, but we do not know whether repeated exposure to appropriate habitat inputs without allowing settling would prime restlessness or migratory flight. In virtually no other organisms have we even addressed the question of which vegetative activities might interact reciprocally with migratory behavior with respect to suppressing (or inhibiting) and priming. Clearly there is much to be done toward integrating the kinds of questions our definition poses into the mainstream of thought and research on migration.

Whether or not this definition is correct in its details, I would argue that it makes one very important conceptual contribution to notions of migration. This is in making explicit a hierarchical distinction between migration as a behavior with a physiological basis and its outcome as part of an organism's population ecology (Kennedy 1985; Gatehouse 1987b). Much of the confusion surrounding studies of migration and of the failure of workers studying different taxa to communicate or understand each other has been due to the failure to distinguish between the behavior of individual migrants and the population consequences of their migratory activities. Once this distinction is recognized, the way should be open for much greater understanding of the ecology and evolution of migration and the other sorts of movements outlined in Chapter 1. Keeping the notion of hierarchy in mind should help to clarify not only the way we view migration, but also the way we make observations and design experiments. The behavioral taxonomy I have proposed will have been especially valuable if it leads to the asking of new and important questions about movement.

Migration as a Syndrome of Traits

Implicit in the above discussion of definition is the notion that migration is not a unitary character but rather consists of a syndrome of interacting physiological and behavioral traits. Physiologically migration requires the mobilization of resources and energy in support of movement behaviors that are not interrupted for foraging and feeding and so demand that considerable extra energy be available before they are undertaken. A number of special adaptations exist to ensure that is the case. Energy pathways are redirected so that instead of mobilizing to put effort into reproduction or growth, fuel is mobilized for movement. This may involve feeding on high-energy sources such as fruit in the case of some bird migrants (Chapter 6) and in any case incorporates hormonal mechanisms to ensure the storage of reserves for the journey and often, in animals, the generation of higher levels of nervous activity. Like the energy pathways they control, hormones, too, are often redirected in their function to support migration instead of maintenance or reproduction. Included in behavioral mechanisms may be special abilities to select optimal wind or current systems in aid of travel, and if lives or journeys are long, there may be highly developed orientation or even navigation mechanisms to ensure that the journey takes place over the proper route to the correct destination.

In addition to behavior and physiological mechanisms, migration syndromes also involve various morphologies. In the most extreme case migrants and nonmigrants may be distinguished by the presence and absence, respectively, of a transporting apparatus. Thus in some insects migrants bear wings and nonmigrants do not, and in several composites some achenes bear pappi for wind transport, while others do not but may possess thick walls to assist survival for long periods. In other migrants morphological contributions to syndromes may be far less dramatic but nevertheless important. It is apparent, for example, that selection has repeatedly acted to make the wings of volant migrants longer than those of their more sedentary counterparts. This is true in both insects and birds. Detailed studies of the wing morphologies of birds, however, also reveal that syndromes are subject to constraints. Birds use their wings for many activities besides migratory flight, and these other uses also contribute to variation in size and shape (Chapter 7). Morphological and physiological constraints apply to all organisms and migrants are no exception. The syndromes we observe result from the interaction of selection and constraint.

The allocation of metabolic resources between migration and reproduction also means that migratory syndromes are a consequence of life-history evolution. Indeed this is virtually guaranteed by the fact that one consequence of migration is the determination of where and when to breed, and so life-history traits will be strongly influenced. Age at first reproduction will be at least in part determined by the duration of the migratory period, while fecundity will be a function of the proportion of resources devoted to reproduction versus the proportion devoted to sustaining migration, such as the maintenance of flight or swimming muscles or the production of silken threads for ballooning. Extensive studies of salmonid fish especially, have already demonstrated the relations between energetic constraints and migratory life histories. If migration is long and in a juvenile stage, large energy reserves in the form of fat (some seeds) or yolk (some plankton) may be required to sustain it. Life-history traits are thus as much a part of migration syndromes as morphology, physiology, and behavior.

What remains still largely an open question is how much of the variation we see in migration syndromes is the result of genetic variation. What we observe is a set of phenotypic correlations, although very few of these have been formally estimated. We are only just beginning to assess the role genetic architecture may play in the expression of these phenotypes. The determination of the amount of genetic variation and covariation allows assessment of how much and how quickly traits can evolve under selection. In addition we could gain additional understanding if we knew how much of the variation we see is due to phenotypic plasticity of one or a few genotypes versus how much is the result of genotypic variation underlying a diversity of syndromes. Finally, if genetic variation does contribute to syndromes, which traits are likely to be genetically correlated and why?

What few studies have been done suggest there is ample genetic variation for quite rapid evolutionary change in migration syndromes. The most dramatic case is undoubtedly that of the blackcap warbler, which in the last few decades has evolved a new migration route between the British Isles and southern Germany. Laboratory studies strongly suggest that the necessary genetic variation exists for the required changes in both duration and direction of migratory flight (Chapter 14). It would thus seem this is not simply a case of phenotypic plasticity. In migratory milkweed bugs genetic correlations exist between life history, morphology, and behavior that clearly define a genetically based migration syndrome; these correlations are absent in nonmigratory populations. In these North American bugs the migration syndrome results from polygenes, but in the Mediterranean lygaeid bug *Horvathiolus gibbicollis*, a Mendelian locus acting pleiotropically influences not only wing morph but also a set of life-history traits that differed between the winged and wingless types (Chapter 13). Physiological studies of wing morph and other aspects of variation in migrants suggest there is hormonal mediation of the interactions among genes, environment, physiology, and migratory behavior (Chapter 6). At present we lack any knowledge of the ratio of polygenic to single-locus syndromes across the span of migrating organisms, yet such a determination would tell us much about how syndromes may have evolved and their potential for future evolutionary change. What the few genetic studies available so far do reveal is that it makes little sense to view single elements of migratory performance in isolation from other aspects of the syndromes of which they are a part, because natural selection is most likely acting on all the traits simultaneously and on the strengths and signs of the correlations among them as well.

Migration as an Adaptation to Shifting or Patchy Environment

Migration is an adaptation that has evolved to allow organisms to cope successfully with environmental heterogeneity in both space and time. In the case of long-lived migrants this allows individuals to pursue a lifetime track through a sequence of habitats no one of which is available for more than part of the life cycle. Those with short life spans pass through habitat sequences over the course of succeeding generations. Long life spans permit round-trip migration; short-lived organisms have time to travel only one way within a generation. Many migrations are seasonal, because seasonal changes are a major factor causing habitats to vary in quality. Seasonal fluctuations that produce flushes of resources exploited by migrants cover a wide spectrum of habitats from arctic tundra to the inundation zones of tropical rivers. Other habitats that change regularly are those associated with the early successional stages of ecological communities, and it has often been noted that denizens of these habitats are usually migratory, especially among plants and insects. Indeed one of the traits characterizing plant species as weedy is their enhanced ability to migrate as seeds and colonize newly opened germination sites. Because rainfall there is not only scarce but also likely to be highly erratic in amount and location, deserts and other arid regions provide some of the most variable of the world's habitats, and they tend to be endowed with a great diversity of migrant species and migratory lifestyles. On a much smaller scale, ephemeral habitats such as carrion or dying trees provide their own habitats uniquely exploitable by migrants. The details of the variety of migratory lifestyles that have evolved to take advantage of the often rich but only briefly available resources of temporary habitats range from the quaint to the spectacular.

If it is true that shifting or patchy habitats select for migratory life histories, then the converse should also be true: denizens of relatively permanent habitats should be sedentary. That this is indeed the case is perhaps best indicated by wing polymorphic insects. In long-lasting habitats such as alpine tundra or deep lakes, a disproportionate segment of the resident insect species lack wings, and are therefore the quintessential nonmigrants (Chapter 13). Among birds stable long-lasting habitats such as tropical rainforest also are home to fewer migrants than occur in habitats with less and more erratic rainfall such as dry forest or savanna. When migrant birds do occur in rainforest, they are likely to require resources that are more patchily distributed, such as fruiting or flowering trees (Chapter 9). The available data thus do support the prediction that organisms of stable environments will evolve sedentary rather than migratory life cycles.

Within and among migrant species, there is much variation in the ways movements are adapted to environmental heterogeneity. Some of this variation reflects the different interests of the organisms concerned. Differential migration of the sexes, for example, may reflect different reproductive strategies. Male birds may migrate less far and so return earlier to territories on the breeding grounds; male eels may travel less far because they can gain by maturing at a small size. In some instances migration may represent the "best of a bad job" and risky movement undertaken only because staying is even riskier in the face of competition or other factors. In still other instances variability in migratory performance may reflect environmental uncertainty or situations where mixed evolutionary strategies may be stable. The very fact of environmental heterogeneity introduces chance and uncertainty into the lives of migrants; movement can reduce the risks, but it cannot eliminate them. Nevertheless, by allowing changes of habitat, migration can be an effective means of reducing the variances that organisms will face.

There is also a practical side to the association between migration and shifting or patchy habitats. By converting large areas to agriculture, we create extensive tracts of exactly the sorts of habitats to which migrant pests are adapted. We then compound the problem by establishing continuous monocultural stands of ruderal plants. It is thus no wonder that a significant proportion of the world's crop pests are migrants; these include fungi, insects, and in some cases birds. It seems all too obvious that if we are to manage such pests successfully, we shall need to understand the biological and meteorological aspects of their migration systems more fully. Some notable efforts demonstrate the truth of this, but there are all too many instances where biological logic has been ignored. The same is true when it comes to the conservation of migrants, the more tragically when, as in the case of Pacific salmon, they are an important human food source. I hope we do better in the future, but the exponential growth of the human population means we have all too little time to correct our mistakes.

Future Directions

Throughout the course of this book I have tried to point out areas where there are gaps in our knowledge or where our understanding is cursory. To fill the gaps and increase our understanding, I advocate a much broader approach to the analysis of migration based on a clear perception of the difference between the behavior of migrating individuals and the outcome of their behavior in terms of the ecology of populations. To do so will, I think, open up new areas of research and end much of the confusion and argument over what is migration and what is not. Such debates have put an unfortunate brake on progress in understanding. On the other hand it is a tribute to the ingenuity and perspicacity of the many biologists who study migration that we have advanced as far as we have. Progress should be even greater in the future.

Concomitant with clarification of concepts should be greater diversity in the organisms we study. To some extent studies have been driven by practical considerations, accounting for the numerous data on pest insects and diadromous fishes, to cite two conspicuous examples. Other organisms like migratory songbirds have attracted attention because of their visibility and esthetic appeal. It is apparent, however, that a distressingly small number of the world's great diversity of migratory species has been studied. In the future I would hope we take much greater advantage of what comparative studies of diverse migrants can tell us. Too often we tend to focus all our attention on our "own" group without regard for what other organisms may do. Thus the repeated emphasis on round-trip movements in the vertebrate literature ignores both the fact that even many vertebrates that are migrants by all other criteria do not make classic round-trips, and the fact that most migrants from other groups of organisms do not make round-trips either. Also missed is the important relation between round-trips and life spans. This then is a plea to study migration as a general biological phenomenon, not simply a trait characteristic of a particular taxon. It would be especially useful to apply similar methods to different organisms, as well as different methods to a few organisms, to gain the most from what comparative data can reveal.

What also seems necessary is a greater application of combined approaches from more than one discipline. I think much of the confusion over characterizing migration could have been avoided had workers more often considered together the physiological mechanisms underlying migratory behavior, the ecological consequences of that behavior, and how those consequences guide its evolution. In addition to being taxon bound, we also tend to be discipline bound, especially in this age of specialization as biology becomes more complex. I suspect, however, that the greatest advances in our understanding of migration will come if we can manage to cross traditional disciplinary boundaries. The rapidly developing interactions between theory and experiment also should be pursued.

As with any active area of scientific inquiry, studies of migration continuously provide new questions for investigation. These involve both proximate mechanisms and evolutionary explanation. With respect to mechanisms we need to know much more concerning the general applicability of the behavioral and physiological criteria for migration that we have defined. How useful is our definition across the kingdoms of living things and in what ways could it or should it be modified to make it more useful? How are metabolic, hormonal, nervous, and sensory pathways specifically mobilized to serve migration? And in an area in which we seem on the verge of really exciting developments, how do the genes mediate those pathways?

The mechanisms that serve migration are organized into arrays or syndromes of traits. These syndromes vary across migrants, and we need to know more about what sorts of variation occur and how natural selection acts to produce particular arrangements of traits. In what ways, for example, are syndromes limited by mechanical, physiological, or phylogenetic constraints? How much genetic variability is present in syndromes, and what guides their further evolution? Implicit in the notion of syndromes is that these assemblages of characters are adaptations to fit certain environments. This leads to the question of which migratory life cycles are best adapted to which environments. What sorts of selection are generated by the different habitats to which migrants have adapted and why in some environments do some organisms migrate while others do not? As with most important questions in biology, providing answers will present interesting challenges, but we can take inspiration from the challenges faced by the migrants themselves. I hope I have provided a useful framework to guide us in the quest.

References

Abele, L. G. and N. Blum. 1977. Ecological aspects of the freshwater decapod crustaceans of the Perlas Archipelago, Panama. Biotropica 9:239–252.

Abele, L. G. and D. B. Means. 1977. *Sesarma jarvisi* and *Sesarma cookei*: montane terrestrial grapsid crabs in Jamaica (Decapoda). Crustaceana 32:91–93.

Able, K. P. 1980. Mechanisms of orientation, navigation, and homing. In: S. A. Gauthreaux, Jr., ed. Animal Migration, Orientation and Navigation. Academic Press, New York, pp. 283–373.

Able, K. P. 1991a. Common themes and variations in animal orientation systems. Amer. Zool. 31:157–167.

Able, K. P. 1991b. The development of migratory orientation mechanisms. In: P. Berthold, ed. Orientation in Birds. Birkhäuser Verlag, Basel, pp. 166–179.

Able, K. P. 1993. Orientation cues used by migratory birds: A review of cue-conflict experiments. Trends Ecol. Evol. 8:367–371.

Achtemeier, G. L., B. Ackerman, L. K. Hendrie, M. E. Irwin, R. Larkin, N. Liquido, E. A. Mueller, R. W. Scott, W. Steiner, and D. J. Voegtlin. 1987. The Pests and Weather Project. Report ILENR/RE-AQ-87/1 of the Illinois Dept. of Energy and Natural Resources. 227 pp.

Ackery, P. R. and R. I. Vane-Wright. 1984. Milkweed Butterflies: Their Cladistics and Biology. Cornell University Press, Ithaca, N.Y.

Adler, K. and J. B. Phillips. 1985. Orientation in a desert lizard (*Uma notata*): time-compensated compass movement and polarotaxis. J. Comp. Physiol. A 156:547–552.

Adler, K. and D. H. Taylor. 1973. Extraocular perception of polarized light by orienting salamanders. J. Comp. Physiol. 87:203–212.

Adriaensen, F. and A. A. Dhondt. 1990. Population dynamics and partial migration of the European robin (*Erithacus rubecula*) in different habitats. J. Anim. Ecol. 59:1077–1090.

Adu-Mensah, K. and R. Kumar. 1977. Ecology of *Oxycarenus* species (Heteroptera: Lygaeidae) in Southern Ghana. Biol. J. Linn. Soc. 9: 349–377.

Aidley, D. J., ed. 1981. Animal Migration. Cambridge University Press.

Aleksiuk, M. 1976. Reptilian hibernation: Evidence of adaptive strategies in *Thamnophis sirtalis parietalis*. Copeia 1976:170–178.

Alerstam, T. 1985. Strategies of migratory flight, illustrated by arctic and common terns, *Sterna paradisea* and *Sterna hirundo*. In: M. A. Rankin, ed. Migration: Mechanisms and Adaptive Significance. Contrib. Marine Sci. 27 (Suppl.): 580–603.

Alerstam, T. 1990. Bird Migration. Cambridge University Press, Cambridge.

Alerstam, T. 1991. Ecological causes and consequences of bird orientation. In: P. Berthold, ed. Orientation in Birds. Birkhäuser, Basel, pp. 202–225.

Alerstam, T. and P. H. Enckell. 1979. Unpredictable habitats and evolution of bird migration. Oikos 33:228–232.

Allen, M. M. and K. B. Snow. 1993. The Monarch Project: a program of practical conservation in California. In: S. B. Malcolm and M. P. Zalucki, eds. Biology and Conservation of the Monarch Butterfly. Los Angeles County Natural History Museum, Science Series, No. 38, pp. 393–394.

Andersen, M. C. 1992. An analysis of variability in seed settling velocities of several wind-dispersed Asteraceae. Am. J. Bot. 79:1087–1091.

Andersson, M. 1980. Nomadism and site tenacity as alternative reproductive tactics in birds. J. Anim. Ecol. 49:175–184.

Antonovics, J. and P. H. van Tiederen. 1991. Ontoecogenophyloconstraints? The chaos of constraint terminology. Trends. Ecol. Evol. 6:166–167.

Arnold, G. P. 1981. Movements of fish in relation to water currents. In: D. J. Aidley, ed. Animal Migration. Cambridge University Press, Cambridge, pp. 55–80.

Arnold, G. P. and P. H. Cook. 1984. Fish migration by selective tidal stream transport: First results with a computer simulation model for the European continental shelf. In: J. D. McCleave, G. P. Arnold, J. J. Dodson, and W. H. Neill, eds. Mechanisms of Migration in Fishes. Plenum, New York, pp. 227–261.

Arnold, S. J. 1992. Constraints on phenotypic evolution. Am. Nat. 140 (Suppl.): s85–s107.

Aschoff, J. 1960. Exogenous and endogenous components in circadian rhythms. Cold Spring Harbor Symp. Quant. Biol. 25:11–28.

Askins, R. A., J. F. Lynch, and R. Greenberg. 1990. Population declines in migratory birds in eastern North America. Current Ornithology 7:1–57.

Atkins, M. D. 1959–61. A study of the flight of the Douglas-fir beetle, *Dendroctonus pseudotsugae* Hopk. (Coleoptera; Scolytidae) I. Flight preparation and response. II. Flight movements. III. Flight capacity. Can. Entomol. 91:283–291; 92:941–954; 93:467–474.

Atkins, M. D. 1966a. Behavioral variation among scolytids in relation to their habitat. Can. Entomol. 98:285–288.

Atkins, M. D. 1966b. Laboratory studies on the behaviour of the Douglas-fir beetle, *Dendroctonus pseudotsugae* Hopkins. Can. Entomol. 98:953–991.

Auburn, J. S. and D. H. Taylor. 1979. Polarized light perception and orientation in larval bull frogs *Rana catesbeiana*. Animal Behav. 27:658–668.

Audet, C., G. J. Fitzgerald, and H. Guderley. 1985. Prolactin and cortisal control of salinity preferences in *Gasterosteus aculeatus* and *Apeltes quadracus*. Behaviour 93:36–55.

Aukema, B. 1991. Fecundity in relation to wing-morph of three closely related species of the *melanocephalus* group of the genus *Calathus* (Coleoptera: Carabidae). Oecologia 87:118–126.

Baggerman, B. 1962. Some endocrine aspects of fish migration. Gen. Comp. Endocrin. Suppl. 1:188–205.

Bairlein, F. 1987. The migratory strategy of the Garden Warbler: A survey of field and laboratory data. Ringing and Migration 8:59–72.

Bairlein, F. 1988. How do migratory songbirds cross the Sahara? Trends Ecol. Evol. 3:191–194.

Bairlein, F. 1990. Nutrition and food selection in migratory birds. In: E. Gwinner, ed. Bird Migration: Physiology and Ecophysiology. Springer-Verlag, Berlin, pp. 198–213.

Bairlein, F. 1992a. Recent prospects on trans-Saharan migration of songbirds. Ibis 134 suppl 1:41–46.

Bairlein, F. 1992b. Migratory strategies of songbirds across the Sahara. Proc. VII Pan-African Ornith. Cong., pp.91–100.

Bairlein, F. and U. Totzke. 1992. New aspects on migratory physiology of trans-Saharan passerine migrants. Ornis Scand. 23:244–250.

Baker, R. R. 1968a. A possible method of evolution of the migratory habit in butterflies. Phil. Trans. Roy. Soc. B 253:309–341.

Baker, R. R. 1968b. Sun orientation during migration in some British butterflies. Proc. R. Ent. Soc. Lond. A 43:89–95.

Baker, R. R. 1978. The Evolutionary Ecology of Animal Migration. Hodder & Stoughton, London.

Baldassare, G. A. and D. H. Fischer. 1984. Food habits of fall migrant shorebirds on the Texas high plains. J. Field Ornithol. 55:220–229.

Baptista, L. F. and L. Petrinovitch. 1984. Social interaction, sensitive phases, and the song template hypothesis in the white-crowned sparrow. Anim. Behav. 32:172–181.

Barker, J. F. and W. S. Herman. 1976. Effect of photoperiod and temperature on reproduction of the monarch butterfly, *Danaus plexippus*. J. Insect Physiol. 22:1565–1568.

Barlow, R. B., M. K. Powers, H. Howard, and L. Kass. 1987. Vision in *Limulus* mating and migration. In: W. F. Herrnkind and A. B. Thistle, eds. Signposts in the Sea. Department of Biological Sciences, Florida State University, Tallahassee, pp. 69–84.

Barnes, R. 1993. Riparian forests: rivers of life. In: C. Peaslee, D. Evans, G. Geupel, and N. Nur, eds. On Behalf of Songbirds: California Planning and Action for the Partners in Flight Initiative. PRBO, Stinson Beach, California, pp. 8–9.

Barry, R. G. and R. J. Chorley. 1987. Atmosphere, Weather & Climate, 5th ed. Methuen, London.

Barton, N. H. and M. Turelli. 1989. Evolutionary quantitative genetics: How little do we know? Ann. Rev. Genet. 23:337–370.

Baskin, Y. 1993. Blue whale population may be increasing off California. Science 260:287.

Bauer, H.-G. and A. Kaiser. 1991. Herbsfangdaten, Verweildauer, Mauser und Biometrie teilziehender Gartenbaumlaufer (*Certhia brachydactyla*) in einem südwestdeutschen Rastgebiet. Die Vogelwarte 36:85–98.

Bazzaz, F. A. 1986. Life history of colonizing plants: some demographic, genetic, and physiological features. In: H. A. Mooney and J. A. Drake, eds. Ecology of Biological Invasions of North America and Hawaii. Ecological Studies, vol. 58:96–110. Springer-Verlag, Berlin.

Beall, G. 1948. The fat content of a butterfly *Danaus plexippus* Linn. as affected by migration. Ecology 29:80–94.

Beamish, F. W. H. 1978. Swimming capacity. Fish Physiology 7:101–187.

Beason, R. C. 1992. You can get there from here: Responses to simulated magnetic equator crossing by the bobolink (*Dolichonyx oryzivorous*). Ethology 91:75–80.

Beason, R. C. and P. Semm. 1991. Neuroethological aspects of avian orientation. In: P. Berthold, ed. Orientation in Birds. Birkhäuser Verlag, Basel, pp. 106–127.

Beck, S. D. 1980. Insect Photoperiodism, 2nd ed. Academic Press, New York.

Bell, W. J. 1991. Searching Behavior: The Behavioral Ecology of Finding Resources. Chapman and Hall, London.

Bell, W. J. and T. R. Tobin. 1982. Chemo-orientation. Biol. Rev. 57:219–260.

Berger, J. 1987. Reproductive fates of dispersers in a harem-dwelling ungulate: the wild horse. In: B. D. Chepko-Sade and Z. Tang-Halpin, eds. Mammalian Dispersal Patterns. University of Chicago Press, Chicago, pp. 41–54.

Berry, R. E. and L. R. Taylor. 1968. High-altitude migration of aphids in maritime and continental climates. J. Anim. Ecol. 37:713–722.

Berthold, P. 1975. Migration: Control and metabolic physiology. Avian Biology 5:77–128.

Berthold, P. 1984a. The endogenous control of bird migration: a survey of experimental evidence. Bird Study 31:19–27.

Berthold, P. 1984b. The control of partial migration in birds: a review. The Ring 10:253–265.

Berthold, P. 1985. Physiology and genetics of avian migration. In: M. A. Rankin, ed. Migration: Mechanisms and Adaptive Significance. Contrib. Marine Sci. 27 (Suppl.), pp. 526–543.

Berthold, P. 1988. Evolutionary aspects of migratory behavior in European warblers. J. Evol. Biol. 1:195–209.

Berthold, P. 1991a. Genetic control of migratory behaviour in birds. Trends Ecol. Evol. 6:254–257.

Berthold, P., ed. 1991b. Orientation in Birds. Birkhäuser Verlag, Basel.

Berthold, P. 1991c. Spatiotemporal programmes and genetics of orientation. In: P. Berthold, ed. Orientation in Birds. Birkhäuser Verlag, Basel, pp. 86–105.

Berthold, P. 1991d. Orientation in birds: A final consideration. In: P. Berthold, ed. Orientation in Birds. Birkhäuser Verlag, Basel, pp. 322–327.

Berthold, P. 1992. Steuerung des Vogelzuges. Biologie in unserer Zeit. 22(1):33–38.

Berthold, P. 1993. Bird Migration: A General Survey. Oxford University Press.

Berthold, P. and U. Querner. 1981. Genetic basis of migratory behavior in European warblers. Science 212:77–79.

Berthold, P. and U. Querner. 1982. Partial migration in birds: experimental proof of polymorphism as a controlling system. Experientia 38:805.

Berthold, P. and S. Terrill. 1988. Migratory behaviour and population growth of blackcaps wintering in Britain and Ireland: some hypotheses. Ringing and Migration 9:153–159.

Berthold, P., J. Griesinger, E. Nowak, and U. Querner. 1991. Satelliten-Telemetrie eines Gänsegeiers (*Gyps fulvus*) in Spanier. J. Ornithol. 132:327–329.

Berthold, P., A. J. Helbig, G. Mohr, and U. Querner. 1992. Rapid microevolution of migratory behaviour in a wild bird species. Nature 360:668–670.

Berthold, P., W. Wiltschko, H. Mildenberger, and U. Querner. 1990. Genetic transmission of migratory behavior into a nonmigratory bird population. Experientia 46:107–108.

Berven, K. A. and T. A. Grudzien. 1990. Dispersal in the wood frog (*Rana sylvatica*): implications for genetic population structure. Evolution 44:2047–2056.

Betts, E. 1976. Forecasting infestations of tropical migrant pests: the Desert Locust and the African Armyworm. Roy. Ent. Soc. Symp. 7:113–134.

Bhakthan, N. M. G., J. H. Borden, and K. K. Nair. 1970. Fine structure of degenerating and regenerating flight muscles in a bark beetle, *Ips confusus*. I. Degeneration. J. Cell. Sci. 6:807–820.

Bhakthan, N. M. G., K. K. Nair, and J. H. Borden. 1971. Fine structure of degenerating and regenerating flight muscles in a bark beetle, *Ips confusus*. II. Regeneration. Canad. J. Zool. 49:85–89.

Bibby, C. J. and R. E. Green. 1981. Autumn migration strategies of reed and sedge warblers. Ornis Scand. 12:1–12.

Biebach, H. 1983. Genetic determination of partial migration in the European robin (*Erithacus rubecula*). Auk 100:601–606.

Biebach, H. 1990. Strategies of trans-Sahara migrants. In: E. Gwinner, ed. Bird Migration: Physiology and Ecophysiology. Springer-Verlag, Berlin, pp. 352–367.

Biebach, H. 1991. Is water or energy crucial for trans-Sahara migrants? Acta XX Cong. Int. Ornith., Vol. II, pp. 773–779.

Biebach, H. 1992. Flight-range estimates for small trans-Sahara migrants. Ibis 134 (Suppl.) 1:47–54.

Biebach, H., W. Friedrich, and G. Heine. 1986. Interaction of bodymass, fat, foraging and stopover period in trans-Sahara migrating passerine birds. Oecologia 69:370–379.

Bigg, M. A., G. Ellis, and K. C. Balcomb. 1986. The photographic identification of individual cetaceans. Whalewatcher (J. Amer. Cetacean Soc.) 20(2):10–12.

Bigg, M. A., G. M. Ellis, J. K. B. Ford, and K. C. Balcomb. 1987. Killer whales: a study of their identification, geneology and natural history in British Columbia and Washington State. Phantom Press, Nanaimo, B.C.

Bilcke, G. 1984. Residence and non-residence in Passerines: dependence on the vegetation structure. Ardea 72:223–227.

Binns, E. S. 1982. Phoresy as migration—some functional aspects of phoresy in mites. Biol. Rev. 57:571–620.

Björnson, B. T., G. Young, R. J. Lin, L. J. Deftos, and H. A. Bern. 1989. Smoltification and seawater adaptation in coho salmon (*Oncorhynchus kisutch*): Plasma calcium regulation, osmoregulation, and calcitonin. Gen. Comp. Endocrinol. 74:346–354.

Black, M. 1993. Recounting a century of failed fishery policy toward Sacramento River salmon and steelhead. Bodega Bay Marine Laboratory, Colloquium VI, Conservation Biology of Endangered Pacific Salmonids: Life History, Genetics and Demography, 8–12 September 1993.

Blackman, R. T. 1975. Photoperiodic determination of the male and female sexual morphs of *Myzus persicae*. J. Insect Physiol. 21:435–453.

Blackmer, J. L. and P. L. Phelan. 1991. Behavior of *Carpophilus hemipterus* in a vertical flight chamber: transition from phototactic to vegetative orientation. Entomol. Exp. Appl. 58:137–148.

Blakemore, R. P. 1975. Magnetotactic bacteria. Science 190:377–379.

Blakers, M., S. J. J. F. Davies, and P. N. Reilly. 1984. The Atlas of Australian Birds. Melbourne Univ. Press, Carlton, Victoria, Australia.

Bland, R. T. 1986. Blackcap. In: P. Lack, ed. The Atlas of Wintering Birds in Britain and Ireland. T. and A. D. Poyser, Ltd., Carlton, U.K., pp. 332–333.

Blem, C. R. 1980. The energetics of migration. In: S. A. Gauthreaux, Jr., ed. Animal Migration, Orientation and Navigation. Academic Press, New York, pp. 175–224.

Blem, C. R. 1990. Avian energy storage. Current Ornithol. 7:59–113.

Bliss, D. E. 1968. Transition from water to land in decapod crustaceans. Amer. Zool. 8:355–392.

Boake, C. R. B., ed. 1994. Quantitative Genetic Studies of Behavioral Evolution. University of Chicago Press, Chicago.

Borden, J. H. and C. E. Slater. 1969. Flight muscle volume change in *Ips confusus* (Coleoptera: Scolytidae). Canad. J. Zool. 47:29–32.

Borowicz, V. A. 1988. Fruit consumption by birds in relation to fat content of pulp. Am. Midl. Nat. 119:121–127.

Botton, M. L. and R. E. Loveland. 1987. Orientation of the horseshoe crab, *Limulus polyphemus*, on a sandy beach. Biol. Bull. 173:289–298.

Bradshaw, W. E. 1986. Pervasive themes in insect life cycle strategies. In: F. Taylor and R. Karban, eds. The Evolution of Insect Life Cycles. Springer, New York, pp.

261–275.

Brady, J., ed. 1982. Biological Timekeeping. Cambridge University Press, Cambridge.

Brosset, A. 1990. A long term study of the rain forest birds of M'Passa (Gabon). In: A. Keast, ed. Biogeography and Ecology of Forest Bird Communities, SPB Academic Publishing, The Hague, pp. 259–274.

Brower, L. P. 1985. New perspectives on the migration biology of the monarch butterfly, *Danaus plexippus* L. In: M. A. Rankin, ed. Migration: Mechanisms and Adaptive Significance. Contrib. Marine Sci. 27 (Suppl.), 748–785.

Brower, L. P. and S. B. Malcolm. 1991. Animal migrations: endangered phenomena. Amer. Zool. 31:265–276.

Brower, L. P., W. H. Calvert, L. H. Hedrick, and J. Christian. 1977. Biological observations on an overwintering colony of monarch butterflies (*Danaus plexippus*) in Mexico. J. Lepidop. Soc. 31:232–242.

Brown, E. S. 1951. The relation between migration rate and type of habitat in aquatic insects with special reference to certain Corixidae. Proc. Zool. Soc. Lond. 121:539–545.

Brown, E. S. 1965. Notes on the migration and direction of flight of *Eurygaster* and *Aelia* species (Hemiptera: Pentatomidae) and their possible bearing on invasions of cereal crops. J. Anim. Ecol. 34:93–107.

Brown, P. 1990. Swift parrot, *Lathamus discolor*. In: J. Brouwer and S. Garnett, eds. Threatened Birds of Australia: An Annotated List. Royal Australasian Ornithologists Union Report Number 68, Melbourne.

Bruggers, R. L. and C. C. H. Elliott, eds. 1989. Quelea quelea: Africa's Bird Pest. Oxford University Press, Oxford.

Buntin, J. D. and E. Ruzycki. 1987. Characteristics of prolactin binding sites in the brain of the ring dove (*Streptopelia risoria*). Gen. Comp. Endocrin. 65:243–253.

Burbidge, A. A. and P. J. Fuller. 1982. Banded stilt breeding at Lake Barlee, Western Australia. Emu 82:212–216.

Butcher, G. S., B. Peterjohn, and C. J. Ralph. 1993. Overview of bird population monitoring programs and databases. In: D. M. Finch and P. W. Stangel, eds. Status and Management of Neotropical Migratory Birds. USDA Rocky Mtn. Forest and Range Exp. Sta., Ft. Collins, Colorado, pp. 192–203.

Caldwell, R. L. 1974. A comparison of the migratory strategies of two milkweed bugs, *Oncopeltus fasciatus* and *Lygaeus kalmii*. In: L. Barton Browne, ed. Experimental Analysis of Insect Behaviour. Springer, New York, pp. 304–316.

Caldwell, R. L. and H. Dingle. 1967. The regulation of cyclic reproductive and feeding activity in the large milkweed bug *Oncopeltus* by temperature and photoperiod. Biol. Bull. 133:510–525.

Caldwell, R. L. and J. P. Hegmann. 1969. Heritability of flight duration in the milkweed bug, *Lygaeus kalmii*. Nature 223:91–92.

Caldwell, R. L. and M. A. Rankin. 1974. Separation of migratory from feeding and reproductive behavior in *Oncopeltus fasciatus*. J. Comp. Physiol. 88:383–394.

Calvert, W. H. and L. P. Brower. 1981. The importance of forest cover for the survival of overwintering monarch butterflies (*Danaus plexippus*, Danaidae) in Mexico. J. Lepidop. Soc. 35:216–225.

Calvert, W. H. and L. P. Brower. 1986. The location of monarch butterfly (*Danaus plexippus* L.) overwintering colonies in Mexico in relation to topography and climate. J. Lepidop. Soc. 40:164–187.

Campbell, E. A. and P. B. Moyle. 1991. Historical and recent population sizes of spring-run chinook salmon in California. In: T. J. Hassler, ed. 1990 Northeast Pacific Chinook and Coho Salmon Workshop Proceedings, American Fisheries Society. Humboldt State University, Arcata, Calif., pp. 155–216.

Carmi, N., B. Pinshow, W. P. Porter, and J. Jaeger. 1992. Water and energy limitations on flight duration in small migrating birds. Auk 109:268–276.

Carr, A. F. 1967. So Excellent a Fishe: A Natural History of Sea Turtles. Scribner, New York.

Carscadden, J. E. and W. C. Leggett. 1975. Life history variations in populations of American shad, *Alosa sapidissima* (Wilson), spawning in tributaries of the St. John River, New Brunswick. J. Fish. Biol. 7:595–609.

Castonguay, M., J. Dutil, C. Audet, and R. Miller. 1990. Locomotor activity and concentration of thyroid hormones in migratory and sedentary juvenile American eels. Trans. Amer. Fish. Soc. 119:946–956.

Cheke, R. 1978. Theoretical rates of increase of gregarious and solitarious populations of the desert locust. Oecologia 35:161–171.

Chen, R.-L., X.-Z. Bao, V. A. Drake, R. A. Farrow, S.-Y. Wang, Y.-J. Sun, and B.-P. Zhai. 1989. Radar observations of the spring migration into northeastern China of the oriental armyworm moth, *Mythimna separata*, and other insects. Ecol. Entomol. 14:149–162.

Clark, L. and J. R. Mason. 1985. Use of nest material as insecticidal and anti-pathogenic agents by the European starling. Oecologia 67:169–176.

Clark, L. and J. R. Mason. 1987. Olfactory discrimination of plant volatiles by the European starling. Anim. Behav. 35:227–235.

Close, R. C. and A. I. Tomlinson. 1975. Dispersal of the grain aphid *Macrosiphum miscanthi* from Australia to New Zealand. N. Z. Entomol. 6:62–65.

Coats, S. A., J. A. Mutchmor, and J. J. Tollefson. 1987. Regulation of migratory flight by juvenile hormone mimic and inhibitor in the western corn rootworm (Coleoptera: Chrysomelidae). Ann. Entomol. Soc. Amer. 80:697–708.

Cochran, W. W. 1987. Orientation and other migratory behaviors of a Swainson's thrush followed for 1,500 km. Animal Behav. 35:927–929.

Cockbain, A. J. 1961. Fuel utilization and duration of tethered flight in *Aphis fabae*. Scop. J. Exp. Biol. 38:163–174.

Cockburn, A. 1992. Habitat heterogeneity and dispersal: environmental and genetic patchiness. In: N. C. Stenseth and W. Z. Lidicker, Jr., eds. Animal Dispersal: Small Mammals as a Model. Chapman & Hall, London, pp. 65–95.

Cody, M. L. 1972. Finch flocks in the Mohave Desert. Theor. Pop. Biol. 2:142–158.

Coemans, M., J. Vos, and J. Nuboer. 1990. No evidence for polarization sensitivity in the pigeon. Naturwiss. 77:138–142.

Cohen, D. 1967. Optimization of seasonal migratory behavior. Amer. Natur. 101:5–17.

Collie, N. L. 1985. Intestinal nutrient transport in coho salmon (*Oncorhynchus kisutch*) and the effects of development, starvation, and seawater adaptation. J. Comp. Physiol. B156:163–174.

Common, I. F. B. 1954. A study of the ecology of the adult bogong moth, *Agrotis infusa* (Boisd.) (Lepidoptera: Noctuidae) with special reference to its behaviour during migration and aestivation. Aust. J. Zool. 2:223–263.

Cox, G. W. 1968. The role of competition in the evolution of migration. Evolution 22:180–192.

Cox, G. W. 1985. The evolution of avian migration systems between temperate and tropical regions of the New World. Am. Nat. 126:451–474.

Cramp, S. and K. E. L. Simmons, eds. 1988. Handbook of the Birds of Europe, the Middle East, and North Africa, Vol. 5. Oxford Univ. Press, London.

Cronin, T. W. and R. B. Forward, Jr. 1979. Tidal vertical migration: an endogenous rhythm in estuarine crab larvae. Science 205:1020–1022.

Cusson, M. and J. N. McNeil. 1989. Involvement of juvenile hormone in the regulation of pheromone release activities in a moth. Science 243:210–212.

Danks, H. V. 1987. Insect Dormancy: An Ecological Perspective. Biol. Survey Canada Monogr. Ser. No. 1, Ottawa.

Danthanarayana, W. 1976a. Diel and lunar flight periodicities in the light brown apple moth, *Epiphyas postvittana* (Walker) (Tortricidae) and their possible adaptive significance. Aust. J. Zool. 24:65–73.

Danthanarayana, W. 1976b. Flight thresholds and seasonal variations in flight activity of the light brown apple moth, *Epiphyas postvittana* (Walker) (Tortricidae). Oecologia 23:171–182.

Danthanarayana, W. 1983. Population ecology of the light brown apple moth, *Epiphyas postvittana* (Lepidoptera: Tortricidae). J. Anim. Ecol. 52:1–33.

Danthanarayana, W., ed. 1986. Insect Flight: Dispersal and Migration. Springer-Verlag, Berlin.

Darling, J. D. and D. J. McSweeney. 1985. Observations on the migrations of North Pacific humpback whales (*Megaptera novaeangliae*). Can. J. Zool. 63:308–314.

Darwin, C. 1859. The Origin of Species by Means of Natural Selection. John Murray, London.

Davey, J. T. 1953. Possibility of movements of the African migratory locust in the solitary phase and the dynamics of its outbreaks. Nature 172:720–721.

Davies, S. J. J. F. 1984. Nomadism as a response to desert conditions in Australia. J. Arid Environ. 7:183–195.

Davis, M. A. 1980. Variation in flight duration among individual *Tetraopes* beetles: implications for studies of insect flight. J. Insect Physiol. 26: 403–406.

Davis, M. A. 1981. The flight capacity of dispersing milkweed beetles, *Tetraopes tetraophthalmus*. Ann. Ent. Soc. Amer. 74:385–386.

Davis, N. T. 1975. Hormonal control of flight muscle histolysis in *Dysdercus fulvoniger*. Ann. Entomol. Soc. Amer. 68:710–714.

Dawbin, W. N. 1966. The seasonal migratory cycle of humpback whales. In: K. S. Norris, ed. Whales, Dolphins and Porpoises. University of California Press, Berkeley and Los Angeles, pp. 145–170.

Deelder, C. L. 1958. On the behavior of elvers (*Anguilla vulgaris*) migrating from the sea into fresh water. J. Conseil 24:135–146.

de Kort, C. A. D. 1969. Hormones and the structural and biochemical properties of the flight muscles in the Colorado beetle. Meded. Land. Wageningen 69:1–63.

de Kort, C. A. D., B. J. Bergot, and D. A. Schooley. 1982. The nature and titre of juvenile hormone in the Colorado potato beetle, *Leptinotarsa decemlineata*. J. Insect Physiol. 28:471–475.

Den Boer, P. J. 1981. On the survival of populations in a heterogeneous and variable environment. Oecologia 50:39–53.

Denno, R. F. 1985. Fitness, population dynamics and migration in planthoppers: the role of host plants. In: M. A. Rankin, ed. Migration: Mechanisms and Adaptive Significance. Contrib. Marine Sci. 27 (Suppl.), pp. 623–640.

Denno, R. F. and G. K. Roderick. 1992. Density-related dispersal in planthoppers: effects of interspecific crowding. Ecology 73:1323–1334.

Denno, R. F., L. W. Douglas, and D. Jacobs. 1985. Crowding and host plant nutrition: environmental determinants of wing form in *Prokelisia marginata*. Ecology 66:1588–1596.

Denno, R. F., G. K. Roderick, K. L. Olmstead, and H. G. Döbel. 1991. Density-related migration in planthoppers (Homoptera: Delphacidae): the role of habitat persistence. Am. Nat. 138:1513–1541.

Denno, R. F. and E. E. Grissell. 1979. The adaptiveness of wing-dimorphism in the salt march-inhabiting planthopper, *Prokelisia marginata* (Homoptera: Delphacidae). Ecology 60:221–236.

Dent, J. N. 1985. Hormonal interactions in the regulation of migratory movements of urodele amphibians. In: B. K. Folett, S. Ishii, and A. Chandala, eds. The Endocrine System and the Environment. Springer-Verlag, Berlin, pp. 79–84.

Derr, J. A. 1980. Coevolution of the life history of a tropical seed feeding insect and its food plants. Ecology 61:881–892.

Derr, J. A., B. A. Alden, and H. Dingle. 1981. Insect life histories in relation to migration, body size, and host plant array: a comparative study of *Dysdercus*. J. Anim. Ecol. 50:181–193.

DeSante, D. 1973. An analysis of the fall occurrence and nocturnal orientations of vagrant wood warblers (Parulidae) in California. Ph.D. Dissertation, Stanford University, California.

DeSante, D. 1983. Annual variability in the abundance of migrant landbirds on Southeast Farallon Island, California. Auk 100:826–852.

DeSante, D. and D. G. Ainley. 1980. The avifauna of the South Farallon Islands, California. Studies Avian Biol. No. 4.

De Vries, P. J. and R. Dudley. 1990. Morphometrics, airspeed, thermoregulation, and lipid reserves of migrating *Urania fulgens* (Uraniidae) moths in natural free flight. Physiol. Zool. 63:235–251.

de Wilde, P. A. W. J. 1973. On the ecology of *Coenobita clypeatus* in Curacao. Stud. Fauna Curacao 44:1–138.

Dhondt, A. A. and E. Matthysen. 1993. Conservation biology of birds: can we bridge the gap between head and heart? Trends Ecol. Evol. 8:160–161.

Diario Oficial. 1986. Organo del los Estados Unidos Mexicanos. México, D. f., Jueves 9 de Octubre de 1986, 398(27):33–41.

Dickison, R. B. B., M. J. Haggis, and R. C. Rainey. 1983. Spruce budworm moth flight and storms: case study of a cold front system. J. Clim. Appl. Meteorol. 22:278–286.

Dingle, H. 1965. The relation between age and flight activity in the milkweed bug, *Oncopeltus*. J. Exp. Biol. 42:269–283.

Dingle, H. 1968. The influence of environment and heredity on flight activity in the large milkweed bug, *Oncopeltus*. J. Exp. Biol. 48:175–184.

Dingle, H. 1969. Ontogenetic changes in phototaxis and thigmokinesis in stomatopod larvae. Crustaceana 16:108–110.

Dingle, H. 1974. Diapause in a migrant insect, the milkweed bug *Oncopeltus fasciatus* (Dallas) (Hemiptera: Lygaeidae). Oecologia 17:1–10.

Dingle, H., ed. 1978a. Evolution of Insect Migration and Diapause. Springer-Verlag, New York.

Dingle, H. 1978b. Migration and diapause in tropical, temperate, and island milkweed bugs. In: H. Dingle, ed. Evolution of Insect Migration and Diapause. Springer, New York, pp. 254–276.

Dingle H. 1980. The ecology and evolution of migration. In: S. A. Gauthreaux, Jr., ed.

Animal Migration, Orientation, and Navigation. Academic Press, New York, pp. 1–101.

Dingle, H. 1982. Function of migration in the seasonal synchronization of insects. Ent. exp. & Appl. 31:36–48.

Dingle, H. 1985. Migration. In: G. A. Kerkut and L. I. Gilbert, eds. Comprehensive Insect Physiology, Biochemistry and Pharmacology, Vol. 9, pp. 375–415.

Dingle, H. 1986. Evolution and genetics of insect migration. In: W. Danthanarayana, ed. Insect Flight: Dispersal and Migration. Springer-Verlag, Berlin, pp. 11–26.

Dingle, H. 1988. Quantitative genetics of life history evolution in a migrant insect. In: G. DeJong, ed. Population Genetics and Evolution. Springer, Berlin, pp. 83–93.

Dingle, H. 1989. The evolution and significance of migratory flight. In: G. J. Goldsworthy and C. H. Wheeler, eds. Insect Flight. CRC Press. Boca Raton, Florida.

Dingle, H. 1991a. Evolutionary genetics of animal migration. Amer. Zool. 31:253–264.

Dingle, H. 1991b. Factors influencing spatial and temporal variation in abundance of the large milkweed bug (Hemiptera: Lygaeidae). Ann. Entomol. Soc. Amer. 84:47–51.

Dingle, H. 1992. Food level reaction norms in size-selected milkweed bugs (*Oncopeltus fasciatus*). Ecol. Entomol. 17:121–126.

Dingle, H. 1994. Genetic analyses of animal migration. In: C. R. B. Boake, ed. Quantitative Genetic Studies of Behavioral Evolution. University Chicago Press, Chicago, pp. 145–164.

Dingle, H. and G. Arora. 1973. Experimental studies of migration in bugs of the genus *Dysdercus*. Oecologia 12:119–140.

Dingle, H. and R. L. Caldwell. 1972. Reproductive and maternal behavior of the mantis shrimp *Gonodactylus bredini* Manning. Biol. Bull. 142:417–426.

Dingle, H. and K. E. Evans. 1987. Responses in flight to selection on wing length in non-migratory milkweed bugs, *Oncopeltus fasciatus*. Entomol. Exp. Appl. 45:289–296.

Dingle, H. and S. A. Gauthreaux, Jr. 1991. Introduction to the Symposium: The maturing of migration. Amer. Zool. 31:153–155.

Dingle, H. and J. B. Haskell. 1967. Phase polymorphism in the grasshopper *Melanoplus differentialis*. Science 155:590–592.

Dingle, H., N. R. Blakley, and E. R. Miller. 1980. Variation in body size and flight performance in milkweed bugs (*Oncopeltus*). Evolution 34:371–385.

Dingle, H., K. E. Evans, and J. O. Palmer. 1988. Responses to selection among life-history traits in a nonmigratory population of milkweed bugs (*Oncopeltus fasciatus*). Evolution 42:79–92.

Dixon, A. F. G., S. Horth, and P. Kindlmann. 1993. Migration in insects: cost and strategies. J. Anim. Ecol. 62:182–190.

Dole, J. W. 1965. Summer movement of adult leopard frogs, *Rana pipiens* Schrieber, in Northern Michigan. Ecology 46:236–255.

Dole, J. W. 1967. Spring movement of adult leopard frogs, *Rana pipiens* Schrieber, in Northern Michigan. Am. Midl. Nat. 78:167–181.

Dolnik, V. R. 1990. Bird migration across arid and mountainous regions of Middle Asia and Kasakhstan. In: E. Gwinner, ed. Bird Migration: Physiology and Ecophysiology. Springer-Verlag, Berlin, pp. 368–386.

Dorst, J. 1962. The Migrations of Birds. Houghton, Boston, Massachusetts.

Døving, K. B., H. Westerberg, and P. B. Johnsen. 1985. Role of olfaction in the behavioral and neuronal responses of Atlantic salmon, *Salmo salar*, to hydrographic stratification. Can. J. Fish. Aquat. Sci. 542:1658–1667.

Downhower, J. F. 1992. The map problem in avian orientation: a re-evaluation. Anim.

Behav. 43:168–169.

Drake, V. A. 1982. Insects in the sea-breeze front at Canberra: A radar study. Weather 37:134–143.

Drake, V. A. 1991. Methods for studying adult movements in Heliothis. In: M. P. Zalucki, ed. Heliothis: Research Methods and Prospects. Spring-Verlag, Berlin, pp. 109–121.

Drake, V. A. and R. A. Farrow. 1988. The influence of atmospheric structure and motions on insect migration. Ann. Rev. Entomol. 33:183–210.

Drake, V. A. and R. A. Farrow. 1989. The 'aerial plankton' and atmospheric convergence. Trends Ecol. Evol. 4:381–385.

Drake, V. A., K. F. Helon, J. L. Readshaw, and D. G. Reid. 1981. Insect migration across Bass Strait during spring: a radar study. Bull. Entomol. Res. 71:449–466.

Drury, W. H. and J. A. Keith. 1962. Radar studies of songbird migration in coastal New England. Ibis 104:449–489.

Drury, W. H. and I. C. T. Nisbet. 1964. Radar studies of orientation of songbird migrants in southeastern New England. Bird Banding 35:69–119.

Dye, C. 1983. Insect movement and fluctuations in insect population size. Antenna 7:174–178.

Dyer, F. C. and J. L. Gould. 1983. Honey bee navigation. Amer. Scient. 71:587–597.

Eastwood, E. 1967. Radar Ornithology. Methuen, London.

Elgood, J. H., C. H. Fry, and R. J. Dowsett. 1973. African migrants in Nigeria. Ibis 115:1–45:375–411.

Ellegren, H. 1991. Stopover ecology of autumn migrating bluethroats (*Luscinia s. svecica*) in relation to age and sex. Ornis Scandinavica 22:340–348.

Ellington, C. P. 1991. Limitations on animal flight performance. J. Exp. Biol. 160:71–91.

Elliott, J. M. 1987. Population regulation in contrasting populations of trout *Salmo trutta* in two Lake District streams. J. Anim. Ecol. 56:83–98.

Elliott, J. M. 1988. Growth, size, biomass and production in contrasting populations of trout *Salmo trutta* in two Lake District streams. J. Anim. Ecol. 57:49–60.

Elliott, J. M. 1989. Growth and size variation in contrasting populations of trout *Salmo trutta*: an experimental study on the role of natural selection. J. Anim. Ecol. 58:45–58.

Ellis, P. E. 1951. The marching behavior of hoppers of the African migratory locust (*Locusta migratoria migratorioides* R. and F.) in the laboratory. Anti-Locust Bull No. 7.

Ellis, P. E., D. B. Carlisle, and D. J. Osborne. 1965. Desert locusts: sexual maturation delayed by feeding on senescent vegetation. Science 149:546–547.

Ellner, S. 1984. Asymptotic behavior of some stochastic difference equation models. J. Math. Biol. 19:169–200.

Ellner, S. 1985. ESS germination rates in randomly varying environments. I. Logistic-type models. II. Reciprocal yield-law models. Theor. Pop. Biol. 28: 50–79, 80–116.

Elton, C. 1927. Animal Ecology. Sidgwick and Jackson, London.

Elton, C. 1930. Animal Ecology and Evolution. Oxford University Press, London.

Emlen, S. T. 1967. Migratory orientation in the Indigo Bunting, *Passerina cyanea*. Part I: The evidence for use of celestial cues. Part II: Mechanisms of celestial orientation. Auk 84:309–342, 463–489.

Emlen, S. T. 1970. The influence of magnetic information on the orientation of the Indigo Bunting, *Passerina cyanea*. Anim. Behav. 18:215–224.

Emlen, S. T. 1975a. Migration: orientation and navigation. Avian Biology 5:129–219.

Emlen, S. T. 1975b. The stellar-orientation system of a migratory bird. Sci. Am. 233(2):102–111.

Emlen, S. T. and J. T. Emlen. 1966. A technique for recording migratory orientation of captive birds. Auk 83:361–367.

Eriksson, L.-O. and H. Lundquist. 1982. Circannual rhythms and photoperiodic regulation of growth and smolting in Baltic salmon (*Salmo salar* L.). Aquaculture 28:113–121.

Faaborg, J. and W. J. Arendt. 1992. Long-term declines of winter resident warblers in a Puerto Rican dry forest: Which species are in trouble? In: J. M. Hagan III and D. W. Johnston, eds. Ecology and Conservation of Neotropical Migrant Landbirds. Smithsonian Institution Press, Washington, D.C., pp. 57–63.

Fairbairn, D. J. 1986. Does alary dimorphism imply dispersal dimorphism in the waterstrider, *Gerris remigis*? Ecol. Entomol. 11: 355–368.

Fairbairn, D. J. 1988. Adaptive significance of wing dimorphism in the absence of dispersal: a comparative study of wing morphs in the waterstrider, *Gerris remigis*. Ecol. Entomol. 13:273–281.

Fairbairn, D. J. and T. C. Butler. 1990. Correlated traits for migration in the Gerridae (Hemiptera, Heteroptera): a field test. Ecol. Entomol. 15: 131–142.

Fairbairn, D. and L. Desranleau. 1987. Flight threshold, wing muscle histolysis, and alary polymorphism: correlated traits for dispersal tendency in the Gerridae. Ecol. Entomol. 11:13–24.

Fairbairn, D. J. and D. A. Roff. 1990. Genetic correlations among traits determining migratory tendency in the sand cricket, *Gryllus firmus*. Evolution 44:1787–1795.

Falconer, D. S. 1989. Introduction to Quantitative Genetics, 3rd. ed. Longman, London.

Farner, D. S. and B. K. Follett. 1979. Reproductive periodicity in birds. In: E. J. W. Barrington, ed. Hormones and Evolution, 2nd ed. Academic Press, London, pp. 829–872.

Farner, D. S. and E. Gwinner. 1980. Photoperiodicity, circannual, and reproductive cycles. In: A. Epple and M. H. Stetson, eds. Avian Endocrinology. Academic Press, London, pp. 331–366.

Farrow, R. A. 1975a. Offshore migration and the collapse of outbreaks of the Australian plague locust (*Chortoicetes terminifera* Walk.) in south-east Australia. Aust. J. Zool. 23:569–595.

Farrow, R. A. 1975b. The African Migratory Locust in its main outbreak area of the Middle Niger: quantitative studies of solitary populations in relation to environmental factors. Locusta, No. 11, 198 pp.

Farrow, R. A. 1984. Detection of transoceanic migration of insects to a remote island in the Coral Sea, Willis Island. Aust. J. Ecol. 9:253–272.

Farrow, R. A. 1990. Flight and migration in acridoids. In: R. F. Chapman and A. Joern, eds. Biology of Grasshoppers. Wiley, New York, pp. 227–314.

Farrow, R. A. and J. E. Dowse. 1984. Method of using kites to carry tow nets in the upper air for sampling migrating insects and its application to radar entomology. Bull. Ent. Res. 74:87–95.

Farrow, R. A. and B. C. Longstaff. 1986. Comparison of the intrinsic rates of natural increase of locusts in relation to the incidence of plagues. Oikos 46:207–222.

Farrow, R. A. and G. McDonald. 1987. Migration strategies and outbreaks of noctuid pests in Australia. Insect Sci. Applic. 8:531–542.

Fenner, M. 1985. Seed Ecology. Chapman and Hall, London and New York.

Fenton, M. B. and D. W. Thomas. 1985. Migrations and dispersal of bats (Chiroptera). In: M. A. Rankin, ed. Migration: Mechanisms and Adaptive Significance. Contrib. Marine Sci. 27 (Suppl.), pp. 407–424.

Fisher, H. I. 1975. The relationship between deferred breeding and mortality in the Laysan albatross. Auk 92:433–441.

Fishpool, L. D. C. and G. B. Popov. 1984. The grasshopper faunas of the savannas of Mali, Niger, Benin and Togo. Bull. Inst. Fondam. Afr. Naire Ser A 43:275–410.

Fitt, G. P. 1989. The ecology of *Heliothis* species in relation to agroecosystems. Ann. Rev. Entomol. 34:17–52.

Fitt, G. P. and J. C. Daly. 1990. Abundance of overwintering pupae and the spring generation of *Helicoverpa* spp. (Lepidoptera: Noctuidae) in northern New South Wales, Australia: implications for pest management. J. Econ. Entomol. 83:1827–1836.

Fletcher, D. J. C. 1978. The African bee, *Apis mellifera adansonii*, in Africa. Ann. Rev. Entomol. 23:151–171.

Foerster, R. E. 1947. Experiments to develop sea-run from land-locked sockeye salmon (*Oncorhynchus nerka kennerlyi*) J. Fish. Res. Board Canada 7:88–93.

Follett, B. K. 1982. Photoperiodic physiology in animals. In: J. Brady, ed. Biological Timekeeping. Cambridge University Press, Cambridge, pp. 83–100.

Follett, B. K. and D. E. Follett, eds. 1981. Biological Clocks in Seasonal Reproductive Cycles. John Wright, Bristol.

Fontaine, M. 1975. Physiological mechanisms in the migration of marine and amphihaline fish. Adv. Mar. Biol. 13:241–355.

Foote, C. J., C. C. Wood, W. C. Clarke, and J. Blackburn. 1992. Circannual cycle of seawater adaptability in *Oncorhynchus nerka*: genetic differences between sympatric sockeye salmon and kokanee. Can. J. Fish. Aquat. Sci. 49:99–109.

Ford, H. A. 1978. The Black Honeyeater: nomad or migrant? S. Aust. Ornithol. 27:263–269.

Ford, H. A. 1989. Ecology of Birds: An Australian Perspective. Surrey Beatty & Sons, Chipping Norton, NSW.

Ford, J. K. B. 1991. Vocal traditions among resident killer whales (*Orcinus orca*) in coastal waters of British Columbia. Can. J. Zool. 69:1454–1483.

Forshaw, J. M. 1981. Australian Parrots, 2nd ed. Landsdowne Editions, Sydney.

Forward, R. B., Jr. 1987. Crustacean larval vertical migration: a perspective. In: W. F. Herrnkind and A. B. Thistle, eds. Signposts in the Sea. Department of Biological Science, Florida State University, Tallahassee, pp. 29–44.

Fosberg, M. A. and M. Peterson. 1986. Modeling airborne transport of gypsy moth (Lepidoptera: Lymantriidae) larvae. Agr. For. Met. 38:1–8.

Foster, W. A. 1978. Dispersal behaviour of an intertidal aphid. J. Anim. Ecol. 47:653–659.

Foster, W. A. and J. E. Treherne. 1978. Dispersal mechanisms in an intertidal aphid. J. Anim. Ecol. 47:205–217.

Fox, W. W. Jr. 1992a. Endangered Species Act, Section 7 Consultation/Conference, Biological Opinion (April 10, 1992). National Marine Fisheries Service, 73 pp. + Appendix.

Fox, W. W. Jr. 1992b. Biological Opinion (February 14, 1992). National Marine Fisheries Service, 48 pp. + maps.

Frazzeta, T. 1975. Complex Adaptations in Evolving Populations. Sinauer, Sunderland, Massachusetts.

Fretwell, S. D. 1972. Populations in a Seasonal Environment. Princeton Univ. Press, Princeton, New Jersey.

Frisch, K. von. 1950. Die Sonne als Kompass im Leben der Bienen. Experientia 6:210–221.

Frisch, K. von. 1967. The Dance Language and Orientation of Bees. Harvard University Press, Cambridge, Massachusetts.

Frith, H. G. 1982. Waterfowl in Australia, rev. ed. Angus and Robertson, Sydney.

Fujisaki, K. 1985. Ecological significance of the wing polymorphism of the oriental chinch bug, *Cavelarius sacchivorous* Okajima (Heteroptera: Lygaeidae). Res. Pop. Ecol. 27:125–136.

Fujisaki, K. 1992. A male fitness advantage to wing reduction in the oriental chinch bug, *Cavelarius saccharivorous* Okajima (Heteroptera: Lygaeidae). Res. Pop. Biol. 34:173–183.

Fullagar, P. J., K. W. Lowe, and S. J. J. F. Davies. 1988. Intracontinental migration of Australian birds. Proc. XIX Int. Ornith. Congr. Vol. I: 791–801.

Fulton, L. 1968. Spawning areas and abundance of chinook salmon *Oncorhynchus tshawytscha* in the Columbia River—past and present. U.S. Fish and Wildlife Service, Special Scientific Report, Fisheries No. 571, 26 pp.

Fuseini, B. A. and R. Kumar. 1975. Ecology of cotton stainers (Heteroptera: Pyrrhococidae) in southern Ghana. Biol. J. Linn. Soc. 7:113–146.

Gabriel, M. H. 1977. Models of insect migration. Ph.D. Thesis, University of Iowa.

Gaines, M. S. and T. R. McClenaghan, Jr. 1980. Dispersal in small mammals. Ann. Rev. Ecol. Syst. 11:163–196.

Ganeshaiah, K. N. and R. Uma Shaanker. 1991. Seed size optimization in a wind-dispersed tree *Butea monosperma*—a trade off between seedling establishment and pod dispersal. Oikos 60:3–6.

Garner, W. W. and H. A. Allard. 1920. Effect of the relative length of the day and night and other factors of the environment on growth and reproduction in plants. J. Agric. Res. 18:553–606.

Garner, W. W. and H. A. Allard. 1923. Further studies in photoperiodism, the response of the plant to the relative length of day and night. J. Agric. Res. 23:871–920.

Gatehouse, A. G. 1986. Migration of the African armyworm *Spodoptera exempta*: genetic determination of migratory capacity and a new synthesis. In: W. Danthanarayana, ed. Insect Flight: Dispersal and Migration, Springer-Verlag, Berlin, pp. 128–144.

Gatehouse, A. G. 1987a. Migration and low population density in armyworm (Lepidoptera: Noctuidae) life histories. Insect Sci. Applic. 8:573–580.

Gatehouse, A. G. 1987b. Migration: A behavioral process with ecological consequences? Antenna 11:10–12.

Gatehouse, A. G. 1989. Genes, environment, and insect flight. In: G. J. Goldsworthy and C. H. Wheeler, eds. Insect Flight. CRC Press. Boca Raton, FL, pp. 115–138.

Gatehouse, A. G. and D. S. Hackett. 1980. A technique for studying flight behaviour of tethered *Spodoptera exempta* moths. Physiol. Entomol. 5:215–222.

Gauthreaux, S. A., Jr. 1978. The ecological significance of behavioral dominance. In: P. P. G. Bateson and R. H. Klopfer, eds. Perspectives in Ethology, Plenum, New York, pp. 17–54.

Gauthreaux, S. A., Jr. 1982. The ecology and evolution of avian migration systems. In: D. S. Farner and J. R. King, eds. Avian Biology, Vol. VI. Academic Press, New York, pp. 93–168.

Gauthreaux, S. A., Jr. 1985a. The temporal and spatial scales of migration in relation to environmental changes in time and space. In: M. A. Rankin, ed. Migration: Mechanisms and Adaptive Significance. Contrib. Marine Sci. 27 (Suppl.), pp. 503–515.

Gauthreaux, S. A., Jr. 1985b. Avian migration mobile research laboratory. In: M. Harwood, ed. Proceedings of Fourth Hawk Migration Conference. Hawk Migration Assoc. of N. America, Rochester, N.Y., pp. 339–346.

Gauthreaux, S. A., Jr. 1991. The flight behavior of migrating birds in changing wind fields:

radar and visual analysis. Amer. Zool. 31:187–204.

Gauthreaux, S. A. 1992. The use of weather radar to monitor long-term patterns of trans-Gulf migration in spring. In: J. M. Hagen III and D. W. Johnston, eds. Ecology and Conservation of Neotropical Migrant Landbirds. Smithsonian Institution Press, Washington, D. C., pp. 96–100.

Gibo, D. L. 1981. Altitudes attained by migrating *Danaus p. plexippus* (Lepidoptera: Danainae), as reported by glider pilots. Can. J. Zool. 59:571–572.

Gibo, D. L. 1986. Flight strategies of migratory monarch butterflies (*Danaus plexippus* L.) in southern Ontario. In: W. Danthanarayana, ed. Insect Flight: Dispersal and Migration. Springer-Verlag, Berlin, pp. 172–184.

Gibo, D. L. and J. A. McCurdy. 1993. Lipid accumulation by migrating monarch butterflies (*Danaus plexippus* L.). Can. J. Zool. 71:76–82.

Gibo, D. L. and M. J. Pallett. 1979. Soaring flight of monarch butterflies, *Danaus plexippus* (Lepidoptera: Danainae) during the late summer migration in southern Ontario. Can. J. Zool. 57:1393–1401.

Gifford, C. A. 1962. Some observations on the general biology of the land crab, *Cardisoma guanhumi* (Latreille) in South Florida. Biol. Bull. 123:207–223.

Gill, D. E. 1978a. Effective population size and interdemic migration rates in a metapopulation of the red-spotted newt (*Notophthalmus viridescens* (Rafinesque). Evolution 32:839–849.

Gill, D. E. 1978b. The metapopulation dynamics of the red-spotted newt, *Notophthalmus viridescens* (Rafinesque). Ecol. Monogr. 48:145–166.

Glick, P. A. 1939. The distribution of insects, spiders, and mites in the air. Tech. Bull. U.S. Dept. Agric. No 673, 150 pp.

Goldsworthy, G. J. 1983. The endocrine control of flight metabolism in locusts. Advances Insect Physiol. 17:149–204.

Goulding, M. 1980. The Fishes and the Forest. University of California Press, Berkeley, California.

Grau, E. G., J. L. Specker, R. S. Nishioka, H. A. Bein, and L. C. Folmar. 1981. Lunar phasing of the thyroxine surge preparatory to seaward migration of salmonid fish. Science 221:607–609.

Greenbank, D. O., G. W. Schaefer, and R. C. Rainey. 1980. Spruce budworm (Lepidoptera: Tortricidae) moth flight and dispersal: new understanding from canopy observations, radar, and aircraft. Mem. Ent. Soc. Canada 110:1–49.

Greenstone, M. H., R. R. Eaton, and C. E. Morgan. 1991. Sampling aerially dispersing arthropods: A high-volume, inexpensive, automobile- and aircraft-borne system. J. Econ. Entomol. 84:1717–1725.

Greenstone, M. H., C. E. Morgan, A.-L. Hultsch, R. A. Farrow, and J. E. Dowse. 1987. Ballooning spiders in Missouri, U.S.A., and New South Wales, Australia: family and mass distributions. J. Arachnol. 15:163–170.

Greenwood, J. J. D. 1993. The ecology and conservation management of geese. Trends Ecol. Evol. 8:307–308.

Greenwood, P. J. and P. H. Harvey. 1982. The natal and breeding dispersal of birds. Ann. Rev. Ecol. Syst. 13:1–21.

Greer Walker, M., F. R. Harden Jones, and G. P. Arnold. 1978. The movements of plaice (*Pleuronectes platessa* L.) tracked in the open sea. J. Conseil 38:58–86.

Gregg, P. C. and A. G. L. Wilson. 1991. Trapping methods for adults. In: M. P. Zalucki, ed. Heliothis: Research Methods and Prospects. Springer-Verlag, Berlin, pp. 30–48.

Groebbels, F. 1928. Zur Physiologie des Vogelzuges. Verh. Ornithol. Ges. Bayern

18:44–74.

Groeters, F. R. 1989. Geographic and clonal variation in the milkweed-oleander aphid, *Aphis nerii* (Homoptera: Aphididae), for winged morph production, life history, and morphology in relation to host plant permanence. Evol. Ecol. 3:327–341.

Groeters, F. R. and H. Dingle. 1987. Genetic and maternal influences on life history plasticity in response to photoperiod by milkweed bugs. Am. Nat. 129:332–346.

Groeters, F. R. and H. Dingle. 1988. Genetic and maternal influences on life history plasticity in milkweed bugs (*Oncopeltus fasciatus*): response to temperature. J. Evol. Biol. 1:317–333.

Gross, M. 1987. The evolution of diadromy in fishes. Am. Fish. Soc. Symp. 1:14–25.

Gross, M. T., R. C. Coleman, and R. McDowall. 1988. Aquatic productivity and the evolution of diadromous fish migration. Science 239:1291–1293.

Gu, H. 1991. Electrophoretic variation in flight-related enzyme loci and its possible association with flight capacity in *Epiphyas postvittana*. Biochem. Genet. 29:345–354.

Gu, H. and W. Danthanarayana. 1992. Quantitative genetic analysis of dispersal in *Epiphyas postvittana*. I. Genetic variation in flight capacity. II. Genetic covariations between flight capacity and life-history traits. Heredity 68:53–60, 61–69.

Gwinner, E. 1986a. Circannual Rhythms. Springer-Verlag, Berlin.

Gwinner, E. 1986b. Circannual rhythms in the control of avian migration. Adv. Study Behav. 16:191–228.

Gwinner, E. 1989. Photoperiod as a modifying and limiting factor in the expression of avian circannual rhythms. J. Biol. Rhythms 4:237–250.

Gwinner, E. 1990a. Circannual rhythms in bird migration: control of temporal patterns and interactions with photoperiod. In: E. Gwinner, ed. Bird Migration: Physiology and Ecophysiology. Springer-Verlag, Berlin, pp. 257–268.

Gwinner, E., ed. 1990b. Bird Migration: Physiology and Ecophysiology. Springer-Verlag, Berlin.

Gwinner, E. and D. Czeschlik. 1978. On the significance of spring migratory restlessness in caged birds. Oikos 30:364–372.

Gwinner, E., M. Zeman, I. Schwabl-Benzinger, S. Jenni-Eierman, L. Jenni, and H. Schwabl. 1992. Corticosterone levels of passerine birds during migratory flight. Naturwissenschaften 79:276–278.

Haber, W. A. 1993. Seasonal migration of monarchs and other butterflies in Costa Rica. In: S. B. Malcolm and M. P. Zalucki, eds. Biology and Conservation of the Monarch Butterfly. No. 38, Science Series, Los Angeles County Natural History Museum, pp. 201–207.

Hagan, J. M. III and D. W. Johnston, eds. 1992. Ecology and Conservation of Neotropical Migrant Landbirds. Smithsonian Institution Press, Washington, D. C.

Hagen, K. S. 1962. Biology and ecology of predaceous Coccinellidae. Ann. Rev. Entomol. 7:289–326.

Hamilton, K. 1985. A study of the variability of the return migration route of Fraser River sockeye salmon (*Oncorhyncus nerka*). Can. J. Zool. 63:1930–1943.

Hamilton, W. D. and R. M. May. 1977. Dispersal in stable habitats. Nature 269:578–581.

Hammock, B. D. 1985. Regulation of juvenile hormone titer: Degradation. In: G. C. Kerkut and L. I. Gilbert, eds. Comprehensive Insect Physiology, Biochemistry and Pharmacology, Vol. 7, pp. 431–472.

Hammond, A. M. and H. W. Fescemyer. 1987. Physiological correlates in migratory noctuids: the velvetbean caterpillar as a model. Insect Sci. Applic. 8:581–589.

Hamner, W. M. 1988. The 'Lost Year' of the sea turtle. Trends Ecol. Evol. 3:116–118.

Han, E.-N. and A. G. Gatehouse. 1991. Genetics of precalling period in the oriental armyworm, *Mythimna separata* (Walker) (Lepidoptera: Noctuidae), and implications for migration. Evolution 45:1502–1510.

Happ, G. M. 1984. Development and reproduction. In: H. E. Evans, ed. Insect Biology. Addison-Wesley, Reading, Massachusetts, pp. 93–113.

Harden Jones, F. R. 1968. Fish Migration. St. Martin's Press, New York.

Harden Jones, F. R. 1981. Fish Migration: strategy and tactics. In: D. J. Aidley, ed. Animal Migration. Cambridge University Press, Cambridge, pp. 139–165.

Harden Jones, F. R. 1984. A view from the ocean. In: J. D. McCleave, G. P. Arnold, J. J. Dodson, and W. H. Neill, eds. Mechanisms of Migration in Fishes. Plenum, New York, pp. 1–26.

Hardie, J. 1980. Juvenile hormone mimics the photoperiodic apterization of the alate gynopara of the aphid, *Aphis fabae*. Nature 286:602–604.

Hardisty, M. W. and I. C. Potter. 1971. Paired species. In: M. W. Hardisty and I. C. Potter, eds. The Biology of Lampreys, Vol. 1. Academic Press, London, pp. 127–206.

Hardy, A. C. 1958. The Open Sea, Its Natural History: The World of Plankton. Houghton, Boston, Massachusetts.

Harris, M. P. 1970. Abnormal migration and hybridization of *Larus argentatus* and *L. fuscus* after interspecies fostering experiments. Ibis 112:488–498.

Hartnoll, R. G. 1963. The freshwater grapsid crabs of Jamaica. Proc. Linn. Soc. Lond. 175:145–169.

Hasler, A. D. 1966. Underwater Guideposts—Homing of Salmon. University of Wisconsin Press, Madison.

Hasler, A. D. and A. T. Scholz. 1983. Olfactory Imprinting and Homing in Salmon. Investigations into the Mechanism of the Imprinting Process. Springer-Verlag, Berlin.

Hasler, A. D. and W. J. Wisby. 1951. Discrimination of stream odors by fishes and relation to parent stream behavior. Am. Nat. 85:223–238.

Hasler, A. D., A. T. Scholz, and R. M. Horrall. 1978. Olfactory imprinting and homing in salmon. Amer. Scient. 66:347–355.

Hassel, M. P. and T. R. E. Southwood. 1978. Foraging strategies of insects. Ann. Rev. Ecol. Syst. 9:75–98.

Heape, W. 1931. Emigration, Migration, and Nomadism. Heffer, Cambridge, U.K.

Hegde, S. G., R. Uma Shaanker, and K. N. Ganeshaiah. 1991. Evolution of seed size in the bird-dispersed tree *Santalum album* L.: a trade off between seedling establishment and dispersal efficiency. Evol. Trends Plants 5:131–135.

Hegmann, J. P. and H. Dingle. 1982. Phenotypic and genetic covariance structure in milkweed bug life history traits. In: H. Dingle and J. P. Hegmann, eds. Evolution and Genetics of Life Histories. Springer, New York, pp. 177–186.

Heimlich-Boran, J. R. 1988. Behavioral ecology of killer whales (*Orcinus orca*) in the Pacific Northwest. Can. J. Zool. 66:565–578.

Helbig, A. J. 1990. Are orientation mechanisms among migratory birds species specific? Trends. Ecol. Evol. 5:365–367.

Helbig, A. J. 1991a. SE- and SW-migrating blackcap (*Sylvia atricapilla*) populations in central Europe: orientation of birds in the contact zone. J. Evol. Biol. 4:657–670.

Helbig, A. J. 1991b. Inheritance of migratory direction in a bird species: a cross-breeding experiment with SE- and SW-migratory blackcaps (*Sylvia atricapilla*). Behav. Ecol. Sociobiol. 28:9–12.

Helfman, G. S., D. E. Facey, L. S. Hales, and E. L. Boyeman. 1987. Reproductive ecology of the American eel. Am. Fish. Soc. Symp. 1:42–56.

Helle, P. and R. J. Fuller. 1988. Migrant passerine birds in European forest successions in relation to vegetation height and geographical position. J. Anim. Ecol. 57:565–579.

Herman, W. S. 1981. Studies on the adult reproductive diapause of the monarch butterfly, *Danaus plexippus*. Biol. Bull. 160:89–106.

Herman, W. S. 1985. Hormonally mediated events in adult monarch butterflies. In: M. A. Rankin, ed. Migration: Mechanisms and Adaptive Significance. Contrib. Marine Sci. (Suppl.), pp. 799–815.

Herrera, C. M. 1978. On the breeding distribution pattern of European migrant birds: MacArthur's theme reexamined. Auk 95:496–509.

Herrera, C. M. 1987. Vertebrate-dispersed plants of the Iberian Peninsula: a study of fruit characteristics. Ecol. Monogr. 57:305–331.

Herrnkind, W. F. 1972. Orientation in shore-living arthropods, especially the sand fiddler crab. In: H. E. Winn and B. Olla, eds. Behavior of Marine Animals, Vol. I., Plenum, New York, pp. 1–59.

Herrnkind, W. F. 1980. Spiny lobsters: patterns of movement. In: J. S. Cobb and B. F. Phillips, eds. Biology and Management of Lobsters, Vol. 1. Academic Press, New York, pp. 349–407.

Herrnkind, W. F. 1985. Evolution and mechanisms of mass single-file migration in spiny lobsters: synopsis. In: M. A. Rankin, ed. Migration: Mechanisms and Adaptive Significance. Contrib. Marine Sci. 27 (Suppl.), pp. 197–211.

Herrnkind, W. F. 1991. Lobsters. In: Fantastic Journeys, The Marvels of Animal Migration. Merehurst, London, pp. 106–111.

Hill, J. K. and A. G. Gatehouse. 1992. Genetic control of the pre-reproductive period in *Autographa gamma* (L.) (silver Y moth) (Lepidoptera: Noctuidae). Heredity 69:458–464.

Hoar, W. S. 1976. Smolt transformation: evolution, behaviour, and physiology. J. Fish. Res. Board Can. 33:1234–1252.

Hoar, W. S. 1988. The physiology of smolting salmonids. Fish Physiology X1B:273–343.

Hobbs, S. E. and W. W. Wolf. 1989. An airborne radar technique for studying insect migration. Bull. Ent. Res. 79:693–704.

Hofer, H. and M. L. East. 1993. The commuting system of Serengeti spotted hyaenas; how a predator copes with migratory prey. I. Social organization. II. Intrusion pressure and commuter's space use. Anim. Behav. 46:547–557, 559–574.

Hoffmann, K. 1954. Versuche zu der im Richtungfinden der Vögel enthaltenen Zeitschät-zung. Z. Tierpsychol. 11:453–475.

Holekamp, K. E. and P. W. Sherman. 1989. Why male ground squirrels disperse. Amer. Scientist 77:232–239.

Holekamp, K. E., H. B. Simpson, and L. Smale. 1985. Endocrine influences on natal dispersal in Belding's ground squirrels (*Spermophilus beldingi*). In: M. A. Rankin, ed. Migration: Mechanisms and Adaptive Significance. Contrib. Marine Sci. 27 (Suppl.), pp. 397–408.

Hollinger, S. E., K. R. Sivier, M. E. Irwin, and S. A. Isard. 1991. A helicopter-mounted isokinetic aerial insect sampler. J. Econ. Entomol. 84:476–483.

Houck, M. A. and B. M. O'Connor. 1991. Ecological and evolutionary significance of phoresy in the Astigmata. Ann. Rev. Entomol. 36:611–636.

Howard, W. E. 1960. Innate and environmental dispersal of individual vertebrates. Amer. Midl. Nat. 63:152–161.

Howe, H. F. 1981. Dispersal of a neotropical nutmeg (*Virola sebifera*) by birds. Auk 98:88–98.

Howe, H. F. and J. Smallwood. 1982. Ecology of seed dispersal. Ann. Rev. Ecol. Syst. 13:201–228.

Howe, H. F. and L. C. Wesley. 1988. Ecological Relationships of Plants and Animals. Oxford University Press, New York.

Huber, H. R. 1987. Natality and weaning success in relation to age at first reproduction in northern elephant seals. Can. J. Zool. 65:1311–1316.

Hughes, L. and M. Westoby. 1992. Effect of diaspore characteristics on removal of seeds adapted for dispersal by ants. Ecology 73:1300–1312.

Humphrey, J. A. C. 1987. Fluid mechanic constraints on spider ballooning. Oecologia 73:469–477.

Hunter, D. M. 1986. Locating infestations of the Australian plague locust *Chortoicetes terminifera* (Walker) in the remote interior of Australia. Proc. 4th Triennial Meeting, Pan Amer. Acridol. Soc. 1985, pp. 185–190.

Hunter, D. M. 1989. The response of Mitchell grasses (*Astrebla* spp.) and button grass (*Dactylotenium radulans* R. Br.) to rainfall and their importance to the survival of the Australian plague locust, *Chortoicetes terminifera* (Walker), in the arid zone. Aust. J. Ecol. 14:467–471.

Hutchings, J. A. and D. W. Morris. 1985. The influence of phylogeny, size and behaviour on patterns of covariation in salmonid life histories. Oikos 45:118–124.

Hutchison, L. V. and B. M. Wenzel. 1980. Olfactory guidance in foraging by procellariiforms. Condor 82:314–319.

Hutto, R. L. 1985. Habitat selection by non-breeding migratory land birds. In: M. L. Cody, ed. Habitat Selection in Birds. Academic Press, London, pp. 455–476.

Idler, D. R. and I. Bitners. 1958. Biochemical studies on sockeye salmon during spawning migration. II. Cholesterol, fat, protein and water in the flesh of standard fish. Can. J. Biochem. Physiol. 36:793–799.

Idler, D. R. and I. Bitners. 1959. Biochemical studies in sockeye salmon during spawning migration. V. Cholesterol, fat, protein and water in the body of the standard fish. J. Fish. Res. Board Can. 16:235–241.

Imboden, C. 1974. Zug, Fremdansiedlung und Brutperiode des Kiebitz *Vanellus vanellus* in Europe. Orn. Beob. 71:5–134.

Inglis, J. M. 1976. Wet season movements of individual wildebeests of the Serengeti migratory herd. East Afr. Wildl. J. 14:17–34.

Irwin, M. E. and J. M. Thresh. 1988. Long-range aerial dispersal of cereal aphids as virus vectors in North America. Phil. Trans. R. Soc. Lond. B 321:421–446.

Isard, S. A., ed. 1993. Alliance for Aerobiology Research workshop report. Alliance for Aerobiology Research Workshop Writing Committee, Champaign, Illinois, 40 pp.

Iwanaga, K. and S. Tojo. 1986. Effects of juvenile hormone and rearing density on wing dimorphism and oocyte development in the brown planthopper, *Nilaparvata lugens*. J. Insect Physiol. 32:585–590.

Jackson, D. J. 1928. The inheritance of long and short wings in the weevil, *Sitona hispidula*, with a discussion of wing reduction among beetles. Trans. Roy. Soc. Edinburgh 55:655–735.

Jaeger, M. M., R. L. Bruggers, B. E. Johns, and W. A. Erickson. 1986. Evidence of itinerant breeding of the red-billed quelea, *Quelea quelea*, in the Ethiopian Rift Valley. Ibis 128:469–482.

Jakob, E. M. 1991. Costs and benefits of group living for pholcid spiderlings: losing food, saving silk. Anim. Behav. 41:711–722.

James, D. G. 1984. Migration and clustering phenology of *Danaus plexippus* (L.)

(Lepidoptera: Nymphalidae) in Australia. J. Aust. Entomol. Soc. 23:199–204.

James, D. G. 1988. Migration and behaviour of nonreproductive *Danaus plexippus* (L.) (Lepidoptera: Nymphalidae) in the Blue Mountains, New South Wales. Aust. Ent. Mag. 15:25–30.

James, D. G. 1993. Migration biology of the monarch butterfly in Australia. In: S. B. Malcolm and M. P. Zalucki, eds. The Biology and Conservation of the Monarch Butterfly. Los Angeles County Natural History Museum, Science Series, No. 38, pp. 189–200.

James, F. C., C. E. McCulloch, and D. A. Wiedenfeld. 1995. New approaches to the analysis of population trends in land birds. Ecology (In press).

Jander, R. 1975. Ecological aspects of spatial orientation. Ann. Rev. Ecol. Syst. 6:171–188.

Järvinen, O. 1976. Migration, extinction, and alary polymorphism in water-striders (*Gerris* Fabr.) Ann. Acad. Sci. Fennici A 206:1–7.

Järvinen, O. and K. Vepsäläinen. 1976. Wing dimorphism as an adaptive strategy in water-striders (*Gerris*). Hereditas 84:61–68.

Jehl, J. R., Jr. 1988. Biology of the eared grebe and Wilson's phalarope in the nonbreeding season: a study of adaptations to saline lakes. Stud. Avian Biol. No. 12, 14 pp.

Jehl, J. R., Jr. 1990. Aspects of the molt migration. In: E. Gwinner, ed. Bird Migration: Physiology and Ecophysiology. Springer-Verlag, Berlin, pp. 102–113.

Johannsen, A. S. 1958. Relation of nutrition to endocrine-reproductive functions in the milkweed bug *Oncopeltus fasciatus* (Dallas) (Heteroptera: Lygaeidae). Nytt Mag. Zool. 7:1–132.

Johnsen, P. B. 1987. New directions in fish orientation studies. In: W. F. Herrnkind and A. B. Thistle, eds. Sign posts in the Sea. Department of Biological Science, Florida State University, Tallahassee, pp. 85–101.

Johnson, C. G. 1960. The basis for a general system of insect migration and dispersal by flight. Nature 186:348–350.

Johnson, C. G. 1963. Physiological factors in insect migration by flight. Nature 198:423–427.

Johnson, C. G. 1969. Migration and Dispersal of Insects by Flight. Methuen, London.

Johnson, C. G. 1976. Lability of the flight system: a context for functional adaptation. Roy. Ent. Soc. Symp. 7:217–234.

Johnson, L. K. 1982. Sexual selection in a brentid weevil. Evolution 36:251–262.

Johnson, S. 1987. Migration and life history strategy of the fall armyworm, *Spodoptera frugiperda* in the Western Hemisphere. Insect Sci. Appl. 8:543–549.

Jones, J. B. 1991. Movements of albacore tuna (*Thunnus alalunga*) in the South Pacific: evidence from parasites. Mar. Biol. 111:1–9.

Jones, P. J. 1989. General aspects of quelea migrations. In: R. L. Bruggers and C. C. H. Elliott, eds. Quelea quelea: Africa's Bird Pest. Oxford University Press, Oxford, pp. 102–112.

Jones, R. E. 1977. Movement patterns and egg distribution in cabbage butterflies. J. Anim. Ecol. 46:195–212.

Jouventin, P. and H. Weimerskirch. 1988. Demographic strategies of southern albatrosses. Acta 19 Cong. Int. Ornith. 1:857–865.

Jouventin, P. and H. Weimerskirch. 1990. Satellite tracking of wandering albatrosses. Nature 343:746–748.

Joyce, R. J. V. 1976. Insect flight in relation to problems of pest control. Symp. Roy. Ent. Soc. Lond. 7:135–155.

Joyce, R. J. V. 1981. The control of migrant pests. In: D. J. Aidley, ed. Animal Migration.

Cambridge Univ. Press, pp. 209–230.

Kaiser, A. 1992. Fat deposition and theoretical flight range of small autumn migrants in southern Germany. Bird Study 39:96–110.

Kaitala, A. 1987. Dynamic life-history strategy of the waterstrider *Gerris thoracicus* as an adaptation to food and habitat variation. Oikos 48:125–131.

Kaitala, A. 1988. Wing muscle dimorphism: two reproductive pathways of the waterstrider *Gerris thoracicus* in relation to habitat instability. Oikos 53:222–228.

Kaitala, A. 1991. Phenotypic plasticity in reproductive behaviour of waterstriders: trade-offs between reproduction and longevity during food stress. Func. Ecol. 5:12–18.

Kaitala, A. and H. Dingle. 1992. Spatial and temporal variation in wing dimorphism of California populations of the waterstrider *Aquarius remigis* (Heteroptera: Gerridae). Ann. Entomol. Soc. Am. 85:590–595.

Kaitala, A. and L. Huldén. 1990. Significance of spring migration and flexibility in flight-muscle histolysis in waterstriders (Heteroptera, Gerridae). Ecol. Entomol. 15:409–418.

Kaitala, A., V. Kaitala, and P. Lundberg. 1993. A theory of partial migration. Amer. Natur. 142:59–81.

Kaitala, V. 1990. Evolutionary stable migration in salmon: a simulation study of homing and straying. Ann. Zool. Fennici 27:131–138.

Kaitala, V., A. Kaitala, and W. M. Getz. 1989. Evolutionarily stable dispersal of a waterstrider in a temporally and spatially heterogeneous environment. Evol. Ecol. 3:283–298.

Kalela, O. 1954. Populationsokologische Geschichtspunkte zur Entstehung des Vogelzuges. Ann. Zool. Fenn. 16:1–30.

Kanciruk, P. and W. F. Herrnkind. 1978. Mass migration of spiny lobster, *Panulirus argus* (Crustacea: Palinuridae): behavior and environmental correlates. Bull. Mar. Sci. 28:601–623.

Kanz, J. E. 1977. The orientation of migrant and nonmigrant monarch butterflies, *Danaus plexippus* (L). Psyche 84:120–141.

Karban, R. 1982. Increased reproductive success at high densities and predator satiation for periodical cicadas. Ecology 63:321–328.

Kareiva, P. M. 1982. Experimental and mathematical analysis of herbivore movement: quantifying the influence of plant spacing and quality of foraging discrimination. Ecol. Monogr. 52:261–282.

Kareiva, P. M. 1983a. Local movements in herbivorous insects: applying a passive diffusion model to mark-recapture field experiments. Oecologia 57:322–327.

Kareiva, P. M. 1983b. Influence of vegetation texture on herbivore populations: resource concentration and herbivore movement. In: R. F. Denno and M. S. McClure, eds. Variable Plants and Herbivores in Natural and Managed Systems. Academic Press, New York, pp. 259–289.

Kareiva, P. M. and N. Shigesada. 1983. Analyzing insect movement as a correlated random walk. Oecologia 56:234–238.

Karr, J. R. 1980. Patterns in the migration systems between the North Temperate Zone and the Tropics. In: A Keast and E. S. Morton, eds. Migrant Birds in the Neotropics. Smithsonian Institution Press, Washington, D.C., pp. 529–543.

Karr, J. R. 1990. Interactions between forest birds and their habitats: A comparative synthesis. In: A. Keast, ed. Biogeography and Ecology of Forest Bird Communities. SPB Academic Publishing, The Hague, pp. 379–386.

Keast, A. 1990. The annual cycle in forest birds relative to latitude and habitat: a synthesis. In: A. Keast, ed. Biogeography and Ecology of Forest Bird Communities. SPB

Academic Publishing, The Hague, pp. 395–401.

Keast, A. and E. A. Morton, eds. 1980. Migrant Birds in the Neotropics. Smithsonian Institution Press, Washington, D.C.

Kennedy, J. S. 1951. The migration of the desert locust, *Schistocerca gregaria* Forsk. I. The behaviour of swarms. II. A theory of long range migrations. Phil. Trans. R. Soc. Lond. Ser. B 235:163–290.

Kennedy, J. S. 1958. The experimental analysis of aphid behaviour and its bearing on current theories of instinct. Proc. 10th Int. Congr. Ent., Montreal, 1956, 2:397–404.

Kennedy, J. S. 1961. A turning point in the study of insect migration. Nature 189: 785–791.

Kennedy, J. S. 1966. Nervous co-ordination of instincts. Cambridge Research 2:29–32.

Kennedy, J. S. 1983. Zigzagging and casting as a programmed response to wind-borne odour: a review. Physiol. Entomol. 8:109–120.

Kennedy, J. S. 1985. Migration, behavioral and ecological. In: M. A. Rankin, ed. Migration: Mechanisms and Adaptive Significance. Contrib. Marine Science 27 (Suppl.), pp. 5–26.

Kennedy, J. S. and C. O. Booth. 1963a. Free flight of aphids in the laboratory. J. Exp. Biol. 40:67–85.

Kennedy, J. S. and C. O. Booth, 1963b. Co-ordination of successive activities in an aphid. The effect of flight on the settling responses. J. Exp. Biol. 40:351–369.

Kennedy, J. S. and C. O. Booth. 1964. Co-ordination of successive activities in an aphid. Depression of settling after flight. J. Exp. Biol. 41:805–824.

Kennedy, J. S. and A. R. Ludlow. 1974. Co-ordination of two kinds of flight activity in an aphid. J. Exp. Biol. 61:173–196.

Kennedy, J. S. and H. L. G. Stroyan. 1959. Biology of aphids. Ann. Rev. Entomol. 4:139–160.

Kennedy, J. S. and M. J. Way. 1979. Summing up the conference. In: R. L. Rabb and G. G. Kennedy, eds. Movements of Highly Mobile Insects: Concepts and methodology in Research. North Carolina State University, Raleigh, pp. 446–456.

Kennedy, J. S., C. O. Booth, and W. J. S. Kershaw. 1961. Host finding by aphids in the field. III. Visual attraction. Ann. Appl. Biol. 49:1–21.

Kerlinger, P. 1989. Flight Strategies of Migrating Hawks. University of Chicago Press, Chicago.

Kerlinger, P. and F. R. Moore. 1989. Atmospheric structure and avian migration. Current Ornith. 6:109–142.

Kerlinger, P., M. R. Lein, and B. J. Sevick. 1985. Distribution and population fluctuations of wintering snowy owls (*Nyctea scandiaca*) in North America. Can. J. Zool. 63:1829–1834.

Ketterson, E. D. and V. Nolan, Jr. 1976. Geographic variation and its climatic correlates in the sex-ratio of eastern-wintering dark-eyed juncos (*Junco hyemalis hyemalis*). Ecology 57:679–693.

Ketterson, E. D. and V. Nolan, Jr. 1979. Seasonal, annual and geographic variation in the sex ratio of wintering populations of dark-eyed juncos (*Junco hyemalis*). Auk 96:532–536.

Ketterson, E. D. and V. Nolan, Jr. 1982. The role of migration and winter mortality in the life-history of a temperate zone migrant, the dark-eyed junco, as determined from demographic analysis of winter populations. Auk 99:243–259.

Ketterson, E. D. and V. Nolan, Jr. 1983. The evolution of differential bird migration. In: R. F. Johnston, ed. Current Ornithology, Vol. I. Plenum, New York, pp. 357–402.

Ketterson, E. D. and V. Nolan, Jr. 1992. Hormones and life histories: an integrative approach. Am. Nat. 140 (Suppl.): s33–s62.

Kevan, D. K. M. 1989. Transatlantic travelers. Antenna 13:12–15.

Kimura, T. and S. Masaki. 1977. Brachypterism and seasonal adaptation in *Orgyia thyallina* Butler (Lepidoptera: Lymantriidae) Kontyû 45:97–106.

Kirkevold, B. C. and J. S. Lockard, eds. 1986. Behavioral Biology of Killer Whales. Alan R. Liss, Inc., New York.

Kirschvink, J. L. and A. K. Kirschvink. 1991. Is geomagnetic sensitivity real? Replication of the Walker-Bitterman magnetic conditioning experiment in honeybees. Amer. Zool. 31:169–185.

Kirschvink, J. L., A. E. Dizon, and J. A. Westphal. 1986. Evidence from strandings for geomagnetic sensitivity in cetaceans. J. Exp. Biol. 120:1–24.

Kirschvink, J. L., D. S. Jones, and B. J. McFadden, eds. 1985. Magnetite Biomineralization and Magnetoreception in Animals: A New Biomagnetism. Plenum, New York.

Kisimoto, R. 1956. Effect of crowding during the larval period on the determination of the wing-form of an adult planthopper. Nature 178:641–642.

Kisimoto, R. 1976. Synoptic weather conditions inducing long-distance immigration of planthoppers, *Sogatella furcifera* Horvath and *Nilaparvata lugens* (Stal). Ecol. Entomol. 1:95–109.

Klausner, E., E. R. Miller, and H. Dingle. 1981. Genetics of brachyptery in a lygaeid bug island population. J. Heredity 72:288–289.

Klimley, A. P. 1985. The areal distribution and autoecology of the white shark, *Carcharodon carcharias*, off the west coast of North America. Mem. S. Calif. Acad. Sci. 9:15 40.

Klinke, R. 1991. Avian bearing mechanisms and performance from infrasound to the mid-frequency range. Acta XX Cong. Int. Ornithol. 1990. Vol. III:1805–1812.

Kramer, G. 1950. Orientierte Zugaktivität gekäfigter Singvögel. Naturwiss. 37:188.

Kramer, G. 1951. Eine neue Methode zur Erforschung der Zugorientierung und die bisker damit ergielten Ergebnisse. Proc X Ornith. Cong., Uppsala, pp. 269–280.

Krebs, J. R. and N. B. Davies, eds. 1984. Behavioural Ecology: An Evolutionary Approach, 2nd ed. Sinauer, Sunderland, Massachusetts.

Kreithen, M. and W. T. Keeton. 1974. Detection of polarized light by the homing pigeon, *Columba livia*. J. Comp. Physiol. 89:83–92.

Krogh, A. and T. Weis-Fogh. 1951. The respiratory exchange of the desert locust (*Schistocerca gregaria*) before, during and after flight. J. Exp. Biol. 28:344–357.

Krogh, A. and T. Weis-Fogh. 1952. A roundabout for studying sustained flight of locusts. J. Exp. Biol. 29:211–219.

Lack, D. 1954. The Natural Regulation of Animal Numbers. Oxford University Press, London.

Lack, D. 1963. Migration across the North Sea studied by radar. Part 5. Movements in August, winter and spring, and conclusion. Ibis 105:461–492.

Lack, D. 1968. Bird migration and natural selection. Oikos 19:1–19.

Lack, D. and G. C. Varley. 1945. Detection of birds by radar. Nature 156: 446.

Lack, P. C. 1990. Palearctic-African systems. In: A. Keast, ed. Biogeography and Ecology of Forest Bird Communities. SPB Academic Publishing, The Hague, pp. 345–356.

Lal, P. 1988. Role of thyroid in sexual and body weight cycles of the migratory red-headed bunting (*Emberiza bruniceps*). Gen. Comp. Endocrin. 70:291–300.

Landreth, H. F. and D. E. Ferguson. 1967. Newts: sun-compass orientation. Science 158:1459–1461.

Lane, B. and A. Jessop. 1985. Tracking of migratory waders in north-western Australia using meteorological radars. Stilt 6:17–28.

Lane, J. 1993. Overwintering monarch butterflies in California: past and present. In: S. B. Malcolm and M. P. Zalucki, eds. Biology and Conservation of the Monarch Butterfly. Los Angeles County Natural History Museum, Science Series, No. 38, pp. 335–344.

Laughlin, R. 1974. A modified Kennedy flight chamber. J. Aust. Ent. Soc. 13: 151–153.

Leach, I. H. 1981. Wintering blackcaps in Britain and Ireland. Bird Study 28:5–14.

Le Boeuf, B. J., D. G. Ainley and T. J. Lewis. 1974. Elephant seals on the Farallones: population structure of an incipient breeding colony. J. Mammal. 55:370–385.

Lees, A. D. 1966. The control of polymorphism in aphids. Adv. Insect Physiol. 3:207–277.

Lees, A. D. 1967. The production of apterous and alate forms in the aphid *Megoura viciae* Buckton, with special reference to the role of crowding. J. Insect Physiol. 13:289–318.

Leggett, W. C. 1977. The ecology of fish migrations. Ann. Rev. Ecol. Syst. 8:285–308.

Leggett, W. C. 1984. Fish migrations in coastal and estuarine environments: a call for new approaches to the study of an old problem. In: J. D. McCleave, G. P. Arnold, J. J. Dodson, and W. H. Neill, eds. Mechanisms of Migration in Fishes. Plenum, New York, pp. 159–178.

Leggett, W. C. 1985. The role of migrations in the life history evolution of fish. In: M. A. Rankin, ed. Migration: Mechanisms and Adaptive Significance. Contrib. Marine Sci. 27 (Suppl.), pp. 277–295.

Lehrer, M. 1991. Bees which turn back and look. Naturwiss. 78:274–276.

Leisler, B. 1990. Selection and use of habitat of wintering migrants. In: E. Gwinner, ed. Bird Migration: Physiology and Ecophysiology. Springer-Verlag, Berlin, pp. 156–174.

Leisler, B. 1992. Habitat selection and coexistence of migrants and Afrotropical residents. Ibis 134 (Suppl.) 1:77–82.

Leisler, B. and H. Winkler. 1985. Ecomorphology. Current Ornithol. 2:155–186.

Leisler, B. and H. Winkler. 1991. Ergebnisse and Konzepte ökomorphologischer Untersuchungen an Vögeln. J. Ornithol. 132:373–425.

Leisler, B., H.-W. Ley, and H. Winkler. 1989. Habitat, behaviour and morphology of *Acrocephalus* warblers: an integrated analysis. Ornis Scand. 20:181–186.

Leong, K. L. H. 1990. Microenvironmental factors associated with the winter habitat of the monarch butterfly (Lepidoptera: Danaidae) in central California. Ann. Entomol. Soc. Amer. 83:907–910.

Leong, K. L. H., D. Frey, G. Brenner, S. Baker, and D. Fox. 1991. Use of multivariate analyses to characterize the monarch butterfly (Lepidoptera: Danaidae) winter habitat. Am. Entomol. Soc. Amer. 84:263–267.

Levey, D. J. and F. G. Stiles. 1992. Evolutionary precursors of long-distance migration: resource availability and movement patterns in neotropical landbirds. Am. Nat. 140:447–476.

Levin, S. A., D. Cohen, and A. Hastings. 1984. Dispersal strategies in patchy environments. Theor. Pop. Biol. 26:165–191.

Li, K.-P., H. Wong, and W. Woo. 1964. Route of the seasonal migration of the oriental armyworm moth in the eastern part of China as indicated by a three-year result of releasing and recapturing of marked moths. Acta Phytophylacica Sinica 3:93–100 (In Chinese with English summary.) (Rev. Appl. Ent. 53:391).

Lidicker, W. Z. and R. L. Caldwell, eds. 1982. Dispersal and Migration. Hutchinson Ross, Stroudsburg, PA.

Lighthill, M. J. 1971. Large-amplitude elongated-body theory of fish locomotion. Proc. Roy. Soc. Lond. B 179:125–138.

Lindauer, M. 1976. Foraging and homing flight of the honey bee: some general problems of orientation. Roy. Ent. Soc. Symp. 7:199–216.

Lindeque, M. and P. M. Lindeque. 1991. Satellite tracking of elephants in northwestern Namibia. Afr. J. Ecol. 29:196–206.

Lindström, Å., and T. Alerstam. 1992. Optimal fat loads in migrating birds: a test of the time minimization hypothesis. Am. Nat. 140:477–491.

Liquido, N. J. and M. E. Irwin. 1986. Longevity, fecundity, change in degree of gravidity and lipid content with adult age, and lipid utilization during tethered flight of alates of the corn leaf aphid, *Rhopalosiphum maidis*. Ann. App. Biol. 108:449–459.

Lockyer, C. 1978. A theoretical approach to the balance between growth and food consumption in fin and sei whales, with special reference to the female reproductive cycle. Twenty-eighth Report of the International Whaling commission, pp. 243–249.

Lockyer, C. H. and S. G. Brown. 1981. The migration of whales. In: D. J. Aidley, ed. Animal migration. Cambridge University Press, Cambridge, pp. 105–137.

Lokesha, R., S. G. Hegde, R. U. Shaanker, and K. N. Ganeshaiah. 1992. Dispersal mode as a selective force in shaping the chemical composition of seeds. Am. Nat. 140:520–525.

Lopez, F., J. M. Serrano, and F. J. Acosta 1994. Parallels between the foraging strategies of ants and plants. Trends Ecol. Evol. 9:150–153.

Lopez Ornat, A. and R. Greenberg. 1990. Sexual segregation by habitat in migratory warblers in Quintana Roo, Mexico. Auk 107:539–543.

Loria, D. L. and F. R. Moore. 1990. Energy demands of migration on Red-eyed Vireos, *Vireo olivaceus*. Behav. Ecol. 1:24–35.

Lowe-McConnell, R. H. 1975. Fish Communities in Tropical Fresh Waters. Longmans, New York.

Lowery, G. H. and R. J. Newman. 1966. A continentwide view of bird migration on four nights in October. Auk 83:547–586.

Lundberg, P. 1985. Dominance behaviour, body weight and fat variations, and partial migration in European blackbirds *Turdus merula*. Behav. Ecol. Sociobiol. 17:185–189.

Lundberg, P. 1988. The evolution of partial migration in birds. Trends Ecol. Evol. 3:172–175.

Lüters, W. and G. Birukow. 1963. Solar-compass orientation of *Apodemus agrarius*. Naturwiss. 50:757–758.

MacArthur, R. H. 1959. On the breeding distribution patterns of North American migrant birds. Auk 76:318–325.

Macauley, E. D. M. 1972. A simple insect flight recorder. Ent. Exp. Appl. 15: 252–254.

MacKay, P. A. and R. J. Lamb. 1979. Migratory tendency in aging populations of the pea aphid *Acyrthosiphon pisum*. Oecologia 39:301–308.

MacKay, P. A. and W. G. Wellington. 1977. Maternal age as a source of variation in the ability of an aphid to produce dispersing forms. Res. Pop. Ecol. 18:195–209.

MacKenzie, D. R., C. S. Barfield, G. G. Kennedy, R. D. Berger, and D. J. Taranto, eds. 1985. The Movement and Dispersal of Agriculturally Important Biotic Agents. Claitor's Publishing, Baton Rouge, LA.

Maclusky, N. J. and F. Naftolin. 1981. Sexual differentiation of the central nervous system. Science 211:1294–1303.

Malcolm, S. B. 1993. Conservation of monarch butterfly migration in North America: an endangered phenomenon. In: S. B. Malcolm and M. P. Zalucki, eds. Biology and Conservation of the Monarch Butterfly. Los Angeles County Natural History Museum, Science Series, No. 38, pp. 357–361.

Malcolm, S. B. and M. P. Zalucki, eds. 1993. Biology and Conservation of the Monarch

Butterfly. Los Angeles County Natural History Museum, Science Series, No. 38.

Malcolm, S. B., B. J. Cockrell, and L. P. Brower. 1993. Spring recolonization of eastern North America by the monarch butterfly: successive brood or single sweep migration? In: S. B. Malcolm and M. P. Zalucki, eds. Biology and Conservation of the Monarch Butterfly. Los Angeles County Natural History Museum, Science Series, No. 38, pp. 253–268.

Mangel, M. and C. W. Clark. 1988. Dynamic Modeling in Behavioral Ecology. Princeton University Press, Princeton, New Jersey.

Marcovitch, S. 1923. Plant lice and light exposure. Science 58:537–538.

Marcovitch, S. 1924. The migration of the Aphididae and the appearance of sexual forms as affected by the relative length of daily light exposure. J. Agric. Res. 27:513–522.

Marden, J. H. 1987. Maximum lift production during takeoff in flying animals. J. Exp. Biol. 130:235–258.

Martin, D. D. and A. H. Meier. 1973. Temporal synergism of corticosterone and prolactin in regulating orientation in the migratory white-throated sparrow, *Zonotrichia albicollis*. Condor 75:369–374.

Martin, G. R. 1991. The question of Polarization. Nature 350:194.

Masaki, S. 1973. Climatic adaptation and photoperiodic response in the band-legged ground cricket. Evolution 26:587–600.

Mascanzoni, D. and H. Wallin. 1986. The harmonic radar: a new method of tracing insects in the field. Ecol. Entomol. 11:387–390.

Masters, A. R., S. B. Malcolm, and L. P. Brower. 1988. Monarch butterfly (*Danaus plexippus*) thermoregulatory behavior and adaptations for overwintering in Mexico. Ecology 69:458–467.

Mathad, S. B. and J. E. McFarlane. 1968. Two effects of photoperiod on wing development in *Gryllodes sigillatus* (Walk.). Canad. J. Zool. 46:57–60.

Mattocks, P. W. Jr. 1976. The role of gonadal hormones in the regulation of the premigratory fat deposition in the white-crowned sparrow, *Zonotrichia leucophrys gambelii*. MS Thesis. University of Washington, Seattle.

Maynard Smith, J. 1976. Evolution and the theory of games. Amer. Scientist 64:41–45.

Maynard Smith, J. 1982. Evolution and the Theory of Games. Cambridge University Press.

Mayr, E. 1957. On the origin of bird migration in the Pacific. Proc. Pacific Sci. Congr. 7, 1949, pp. 387–394.

McAnelly, M. L. 1985. The adaptive significance and control of migratory behavior in the grasshopper *Melanoplus sanguinipes*. In: M. A. Rankin, ed. Migration: Mechanisms and Adaptive Significance. Contrib. Marine Sci. 27 (Suppl.), pp. 687–703.

McAnelly, M. L. 1986. Migration in the grasshopper *Melanoplus sanguinipes* (Fab.). I. The capacity for flight in non-swarming populations. Biol. Bull. 170:368–377.

McAnelly, M. L. and M. A. Rankin. 1986. Migration in the grasshopper *Melanoplus sanguinipes* (Fab.). II. Interactions between flight and reproduction. Biol. Bull. 170:378–392.

McCleave, J. D. 1987. Migration of *Anguilla* in the ocean: Signposts for adults! Signposts for leptocephali? In: W. F. Herrnkind and A. B. Thistle, eds. Signposts in the Sea. Department of Biological Science, Florida State University, Tallahassee, pp. 102–117.

McCleave, J. D. and R. C. Kleckner. 1985. Oceanic migrations of Atlantic eels (*Anguilla* spp.): adults and their offspring. In: M. A. Rankin, ed. Migration: Mechanisms and Adaptive Significance. Contrib. Marine Sci. 27 (Suppl.), pp. 316–337.

McCleave, J. D. and R. C. Kleckner. 1987. Distribution of leptocephali of the catadromous *Anguilla* species in the western Sargasso Sea in relation to water circulation and migration. Bull. Mar. Sci. 41:789–806.

McCleave, J. D., G. P. Arnold, J. J. Dodson, and W. H. Neill, eds. 1984. Mechanisms of Migration in Fishes. Plenum, New York.

McConaugha, J. R. 1992. Decapod larvae: dispersal, mortality, and ecology. A working hypothesis. Amer. Zool. 32:512–523.

McConnell, B. J., C. Chambers, K. S. Nicholas, and M. A. Fedak. 1992. Satellite tracking of grey seals (*Halichoerus grypus*). J. Zool. 226:271–282.

McCook, H. C. 1890. American Spiders and Their Spinning Work, Vol. 2. Academy of Natural Sciences, Philadelphia.

McCullogh, D. R. 1985. Long-range movements of large terrestrial mammals. In: M. A. Rankin, ed. Migration: Mechanisms and Adaptive Significance. Contrib. Marine Sci. 27 (Suppl.), pp. 444–465.

McDonald, G. and P. G. Cole. 1991. Factors influencing oocycte development in *Mythimna convecta* (Lepidoptera: Noctuidae) and their possible impact on migration in eastern Australia. Bull. Entomol. Res. 81:175–184.

McDowall, R. M. 1988. Diadromy in Fishes: Migrations between Freshwater and Marine Environments. Croon Helm, London.

McEvoy, P. B. and C. S. Cox. 1987. Wind dispersal distances in dimorphic achenes of ragwort, *Senecio jacobaea*. Ecology 68:2006–2015.

McFarlane, G. A. and R. J. Beamish. 1986. A tag suitable for assessing long-term movements of spiny dogfish and preliminary results from use of this tag. North Amer. J. Fish. Mgmt. 6:69–76.

McFarlane, J. E. 1966. Studies on group effects in crickets. III. Wing development of *Gryllodes sigillatus* (Walk.). Canad. J. Zool. 44:1017–1021.

McKean, J. L. and L. W. Braithwaite. 1976. Moult, movements, age and sex composition of mountain duck, *Tadorma tadornoides*, banded at Lake George, N.S.W. Aust. Wildl. Res. 3:173–179.

McNeil, J. N. 1987. The true armyworm *Pseudaletia punctata*: a victim of the pied piper or a seasonal migrant? Insect Sci. Appl. 8:591–597.

McNeil, J. N. and B. D. Roitberg. 1985. Experimental analysis of individual behavior as a means to understanding population dispersal. In: D. R. MacKenzie, C. S. Barfield, G. G. Kennedy, R. D. Berger, and D. J. Taranto, eds. The Movement and Dispersal of Agriculturally Important Biotic Agents. Claitor's Publishing, Baton Rouge, Louisiana, pp. 145–150.

McNeil, J. N., M. Cusson, I. Orchard, and S. S. Tobe. 1995. Physiological integration of migration in Lepidoptera. In: V. A. Drake and A. G. Gatehouse, eds. Insect Migration: Physical Factors and Physiological mechanisms. (In press)

Meier, A. H. and A. G. Fivizanni. 1980. Physiology of migration. In: S. A. Gauthreaux, Jr., ed. Animal Migration, Orientation and Navigation. Academic Press, New York, pp. 225–282.

Meier, A. H. and J. M. Wilson. 1985. Resetting annual cycles with neurotransmitter-affecting drugs. In: B. K. Follett, S. Ishii and A. Chandola, eds. The Endocrine System and the Environment. Japan Science Press, Tokyo/Springer, Berlin, pp. 149–157.

Merkel, F. W. and W. Wiltschko. 1965. Magnetismus und Richtungsfinden zugenruhiger Rotkehlchen (*Erithacus rubecula*). Vogelwarte 23:71–77.

Messina, F. J. 1987. Genetic contribution to the dispersal polymorphism of the cowpea weevil (Coleoptera: Bruchidae). Ann. Entomol. Soc. Am. 80:12–16.

Meylan, A. B., B. W. Bowen, and J. C. Avise. 1990. A genetic test of the natal homing versus social facilitation models of green turtle migration. Science 248:724–727.

Mikkola, K. 1967. Immigrations of Lepidoptera, recorded in Finland in the years 1946–1966, in relation to aircurrents. Ann. Zool. Fenn. 2:124–139.

Mikkola, K. 1986. Direction of insect migrations in relation to the wind. In W. Danthanarayana, ed. Insect Flight: Dispersal and Migration. Springer-Verlag, Berlin, pp. 152–171.

Mönkkönen, M. and P. Helle. 1989. Migratory habits of birds breeding in different stages of forest succession: a comparison between the Palearctic and the Nearctic. Ann. Zool. Fennici 26:323–330.

Moore, F. R. 1986. Sunrise, skylight polarization, and the early morning orientation of night-migrating warblers. Condor 88:493–498.

Moore, F. R. 1987. Sunset and the orientation behaviour of migrating birds. Biol. Rev. 62:65–86.

Moore, F. R. 1991. Ecophysiological and behavioral response to energy demand during migration. Acta XX Cong. Int. Ornith. Vol. II:753–760.

Moore, F. R. and J. B. Phillips. 1988. Sunset, skylight polarization and the migratory orientation of yellow-rumped warblers, *Dendroica coronata*. Anim. Behav. 36:1770–1778.

Moran, N., J. Seminoff, and L. Johnstone. 1993. Genotypic variation in propensity for host alternation within a population of *Pemphigus betae* (Homoptera: Aphididae). J. Evol. Biol. 6:691–705.

Moreau, R. E. 1951. The migration system in perspective. Proc. Int. Ornithol. Cong. 10, 1954, pp. 245–248.

Moreau, R. E. 1961. Problems of Mediterranean-Sahara migration. Ibis 103:373–421, 580–623.

Moreau, R. E. 1972. The Palearctic-African Bird Migration Systems. Academic Press, New York.

Mori, K. and F. Nakasuji. 1990. Genetic analysis of the wing-form determination of the small brown planthopper, *Laodelphax striatellus* (Hemiptera: Delphacidae). Entomol. Exp. Appl. 28:47–53.

Moriya, T. and J. N. Dent. 1986. Hormonal interaction in the mechanism of migratory movement in the newt, *Notophthalmus viridescens*. Zool. Sci (Tokyo) 3:669–676.

Morse, A. N. C. 1991. How do planktonic larvae know where to settle? Am. Scientist 9:154–167.

Morse, D. H. 1971. The insectivorous bird as an adaptive strategy. Ann. Rev. Ecol. Syst. 2:177–200.

Morton, E. S. 1990. The evolution of habitat segregation by sex in the hooded warbler: experiments on proximate causation. Am. Nat. 135:319–333.

Morton, E. S. 1992. What do we know about the future of migrant landbirds? In: J. M. Hagan III and D. W. Johnston, eds. Ecology and Conservation of Neotropical Migrant Landbirds. Smithsonian Institution Press, Washington, D.C., pp. 579–589.

Morton, E. S., J. F. Lynch, K. Young, and P. Melhop. 1987. Do male hooded warblers form nonbreeding territories in tropical forest? Auk 104:133–135.

Mousseau, T. A. and H. Dingle. 1991. Maternal effects in insect life histories. Ann. Rev. Entomol. 36:511–534.

Mousseau, T. A. and D. A. Roff. 1987. Natural selection and the heritability of fitness components. Heredity 59:181–197.

Moyle, P. B. 1976. Inland Fishes of California. University of California press, Berkeley.

Moyle, P. B., J. D. Williams, and E. D. Wikramanayake. 1989. Fish species of special concern of California. Final Report to California Dept. Fish and Game, Inland

Fisheries, Rancho Cordova, Calif., October, 1989, 222 pp.

Mrosovsky, N. and S. F. Kingsmill. 1985. How turtles find the sea. Z. Tierpsychol. 67:237–256.

Mulvihill, R. S. and C. R. Chandler. 1990. The relationship between wing shape and differential migration in the dark-eyed junco. Auk 107:490–499.

Munro, U. and W. Wiltschko. 1992. Orientation studies on Yellow-faced Honeyeaters *Lichenostomus chrysops* (Meliphagidae) during autumn migration. Emu 92:181–184.

Munro, U. and W. Wiltschko. 1993. Magnetic compass orientation in the Yellow-faced Honeyeater, *Lichenostomus chrysops*, a day migrating bird from Australia. Behav. Ecol. Sociobiol. 32:141–145.

Munro, U., W. Wiltschko, and H. A. Ford. 1993. Changes in migratory direction of Yellow-faced Honeyeaters *Lichenostromus chrysops* (Meliphagidae) during autumn migration. Emu 93:59–62.

Murai, M. 1977. Population studies of *Cavelarius sacchivorous* Okajima (Heteroptera: Lygaeidae): adult dispersal in relation to density. Res. Pop. Ecol. 18:147–159.

Murray, B. G. 1989. A critical review of the transoceanic migration of the blackpoll warbler. Auk 106:8–17.

Myers, J. H. and C. J. Krebs. 1971. Genetic, behavioral and reproductive attributes of dispersing field voles, *Microtus pennsylvanicus* and *Microtus ochrogaster*. Ecol. Monogr. 41:53–78.

Nagano, C. D., W. H. Sakai, S. B. Malcolm, B. J. Cockrell, J. P. Donahue, and L. P. Brower. 1993. Spring migration of monarch butterflies in California. In: S. B. Malcolm and M. P. Zalucki, eds. Biology and Conservation of the Monarch Butterfly. Los Angeles County Natural History Museum, Science Series, No. 38, pp. 219–232.

Nair, C. R. M. and V. K. K. Prabhu. 1985. Entry of proteins from degenerating flight muscles into oocyctes in *Dysdercus cingulatus* (Heteroptera: Pyrrhocoridae). J. Insect Physiol. 31:383–387.

Nehlson, W., J. E. Williams, and J. A. Lichatowich. 1991. Pacific salmon at the crossroads: stocks at risk from California, Oregon, Idaho, and Washington. Fisheries 16(2):4–21.

Nelson, D. R. 1987. On the use of ultrasonic tracking in orientation studies. In: W. F. Herrnkind and A. B. Thistle, eds. Signposts in the Sea. Department of Biological Science, Florida State University, Tallahassee, pp. 118–129.

Newton, I. 1972. Finches. Collins, London.

Nielsen, E. T. 1961. On the habits of the migratory butterfly, *Ascia monuste* L. Biol. Meddr. 23:1–81.

Nikolsky, G. V. 1963. The Ecology of Fishes. Academic Press, New York.

Nisbet, I. C. T. and W. H. Drury. 1967. Orientation of spring migrants studied by radar. Bird Banding 38:173–186.

Nishiwaki, M. 1966. Distribution and migration of the larger cetaceans of the North Pacific as shown by Japanese whaling results. In: K. S. Norris, ed. Whales, Dolphins, and Porpoises. Univ. of California Press, Berkeley.

Nix, H. A. 1976. Environmental control of breeding, post-breeding dispersal and migration of birds in the Australian region. Proc. 16th Int. Ornith. Cong., 1974, pp. 272–305.

Noda, T. and K. Kiritani. 1989. Landing places of migratory planthoppers *Nilaparvata lugens* (Stal) and *Sogatella furcifera* (Horvath) (Homoptera: Delphacidae) in Japan. Appl. Ent. Zool. 24:59–65.

Norse, E. A. 1977. Aspects of the zoogeographic distribution of *Callinectes* (Brachyura: Portunidae). Bull. Mar. Sci. 27:440–447.

Nowak, E., P. Berthold, and U. Querner. 1990. Satellite tracking of migrating Bewick's swans. A European pilot study. Naturwiss. 77:549–550.

Nur, N. 1993. A sharper focus on songbird populations. In: C. Peaslee, D. Evans, G. Geupel, and N. Nur, eds. On Behalf of Songbirds: California Planning and Action for Partners in Flight Initiative. PRBO, Stinson Beach, California, pp. 7–16.

Obrecht, H. H., C. J. Pennycuick, and M. R. Fuller. 1988. Wind tunnel experiments to assess the effect of back-mounted radio transmitters on bird body drag. J. Exp. Biol. 135:265–273.

Ogarrio, R. 1993. Conservation actions taken by Monarca, A. C., to protect the overwintering sites of the monarch butterfly in Mexico. In: S. B. Malcolm and M. P. Zalucki, eds. Biology and Conservation of the Monarch Butterfly. Los Angeles County Natural History Museum, Science Series, No. 38, pp. 377–378.

Ogden, J. C. and T. P. Quinn. 1984. Migration in coral reef fishes: ecological significance and orientation mechanisms. In: J. D. McCleave, G. P. Arnold, J. J. Dodson, and W. H. Neill, eds. Mechanisms of Migration in Fishes. Plenum, New York, pp. 293–306.

Oku, T. 1980. Evening activity of the pre-reproductive adults of two Euxoa cutworms (Lepidoptera: Noctuidae). Appl. Entomol. Zool. 15:344–347.

Oku, T. 1983a. Aestivation and migration in noctuid moths. In: V. K. Brown and I. Hodek, eds. Diapause and Life Cycle Strategies in Insects. Junk, The Hague, pp. 219–231.

Oku, T. 1983b. Annual and geographical distribution of crop infestation in northern Japan by the Oriental armyworm in special relation to the migration phenomenon. (In Japanese with English summary). Misc. Publ. Tohoku Nat. Agric. Exp. Station No. 3: 1–49 (Rev. Appl. Ent. A 71:854–855).

Olson, D. B., F. A. Schott, R. J. Zantopp, and K. D. Leaman. 1984. The mean circulation east of the Bahamas as determined from a recent measurement program and historical XBT data. J. Phys. Ocean. 14:1470–1487.

Orr, R. T. 1970. Animals in Migration. Macmillan, New York.

Palmer, J. O. and H. Dingle. 1986. Direct and correlated responses to selection among life-history traits in milkweed bugs (Oncopeltus fasciatus). Evolution 40:767–777.

Palmer, J. O. and H. Dingle. 1989. Responses to selection on flight behavior in a migratory population of milkweed bug (Oncopeltus fasciatus). Evolution 43:1805–1808.

Papi, F. 1990. Olfactory navigation in birds. Experientia 46:352–363.

Papi, F. 1991. Olfactory navigation. In: P. Berthold, ed. Orientation in Birds. Birkhäuser Verlag, Basel, pp. 52–85.

Parker, W. E. and A. G. Gatehouse. 1985a. The effect of larval rearing conditions on flight performance in females of the African armyworm, Spodoptera exempta (Walker) (Lepidoptera: Noctuidae). Bull. Ent. Res. 75:35–47.

Parker, W. E. and A. G. Gatehouse. 1985b. Genetic factors controlling flight performance and migration in the African armyworm moth, Spodoptera exempta (Walker) (Lepidoptera: Noctuidae). Bull. Ent. Res. 75:49–63.

Pashley, D. and G. L. Bush. 1979. The use of allozymes in studying insect movement with special reference to the codling moth, Laspeyresia pomonella (L.) (Olethreutidae). In: R. L. Rabb and G. G. Kennedy, eds. Movement of Highly Mobile Insects: Concepts and Methodology in Research. North Carolina State University, Raleigh, pp. 333–341.

Pathak, V. K. and A. Chandola. 1982a. Involvement of thyroid gland in the development of migratory disposition in the red-headed bunting, Emberiza bruniceps. Horm. Behav. 16:46–59.

Pathak, V. K. and A. Chandola. 1982b. Seasonal variations in extra thyroidal conversion of thyroxine to triiodothyronine and migratory disposition in red-headed bunting. Gen.

Comp. Endocrin. 47:433–440.

Pearce, A. 1981. A brief introduction to descriptive physical oceanography. Australia CSIRO, Div. of Fisheries and Oceanography Report 132, 51 pp.

Peaslee, C., D. Evans, G. Geupel, and N. Nur, eds. 1993. On Behalf of Songbirds: California Planning and Action for the Partners in Flight Initiative. PRBO, Stinson Beach, California.

Pedgley, D. E. 1982. Windborne Pests and Diseases. Ellis Horwood, Chichester, U.K.

Pedgley, D. E., W. W. Page, A. Mushi, P. Odiyo, J. Amisi, C. F. Dewhurst, W. R. Dunstan, L. D. C. Fishpool, A. W. Harvey, T. Megenasa, and D. J. W. Rose. 1989. Onset and spread of an African armyworm upsurge. Ecol. Entomol. 14:311–333.

Pennycuick, C. J. 1972a. Soaring behaviour and performance of some East African birds, observed from a motor glider. Ibis 114:178–218.

Pennycuick, C. J. 1972b. Animal Flight. Edward Arnold, London.

Pennycuick, C. J. 1983. Thermal soaring compared in three dissimilar tropical bird species, *Fregata magnifiscens, Pelecanus occidentalis* and *Coragyps atratus*. J. Exp. Biol. 102:307–325.

Pennycuick, C. J. 1989. Bird Flight Performance: A Practical Calculation Manual. Oxford University Press, Oxford.

Pennycuick, C. J., T. Alerstam, and B. Larson. 1979. Soaring migration of the Common Crane, *Grus grus* observed by radar and from an aircraft. Ornis Scand. 10:241–251.

Pennycuick, L. 1975. Movements of the migratory wildebeest population in the Serengeti area between 1960 and 1973. East Afr. Wildl. J. 13:65–87.

Perdeck, A. C. 1958. Two types of orientation in migrating *Sturnus vulgaris* and *Fringilla coelebs* as revealed by displacement experiments. Ardea 46:1–37.

Perrin, N. and J. Travis. 1992. On the use of constraints in evolutionary biology and some allergic reactions to them. Func. Ecol. 6:361–363.

Phillips, B. F. 1981. The circulation of the southeastern Indian Ocean and the planktonic life of the Western Rock Lobster. Oceanogr. Mar. Biol. Ann. Rev. 19:11–39.

Phillips, B. F. and L. Olsen. 1975. Swimming behaviour of the Puerulus larvae of the western rock lobster. Aust. J. Mar. Freshwater Res. 26:415–417.

Phillips, J. B. 1985. Magnetic compass orientation in the eastern red-spotted newt, *Notophthalmus viridescens*. J. Comp. Physiol. A 158:103–109.

Phillips, J. B. 1986. Two magnetoreception pathways in a migratory salamander. Science 233:765–767.

Phillips, J. B. 1987. Laboratory studies of homing orientation in the eastern red-spotted newt, *Notophthalmus viridescens*. J. Exp. Biol. 131:215–229.

Phillips, J. B. and S. C. Borland. 1992. Magnetic compass information is eliminated under near-infrared light in the eastern red-spotted newt *Notophthalmus viridescens*. Anim. Behav. 44:796–797.

Phillips, J. B. and F. R. Moore. 1992. Calibration of the sun compass by sunset polarized light patterns in a migratory bird. Behav. Ecol. Sociobiol. 31:189–193.

Pittendrigh, C. S. 1960. Circadian rhythms and the circadian organization of living systems. Cold Spring Harbor Symp. Quant. Biol. 25:159–184.

Popov, G. B. 1965. Review of the work of the Desert Locust Ecological Survey June 1958–March 1964 and considerations and conclusions arising from it. Prog. Rept. UNSF/DL/ES/8. FAO, Rome.

Popov, G. B. and M. Ratcliffe. 1968. The sahelian tree locust, *Anacridium melanorhodon*. Anti-Locust Mem. No. 9.

Potter, I. C. 1980. The Petromyzoniformes with particular reference to paired species. Can.

J. Fish. Aquat. Sci. 37:1595–1615.

Potts, D. C. 1984. Generation times and the Quaternary evolution of reef-building corals. Paleobiology 10:48–58.

Power, J. H. and J. D. McCleave. 1983. Simulation of the North Atlantic Ocean drift of *Anguilla* leptocephali. Fish. Bull. 81:483–500.

Pringle, J. D. 1987. The Shorebirds of Australia. Angus and Robertson, North Ryde, NSW, Australia.

Prokopy, R. J. and B. D. Roitberg. 1984. Foraging behavior of true fruit flies. Am. Scientist 72:41–49.

Prunet, P., G. Boeuf, J. P. Bolton, and G. Young. 1989. Smoltification and seawater adaptation in the Atlantic salmon (*Salmo salar*): plasma prolactin, growth hormone, and thyroid hormones. Gen. Comp. Endocrinol. 74:355–364.

Purchase, D. 1985. Bird-banding and the migration of yellow-faced and white-naped honeyeaters through the Australian Capital Territory. Corella 9:59–62.

Quinn, J. F. and A. Hastings. 1987. Extinction in subdivided habitats. Conserv. Biol. 1:198–208.

Quinn, T. P. 1980. Evidence for celestial and magnetic compass orientation in lake migrating sockeye salmon fry. J. Comp. Physiol. A 137:243–248.

Quinn, T. P. 1984. Homing and straying in Pacific salmon. In: J. D. McCleave, G. P. Arnold, J. J. Dodson, and W. H. Neill, eds. Mechanisms of Migration in Fishes. Plenum, New York, pp. 357–362.

Quinn, T. P. 1985. Homing and the evolution of sockeye salmon (*Oncorhynchus nerka*). In: M. A. Rankin, ed. Migration: Mechanisms and Adaptive Significance. Contrib. Marine Sci. 27 (Suppl.), pp. 351–366.

Quinn, T. P. and E. L. Brannon. 1982. The use of celestial and magnetic cues by orienting sockeye salmon smolts. J. Comp. Physiol. A 147:547–552.

Quinn, T. P. and R. D. Brodeur. 1991. Intra-specific variations in the movement patterns of marine animals. Amer. Zool. 31:231–242.

Rabb, R. L. and G. G. Kennedy, eds. 1979. Movement of Highly Mobile Insects: Concepts and methodology in Research. North Carolina State University, Raleigh.

Rabb, R. L. and R. E. Stinner. 1978. The role of insect dispersal and migration in population processes. In: C. R. Vaughan, W. Wolf, and W. Klassen, eds. Radar, Population Ecology, and Pest Management. NASA, Wallops Island, Virginia, pp. 3–16.

Rainey, R. C. 1951. Weather and the movements of locust swarms: a new hypothesis. Nature 168:1057–1060.

Rainey, R. C. 1978. The evolution and ecology of flight: the "oceanographic" approach. In: H. Dingle, ed. The Evolution of Insect Migration and Diapause. Springer, New York, pp. 33–48.

Rainey, R. C. 1989. Migration and Meteorology: Flight Behaviour and the Atmospheric Environment of Locusts and Other Migrant Pests. Clarendon Press, Oxford.

Rainey, R. C., K. A. Browning, R. A. Cheke, and M. J. Haggis, eds. 1990. Migrant Pests: Progress, Problems, and Potentialities. Royal Society (Proceedings), London.

Ramenofsky, M. 1990. Fat storage and fat metabolism in relation to migration. In: E. Gwinner, ed. Bird Migration: Physiology and Ecophysiology. Springer-Verlag, Berlin, pp. 214–231.

Rankin, M. A. 1978. Hormonal control of insect migratory behavior. In: H. Dingle, ed. The Evolution of Insect Migration and Diapause. Springer, New York, pp. 5–32.

Rankin, M. A., ed. 1985. Migration: Mechanisms and Adaptive Significance. Contrib. Marine Science 27 (Suppl.). 868 pp.

Rankin, M. A. 1985b. Introduction. In: M. A. Rankin, ed. Migration: Mechanisms and Adaptive Significance. Contrib. Marine Science 27 (Suppl.), pp. 1–4.

Rankin, M. A. 1989. Hormonal control of flight. In: G. J. Goldsworthy and C. H. Wheeler, eds. Insect Flight. CRC Press. Boca Raton, Florida, pp. 139–163.

Rankin, M. A. 1991. Endocrine effects on migration. Amer. Zool. 31:217–230.

Rankin, M. A. and J. C. A. Burchsted. 1992. The cost of migration in insects. Annu. Rev. Entomol. 37:533–559.

Rankin, M. A. and S. M. Rankin. 1980a. Some factors affecting presumed migratory flight activity of the convergent ladybeetle, *Hippodamia convergens* (Coccinellidae: Coleoptera). Biol. Bull. 158:336–369.

Rankin, S. M. and M. A. Rankin. 1980b. The hormonal control of migratory flight behaviour in the convergent ladybird beetle *Hippodamia convergens*. Physiol. Entomol. 5:175–182.

Rankin, M. A. and L. M. Riddiford. 1978. Hormonal control of migratory flight in *Oncopeltus fasciatus*: the effects of the corpus cardiacum, corpus allatum, and starvation on migration and reproduction. Gen. Comp. Endocrin. 33:309–321.

Rankin, M. A., M. L. McAnelly, and J. E. Bodenhamer. 1986. The oogenesis-flight syndrome revisited. In: W. Danthanarayama, ed. Insect Flight: Dispersal and Migration. Springer-Verlag, Berlin, pp. 27–48.

Rappole, J. H. and D. W. Warner. 1976. Relationships between behavior, physiology and weather in avian transients at migration stopover sites. Oecologia 26:193–212.

Rappole, J. H., M. A. Ramos, and K. Winker. 1989. Wintering wood thrush movements and mortality in southern Veracruz. Auk 106:402–410.

Rasmuson, B., M. Rasmuson, and J. Nygren. 1977. Genetically controlled differences in behavior between cycling and noncycling populations of field voles (*Microtus agrestis*). Hereditas 87:33–42.

Raulston, J. R., S. D. Pair, F. A. Pedraga Martinez, J. Westbrook, A. N. Sparks, and V. M. Sanchez Valdez. 1986. Ecological studies indicating the migration of *Heliothis zea*, *Spodoptera frugiperda*, and *Heliothis virescens* from Northeastern Mexico and Texas. In: W. Danthanarayana, ed. Insect Flight: Dispersal and Migration, Springer-Verlag, Berlin, pp. 204–220.

Rayner, J. M. V. 1988. Form and function in avian flight. Current Ornith. 5:1–66.

Rayner, J. M. V. 1990. The mechanics of flight and bird migration performance. In: E. Gwinner, ed. Bird Migration: Physiology and Ecophysiology. Springer-Verlag, Berlin, pp. 283–299.

Real, L. A. 1980. Fitness, uncertainty, and the role of diversification in evolution and behavior. Amer. Natur. 115:623–638.

Rectenwald, H. 1989. California Dept. Fish and Game memorandum (8/16/89) to Dick Daniel, Environmental Services Division, Status of Winter-run Chinook Salmon Prior to Construction of Shasta Dam. 2 pp. w/ attachments.

Reese, E. S. 1989. Orientation behavior of butterflyfishes (family Chaetodontidae) on coral reefs: spatial learning of route specific landmarks and cognitive maps. Environ. Biol. Fishes 25:79–86.

Reffalt, W. C. 1985. A nationwide survey: wetlands in extremis. Wilderness 49:28–41.

Reid, R. W. 1962. Biology of the mountain pine beetle, *Dendroctonus monticolae* Hopkins in the East Kootenay Region of British Columbia. I. Life cycle, brood development, and flight periods. Canad. Entomol. 94:531–538.

Reisner, M. 1993. Cadillac Desert: The American West and Its Disappearing Water. Revised and Updated Edition. Penguin Books, New York.

Reiter, J., K. J. Panken, and B. L. Le Boeuf. 1981. Female competition and reproductive success in northern elephant seals. Anim. Behav. 29:670–687.

Reznick, D. 1981. "Grandfather effects": the genetics of interpopulation differences in offspring size in the mosquito fish. Evolution 35:941–953.

Richardson, W. J. 1979. Southeastward shorebird migration over Nova Scotia and New Brunswick in autumn: a radar study. Can. J. Zool. 57:107–124.

Richardson, W. J. 1980. Autumn landbird migration over the western Atlantic Ocean as evident from radar. Acta 17th Int Congr. Ornith. Berlin, pp. 501–506.

Richardson, W. J. 1985. The influence of weather on orientation and numbers of avian migrants over eastern Canada: a review. In: M. A. Rankin, ed. Migration: Mechanisms and Adaptive Significance. Contrib. Marine Sci. 27 (Suppl.), pp. 604–617.

Richardson, W. J. 1990. Timing of bird migration in relation to weather: A review. In E. Gwinner, ed. Bird Migration: Physiology and Ecophysiology. Springer-Verlag, Berlin, pp. 78–101.

Richardson, W. J. 1991. Wind and orientation of migrating birds: a review. In: P. Berthold, ed. Orientation in Birds. Birkhäuser Verlag, Basel, pp. 226–249.

Richman, N. H. III, and W. W. Zaugg. 1987. Effects of cortisol and growth hormone on osmoregulation in pre- and desmoltified coho salmon (*Oncorhynchus kisutch*). Gen. Comp. Endocrinol. 65:189–198.

Richter, C. J. J. 1970. Aerial dispersal in relation to habitat in eight wolf spider species (*Pardosa, Araneae, Lycosidae*). Oecologia 5:200–214.

Ricklefs, R. E. 1992. The megapopulation: a model of demographic coupling between migrant and resident landbird populations. In: J. M. Hagan III and D. W. Johnston, eds. Ecology and Conservation of Neotropical Migrant Landbirds. Smithsonian Institution Press, Washington, D.C., pp. 537–548.

Riley, J. R. 1978. Quantitative analysis of radar returns from insects. In: C. R. Vaughan, W. Wolf, and W. Klassen, eds. Radar, Insect Population Ecology, and Pest Management. NASA Conf. Publ. 2070, Wallops Island, Virginia.

Riley, J. R. 1985. Radar cross-sections of insects. Proc. Inst. Elect. Electron. Engrs. 73:228–232.

Riley, J. R. 1989. Remote sensing in entomology. Ann. Rev. Entomol. 34:247–271.

Riley, J. R. 1992. A millimetric radar to study the flight of small insects. Electr. Comm. Engin. J., February, 1992, pp. 43–47.

Riley, J. R. and D. R. Reynolds. 1979. Radar-based studies of the migratory flight of grasshoppers in the middle Niger area of Mali. Proc. R. Soc. Lond. Ser. B 204:67–82.

Riley, J. R., D. R. Reynolds, and M. J. Farmery. 1983. Observations of the flight behaviour of the armyworm moth, *Spodoptera exempta*, at an emergence site using radar and infra-red optical techniques. Ecol. Entomol. 8:395–418.

Riley, J. R., D. R. Reynolds, and R. A. Farrow. 1987. The migration of *Nilaparvata lugens* (Stal) (Delphacidae) and other Hemiptera associated with rice during the dry season in the Philippines: a study using radar, visual observations, aerial netting and ground trapping. Bull. Ent. Res. 77:156–169.

Riley, J. R., X.-N. Cheng, X.-X. Zhang, D. R. Reynolds, G.-M. Xu, A. D. Smith, J.-Y. Cheng, A.-D. Bao, and B.-P. Zhai. 1991. Long-distance migration of *Nilaparvata lugens* (Stal) (Delphacidae) in China: radar observations of mass return flight in the autumn. Ecol. Entomol. 16:471–489.

Rimmer, D. W. and B. F. Phillips. 1979. Diurnal migration and vertical distribution of phyllosoma larvae of the western rock lobster *Panulirus cygnus*. Mar. Biol. 54:109–124.

Ritchie, M. and D. Pedgley. 1989. Desert locusts cross the Atlantic. Antenna 13:10–12.

Robbins, C. S., D. Bystrak, and P. H. Geissler. 1986. The breeding bird survey: its first 15 years, 1965–1979. U.S. Dept. Interior Fish Wildl. Serv. Res. Pub. 157:1–196.

Robbins, C. S., J. W. Fitzpatrick, and P. B. Hamel. 1992. A warbler in trouble: *Dendroica cerulea*. In: J. M. Hagen III and D. W. Johnston, eds. Ecology and Conservation of Neotropical Migrant Landbirds. Smithsonian Institution Press, Washington, D.C., pp. 549–562.

Robbins, C. S., J. R. Sauer, R. Greenberg, and S. Droege. 1989. Population declines in North American birds that migrate to the Neotropics. Proc. Natl. Acad. Sci. 86:7658–7662.

Robert, Y. and E. Choppin de Janvry. 1977. Sur l'interêt d'implanter en France un réseau de piêgeage pour améliorer la lutte contre les pucerons. Bull. Tech. Information 323:559–568.

Robinson, T. and C. Minton. 1989. The enigmatic banded stilt. Birds International 1(4):72–85.

Rodda, G. H. 1984. The orientation and navigation of juvenile alligators: evidence of magnetic sensitivity. J. Comp. Physiol. A 154:649–658.

Rodda, G. H. and J. B. Phillips. 1992. Navigational systems develop along similar lines in amphibians, reptiles, and birds. Ethol. Ecol. Evol. 4:43–51.

Roff, D. A. 1974a. Spatial heterogeneity and the persistence of populations. Oecologia 15:245–258.

Roff, D. A. 1974b. The analysis of a population model demonstrating the importance of dispersal in a heterogeneous environment. Oecologia 15:259–275.

Roff, D. A. 1975. Population stability and the evolution of dispersal in a heterogeneous environment. Oecologia 19:217–237.

Roff, D. A. 1986. The evolution of wing dimorphism in insects. Evolution 40:1009–1020.

Roff, D. A. 1988. The evolution of migration and some life history parameters in marine fishes. Environ. Biol. Fish. 22:133–146.

Roff, D. A. 1990. The evolution of flightlessness in insects. Ecol. Monogr. 60:389–421.

Roff, D. A. 1991. Life history consequences of bioenergetic and biomechanical constraints on migration. Amer. Zool. 31:205–215.

Roff, D. A. 1994. Habitat persistence and the evolution of wing dimorphism in insects. Am. Nat. 144:772–798.

Roff, D. A. and D. J. Fairbairn. 1991. Wing dimorphisms and the evolution of migratory polymorphisms among the Insecta. Amer. Zool. 31:243–251.

Roff, D. A. and T. A. Mousseau. 1987. Quantitative genetics and fitness: lessons from *Drosophila*. Heredity 58:103–118.

Roffey, J. 1985. The FAO desert locust programme. In: D. R. MacKenzie, C. S. Barfield, G. G. Kennedy, R. D. Berger, and D. J. Taranto, eds. The Movement and Dispersal of Agriculturally Important Biotic Agents. Claitor's Publishing, Baton Rouge, Louisiana, pp. 533–540.

Rogers, D. 1983. Pattern and process in large scale animal movement. In: I. R. Swingland and P. J. Greenwood, eds. The Ecology of Animal Movement. Oxford University Press, Oxford, pp. 159–180.

Rogers, L. E., R. L. Buschbomb, and C. R. Watson. 1977. Length-weight relationships of shrub-steppe invertebrates. Ann. Ent. Soc. Amer. 70:51–53.

Root, R. B. and P. M. Kareiva. 1984. The search for resources in cabbage butterflies (*Pieris rapae*): ecological consequences and adaptive significance of Markovian movements in a patchy environment. Ecology 65:147–165.

Rose, D. J. W., C. F. Dewhurst, W. W. Page, and L. D. C. Fishpool. 1987. The role of migration in the life system of the African armyworm, *Spodoptera exempta*. Insect. Sci. Applic. 8:561–569.

Routledge, R. D. and T. B. Swartz. 1991. Taylor's Power Law re-examined. Oikos 60:107–112.

Rowan, W. 1925. Relation of light to bird migration and development changes. Nature 115:494–495.

Rowan, W. 1932. Experiments on bird migration. III. The effects of artificial light, castration and certain extracts on the autumn movements of the American crow (*Corvus brachyrhynchos*). Proc. Nat. Acad. Sci. 18:639–654.

Rowley, W. A. and C. L. Graham. 1968. The effect of age on the flight performance of female *Aedes aegypti* mosquitoes. J. Insect Physiol. 14:719–728.

Rudloe, A. and W. F. Herrnkind. 1976. Orientation of *Limulus polyphemus* in the vicinity of breeding beaches. Mar. Behav. Physiol. 4:75–89.

Ryker, L. C. 1984. Acoustic and chemical signals in the life cycle of a beetle. Scient. Amer. 250(6):112–123.

Sacktor, B. 1974. Biological oxidations and energetics in insect mitochondria. In: M. Rockstein, ed. The Physiology of Insects, Vol. 4. Academic Press, London, pp. 271–353.

Salmon, M. and J. Wyneken. 1987. Orientation and swimming behavior of hatchling loggerhead turtles. *Caretta caretta* L. during their offshore migration. J. Exp. Mar. Biol. Ecol. 109:137–153.

Salmon, M., J. Wyneken, E. Fritz, and M. Lucas. 1992. Seafinding by hatchling sea turtles: role of brightness, silhouette, and beach slope as orientation cues. Behaviour 122:56–77.

Salomonsen, F. 1955. The evolutionary significance of bird migration. Biol. Medd. Dan. Vid. Selsk. 22:1–62.

Salomonsen, F. 1967. Migratory movements of the Arctic Tern (*Sterna paradisaea*) in the Southern Ocean. Biol. Medd. Dan. Vid. Selsk. 24:1.

Salomonsen, F. 1968. The moult migration. Wildfowl 19:5–24.

Santschi, F. 1911. Observations et remarques critiques sur le mécanisme de l'orientation chez les fourmis. Rev. Suisse Zool. 19:305–338.

Santschi, F. 1913. A propos de l'orientation virtuelle chez les fourmis. Bull. Soc. Hist. Nat. Afr. Nord. 5:231–235.

Sappington, T. W. and W. B. Showers. 1992. Reproductive maturity, mating status, and long duration flight behavior of *Agrotis ipsilon* (Lepidoptera: Noctuidae) and the conceptual misuse of the oogenesis-flight syndrome by entomologists. Environ. Entomol. 21:677–688.

Sato, T. 1977. Life history and diapause of the white spotted tussock moth *Orgyia thyelina* Butter (Lepidoptera: Lymantriidae) Jap. J. Appl. Entomol. Zool. 21:6–14.

Sauer, F. 1957. Die Sternorientierung Nächtlich Ziehender Grasmücken, *Sylvia atricapilla* und *carruca*. Z. Tierpsychol. 14:29–70.

Sauer, J. R. and S. Droege. 1992. Geographic patterns in population trends of Neotropical migrants in North America. In: J. M. Hagan III and D. W. Johnston, eds. Ecology and Conservation of Neotropical Migrant Landbirds. Smithsonian Institution Press, Washington, D.C., pp. 26–42.

Saunders, D. S. 1982. Insect Clocks. 2nd. ed. Pergamon Press, Oxford.

Schaefer, G. W. 1969. Radar studies of locust, moth, and butterfly migration in the Sahara. Proc. R. Entomol. Soc. Ser. C 34:33, 39–40.

Schaefer, G. W. 1976. Radar observations of insect flight. Symp. Roy. Entomol. Soc. London 7:157–197.

Schaffer, W. M. 1979. The theory of life-history evolution and its application to Atlantic salmon. Symp. Zool. Soc. Lond. 44:307–326.

Schaffer, W. M. and P. F. Elson. 1975. The adaptive significance of variations in life history among local populations of Atlantic salmon in North America. Ecology 56:577–590.

Schmidt, B. R. 1993. Are hybridogenetic frogs cyclical parthenogens? Trends Ecol. Evol. 8:271–273.

Schmidt, J. 1912. The reproduction and spawning places of the fresh-water eel (*Anguilla vulgaris*). Nature 89:633–636.

Schmidt, J. 1922. The breeding places of the eel. Phil. Trans. Roy. Soc. Lond. B 211:179–208.

Schmidt-Koenig, K. 1973. Über die Navigation der Vögel. Naturwiss. 60:88–94.

Schmidt-Koenig, K. 1985. Migration strategies of Monarch Butterflies. In: M. A. Rankin, ed. Migration: Mechanisms and Adaptive Significance. Contrib. Marine Sci. 27 (Suppl.), pp. 786–798.

Schmidt-Koenig, K. 1987. Bird navigation: Has olfactory orientation solved the problem? Quart. Rev. Biol. 62:31–47.

Schmidt-Koenig, K., J. U. Ganzhorn, and R. Ranvaud. 1991. The sun compass. In: P. Berthold, ed. Orientation in Birds. Birkhäuser Verlag, Basel, pp. 1–15.

Schmidt-Nielsen, K. 1975. Animal Physiology. Cambridge University Press, Cambridge.

Schmitt, J., D. Ehrhardt, and D. Swartz. 1985. Differential dispersal of self-fertilized and outcrossed progeny in jewelweed (*Impatiens capensis*). Am. Nat. 126:570–575.

Schneider, D. C. and B. A. Harrington. 1981. Timing of shorebird migration in relation to prey depletion. Auk 98:801–811.

Schodde, R., G. F. van Tets, C. R. Champion, and G. S. Hope. 1975. Observations on birds at glacial altitudes on the Carstenz Massif, western New Guinea. Emu 75:65–72.

Schwabl, H. 1983. Ausprögung und Bedeutung des Teilzugverhaltens einer südwest-deutschen Population der Amsel, *Turdus merula*. J. Ornithol. 124:101–115.

Schwabl, H. and D. S. Farner. 1989. Dependency on testosterone of photoperiodically-induced vernal fat deposition in female white-crowned sparrows. Condor 91:108–112.

Schwabl, H. and B. Silverin. 1990. Control of partial migration and autumnal behavior. In: E. Gwinner, ed. Bird Migration: Physiology and Ecophysiology. Springer-Verlag, Berlin, pp. 144–155.

Schwabl, H., I. Schwabl-Benzinger, A. R. Goldsmith, and D. S. Farner. 1988. Effects of ovariectomy on long day-induced premigratory fat deposition, plasma levels of luteinizing hormone and prolactin, and molt in white-crowned sparrows, *Zonotrichia leucophrys gambelii*. Gen. Comp. Endocrin. 71:398–405.

Schwarz, H. H. and J. K. Müller. 1992. The dispersal behaviour of the phoretic mite *Poecilochirus carabi* (Mesostigmata; Parasitidae): adaptation to the breeding biology of its carrier *Necrophorus vespilloides* (Coleoptera, Silphidae). Oecologia 89:487–493.

Scorer, R. S. 1978. Environmental Aerodynamics. Ellis Horwood, Chickester, U.K.

Scott, R. W. and G. L. Achtemeier. 1987. Estimating pathways of migrating insects carried in atmospheric winds. Environ. Entomol. 16:1244–1254.

Seeley, T. D. 1985. Honeybee Ecology. Princeton University Press, Princeton, New Jersey.

Semlitsch, R. D. 1993. Asymmetric competition in mixed populations of tadpoles of the hybridogenetic *Rana esculenta* complex. Evolution 47:510–519.

Semlitsch, R. D. and H.-U. Reyer. 1992. Performance of tadpoles from the hybridogenetic

Rana esculenta complex: interactions with pond drying and interspecific competition. Evolution 46:665–676.

Shapiro, A. M. 1976. Seasonal polyphenism. Evol. Biol. 9:259–333.

Shaw, M. J. P. 1970. Effects of population density on alienicolae of *Aphis fabae* Scop. I. The effect of crowding on the production of alatae in the laboratory. Ann. Appl. Biol. 65:191–196.

Sheridan, M. A. 1986. Effects of thyroxin, cortisol, growth hormone, and prolactin on lipid metabolism of coho salmon, *Oncorhynchus kisutch*, during smoltification. Gen. Comp. Endocrinol. 64:220–238.

Sherrington, C. S. 1906. The Integrative Action of the Nervous System. Yale University Press, New Haven.

Sherry, T. W. and R. T. Holmes. 1993. Limits for migrant landbirds year-round. In: C. Peaslee, D. Evans, G. Geupel, and N. Nur, eds. On Behalf of Songbirds: California Planning and Action for the Partners in Flight Initiative. PRBO, Stinson Beach, California, pp. 4–5.

Shoubridge, E. A. 1978. Genetic and reproductive variation in American shad. M.Sc. Thesis, McGill University, Montreal, 73 pp.

Showers, W. B., F. Whitford, R. B. Smelser, A. J. Keaster, J. F. Robinson, J. D. Lopez, and S. E. Taylor. 1989. Direct evidence for meteorologically driven long-range dispersal of an economically important moth. Ecology 70:987–992.

Silvertown, J. W. 1984. Phenotypic variety in seed germination behavior: the ontogeny and evolution of somatic polymorphism in seeds. Am. Nat. 124:1–16.

Simons, P. 1992. The Action Plant. Blackwells, Oxford.

Sinclair, A. R. E. 1983. The function of distance movements in vertebrates. In: I. R. Swingland and P. J. Greenwood, ed. The Ecology of Animal Movement. Oxford, pp. 240–258.

Sinsch, U. 1992. Sex-biased site fidelity and orientation behavior in reproductive natterjack toads (*Bufo calamita*). Ethol. Ecol. Evol. 4:15–32.

Skiles, D. D. 1985. The geomagnetic field: its nature, history, and biological relevance. In: J. L. Kirschvink, D. S. Jones, and B. J. McFadden, eds. Magnetite Biomineralization and Magnetoreception in Organisms. Plenum, New York, pp. 43–102.

Skolnik, M. I. 1980. Introduction to Radar Systems. McGraw-Hill, Kogakushe, Tokyo.

Slansky, F. 1980. Food consumption and reproduction as affected by tethered flight in female milkweed bugs (*Oncopeltus fasciatus*). Entomol. Exp. Appl. 28:277–286.

Slater, J. A. 1964. A catalogue of the Lygaeidae of the world. Univ. of Connecticut Press, Storrs.

Slater, J. A. 1975. On the biology and zoogeography of Australian Lygaeidae (Hemiptera: Heteroptera) with special reference to the Southwest fauna. J. Aust. Entomol. Soc. 14:47–64.

Slater, J. A. 1977. On the incidence and evolutionary significance of wing polymorphism in lygaeid bugs with particular reference to those of South Africa. Biotropica 9:217–229.

Smith, N. G. 1980. Hawk and vulture migrations in the Neotropics. In: A. Keast and E. S. Morton, eds. Migrant Birds in the Neotropics. Smithsonian Institution, Washington, D.C., pp. 51–65.

Smithers, C. N. 1977. Seasonal distribution and breeding status of *Danaus plexippus* (L.) (Lepidoptera: Nymphalidae) in Australia. J. Aust. Ent. Soc. 16:175–184.

Snook, L. C. 1993. Conservation of the monarch butterfly reserves in Mexico: focus on the forest. In: S. B. Malcolm and M. P. Zalucki, eds. Biology and Conservation of the

Monarch Butterfly. Los Angeles County Natural History Museum, Science Series, No. 38, pp. 363–375.

Snyder, R. J. 1988. Migration and evolution of life histories of the threespine stickleback (*Gasterosteus aculeatus* L.). Ph.D. Thesis, University of California, Davis.

Snyder, R. J. 1991. Migration and life histories of the threespine stickleback: evidence for adaptive variation in growth rate between populations. Environ. Biol. Fish. 31:381–388.

Snyder, R. J. and H. Dingle. 1989. Adaptive, genetically based differences in life history between estuary and freshwater threespine sticklebacks (*Gasterosteus aculeatus* L.). Can. J. Zool. 67:2448–2454.

Snyder, R. J., B. A. McKeown, and L. B. Holtby. 1995. Changes in plasma prolactin, thyroxine, and growth hormone during smoltification in naturally occurring coho salmon (*Oncorhynchus kisutch* W.). Manuscript.

Solbreck, C. 1971. Displacement of marked *Lygaeus equestris* (L.) (Het., Lygaeidae) during pre- and posthibernation migrations. Acta. Ent. Fenn. 28:74–83.

Solbreck, C. 1972. Sexual cycle, and changes in feeding activity and fat body size in relation to migration in *Lygaeus equestris* (L.) (Het., Lygaeidae). Ent. Scand. 3:267–274.

Solbreck, C. 1974. Maturation of post-hibernation flight behaviour in the coccinellid *Coleomegilla maculata* (DeGeer). Oecologia 17:265–275.

Solbreck, C. 1976. Flight patterns of *Lygaeus equestris* (Heteroptera) in spring and autumn with special reference to the influence of weather. Oikos 27:134–143.

Solbreck, C. 1978. Migration, diapause, and direct development as alternative life histories in a seed bug, *Neacoryphus bicrucis*. In: H. Dingle, ed. Evolution of Insect Migration and Diapause. Springer, New York, pp. 195–217.

Solbreck, C. 1985. Insect migration strategies and population dynamics. In: M. A. Rankin, ed. Migration: Mechanisms and Adaptive Significance. Contrib. Marine Sci. 27 (Suppl.), pp. 641–662.

Solbreck, C. 1986. Wing and flight muscle polymorphism in a lygaeid bug, *Horvathiolus gibbicollis*: determinants and life history consequences. Ecol. Entomol. 11:435–444.

Solbreck, C. and D. B. Anderson. 1989. Wing reduction; its control and consequences in a lygaeid bug, *Spilostethus pandurus*. Hereditas 111:1–6.

Solbreck, C. and O. Kugelberg. 1972. Field observations on the seasonal occurrence of *Lygaeus equestris* (L.). (Het. Lygaeidae) with special reference to food plant phenology. Ent. Scand. 3:189–210.

Solbreck, C. and I. Pehrson. 1979. Relations between environment, migration and reproduction in a seed bug, *Neacoryphus bicrucis* (Say) (Heteroptera: Lygaeidae). Oecologia 43:51–62.

Solbreck, C. and B. Sillén-Tullberg. 1981. Control of diapause in a "monovoltine" insect, *Lygaeus equestris* (Heteroptera). Oikos 36:68–74.

Solbreck, C., D. B. Anderson, and J. Förare. 1990. Migration and the coordination of life cycles as exemplified by Lygaeinae bugs. In: F. Gilbert, ed. Insect Life Cycles: Genetics, Evolution, and Coordination. Springer-Verlag, London, pp. 197–214.

Sorensen, A. E. 1978. Somatic polymorphism and seed dispersal. Nature 276:174–176.

Soulé, M. E., ed. 1986. Conservation Biology. Sinauer Associates, Inc., Sunderland, Massachusetts.

Southwood, T. R. E. 1961. A hormonal theory of the mechanism of wing polymorphism in Heteroptera. Proc. Roy. Entomol. Soc. A 36:63–66.

Southwood, T. R. E. 1962. Migration of terrestrial arthropods in relation to habitat. Biol. Rev. 37:171-214.

Southwood, T. R. E. 1971. The role and measurement of migration in the population system of an insect pest. Tropical Science 13:275-278.

Southwood, T. R. E. 1977. Habitat, the templet for ecological strategies? J. Anim. Ecol. 46:337-365.

Southwood, T. R. E. 1981. Ecological aspects of insect migration. In: D. J. Aidley, ed. Animal Migration. Cambridge Univ. Press, Cambridge, pp. 196-208.

Sparks, A. N. 1979. An introduction to the status, current knowledge, and research on movement on selected Lepidoptera in southeastern United States. In: R. L. Rabb and G. G. Kennedy, eds. Movement of Highly Mobile Insects: Concepts and Methodology in Research. North Carolina State University Press.

Stearns, S. C. 1989. The evolutionary significance of phenotypic plasticity. Bioscience 39:436-445.

Stengel, M. M. C. 1974. Migratory behavior of the female of the common cockchafer *Melolontha melolontha* L. and its neuroendocrine regulations. In: L. Barton Browne, ed. Experimental Analysis of Insect Behaviour. Springer, New York, pp. 297-316.

Stenseth, N. C. and W. Z. Lidicker, eds. 1992a. Animal Dispersal. Small Mammals as a Model. Chapman & Hall, London.

Stenseth, N. C. and W. Z. Lidicker. 1992b. The study of dispersal: a conceptual guide. In: N. C. Stenseth and W. Z. Lidicker, Jr., eds. Animal Dispersal. Small Mammals as a Model. Chapman & Hall, London, pp. 5-20.

Stephens, D. W. and J. R. Krebs. 1986. Foraging Theory. Princeton Univ. Press, Princeton, New Jersey.

Stinner, R. E. and M. Saks. 1985. Within region movement models and their problems. In: D. R. MacKenzie, C. S. Barfield, G. G. Kennedy, R. D. Berger, and D. J. Taranto, eds. The Movement and Dispersal of Agriculturally Important Biotic Agents. Claitor's Publishing, Baton Rouge, Louisiana, pp. 259-264.

Stinner, R. E., R. L. Rabb, and J. R. Bradley. 1977. Natural factors operating in the population dynamics of *Heliothis zea* in North Carolina. Proc. 15th Int. Cong. Entomol., Washington, D. C., pp. 622-642.

Stinner, R. E., M. Saks, and L. Dohse. 1986. Modelling of agricultural pest displacement. In: W. Danthanarayana, ed. Insect Flight: Dispersal and Migration. Springer-Verlag, Berlin, pp. 235-241.

Stoddard, P. K., J. E. Marsden, and T. C. Williams. 1983. Computer simulation of autumnal bird migration over the western North Atlantic. Anim. Behav. 31:173-180.

Strathman, R. R. 1982. Selection for retention or export of larvae in estuaries. In: V. Kennedy, ed. Estuarine Comparisons. Academic Press, New York, pp. 521-536.

Suarez, R. K., J. R. B. Lighton, C. D. Moyes, G. S. Brown, C. L. Gass, and P. W. Hochachka. 1990. Fuel selection in rufous hummingbirds: ecological implications of metabolic biochemistry. Proc. Natl. Acad. Sci. USA 87:9207-9210.

Sutherland, O. R. W. 1969. The role of crowding in the production of winged forms by two strains of the pea aphid, *Acyrthosiphon pisum*. J. Insect Physiol. 15:1385-1410.

Swingland, I. R. 1983. Intraspecific differences in movement. In: I. R. Swingland and P. J. Greenwood, eds. The Ecology of Animal Movement. Clarendon Press, Oxford, pp. 102-114.

Swingland, I. R. and C. M. Lessels. 1979. The natural regulation of giant tortoise populations on Aldabra Atoll. Movement polymorphism, reproductive success and mortality. J. Anim. Ecol. 48:639-654.

Sydeman, W. J., H. R. Huber, S. D. Emslie, C. A. Ribic, and N. Nur. 1991. Age-specific weaning success of northern elephant seals in relation to previous breeding experience. Ecology 72:2204–2217.

Talbot, L. M. and M. H. Talbot. 1963. The wildebeest in western Masailand, East Africa. Wildlife Monogr. 12:1–88.

Tanaka, S. and H. Wolda. 1987. Seasonal wing length dimorphism in a tropical seed bug: ecological significance of the short-winged form. Oecologia 73:559–565.

Tauber, M. J., C. A. Tauber, and S. Masaki. 1986. Seasonal Adaptations of Insects. Oxford University Press, New York.

Taylor, E. B. 1990. Phenotypic correlates of life-history variation in juvenile chinook salmon, *Oncorhynchus tshawytscha*. J. Anim. Ecol. 59:455–468.

Taylor, L. R. 1958. Aphid dispersal and diurnal periodicity. Proc. Linn. Soc. Lond. 169:67–73.

Taylor, L. R. 1961. Aggregation, variance, and the mean. Nature 189:732–735.

Taylor, L. R. 1979. The Rothamsted Insect Survey—an approach to the theory and practice of synoptic pest forecasting in agriculture. In: R. L. Rabb and G. G. Kennedy, eds. Movement of Highly Mobile Insects: Concepts and Methodology in Research. North Carolina State University, Raleigh, pp. 148–185.

Taylor, L. R. 1986a. Synoptic ecology, migration of the second kind, and the Rothamsted Insect Survey. Presidential address. J. Anim. Ecol. 55:1–38.

Taylor, L. R. 1986b. The four kinds of migration. In: W. Danthanarayana, ed. Insect Flight: Dispersal and Migration. Springer-Verlag, Berlin, pp. 265–280.

Taylor, L. R. and R. A. J. Taylor. 1977. Aggregation, migration, and population mechanics. Nature 265:415–421.

Taylor, L. R. and R. A. J. Taylor. 1983. Insect migration as a paradigm for survival by movement. In: I. R. Swingland and P. J. Greenwood, eds. The Ecology of Animal Movement, Clarendon Press, Oxford, pp. 181–214.

Taylor, R. A. J. and D. Reling. 1986a. Density/height profile and long-range dispersal of first-instar gypsymoth (Lepidoptera: Lymantriidae). Environ. Entomol. 15:431–435.

Taylor, R. A. J. and D. Reling. 1986b. Preferred wind direction of long distance leaf hopper (*Empoasca fabae*) migrants and its relevance to the return of small insects. J. Anim. Ecol. 55:1103–1114.

Taylor, V. A. 1978. A winged elite in a subcortical beetle as a model for a prototermite. Nature 276:73–75.

Tchernavin, V. 1939. The origin of salmon—is its ancestry marine or fresh water? Salmon and Trout Mag. 95:120–140.

Telenius, A. and P. Torstensson. 1991. Seed wings in relation to seed size in the genus *Spergularia*. Oikos 61:216–222.

Terborgh, J. 1989. Where Have All the Birds Gone? Princeton University Press, Princeton, New Jersey.

Terborgh, J. 1992. Perspectives on the conservation of Neotropical migrant landbirds. In: J. M. Hagan III and D. W. Johnston, eds. Ecology and Conservation of Neotropical Migrant Landbirds. Smithsonian Institution Press, Washington, D.C., pp. 7–12.

Terrill, S. B. 1991. Evolutionary aspects of orientation and migration in birds. In: P. Berthold, ed. Orientation in Birds. Birkhäuser Verlag, Basel, pp. 180–201.

Terrill, S. B. and P. Berthold. 1990. Ecophysiological aspects of rapid population growth in a novel migratory blackcap (*Sylvia atricapilla*) population: an experimental approach. Oecologia 85:266–270.

Thapliyal, J. P. and P. Lal. 1984. Light, thyroid, gonad, and photorefractory state in the migratory red-headed bunting, *Emberiza bruniceps*. Gen. Comp. Endocrin. 56:41–52.

Thapliyal, J. P. Lal, A. K. Pati, and B. B. P. Gupta. 1983. Thyroid and gonad in the oxidative metabolism, erythropoiesis, and light response of the migratory red-headed bunting, *Emberiza bruniceps*. Gen. Comp. Endocrin. 51:444–453.

Thomas, D. W. 1983. The annual migrations of three species of West African fruit bats (Chiroptera: Pteropodidae). Canad. J. Zool. 61:2266–2272.

Thompson, J. N. and O. Pellmyr. 1991. Evolution of oviposition behavior and host preference in Lepidoptera. Ann. Rev. Entomol. 36:65–89.

Thomson, A. L. 1926. Problems of Bird Migration. Witherby, London.

Thorarinsson, K. 1986. Population density and movement: a critique of Δ - models. Oikos 46:70–81.

Thorsen, G. 1964. Light as an ecological factor in the dispersal and settlement of larvae of marine bottom invertebrates. Ophelia 1:167–208.

Tinbergen, N. 1953. The Study of Instinct. Methuen, London.

Tolman, E. C. 1948. Cognitive maps in rats and men. Psychol. Rev. 55:189–208.

Toonen, R. J. and J. R. Pawlik. 1995. Settlement of the gregarious tube worm *Hydroides dianthus* (Polychaeta: Serpulidae): I. All larvae are not created equal. Marine Biol. In press.

Tucker, V. A. 1968. Respiratory exchange and evaporative water loss in the flying budgerigar. J. Exp. Biol. 48:67–87.

Tucker, V. A. 1972. Metabolism during flight in the laughing gull, *Larus atricilla*. Am. J. Physiol. 222:237–245.

Tunner, H. G. 1991. Migratory behaviour of *Rana lessonae* and its hybridogenetic associate *Rana esculenta* of Lake Neusiedl (Austria, Hungary). Proc. 6th Ord. Gen. Meeting Eur. Soc. Herp. Budapest, p. 91.

Tunner, H. G. and H. Nopp. 1979. Heterosis in the common European water frog. Naturwiss. 66:268.

Ugolini, A. and L. Pardi. 1992. Equatorial sandhoppers do not have a good clock. Naturwiss. 79:279–281.

Unnithan, G. C. and K. K. Nair. 1977. Ultrastructure of juvenile hormone induced degenerating flight muscles in a bark beetle, *Ips paraconfusus*. Cell Tissue Res. 185:481–486.

Unnithan, G. C., K. K. Nair, and W. S. Bowers. 1977. Precocene induced degeneration of the corpus allatum of adult females of the bug *Oncopeltus fasciatus*. J. Insect Physiol. 23:1081–1094.

Urquhart, F. A. 1960. The Monarch Butterfly. University of Toronto Press.

Urquhart, F. A. 1987. The Monarch Butterfly: International Traveler. Nelson-Hall, Chicago.

Uvarov, B. P. 1921. A revision of the genus *Locusta* L. (*Pachytylus* Fab.) with a new theory as to the periodicity and migrations of locusts. Bull. Entomol. Res. 12:135–163.

Uvarov, B. P. 1966. Grasshoppers and Locusts, Vol. 1. Cambridge University Press, Cambridge.

Uvarov, B. P. 1977. Grasshoppers and Locusts, Vol. 2. Center for Overseas Pest Research, London.

van der Pijl, L. 1969. Principles of Dispersal in Higher Plants. Springer-Verlag, Berlin.

Vander Wall, S. B. 1992. The role of animals in dispersing a "wind-dispersed" pine. Ecology 73:614–621.

van Noordwijk, A. J. 1987. On the implications of genetic variation for ecological research. Ardea 75:13–19.

van Noordwijk, A. J. 1990. The methods of genetical ecology applied to the study of evolutionary change. In: K. Wöhrmann and S. Jain, eds. Population Biology: Ecological and Evolutionary Viewpoints. Springer, Berlin, pp. 291–319.

van Noordwijk, A. J. and M. Gebhardt. 1987. Reflections on the genetics of quantitative traits with continuous environmental variation. In: V. Loeschke, ed. Genetic Constraints on Adaptive Evolution. Springer, Berlin, pp. 73–90.

Van Valen, L. 1971. Group selection and the evolution of dispersal. Evolution 25:591–598.

Vaughn, C. R. 1978. The NASA radar entomology program at Wallops Flight Center. In: C. R. Vaughan, W. Wolf, and W. Klassen, eds. Radar, Population Ecology, and Pest Management. NASA, Wallops Island, Virginia, pp. 161–169.

Venable, D. L. 1985. The evolutionary ecology of seed heteromorphism. Am. Nat. 126:577–595.

Venable, D. L. and J. S. Brown. 1988. The selective interactions of dispersal, dormancy, and seed size as adaptations for reducing risk in variable environments. Am. Nat. 131:360–384.

Venable, D. L. and L. Lawlor. 1980. Delayed germination and dispersal in desert annuals: escape in space and time. Oecologia 46:272–282.

Venable, D. L. and D. A. Levin. 1983. Morphological dispersal structures in relation to growth habit in the Compositae. Plant. Syst. Evol. 143:1–16.

Venable, D. L. and D. A. Levin. 1985. Ecology of achene dimorphism in *Heterotheca latifolia* I. Achene structure, germination and dispersal. J. Ecol. 73:133–145.

Vepsäläinen, K. 1971. The role of gradually-changing daylength in the determination of wing-length, alary dimorphism and diapause in a *Gerris odontogaster* population (Gerridae, Heteroptera) in south Finland. Ann. Acad. Sci. Fenn. 183:1–25.

Vepsäläinen, K. 1973. The distribution and habitats of *Gerris* Fabr. species (Heteroptera: Gerridae) in Finland. Ann. Zool. Fennici 10:419–444.

Vepsäläinen, K. 1974. The life cycles and wing lengths of Finnish *Gerris* Fabr. species (Heteroptera: Gerridae). Acta Zool. Fennica 141:1–73.

Vepsäläinen, K. 1978. Wing dimorphism and diapause in *Gerris*: determination and adaptive significance. In: H. Dingle, ed. Evolution of Insect Migration and Diapause. Springer, New York, pp. 218–253.

Via, S. and R. Lande. 1985. Genotype-environment interaction and the evolution of phenotypic plasticity. Evolution 39:505–523.

Villard, M.-A., K. Freemark, and G. Merriam. 1992. Metapopulation theory and Neotropical migrant birds in temperate forests: an empirical investigation. In: J. M. Hagan III and D. W. Johnston, eds. Ecology and Conservation of Neotropical Migrant Landbirds. Smithsonian Institution Press, Washington, D.C., pp. 474–482.

Vince-Prue, D. 1975. Photoperiodism in Plants. McGraw-Hill, New York.

Vleck, C. M., J. C. Wingfield, and D. S. Farner. 1980. No temporal synergism of prolactin and corticosterone influencing reproduction in white-crowned sparrows. Am. Zool. 20:899.

Wagner, D. L. and J. K. Liebherr. 1992. Flightlessness in insects. Trends Ecol. Evol. 7:216–220.

Walcott, C. 1991. Magnetic maps in pigeons. In: P. Berthold, ed. Orientation in Birds. Birkhäuser, Basel, pp. 38–51.

Waldvogel, J. A. 1989. Olfactory orientation by birds. Current Ornithology 6:269–321.

Walker, J. M. and W. A. Venables. 1990. Weather and bird migration. Weather 45:47–56.

Walker, M. M. 1984. Learned magnetic field discrimination in yellowfin tuna, *Thunnus albacares*. J. Comp. Physiol. A 155:673–679.

Walker, T. J. 1980. Migrating Lepidoptera: Are butterflies better than moths? Fla. Entomol. 63:79–98.

Walker, T. J. 1985a. Butterfly migration in the boundary layer. In M. A. Rankin, ed. Migration: Mechanisms and Adaptive Significance. Contrib. Marine Sci. 27 (Suppl.), pp. 704–722.

Walker, T. J. 1985b. Permanent traps for monitoring butterfly migration: tests in Florida, 1979–84. J. Lepid. Soc. 39:313–320.

Walker, T. J. 1991. Butterfly migration from and to peninsular Florida. Ecol. Entomol. 16:241–252.

Wallington, C. E. 1977. Meteorology for Glider Pilots, 3rd ed. John Murray, London.

Wallraff, H. G. 1991. Conceptual approaches to avian navigation systems. In: P. Berthold, ed. Orientation in Birds. Birkhäuser Verlag, Basel, pp. 128–165.

Waloff, N. 1983. Absence of wing polymorphism in the arboreal, phytopagous species of some taxa of temperate Hemiptera: an hypothesis. Ecol. Entomol. 2:229–232.

Ward, P. 1971. The migration patterns of *Quelea quelea* in Africa. Ibis 113:275–297.

Ward, P. and P. J. Jones. 1977. Pre-migratory fattening in three races of the red-billed quelea, *Quelea quelea* (Aves: Ploceidae), an intra-tropical migrant. J. Zool. 181:43–56.

Washburn, J. O. and L. Washburn. 1984. Active aerial dispersal of minute wingless arthropods: exploitation of boundary-layer velocity gradients. Science 223:1088–1089.

Watson, A. 1992. A red grouse perspective on dispersal in small mammals. In: N. C. Stenseth and W. Z. Lidicker, eds. Animal Dispersal. Small Mammals as a Model. Chapman & Hall, London, pp. 260–273.

Wehner, R. 1983. Celestial and terrestrial navigation: human strategies—insect strategies. In: F. Huber and H. Markl, eds. Neuroethology and behavioral physiology. Springer-Verlag, Berlin.

Wehner, R. 1989. Neurobiology of polarization vision. Trends Neurosci. 12:353–359.

Wehner, R. 1990. On the brink of introducing sensory ecology: Felix Santschi (1872–1940)–Tabib-en-Neml. Behav. Ecol. Sociobiol. 27:295–306.

Wehner, R. and R. Menzel. 1990. Do insects have cognitive maps? Ann. Rev. Neurosci. 13:403–414.

Wehner, R. and S. Wehner. 1990. Insect navigation: use of maps or Ariadne's thread? Ethol. Ecol. Evol. 2:27–48.

Weihs, D. 1984. Bioenergetic considerations in fish migration. In: J. D. McCleave, G. P. Arnold, J. J. Dodson, and W. H. Neill, eds. Mechanisms of Migration in Fishes. Plenum, New York, pp. 487–508.

Welham, C. V. J. 1994. Flight speeds of migrating birds: a test of maximum range speed predictions from three aerodynamic equations. Behav. Ecol. 5:1–8.

Wellington, W. G. 1945. Conditions governing the distribution of insects in the free atmosphere. Introduction. Can. Ent. 77:7–15.

Wellington, W. G. 1977. Returning the insect to insect ecology: some consequences for pest management. Environ. Entomol. 6:1–8.

Wells, D. R. 1990. Migratory birds and tropical forest in the Sunda region. In: A. Keast, ed. Biogeography and Ecology of Forest Bird Communities. SPB Academic Publishers, The Hague, pp. 357–369.

Wenner, A. M. and A. M. Harris. 1993. Do California monarchs undergo long-distance directed migration? In: S. B. Malcolm and M. P. Zalucki, eds. Biology and Conservation of the Monarch Butterfly. Los Angeles County Natural History Museum, Science Series, No. 38, pp. 209–218.

Westerberg, H. 1984. The orientation of fish and the vertical stratification at fine- and micro-structure scales. In: J. D. McCleave, G. P. Arnold, J. J. Dodson, and W. H. Neill, eds. Mechanisms of Migration in Fishes. Plenum, New York, pp. 179–203.

Weygoldt, P. 1969. The Biology of Pseudoscorpions. Harvard University Press, Cambridge, Massachusetts.

Wheeler, C. H. 1989. Mobilization and transport of fuels to the flight muscles. In: G. J. Goldsworthy and C. H. Wheeler, eds. Insect Flight. CRC Press. Boca Raton, Florida, pp. 273–303.

Whitesel, T. A. 1992. Plasma thyroid hormone levels in migratory and lake-resident coho salmon juveniles from the Karluk River system, Alaska. Trans. Am. Fish. Soc. 121:199–205.

Wilcove, D. 1985. Nest predation in forest tracts and the decline of migratory songbirds. Ecology 66:1211–1214.

Williams, C. B. 1930. The Migration of Butterflies. Oliver and Boyd, Edinburgh.

Williams, C. B. 1958. Insect Migration. Collins, London.

Williams, T. C. and J. M. Williams. 1990. The orientation of transoceanic migrants. In E. Gwinner, ed. Bird Migration: Physiology and Ecophysiology. Springer-Verlag, Berlin, pp. 7–21.

Williams, T. C., J. M. Williams, L. C. Ireland, and J. M. Teal. 1978. Estimated flight time for transatlantic autumnal migrants. Am. Birds 32:275–280.

Williamson, K. 1975. Birds and climatic change. Bird Study 22:143–164.

Willson, M. F. 1976. The breeding distribution of North American migrant birds: A critique of MacArthur (1959). Wilson Bull. 88:582–587.

Wilson, E. O., ed. 1988. Biodiversity. National Academy Press, Washington, D.C.

Wilson, E. O. and W. Bossert. 1963. Chemical communication among animals. Recent Progress in Hormone Research 19:673–716.

Wilson, K. and A. G. Gatehouse. 1992. Migration and genetics of pre-reproductive period in the moth, *Spodoptera exempta* (African armyworm). Heredity 69:255–262.

Wilson, K. and A. G. Gatehouse. 1993. Seasonal and geographical variation in the migratory potential of outbreak populations of the African armyworm moth, *Spodoptera exempta*. J. Anim. Ecol. 62:169–181.

Wilson, W. H. 1991. The foraging ecology of migratory shorebirds in marine soft-sediment communities: The effects of episodic predation on prey populations. Amer. Zool. 31:840–848.

Wiltschko, R. 1991. The role of experience in avian navigation and homing. In: P. Berthold, ed. Orientation in Birds. Birkhäuser Verlag, Basel, pp. 250–269.

Wiltschko, R. 1992. Das verhalten verfrachteter vögel. Die Vogelwarte 36:249–310.

Wiltschko, R. and W. Wiltschko. 1978. Relative importance of stars and magnetic field for the accuracy of orientation in night-migrating birds. Oikos 30:195–206.

Wiltschko, R., M. Schops, and U. Kowalski. 1989. Pigeon homing: wind exposition determines the importance of olfactory input. Naturwiss. 76:229–231.

Wiltschko, W. 1968. Uber den Einfluss statischer Magnetfelder auf die Zugorientierung der Rotkelchen (*Erithacus rubecula*). Z. Tierpsychol. 25:537–558.

Wiltschko, W. and R. P. Balda. 1989. Sun compass orientation in seed-caching scrub jays (*Aphelocoma coerulescens*). J. Comp. Physiol. A 164:717–721.

Wiltschko, W. and R. Wiltschko. 1972. Magnetic compass of European robins. Science 176:62–64.

Wiltschko, W. and R. Wiltschko. 1976. Interrelations of magnetic compass and star orientation in night-migrating birds. J. Comp. Physiol. A 109:91–99.

Wiltschko, W. and R. Wiltschko. 1981. Disorientation of inexperienced young pigeons after transportation in total darkness. Nature 291:433–434.

Wiltschko, W. and R. Wiltschko. 1988. Magnetic orientation in birds. Current Ornithology 5:67–121.

Wiltschko, W. and R. Wiltschko. 1991. Magnetic orientation and celestial cues in migratory orientation. In: P. Berthold, ed. Orientation in Birds. Birkhäuser Verlag, Basel, pp. 16–37.

Wiltschko, W. and R. Wiltschko. 1992. Migratory orientation: Magnetic compass orientation of garden warblers (*Sylvia borin*) after a simulated crossing of the magnetic equator. Ethology 91:70–74.

Wiltschko, W., P. Daum, A. Fergenbauer-Kimmel, and R. Wiltschko. 1987. The development of the star compass in garden warblers, *Sylvia borin*. Ethology 74:285–292.

Wingfield, J. C., H. Schwabl, and P. W. Mattocks, Jr. 1990. Endocrine mechanisms of migration. In: E. Gwinner, ed. Bird Migration: Physiology and Ecophysiology. Springer-Verlag, Berlin, pp. 232–256.

Winkler, H. and B. Leisler. 1985. Morphological aspects of habitat selection in birds. In: M. L. Cody, ed. Habitat Selection in Birds. Academic Press, London, pp. 415–434.

Winkler, H. and B. Leisler. 1992. On the ecomorphology of migrants. Ibis 134 (Suppl.) 1:21–28.

Winston, M. L., G. W. Otis, and O. R. Taylor. 1979. Absconding behaviour of the Africanized honeybee in South America. J. Apic. Res. 18:85–94.

Winter, J. D., V. B. Kuechle, D. B. Siniff, and J. R. Tester. 1978. Equipment and methods for radio tracking freshwater fish. Univ. Minnesota Agr. Exp. Station, Misc. Report 152-1978, pp. 3–18.

Wippelhauser, G. S. and J. D. McCleave. 1988. Rhythmic activity of migrating juvenile American eels *Anguilla rostrata*. J. Mar. Biol. Assoc. U.K. 68:81–91.

Wolcott, T. G. and D. L. Wolcott. 1982. Larval loss and spawning behavior in land crabs, *Gecarcinus lateralis* Freminville. J. Crust. Biol. 2:477–485.

Wolcott, T. G. and D. L. Wolcott. 1985. Factors influencing the limits of migratory movements in terrestrial crustaceans. In: M. A. Rankin, ed. Migration: Mechanisms and Adaptive Significance. Contrib. Marine Sci. 27 (Suppl.), pp. 255–273.

Woodrow, K. P., A. G. Gatehouse, and D. A. Davies. 1987. The effect of larval phase on flight performance of African armyworm moths, *Spodoptera exempta* (Walker) (Lepidoptera: Noctuidae). Bull. Ent. Res. 77:113–122.

Wooller, R. D., J. S. Bradbury, I. J. Skira, and D. L. Serventy. 1990. Reproductive success of short-tailed shearwaters, *Puffinus tenuirostris*, in relation to their age and breeding experience. J. Anim. Ecol. 59:161–170.

Wootton, R. J. 1976. The Biology of Sticklebacks. Academic Press, New York.

Wright, S. 1932. The roles of mutation, inbreeding, crossbreeding, and selection in evolution. Proc. Sixth Int. Congr. Genetics 1(1932):356–366.

Wyndham, E. 1982. Movements and breeding seasons of the Budgerigar. Emu 82:276–282.

Wyneken, J. and M. Salmon. 1992. Frenzy and postfrenzy swimming activity in loggerhead, green, and leatherback hatchling sea turtles. Copeia 1992:478–484.

Wynne-Edwards, V. C. 1935. On the habits and distribution of birds on the North Atlantic. Proc. Boston Soc. Nat. Hist. 40:233–346.

Yokoyama, K. 1976. Hypothalamic and hormonal control of photoperiodically induced vernal functions in the white-crowned sparrow, *Zonotrichia leucophrys gambelii*. I. The effects of hypothalamic lesions and exogenous hormones. Cell Tissue Res. 174:391–416.

Yokayama, K. and D. S. Farner. 1978. Induction of zugenruhe by photostimulation of encephalic receptors in white-crowned sparrows. Science 201:76–79.

Young, G., B. Th. Björnsson, P. Prunet, R. J. Lin, and H. A. Bern. 1989. Smoltification and seawater adaptation in coho salmon (*Oncorhynchus kisutch*): plasma production, growth hormone, thyroid hormones, and cortisal. Gen. Comp. Endocrin. 74:335–345.

Young, J. A. 1991. Tumbleweed. Scient. Amer. 264(3):58–63.

Zalucki, M. P. and R. L. Kitching. 1982. The analysis and description of movement in adult *Danaus plexippus* L. (Lepidoptera: Danainae) Behaviour 80:174–198.

Zalucki, M. P., G. Daglish, S. Firempong, and P. Twine. 1986. The biology and ecology of *Heliothis armigera* (Hübner) and *H. punctigera* Wallengren (Lepidoptera: Noctuidae) in Australia: What do we know? Aust. J. Zool. 34:779–814.

Zeh, D. W. and J. A. Zeh. 1992. On the function of harlequin beetle-riding in the pseudoscorpion *Cordylochernes scorpioides* (Pseudoscorpionida: Chernetidae). J. Arachnol. 20:47–51.

Zeng, Z. 1988. Long-term correlated response, interpopulation covariation, and interspecific allometry. Evolution 42:363–374.

Zera, A. J. 1981. Extensive variation at the a-glycerophosphate dehydrogenase locus in species of water striders (Gerridae: Hemiptera). Biochem. Genet. 19:797–812.

Zera, A. J. and K. C. Tiebel. 1988. Brachypterizing effect of group rearing, juvenile hormone-III and methoprene on winglength development in the wing-dimorphic cricket, *Gryllus rubens*. J. Insect Physiol. 34:489–498.

Zera, A. J. and K. C. Tiebel. 1989. Differences in juvenile hormone esterase activity between presumptive macropterous and brachypterous *Gryllus rubens*: implications for the hormonal control of wing polymorphism. J. Insect Physiol. 35:7–17.

Zera, A. J. and K. C. Tiebel. 1991. Photoperiodic induction of wing morphs in the waterstrider *Limnoporus canaliculatus* (Gerridae: Hemipera). Ann. Entomol. Soc. Am. 84:508–516.

Zera, A. J. and S. S. Tobe. 1990. Juvenile hormone-III biosynthesis in presumptive long-winged and short-winged *Gryllus rubens*: implications for the endocrine regulation of wing dimorphism. J. Insect Physiol. 36:271–280.

Zera, A. J., C. Strambi, K. C. Tiebel, A. Strambi, and M. A. Rankin. 1989. Juvenile hormone and ecdysteroid titres during critical periods of wing morph determination in *Gryllus rubens*. J. Insect Physiol. 35:501–511.

Zhao, S.-J. 1986. Relation between long-distance migration of oriental armyworms and seasonal variation of general circulation over east Asia. Advances Atmos. Sci. 3:215–226.

Ziegler, J. R. 1976. Evolution of the migration response: emigration by *Tribolium* and the influence of age. Evolution 30:579–592.

Index